Understanding Global Climate Change

Understanding Global Climate Change

Modelling the Climatic System and Human Impacts

Second Edition

Arthur P. Cracknell

Costas A. Varotsos

CRC Press
Taylor & Francis Group
Boca Raton London New York

CRC Press is an imprint of the
Taylor & Francis Group, an **informa** business

Second edition published 2022
by CRC Press
6000 Broken Sound Parkway NW, Suite 300, Boca Raton, FL 33487-2742

and by CRC Press
2 Park Square, Milton Park, Abingdon, Oxon, OX14 4RN

© 2022 Taylor & Francis Group, LLC

First edition published by CRC Press 1998

CRC Press is an imprint of Taylor & Francis Group, LLC

ISBN: 978-0-367-19591-5 (hbk)
ISBN: 978-1-032-00161-6 (pbk)
ISBN: 978-0-429-20332-9 (ebk)

Typeset in Times
by SPi Global, India

Contents

Preface...xi
Acknowledgements...xvii
Biography...xix
Abbreviations...xxi

Chapter 1   The great global warming scare 1

1.1   Climate change: The background 1
    1.1.1   The definition of climate 1
    1.1.2   The scale in time and space..5
    1.1.3   Chaos...6
    1.1.4   The causes of climate change...9
        1.1.4.1   Events that occur outside the earth 9
    1.1.5   Natural events on the earth 10
        1.1.5.1   Plate tectonics 10
        1.1.5.2   The cryosphere..12
        1.1.5.3   The lithosphere, volcanic eruptions 14
        1.1.5.4   Ocean circulation 14
        1.1.5.5   The biosphere...16
    1.1.6   Human activities..18
        1.1.6.1   Carbon dioxide CO_2..18
        1.1.6.2   Ozone 18
        1.1.6.3   Development of land areas....................................26
        1.1.6.4   Other greenhouse gases 29
        1.1.6.5   Human activities and the greenhouse effect 31
1.2   The emergence of the modern environmental movement
    (alias the green movement)...33
    1.2.1   Back to nature..33
    1.2.2   Acid rain...35
    1.2.3   The club of Rome, limits to growth, the Stockholm
        conference 1972, only one earth, blueprint for survival.....................37
    1.2.4   Cold war cooperation between East and West....................40
    1.2.5   Some more conferences..41
    1.2.6   The intergovernmental panel on climate change.................43
    1.2.7   Why we disagree about climate change (Hulme 2009).....................45
1.3   Sustainability, survival...47
    1.3.1   Introduction 47
    1.3.2   Limits to growth, the Club of Rome....................................47
    1.3.3   Defining sustainability...50
    1.3.4   Our way of life 53
    1.3.5   The end of fossil fuels and other minerals 54
    1.3.6   Can the party continue?..57
        1.3.6.1   Gas 58
        1.3.6.2   Coal...58
        1.3.6.3   Nuclear..59
        1.3.6.4   Renewables 59

          1.3.6.5   Hydropower and geothermal 59
          1.3.6.6   Wind.. 62
          1.3.6.7   Solar 62
          1.3.6.8   Biofuels 63
          1.3.6.9   Waves, tidal systems, etc 65
          1.3.6.10   Hydrogen and fuel cells 65
          1.3.6.11   Cement/Concrete 65
     1.3.7   Sustainable energy without the hot air 66
     1.3.8   Population 68
     1.3.9   The collapse of former civilisations 69
     1.3.10   Easter island 70
     1.3.11   Incomplete collapse: Example the end of the roman empire 71
     1.3.12   Current environmental threats 71

Chapter 2   The atmosphere 75

     2.1   Introduction 75
     2.2   Composition of the atmosphere 76
     2.3   The Earth's radiation budget 80
          2.3.1   Global energy flows 80
          2.3.2   Earth's radiation budget and climate 82
     2.4   Optically active minor gaseous components 83
     2.5   Aerosols 93
          2.5.1   Natural and anthropogenic aerosols in the atmosphere 93
          2.5.2   Aerosol optical properties 93
          2.5.3   Sulphate aerosols 94
          2.5.4   The spatial and temporal distribution of aerosols 95
     2.6   Clouds 96
          2.6.1   Formation of clouds 96
          2.6.2   Types of clouds 97
          2.6.3   Cloudiness and radiation 98

Chapter 3   The hydrosphere 107

     3.1   Introduction 107
     3.2   The hydrological cycle 109
     3.3   The Oceans 112
          3.3.1   The thermohaline circulation 112
          3.3.2   Studies of heat and water balances in the world's oceans 114
          3.3.3   Nonlinearities in oceans and climate feedbacks 117
     3.4   The Lakes 118
     3.5   The Rivers 121
     3.6   Case-studies of hydrological applications 124
          3.6.1   Modelling the state of the Okhotsk Sea ecosystems 125
               3.6.1.1   The Aral Sea 128
               3.6.1.2   The future of the Aral Sea 133
               3.6.1.3   The water balance of the Aral Sea, a new recovery
                         scenario 133
               3.6.1.4   Other threatened seas 137

Chapter 4 The biosphere, lithosphere, and cryosphere 141

    4.1   Background.. 141
    4.2   The biosphere 142
          4.2.1   The cumulative carbon transfers since the industrial revolution 143
          4.2.2   The carbon cycle: annual data 146
    4.3   The lithosphere 154
    4.4   The cryosphere 157
          4.4.1   The Arctic and the Antarctic... 160
          4.4.2   Glaciers.. 165
          4.4.3   Observations of the cryosphere 166
          4.4.4   Thermal radiation and the snow cover: exploration tools 169

Chapter 5 Energy, the driver of the climate system 171

    5.1   Energy.. 171
          5.1.1   Why discuss energy?.. 171
          5.1.2   Renewable energy 171
    5.2   A digression: What is energy?... 173
    5.3   The Green Agenda and energy, Energiewende.. 175
    5.4   The curse of intermittency 182
    5.5   Nuclear energy and the generation of electricity..................................... 189
    5.6   Energy resources and the environment.. 192
    5.7   Decarbonisation potential in the global energy system 193
    5.8   Radioactive environmental contamination from nuclear energy 195
          5.8.1   Nuclear accidents 195
          5.8.2   Nuclear waste management.. 197

Chapter 6 Climate data, analysis, modelling 201

    6.1   Introduction 201
    6.2   Meteorological data.. 202
    6.3   Meteorological satellites... 203
    6.4   Satellites and climate modelling... 211
    6.5   The NASA Earth Observing System (EOS).. 212
    6.6   Trace gases and pollutants.. 217
    6.7   Earth radiation budget, clouds, and aerosol.. 221
          6.7.1   The stratosphere 223
          6.7.2   The troposphere.. 224
    6.8   Volcanic eruptions 226
          6.8.1   The eruption of circa AD 535-6 226
          6.8.2   Ice core records 228
          6.8.3   Radiocarbon dating... 232
          6.8.4   Tree Rings 232
    6.9   Aerosols and climate 234
          6.9.1   Introduction 234
          6.9.2   Modelling aerosol properties.. 235
          6.9.3   Climatic impacts of aerosols 236
          6.9.4   Tropospheric aerosols.. 237

       6.9.5   Sulphate aerosols 237
       6.9.6   Stratospheric aerosols 238
   6.10 Remote sensing and volcanic eruptions 242
       6.10.1 Satellite observations 248
       6.10.2 Aircraft observations 250
       6.10.3 Balloon observations 252
       6.10.4 Surface observations 253
       6.10.5 Climatic consequences of the Mount Pinatubo eruption 255
   6.11 Paleoclimatology 257
       6.11.1 Desertification 257
       6.11.2 Land degradation 261
       6.11.3 The concepts of land cover and land use 262
       6.11.4 Palaeoclimatic information: catastrophic changes 264
   6.12 Models 265
       6.12.1 Weather forecast models 266
       6.12.2 Climate-forecast models 268
       6.12.3 The surface boundary conditions 270
       6.12.4 Feedback mechanisms 273
       6.12.5 The use of general: circulation models 274
   6.13 Carbon dioxide and climate 277
       6.13.1 Numerical modelling for CO_2 increase 277

Chapter 7   The IPCC and its recommendations 281

   7.1   Introduction 281
   7.2   Soviet Climatology in the Second Half of the 20th Century 281
       7.2.1   The cold war period 281
       7.2.2   Soviet climate change dialogue with the West 288
       7.2.3   US-USSR climate science collaboration 288
       7.2.4   Soviet involvement in the activities of the WMO and IPCC 289
       7.2.5   The distinctive soviet contribution 292
   7.3   The IPCC Reports 293
       7.3.1   Background 293
       7.3.2   Why are the IPCC assessments so important? 294
       7.3.3   The IPCC first assessment 295
       7.3.4   The second and subsequent assessment reports 297
            7.3.4.1 Concerns about the IPCC's climate models 301
            7.3.4.2 Political manipulation 302
            7.3.4.3 Himalayan glaciers 302
       7.3.5   Predicted consequences of climate change 303
            7.3.5.1 Sea-level rise 304
            7.3.5.2 Freshwater resources 305
            7.3.5.3 Desertification 306
            7.3.5.4 Agriculture and food supply 306
            7.3.5.5 Natural ecosystems 308
            7.3.5.6 Impact on human health 309
            7.3.5.7 Costs 310
            7.3.5.8 Consensus and Validation 311
   7.4   The concept of global ecology 312

7.5   Academician Kirill Kondratyev and the Intergovernmental Panel
        on Climate Change (IPCC)..319
7.6   The UNFCCC and the Kyoto Protocol...323
7.7   Climate predictions..328
7.8   Cooling off on global warming...330
        7.8.1   The soviet climatologists..330
        7.8.2   Human-induced global warming sceptics 330
        7.8.3   How do we define mean global (or global mean)
                  temperature?..334
7.9   The Nongovernmental International Panel on
        Climate Change (NIPCC)...337
        7.9.1   The origins of the NIPCC...337
        7.9.2   The NIPCC report of 2008 338
                  7.9.2.1   No consensus 340
                  7.9.2.2   Why scientists disagree...340
                  7.9.2.3   Scientific method vs. political science...................340
                  7.9.2.4   Flawed projections 341
                  7.9.2.5   False postulates 341
                  7.9.2.6   Unreliable circumstantial evidence........................341
                  7.9.2.7   Policy implications 341
7.10   Politics, Margaret Thatcher and James Hansen....................................342
7.11   The green movement and human-induced global warming 343
7.12   Consensus 348
        7.12.1   Scientific consensus..348
        7.12.2   Surveys allegedly supporting consensus 350
        7.12.3   Evidence of lack of consensus...351
                  7.12.3.1   Controversies 353
                  7.12.3.2   Climate impacts 353
                  7.12.3.3   Why scientists disagree...354
                  7.12.3.4   Appeals to consensus 354
7.13   Climate Models, C.P., SCC, and IAMs 354
        7.13.1   C.P. or ceteris paribus 354
        7.13.2   SCC, the social cost of carbon, carbon tax........................356

Chapter 8   Climate change: Energy resources–nuclear accidents....................361

8.1   Background...361
8.2   Nuclear war and climate 361
8.3   Nuclear energy, nuclear winter 363
        8.3.1   Nuclear energy 363
        8.3.2   Nuclear winter...364
8.4   The big mistake surrounding the IPCC and the UNFCCC...................365
        8.4.1   The Montreal Protocol and the Kyoto Protocol 366
        8.4.2   The United Nations and the Montreal: Kyoto Protocols 368
8.5   What can I do?..369
        8.5.1   The Anthropocene 371
        8.5.2   Greta Thunberg; extinction rebellion 373
        8.5.3   Aims 376
        8.5.4   Principles...377

8.6   Sustainability 379
        8.6.1   The United Nations Sustainable Development Goals (SDGs) 382
8.7   The World's most dangerous animal 383
8.8   Economic growth is not our salvation 383
8.9   Recent additions to the Climate Change Literature 388
8.10   A new tool for environmental risk assessment: Natural time analysis 392

References 397
Index 419

Preface

Who are these guys? Two humble foot soldiers in the large army of people who earn their daily bread as working scientists in academic institutions, in some countries in national academies of sciences, in industrial research laboratories, etc. We have never made any Earth-shattering scientific discoveries and will never win a Nobel prize. There is only a rather limited, but of course unknown, number of major discoveries to be made (such as the DNA double helix for example) and these are usually the achievements of a large cohort of workers. The people who make the big discoveries are, of course, brilliant people but they also happen to be from the right background and in the right place at the right time. We should not be jealous of them.

As to ourselves, we just happen to have had the very great privilege to have worked with the late Russian academician Kirill Kondratyev, in his last few years after he had rejected the political manoeuvrings behind the formation of the IPCC (Intergovernmental Panel of Climate Change) and the IPCC's emphasis on almost blind adherence to computer models and to an extreme extent its rejection of historical evidence and of basic human common sense regarding the climate. Kondratyev was ostracised by the international meteorology and climatology community and went on to spend the last 15 or so years of his life very productively establishing the science of global ecology.

This book came about because in 2017 Irma Britton, of the publishers CRC Press, raised the question of producing a new edition of "Observing Global Climate Change" by K. Ya. Kondratyev and A.P. Cracknell which was published in 1998. "Observing Global Climate Change" was one of the plagues of unexciting books on climate change/global warming that never made a fortune for its authors and probably might as well never have been published. Kirill Kondratyev had passed away in 2006 and so in answer to Irma's suggestion we proposed the preparation of a new edition as a collaboration between Arthur P. Cracknell, formerly of Dundee University, and Costas A. Varotsos, currently of the National and Kapodistrian University of Athens. We had already worked together for about 25 years on atmospheric physics, principally on ozone depletion, and Costas A. Varotsos had also worked with Kirill Kondratyev for several years.

What was the justification for producing a revision of a 20-year-old rather undistinguished book on climate change when there are already so many other books already published and when climate change is such a highly political issue these days? Why do we dare to put on to the market yet another book on climate change? The reason is that we are putting forward a different view of climate change from the conventional one that, until recently at least, climate change has largely been synonymous with anthropogenically induced global warming due to increased CO_2 emissions from the combustion of fossil fuels. We take the view that human-induced global warming is indeed a threat to our way of life, but that it is not the only threat and it may well not even be the most serious threat. This view is not new, but we believe that it is terribly neglected. We had already taken this view over ten years ago in our "Global Ecology and Ecodynamics – Anthropogenic changes to Planet Earth " (A.P. Cracknell et al. 2009, Springer, Praxis) in which we put forward, in Chapter 18 on sustainability, the idea (which was not new even then but owed a great deal to Jared Diamond several years before) that there were at least 12 environmental threats to our survival and the "generation of greenhouse gases and ozone destroying chemicals" was only one of them. And moreover that any one of these 12 problems of non-sustainability would suffice to limit our lifestyle within the next few decades. There is nothing magical in this context in the figure of 12; one could make other lists. We quote Jared Diamond (2005):

"These 12 sets of problems are not really separate from each other. They are linked and one problem exacerbates another or makes its solution more difficult. But any one of these 12 problems of non-sustainability would suffice to limit our lifestyle within the next several decades. As Diamond says "They are like time bombs with fuses of less than 50 years. ...... People often ask, 'What is the single most important environental/population problem facing the world today?' A flip answer would be, 'The single most important problem is our misguided focus on identifying the single most important problem!' That flip answer is essentially correct, because any of the dozen problems if unsolved would do us grave harm, and because they all interact with each other. If we solved 11 of the problems, but not the 12th, we would still be in trouble, whichever was the problem that remained unsolved. We have to solve them all."

Kirill Kondratyev was an early exponent of this view, but he was a lone voice and he was ostracised by the highly politicised climate change research community in the closing years of the 20th century and the early years of the 21st century. He spent the closing years of his life refining the ideas of global ecology. In the decades following the publication of the IPCC's First Report, a sharp division developed between two groups of antagonists. On the one hand, there were those who held the view that human-induced global warming, as a result of the increasing concentrataion of CO_2 in the atmosphere, was a major environmental threat of cataclysmic proportions. On the other hand, there were those who can loosely be described as sceptics who dissented from that view; Kirill Kondratyev was one of the first of these. Over the years, the arguments became bitter, political, and emotional and it has almost reached the level of a relgious dispute where each side tended not to listen to the arguments put forward by the other side.

It just so happens that we have been putting the finishing touches to the manuscript of this book during the global lockdown for the coronavirus or Covid-19 pandemic of 2020 and this highlighted – for us at least – the fragility of our way of life and reinfornced Kirill Kondratyev's and Jared Diamond's view that while human-induced global warming is one threat to our way of life it is not necessarily the most serious such threat. Covid-19 is not just a threat to our health or even to some of our lives, but it could cause economic disaster, civil disruption, and even war – we shall see.

This is not a textbook on climate or climate change, of which there are already many available; we just mention two by Mark Kaslin "Climate - a very short introduction (2013)" and "Climate Change - a very short introduction (2014)." Nor is it an encyclopaedia of the current state of climate change research. The IPCC and its assessment reports (but not the politically controlled and contaminated executive summaries) have done a great job in gathering together and evaluating the evidence on human-induced global warming. Climate change has become interwoven with questions of the sustainability of our present way of life and much has been written on this. Ultimately our way of life is unsustainable. While we were hunter gatherers, several millennia ago, the question of sustainability did not arise. The luxury of considering the question of sustainability, and all the other intellectual activities in which we now indulge, only became a possibility once we (or at least the privileged among us) had roofs over our heads and did not have to worry too much about where our next meal was coming from.

One important thing to understand about climate change is that climate change is not the only or even the most important thing that affects our well-being here on Earth, the future for our children, our grandchildren, etc., or our survival as a species, *homo sapiens*. The present interest in climate change has been (a) stirred up by what we loosely call "green" interests and (b) maintained by the media who have products to sell and (c) and by scientists who run big climate computer models and who derive their livelihood and their pleasure from running those models. This influence has been achieved, to a considerable extent via the Intergovernmental Panel on Climate Change (the IPCC), and we shall of course devote some considerable attention to this organisation and its activities. We do not, of course, deny that climate and climate change are important, but the first thing we would ask of our readers is that you try to see this in perspective, in the context of all the factors that affect our lives and pose threats to our lives.

In this book, we have tried to be careful not to take sides in the arguments between the IPCC protagonists and sceptics. It is therefore very interesting to read a book published in 2009 by Mike Hulme entitled "Why we disagree about climate change" (Hulme 2009). The publicity for his book starts off:

"Climate change is not 'a problem' waiting for 'a solution'. It is an environmental, cultural and political phenomenon which is reshaping the way we think about ourselves, our societies and humanity's place on Earth."

If you don't have time to read the whole of Mike Hulme's book, the Preface is a "must," no-one should dare to talk or write about climate change without having to read that Preface (Hulme, 2009); it summarises the whole book and also includes a potted biography of the author as a working climate scientist as well as a writer and popular speaker on, dare we say it, "climate change." One might argue that there is nothing left for us to say. However, ten years have passed since 2009 and things have changed. They have changed in two ways. First in terms of our understanding of survival or sustainability and the UNFCCC has now produced its list of Sustainable Development Goal (SDGs) and hopefully is not allowing itself to be totally dominated by the advice of the IPCC and human-induced global warming (Varotsos and Cracknell, 2020). Secondly in terms of the developments of the Gadarene swine phenomenon of various governments rushing towards the cliff edge following the example of Germany, and its Energiewende in particular, as a result of the green political agenda exploiting the influence of the IPCC and the Kyoto Protocol. (The Gadarene swine were the pigs into which Jesus cast the demons that had possessed a madman and as a result the pigs all ran over a steep cliff and fell into the sea and were drowned.)

One thing we learned from Mike Hulme's book is that any writer should declare his or her position at the outset. Our interests, as working scientists, have been in a small area of the use of artificial satellites in studying the Earth and particularly the concentration of ozone in the atmosphere. People are, by and large, familiar with ozone depletion and the Montreal protocol which aims to protect the ozone layer (Cracknell and Varotsos, 2012; Varotsos, 2002). This has sometimes been regarded, unsuccessfully, as a model for the Kyoto Protocol regarding carbon dioxide emissions and we shall examine this issue in Chapter 8.

The history of this present book begins with a book written in Russian by Kirill Kondratyev who produced a rough English translation of that book. This was revised and polished to produce "Understanding Global Climate Change" which was published by Taylor & Francis in 1998. We (Costas A. Varotsos and Arthur P. Cracknell) both first met Kirill Kondratyev in the summer of 1992 at a postgraduate remote sensing summer school on "Remote Sensing and Global Climate Change" organised in Dundee (R.A. Vaughan and A.P. Cracknell, 1994, Springer, Berlin). We both worked with Kirill from then until his death in 2006 and we came to understand quite well his approach to climate change. Our aim in this book is to try to apply his critical attitude to climate change to the present state of knowledge of the subject. Basically the human-induced increase in the concentration of atmospheric carbon dioxide is indeed an important factor in present climate change, but there are many others which we should not fail to take into account. Eventually, on the timescale of geology, we as a species, *homo sapiens,* will become extinct while life on Earth for many other species will continue, but ultimately life itself will become extinct.

We should like to make a point about some of the terminology used in the book to describe the phenomenon of climate change. The three most commonly used shorthand descriptions are "climate change," "global warming," and the "greenhouse effect." We take the word "climate" to mean "the general weather conditions prevailing in an area over a long period" (how long is long?). The climate varies from place to place on the Earth. It also varies at a given place as a function of time. We use the expression "climate change" to refer to the variation of the climate with time at a given place. Where a more precise association of physical cause and effect is required, we preface climate change with the adjectives "natural" or "human-induced"/"anthropogenic." The important thing to understand is that climate change occurs as a result of natural causes but it is modified as a result of human activities and there is no simple way to separate these causes. So if we, as a species, wish to control the climate for our benefit, we need to understand the natural causes and how to interfere with those causes and we need to understand what the consequences of human intervention are likely to be. The other thing that we need to understand is how political activities have seized control of the climate agenda.

The term global warming is widely used in discussions of the climate; this should be used in a purely neutral way to indicate that the Earth's surface temperature is rising (or not), irrespective of the extent to which this is due to natural causes or as a result of human activities.. The term "greenhouse effect" is used to describe the fact that the atmosphere provides a shield so that the Earth is very much warmer than it would be without this effect being present, specifically by about 33 degrees C at present. In simple terms, the greenhouse effect is good in that without it much of life on Earth would be impossible. The term "human-induced greenhouse effect" is used to describe the extra warming that arises because of the combustion of fossil fuels and the consequent increase in the concentration of CO_2 in the atmosphere as well as the presence of a number of other natural and artificial gases into the atmosphere as well.

Until very recently, our knowledge of the global climate and of the way it is changing was rather restricted. Two things have changed that. The first was the launching of remote sensing Earth observing (weather) satellites which means that weather observations are available uniformly over all the globe rather than at a very limited number of surface weather stations or ships of opportunity. One can be fairly precise about dates, first there was the date of launch of Sputnik in 1957 and secondly the launch of first meteorological satellite, TIROS-1, in 1960 (Varotsos and Krapivin, 2020). The second cause of change lies in the rapid development of computers and data storage media. It is more difficult to give a precise date for this, but it was somewhere around 1970. This enabled computer models of the weather and of the climate to be developed and the vast amount of satellite data to be processed and stored. An excellent idea of what we might call classical climatology, i.e. before the occurrence of these two developments, can be gained from Hubert Lamb's "Climate History and the Modern World" (H.H. Lamb, Routledge, 1982, 1995) although the beginnings of these developments did find their way into that book.

Finally, to any potential reviewers – if you have read this far – there is an accusation you could lay against us, namely that there is not a single new or original result, idea, or thought within the pages of this book. That could be held to be true, but what we have tried to do is to distil the wisdom of various recent authors and to quote their work directly on occasion to try to restore a sense of balance into people's discussions of climate change. "Observing Global Climate Change" which was published in 1998 contained 57 pages of references amounting to about 6000 individual references. Following the very wise and convenient advice of our editor Irma Britton at CRC, we have tried to reduce the list here. There are two reasons (a) people these days can so easily use a search engine to find further information on a topic and so we have tried to make sure that our text is liberally sprinkled with useful searchable keywords, and (b) our laziness. Accepting that the number of readers who are likely to want to look up references that were cited "Observing Global Climate Change" in 1998, we have merely cited such references as "KC98" in the hope that people may be able to find a copy of that book.

We finally conclude this Preface with three points which we have pulled out from the book itself and which seem to us to be particularly topical or important just now. The first is the phenomenon of more than 20 oscillations, known as Dansgaard-Oeschger oscillations (see Section 6.11.4), which have been identified from the Greenland ice-cores record of the last ice age (between 110,000 and 23,000 years before the present). In each of these Dansgaard-Oeschger oscillations, there was a sudden sharp rise in temperature of between 2 and 10 °C over a period of a decade or so and this was followed by a slow cooling over several centuries, taking on average about 1,500 years. To put this in perspective, at the time of the lecture mentioned in Chapter 1 (Section 1.2.6) by Prof. Hoesung Lee, the "Chair" of the IPCC in October 2018 at the Royal Society of Edinburgh which was given a week after the publication of the IPCC Special Report on preventing global warming greater than 1.5 °C on the grounds of all the terrible things that would happen to us all if the mean global temperature rose by more than 1.5 °C, oblivious of the fact that these Dansgaard-Oeschger oscillations have occurred many times in the past and appear to have involved rises of nearer 10 °C than 1.5 °C. There is a school of thought that argues that it is futile to try to fight against nature (See King Canute Fig. 1.1 and James Lovelock's books warning about Gaia). In our view, it would be far more beneficial to invest some more serious research effort into studying what caused the Dansgaard-Oeschger oscillations than investing millions of dollars in computer modelling of the climatic effects of human-induced emissions of CO_2.

The second is to quote perhaps the best example that we have of a disastrous climate change, described as the world's worst ecological disaster, namely the vanishing of the fruitful Aral Sea and turning it and its flourishing surroundings into an arid desert. This was nothing whatsoever to do with the burning of fossil fuels but was due to human greed and the diversion of the water from the rivers that flowed into this enclosed Sea.

And the third example is of a change in social and political attitudes to the extent that "green" activists a few years ago sought to mobilise public opinion to persuade governments to support legal action and political pressure to enforce reductions of CO_2 emissions by curbing the burning of fossil fuels and activists like Greta Thunberg still try to follow this line. However, several of the recent books (Mackay 2009, Berners-Lee, 2010, 2019, Rees, 2003) now take the line that individuals should curb their greed and act in various ways to protect the environment on the grounds that, in the words of a well-known UK supermarket chain "Every little helps," though it may not help enough to prevent disaster sooner or later. These changes in social attitudes appear to have occurred rather suddenly in the past, over a few years or perhaps a decade or two and certainly on a different timescale from the 3 or 4 degrees per century predicted by climate models. While the evidence for abrupt changes is quite clear, the mechanisms driving these changes are less clear and are still the subject of very active research. Even if the causes of these changes were known it seems unlikely that computer models would ever predict sudden changes.

Arthur P. Cracknell and Costas A. Varotsos
August 2020

Acknowledgements

We should like to acknowledge the contribution of Mrs Pauline Lovell in editing the text of the book; and Irma Britton and Rebecca Pringle of Taylor & Francis Group for their editorial support.

Biography

Arthur P. Cracknell graduated in physics from Cambridge University in 1961 and then did his DPhil at Oxford University on "Some band structure calculations for metals". He worked as a lecturer in physics at Singapore University (now the National University of Singapore) from 1964-1967 and at Essex University from 1967-1970, before moving to Dundee University in 1970, where he became a professor in 1978. He retired from Dundee University in 2002 and now holds the title of emeritus professor there. He is a Fellow of the Institute of Physics (UK), a Fellow of the Royal Society of Edinburgh (Scotland), and a Fellow of the Remote Sensing and Photogrammetry Society (UK). He is currently working on various short-term contracts in several universities and research institutes in China and Malaysia.

After several years of research work on the study of group-theoretical techniques in solid state physics, in the late 1970s Prof. Cracknell turned his research interests to remote sensing and has been the editor of the International Journal of Remote Sensing and the Remote Sensing Letters. His particular research interests include the extraction of the values of various geophysical parameters from satellite data and the correction of remotely-sensed images for atmospheric effects. He and his colleagues and research students have published more than 250 research papers and he is the author or co-author of several books, both on theoretical solid state physics and on remote sensing. He also pioneered the MSc course in remote sensing at Dundee University, which has now been running since the 1980s. His latest books include *The Advanced very High Resolution Radiometer (AVHRR)* (Taylor & Francis, 1997), *Visible Infrared Imager Radiometer Suite: A new operational cloud imager* written with Keith Hutchison (CRC/Taylor & Francis, 2006) a second edition of *Introduction to Remote Sensing* written with Ladson Hayes (CRC/Taylor & Francis, 2007), *Observing Global Climate Change* written with Kirill Kondratyev (CRC/Taylor & Francis, 1998), *Remote Sensing and Climate Change: Role of Earth Observation* (Springer Praxis, 2001), *Global Climatology and Ecodynamics: Anthropogenic Changes to Planet Earth* written with Costas A. Varotsos and Vladimir Krapivin (Springer Praxis, 2008) and *Remote Sensing and Atmospheric Ozone: Human Activities versus Natural Variability* written with Costas A. Varotsos (Springer Praxis, 2012).

Costas A. Varotsos is Full Professor in Atmospheric Physics at the Dept. of Environmental Physics and Meteorology of the Faculty of Physics and former Dean of the School of Science of the National and Kapodistrian University of Athens (UoA). Since 1989 he teaches has been teaching several subjects of the Environmental and Climate Dynamics which are also the main topics of his research interests (e.g. Remote Sensing, Climate Dynamics, Atmospheric Physics & Chemistry, Environmental Change, Non-linear Processes).

He has published 15 Books-monographs (mainly by SPRINGER) and more than 300 research papers in the fields of Geophysics Remote Sensing, Climate/Biosphere, Atmospheric Physics & Chemistry, and Environmental Change. Also, he has published five university Books-monographs in Greek in the same fields.

He is Honorary Professor of the Russian Academy of Natural Sciences, the National Academy of Sciences of Armenia, Fellow of the Royal Meteorological Society (Oxford, UK) and full Member of the World Academy of Sciences and of the European Academy of Natural Sciences (Hanover, Germany). He is a regular or complimentary member of several scientific societies including the

American Meteorological Society, the American Geophysical Union and the European Geophysical Union.

He has been awarded the gold A.S. Popov medal of the Russian A.S. Popov Society, the VI Vernadsky medal of the Russian Academy of Natural Sciences, the RADI Award of Chinese Academy of Sciences and the MOST/ESA Award.

He has been appointed as Specialty Chief Editor of "Frontiers in Environmental Science" of the NATURE Publishing Group (2013-2019), Editor-in-Chief of the "Remote Sensing Letters" of T&F, Reviews Editor of the "Water, Air & Soil Pollution" of Nature-Springer, Editor of the "International Journal of Environmental Research and Public Health" - Section Environmental Engineering and Public Health of MDPI, Editor of the "Big Earth Data", Editor of "Remote Sensing" of MDPI, Associate Editor of the "International Journal of Remote Sensing", Advisor of the "Environmental Science and Pollution Research" and Editorial board member or Guest Editor in several Journals indexed in WoS.

His research papers have received more than 6,500 citations from other scientists, with H-index=58 in Web of Science (All databases), Thomson-ISI. Considering his Books-monographs then the citations are more than 17,000 and H-index=71.

Abbreviations

AAE	All-aerosol-effect
ABL	Atmospheric boundary layer
ACC	Antarctic circumpolar current
ACP	Applied Climatology Programme
ACRIM	Active Cavity Radiometer Irradiance Monitor
ACRIMSAT	Active Cavity Radiometer Irradiance Monitor Satellite
ADEOS	Advanced Earth Observation Satellite
AGC	Atmospheric general circulation
AGCM	Atmospheric general circulation model
AGGG	Advisory Group on Greenhouse Gases
AHT	Atmosphere heat transport
AIE	Aerosol-indirect-effect
AIRS	Atmospheric Infra-red Sounder
ALT	Altimeter radar
AMAP	Arctic Monitoring and Assessment Programme
AMSR	Advanced Microwave Scanning Radiometer
AMSU-A	Advanced Microwave Sounder
AMW	Active Microwave
AOT	Aerosol optical thickness
APE	Available potential energy
AQUA	NASA Earth Science satellite mission for the Earth's water cycle
ASR	Absorbed solar radiation
ASTER	Advanced Radiometer to Measure Thermal Emission/Reflection
ATS	Advanced Technology Satellite
ATSR	Along-Track Scanning Radiometer
AVHRR	Advanced Very High Resolution Radiometer
BATS	Biosphere–Atmosphere Transport Scheme
CAENEX	Complex Atmospheric Energetic Experiment
CCGT	Combined cycle gas turbine
CCM	Community Climate Model
CCS	Carbon capture and storage
CCSA	Carbon Capture and Storage Association
CDIC	Carbon Dioxide Information Center
CDP	Climatic Data Programme
CERES	Clouds and the Earth's Radiant Energy System
CFCs	Chlorofluorocarbons
CGMS	Co-ordination of Geostationary Meteorological Satellites
CHAMMP	Computer Hardware Advanced Mathematics & Model Physics
CLiC	Climate and Cryosphere
CLIVAR	Climate and Ocean – Variability and Change
CN	Condensation nuclei
CNTL-C	Control simulation with the cloud scheme
CNTL-SA	Control simulation with saturation adjustment
COADS	Comprehensive ocean data set
COP	Conference of the Parties
COSPAR	Committee on Space Research
CPR	Cloud Profiling Radar

CRF	Cloud-radiation forcing
CSMP	Climate System Modelling Programme
CSP	Concentrated solar power
CTT	Cloud top temperature
CUEs	Critical-use exemptions
CZCS	Coastal Zone Colour Scanner
d	Day
DDT	Dichloro-diphenyl-trichloro-ethane
DMS	Dimethyl sulphide
DMSP	Defense Meteorological Satellite Program
DMSPSSM	Defense Meteorological Satellite Program Special Sensor Microwave
DTE	Downward thermal emission
DVI	Dust veil index
EAZO	Energetically active zones of the ocean
EC	Extinction coefficient
ECHAM	European Center + HAMburg
ECMWF	European Centre for Medium Range Weather Forecasts
EEG	Erneuerbare-Energien-Gesetz (Renewable Energy Act)
ELA	Equilibrium line altitudes
EMMA	Electricity Market Model
ENSO	El Niño / Southern Oscillation
ENVISAT	Environmental Satellite
EOS	Earth Observing System
EOSPSO	EOS Project Science Office
EPA	Environmental Protection Agency
ERB	Earth radiation budget
ERBE	Earth Radiation Budget Experiment
ERBS	Earth radiation budget satellite
EROEI	Energy Return on Energy Invested
ERS	European Remote Sensing
ESA	European Space Agency
ESC	Extended stratus cloudiness
ESMR	Electronic Scanning Multi-channel Radiometer
ExMOP	Extraordinary Meeting of the Parties
FGGE	First GARP Global Experiment
FIFE	First ISLSCP Field Experiment
Flops	Floating-point operations per second
FOV	Field of view
FSE	Fuel switching and efficiency
FY	FēngYún satellites
GAAREX	Global Atmosphere Aerosol Radiation Experiment
GAC	Global area coverage
GARP	Global Atmospheric Research Programme
GATE	GARP Atlantic Tropical Experiment
GAW	Global Atmosphere Watch
GCIP	GEWEX Continental Scale International Project
GCM	General circulation model
GCOS	Global Climate Observing System
GEWEX	Global Energy and Water Cycle Experiment
GEO	Geosynchronous
Geosat	Geodetic Satellite

GFDL	Geophysical Fluid Dynamics Laboratory
GG	Greenhouse gases
GGI	Instruments for coordinate and geodetic information
GHGs	Greenhouse gases
GISS	Goddard Institute for Space Studies
GLAS	Goddard Laboratory for Atmospheric Sciences
GLOF	Glacier lake outburst flood
GLRS-A	Geodynamic laser range sounder
GMES	Global Monitoring for Environment and Security
GOES	Geostationary Observational Earth Satellite
GOEZS	Project to study processes in the euphotic zone
GOMS	Geostationary Operational Meteorological Satellite
GOOS	Global Ocean Observing System
GOS	Geomagnetic complex instrument
GPCP	Global Precipitation Climatology Project
Gt	Gigatonne
GTOS	Global Terrestrial Observing System
HAPEX	Hydrological Atmospheric Pilot Experiment
HCFCs	Hydrochlorofluorocarbons
HIRDLS	High-resolution dynamics limb sounder
HIRIS	High-Resolution Imaging Spectrometer
HKH	Hindu Kush Himalaya
HLW	High-level waste
HSB	Humidity Sounder for Brazil
hPa	Hectopascal (100 Pa = 1 mbar)
IASA	International Association for Sustainable Aviation
ICSU	International Council of Scientific Unions
IGBP	International GeosDhere–Biosnhere Programme
IIASA	International Institute for Applied Systems Analysis
iLEAPS	integrated land ecosystem-atmosphere processes study
INES	International Nuclear and Radiological Event
INSAT	Indian National Satellite
IOC	Intergovernmental Oceanographic Commission
IPCC	Intergovernmental Panel on Climate Change
IPEI	Ionospheric Plasma and Electrodynamics Instrument
IRSR	Infra-red Scanning Radiometer
ISC	International Science Council
ISCCP	International Satellite Cloud Climatology Project
ISLSCP	International Satellite Land Surface Climatology Project
ITCZ	Inter-tronical Convergence Zone
IWC	Ice water content
IWP	Ice water path
JASON	Joint Altimetry Satellite Oceanography Network
JERS-1	Japanese Earth Resource Satellite
JGOFS	Joint Global Ocean Flux Study
K	Kelvin (used for both points and differences on the kelvin temperature scale)
KC98	Kondratyev and Cracknell (1998)
kt	Kilotonne
LAC	Local area coverage
LAWS	Lidar wind sounder
LEO	Low Earth Orbit

LIS	Lightning video sensor
LWC	Liquid water content
MEO	Medium Earth orbit
MetOp	Meteorological Operational
MGACs	Minor gaseous and aerosol components
MGCs	Minor gaseous components
MHS	Microwave Humidity Sounder
MIMR	Multi-spectral Video Microwave Radiometer
MISR	Multi-angle Spectro-radiometer
MLF	Multilateral Fund
MLS	Microwave limb sounder
MODIS	Moderate Resolution Imaging Spectroradiometer
MOP	Meetings of Parties
MOPITT	Measurement of Pollution in the Troposphere
MOS-1	Japanese oceanographic satellite
MSA	Methane-sulphonate aerosol
MSD	Mean standard deviation
MSISE MSR	Mass Spectrometer - Incoherent Scatter multi-channel scanning radiometer
MSU	Microwave Sounding Unit
Mt	Megatonne
NCAR	National Center for Atmospheric Research
NDYI	Normalised difference vegetation index
NEA	Nuclear Energy Agency
NEMD	North East Monsoon Drift
NGOs	Nongovernmental organisations
NH	Northern hemisphere
NOAA	National Oceanic and Atmospheric Administration
NRBS	Normalised backscattering
NPOESS	National Polar-orbiting Operational Environmental Satellite System
NPP	Net primary productivity
NTA	Natural Time Analysis
OGC	ocean general circulation
ODS	Ozone-depleting chemical
OECD	Organization for Economic Co-operation and Development
OHT	Ocean heat transport
OLR	Outgoing longwave radiation
OMI	Ozone Monitoring Instrument
OSEM	Okhotsk Sea Ecosystem Model
OSU	Oregon State University
PMW	Passive Microwave
POA	Primary organic aerosols
POM	Particulate organic matter
ppb(v)	Parts per billion (by volume)
ppm(v)	Parts per million (by volume)
ppt (v)	Parts per trillion (by volume)
PV	Photovoltaic
RH	Relative humidity
RCP	Representative Concentration Pathways
SAFIRE	Submillimeter and far infrared experiment
SAGE	Stratospheric Aerosol and Gas Experiment

SAR	Synthetic aperture radar
SARAL	Satellite with ARgos and ALtiKa
SAT	Surface air temperature
SBUV	Solar backscattered ultraviolet
SC	Solar constant
SD	Snow depth
SeaWiFS	Sea-Viewing Wide Field-of-View Sensor
SH	Southern hemisphere
SIRS	Satellite Infra-red Spectrometer
SLR	Side-looking radar
SMM(1)	Stratospheric Measurement Mission
SMM(1)	Solar-Maximum Mission
SMMR	Scanning Multi-channel Microwave Radiometer
S-NPP	Suomi National Polar-orbiting Partnership
SOA	Secondary organic aerosols
SOLSTICE	an instrument to measure extra-atmospheric UV solar radiation
SORCE	Solar Radiation and Climate Experiment
SPARC	Stratosphere-troposphere Processes And their Role in Climate
SPOT	Système pour l'Observation de la Terre
SRB	Surface radiation budget
SPD	Sozialdemokratische Partei Deutschlands (Social Democratic Party of Germany
SSM/I	Special Sensor Microwave Imager
SST	Sea surface temperature
SSU	Stratospheric Sounding Unit
STEM	Science, Technology, Math and Engineering
SWIRLS	Stratospheric wind infrared limp sounder
SWMD	South West Monsoon Drift
TC	Temperature cubed
TEAP	Technology and Economic Assessment Panel
TERRA	formerly known as EOS/AM-1
TES	Tropospheric emission spectrometer
THC	Thermohaline circulation
TIMS	Thermal Infrared Mapping Spectrometer
TIROS	Television Infrared Observation Satellite
TLR	Temperature lapse rate
TM	Landsat thematic mapper
TOA	Top of the atmosphere
TOC	Total ozone content
TOGA	Tropical Ocean and Global Atmosphere
TOMS	Total Ozone Monitoring Spectrometer
TOPEX/POSEIDON	a French/American satellite
TOVS	TIROS Operational Vertical Sounder
TRMM	Tropical Rain Measuring Mission
TS	Temperature Squared
TSIS	Total and Spectral Solar Irradiance Sensor
UARS	Upper Atmosphere Research Satellite
UCAR	University Corporation on Atmospheric Research
UKMO	UK Meteorological Office
UNEP	United Nations Environment Programme
UV	Ultra violet

VISR	Visible Infra-red Scanning Radiometer
VRE	Variable renewable energy
WC	Water content
WCAP	World Climate Application Programme
WCC-1	First World Climate Conference
WCC-2	Second World Climate Conference
WCP	World Climate Programme
WCRP	World Climate Research Programme
WEP	watering / evaporation / precipitation
WMO	World Meteorological Organization
WOCE	World Ocean Circulation Experiment
WWW	World Weather Watch
XIE	X-ray imaging experiment
yr	Year
ZAPE	Zonal available potential energy

1 The great global warming scare

1.1 CLIMATE CHANGE: THE BACKGROUND

It seems appropriate to begin this book with the story of King Canute, whose date of birth is not known but he seems to have been involved in the Scandinavian/Viking conquest of England in 1013/1014. This story is an apocryphal anecdote illustrating the piety or humility of King Canute the Great, recorded in the 12th century by Henry of Huntingdon. In Huntingdon's account, Canute set his throne by the sea shore and commanded the incoming tide to halt and not wet his feet and robes. Yet continuing to rise as usual the tide dashed over his feet and legs without respect to his royal person. In the story, Canute demonstrates to his flattering courtiers that he has no control over the elements (the incoming tide), explaining that secular power is vain compared to the supreme power of God, see Figure 1.1.

In modern terms, we can consider the concept of Gaia, as expounded by James Lovelock; the most relevant of his many books is "The Revenge of Gaia" (Lovelock 2006). We quote from the beginning of Chapter 8 "The concept of Gaia, a living planet, is for me the essential basis of a coherent and practical environmentalism; it counters the persistent belief that the Earth is a property, an estate, there to be exploited for the benefit of humankind. This false belief that we own the Earth, or are its stewards, allows us to pay lip service to environmental policies and programmes but to continue with business as usual. A glance at any financial newspaper confirms that our aim is still growth and development. We cheer at any new discovery of gas or oil deposits and regard the current rise in petroleum prices as a potential disaster, not a welcome curb on pollution. Few, even among climate scientists and ecologists, seem yet to realize fully the potential severity, or the imminence, of catastrophic global disaster; understanding is still in the conscious mind alone and not yet the visceral reaction of fear. We lack an intuitive sense, an instinct, that tells us when Gaia is in danger."

1.1.1 THE DEFINITION OF CLIMATE

We take as our starting point the definitions of the two words "weather" and "climate" in the *Concise Oxford Dictionary:*

WEATHER: Atmospheric conditions prevailing at a place and time, combination, produced by heat or cold, clearness or cloudiness, dryness or moisture, wind or calm, high or low pressure and electrical state, of local air and sky.

CLIMATE: (Region with certain) conditions of temperature, dryness, wind, light, etc. Another dictionary definition is "the general weather conditions prevailing in an area over a long period."

As scientists, we would probably prefer to use the words rainfall or humidity, rather the dryness, because these are quantities that we can measure.

We can illustrate the difference between the weather and the climate. The weather refers to a certain place and a certain time. It may be bright and sunny here just now but 50 or 100 km away it

FIGURE 1.1   *Canute Rebukes His Courtiers* (credit to "Alphonse-Marie-Adolphe de Neuville").

might be wet and windy there at this time. It may be bright, warm, and sunny here today but it may be cold, wet, and windy here tomorrow. However, we would use the word climate to represent a temporal average of the weather conditions, particularly temperature and rainfall, see Figure 1.2 but also a few descriptive comments about the regional variations in things such as the humidity, winds, etc. and it could be specified in terms of the season.

We might use it to refer to a reasonably large geographical area, such as mainland Greece for example. If we were to talk of the climate of a small area, e.g. the city of Athens which differs significantly from that of the larger region in which it is situated, particularly because it is surrounded by an arc of low hills, we would speak of a regional or local climate. For a very small area that differed in some way from its surroundings, we might speak of a micro-climate.

The reason why climate change is topical just now is that it is part of the current environmental debate. We shall trace the development of modern-day environmentalism and the development of green politics in Section 1.2. Throughout the world, people are becoming very conscious of mankind's effect on the environment and people's awareness is being maintained and refreshed from time to time by media interest in the big international conferences that are held and stage managed on environmental issues. However, the first point that needs to be made is that although at the present time we are conscious of, and largely concerned with, possible anthropogenic changes in the climate, there are very large natural changes in the climate anyway, see Figure 1.3.

At one time, northern Europe was covered by glaciers for example. The question is how significant are these changes and what, can we do, or should we do, to attempt to control the anthropogenic changes or, indeed, to control the natural changes as well. The second point that needs to be made is that climate is a temporal average taken over a long timescale, whereas reliable detailed observations of the weather (from conventional records and from satellite data) have only been made over a relatively short and recent period of time. Archaeological or geological evidence that we have for longer time periods does exist but is of a much less detailed nature, see Section 5.3.

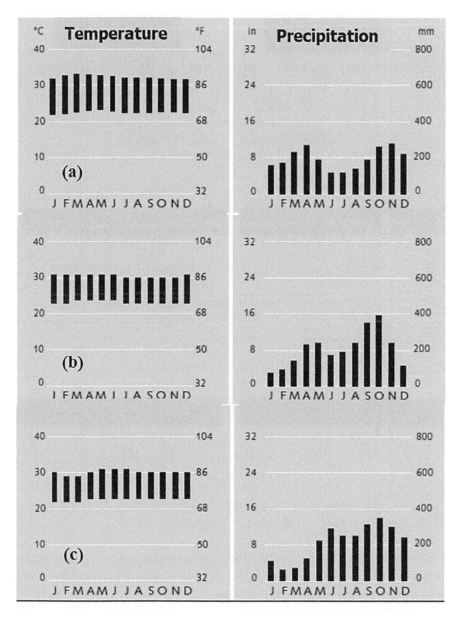

FIGURE 1.2 Seasonal average and extreme values of temperature and precipitation at (a) Kuala Lumpur, (b) Georgetown, and (c) Kota Kinabalu.

By definition, climatologists study the climate. Study of past climate may be of recent climate or of more distant past climate and it is largely an academic subject in that it is of no practical relevance, or we might say it is not useful knowledge. Consequently, it is difficult to determine the extent to which mankind's activities over the last century or two are actually already affecting the climate significantly. In view of this, then, the third point is that given a shortage of detailed experimental evidence about the climate itself and a lack of unequivocal direct evidence regarding the extent to which mankind's activities are already affecting the climate, a great deal of effort has gone into constructing models in which one makes use of the environmental parameters (such as

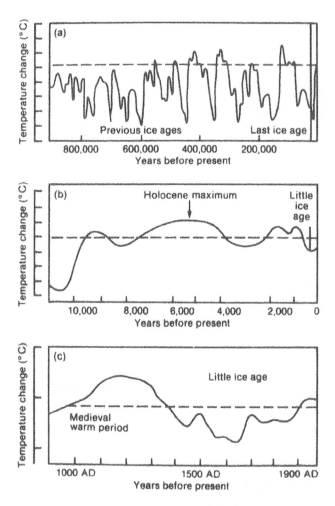

FIGURE 1.3   Global temperature variation since the Pleistocene on three timescales, (a) the last million years, (b) the last 10,000 years, and (c) the last 1000 years (Houghton 1991).

the CO_2 concentration) that can be quantified and use them to calculate predicted effects on the climate. Constructing such models is possible, but the really serious difficulties begin to arise when one tries to use these models. This is because the atmosphere/earth system is very complicated and one is trying to extrapolate its behaviour over a very long time period, when we cannot even predict the weather more than a few days in advance with any degree of reliability! It is, perhaps, not surprising that different groups of people working with different models often produce different predictions of climate change.

One can study the present climate, in which case the work is of the nature of recording the facts of the present climate. Though the question arises of how do we define the "present." Given that the climate is a (temporal) average of weather conditions, one needs to define the period over which one is averaging. It is a bit like trying to teach the elements of differential calculus. One introduces the idea of an interval, defines the slope of a chord joining the points of the variable at the two ends of the interval it vanishes when the limiting value of the slope defines the value of the instantaneous rate of change of the variable. But this concept only applies to what we describe as a differentiable function which basically means a smooth or steadily varying function. But if we take a climatic variable, it is not like that. Consider the temperature at a certain point then it is

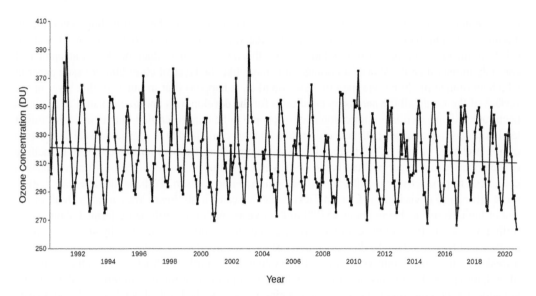

FIGURE 1.4 Looking for a long-term trend in a fluctuating variable. Example of total ozone temporal evolution over Athens from the Athens Dobson spectrophotometer (update of Varotsos and Cracknell 1993).

varying rapidly with time and we are looking for a very slow trend in this rapidly varying function, see Figure 1.4.

This figure shows the temperature variation over a very short period (of 12 years), whereas we are interested in trying to detect a small trend over a long period of, say, a few years. While there are mathematical tools for doing this, it is rather like looking for a needle in a very large haystack. A climatologist is probably not interested in a variable, e.g. the temperature, at a particular point but in the average temperature over an area, it may be a local area, a regional area, or it may be the whole Earth, i.e. global. But a climatologist cannot study the future climate. This is because it has not happened yet. He/she can only predict, i.e. guess possibly making an informed guess. If we knew everything about all the things – not only their initial values but also how they varied through the period in question – which contribute to the climate and were to apply the laws of physics to these things (quantities, variables), then we ought to be able to calculate what the future climate would be. This is what a climate model attempts to do, but it cannot do this exactly because the present values are not all known, the physical laws may or may not be fully known, and the mathematics may be intractable. Any predictions made by the model will be liable to errors and possibly they may even be meaningless, on the principle of "garbage in, garbage out." And there is no way to test the accuracy of the model because we don't know what the actual climate will be. So how can the model be tested? What can be done, and is done, is to take what is known about the climate at some time t_1 in the past and run the model forward to predict the climate at some later time t_2 and compare the prediction with the known climate at t_2. And that is what modellers do and the result should enable them to have a certain degree of confidence in their predictions of the future climate.

1.1.2 THE SCALE IN TIME AND SPACE

It is useful to appreciate the extremely fragile nature of life. We occupy a very thin spherical shell on the surface of a small and insignificant planet. In an aircraft on a commercial flight at, say, an altitude of 10,000 m (10 km), the aircraft cabin has to be pressurised and heated; at the outside pressure and temperature, the crew and passengers would die very quickly. Similarly, if we go very far down below the surface of the sea, we need to be protected from the surrounding environment. Taking this 10 km thickness as an order of magnitude, it corresponds to 10/6400, i.e. 1/640 or about 0.16% of the radius of the Earth. In relative terms, this layer is thinner than the skin of an apple and certainly thinner than the skin of an orange; it is more like the layer of cling film or tissue paper in which an orange might be wrapped in a supermarket. It happens that in our very thin layer at the surface of the Earth, all the necessary conditions for the development and maintenance of life have been satisfied and so it is not unreasonable for people to be concerned that human activities might lead to quite small changes in the conditions and that this could lead to the partial or total destruction of life. A change of only a few degrees in average temperature leads either to ice ages or to melting of the polar ice and either of these could destroy much of our present civilisation. Smaller changes could lead to productive agricultural land being turned into a desert. Perhaps the timescale is even more difficult to envisage. Sir Crispin Tickell (1977) scales it all down. We currently think of the age of the universe as being about 15,000 million years, with the Earth being formed about 4,600 million years ago. If we remove all the zeros and represent 4,600 million years by 46 years (almost an average life span), then the dinosaurs died just over six months ago, *homo sapiens* emerged about a week ago, the zero of our western counting system, at the birth of Christ, was less than a quarter of an hour ago and the industrial revolution has lasted just over a minute. In terms of the climatic situation, on the same timescale, there were major ice ages on the Earth about 9, 7, 6, 4, and 3 years ago, the most recent series of glaciations began less than a week ago and the last glaciers retreated about an hour ago. As Tickell puts it ".... we live in a tiny, damp, curved space at a pleasantly warm moment." Notice from Figure 1.3 that the temperature variations are quite small; however, a rise or fall of only a few tenths of a degree may correspond to quite a large change in the weather conditions. The difference in average temperature of about 1 °C between A.D. 1200 and A.D. 1600 corresponds to some quite significant differences in lifestyle, vineyards in southern England, and fairs on the frozen River Thames in London, respectively. The range of temperature corresponding to what we would regard as extreme conditions is also quite small. At the height of the last ice age, when most of Britain was covered by ice, the mean global temperature was only about 6 °C below the present value.

1.1.3   CHAOS

The term chaos theory is now used with a rather specific meaning. Chaos theory is a part of mathematics. It looks at certain systems that are very sensitive. A very small change may make the system behave completely differently. Very small changes in the starting position of a chaotic system make a big difference after a while. In the early days, it was developed in connection with various mathematical problems, which in some cases were inspired by examples from physics. It was later applied very widely in biology, cryptology, robotics, etc. A good account of some of these will be found in the series of articles from the New Scientist collected together and edited by Hall (1991). Its study in the context of meteorology was developed by Lorenz (1963). The story is told in Wikipedia. Edwards N. Lorenz was using an early computer to run a weather forecast model, which consisted of a set of three ordinary non-linear equations, now known as the Lorenz equations and which have now been extensively studied in various contexts (Sparrow 1982):

$$\frac{dx}{dt} = \sigma\left(y - x\right) \tag{1.1}$$

$$\frac{dy}{dt} = x\left(\rho - z\right) - y \tag{1.2}$$

$$\frac{dz}{dt} = xy - \beta z \tag{1.3}$$

The equations relate the properties of a two-dimensional fluid layer uniformly warmed from below and cooled from above, i.e. a very simple model of the atmosphere. The variable x is proportional to the rate of convection, y to the horizontal temperature variation, and z to the vertical temperature variation. The constants σ, ρ, and β are system parameters proportional to the Prandtl number, the Rayleigh number, and certain physical dimensions of the layer itself. These equations also arise in simplified models of various other systems. Lorenz wanted to see a sequence of data again, and to save time, he started the simulation in the middle of its course. He did this by entering a printout of the data that corresponded to conditions in the middle of the original simulation. To his surprise, the weather the computer began to predict was completely different from the previous calculation. He tracked this down to the computer printout. The computer worked with six-digit precision, but the printout had rounded the variables to three digits. This difference is small and one might have expected that it would have no practical effect. However, what Lorenz discovered was that small changes in initial conditions produced large changes in long-term outcome. This is now a well-known feature of the Lorenz equations. The equations are nonlinear, non-periodic, and three-dimensional. They are deterministic, i.e. for a given set of values of the parameters σ, ρ, and β, there is a unique solution. But for very tiny changes in the values of these parameters, the solution will be very different.

Lorenz's discovery, which gave its name to Lorenz attractors, showed that even detailed atmospheric modelling cannot, in general, make precise long-term weather predictions. It is the critical dependence of a system's behaviour on the precise starting conditions which is known as chaotic behaviour. In a weather system, many events that occur are very small and localised but they may build up to have large effects. This is typified by the title of a lecture given by Lorenz in 1979 "Predictability: does the flap of a butterfly's wings in Brazil set off a tornado in Texas?"

Weather forecasting has come a long way in the last 50 or so years. One stage came with the advent of polar-orbiting weather satellites in 1960 with the launch of TIROS-1 (Television Infrared Observation Satellite) and the geostationary weather satellite GOES-1 (Geostationary Observational Environmental Satellite) in 1975 and all their successors, see Section 6.3. These satellites generated images showing the cloud patterns associated with developing weather systems in a way that had not been possible before and contributed to the tools available to the forecaster. The increasing availability of bigger and faster computers meant that weather forecasting models came to be developed and run. As time went on, these forecasts have become more and more reliable, see Figure 1.5, although it is realised that, in accordance with the work of Lorenz which we have already mentioned, there are practical limits to their reliability which decreases as one tries to make longer-term forecasts.

Climate forecast models developed from weather forecast models and began to be taken seriously when the IPCC was set up in 1988 and began to produce its series of Reports (see Section 1.2.4). It should not be surprising to find that the difficulties encountered by Lorenz would also apply to computer climate models. The last few decades have seen an enormous amount of work go into developing climate models and we shall devote some attention to this subject later (see Chapters 6 and 7).

Scientists have spent many centuries searching for simple rules or laws, that govern the Universe. For example, Newton succeeded in explaining how the planets appeared to move across the sky with his simple laws of motion and theory of gravitation. At the beginning of the 19th century, Laplace believed firmly in a Newtonian universe that worked on clockwork principles. He proposed that if you knew the position and velocities of all the particles in the Universe at a given time, you could predict its future for all time. Such a deterministic view received its first blow with the development of quantum mechanics in the 1920s to describe the world of the very small. Quantum mechanics is used to describe the behaviour of fundamental particles; it is a statistical theory based on probability. According to the uncertainty principle in quantum mechanics, it is impossible to measure the position and momentum (or the position and velocity) of a particle at precisely the same time. In spite of the statistical aspects of quantum mechanics, physicists have used quantum mechanics to construct

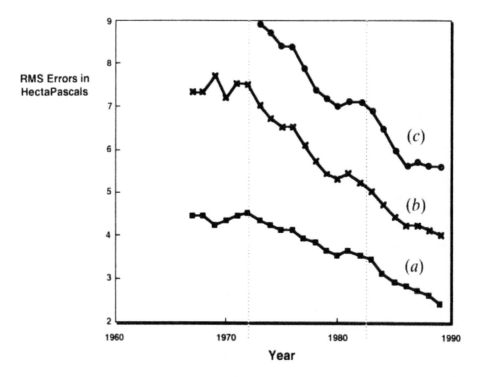

FIGURE 1.5 Errors (r.m.s. differences of forecasts of surface pressure compared with analyses) of UK Meteorological Office forecasting models since 1966 for (a) 24-hour, (b) 48-hour, and (c) 72-hour forecasts (Houghton 1991).

a reasonably robust theoretical framework for describing the fundamental properties of matter and the forces at work in the Universe. Thus, one may hope to explain how the Universe has evolved and indeed even how it came into existence. The reductionist view is that once one has formulated the theory completely, it should be possible to explain the more complicated natural phenomena – how atoms and molecules behave (chemistry) and how they organise into self-replicating entities (biology). In principle, it is just a matter of time and effort, or of computing power. Since human beings are subject to the same laws of nature as the galaxies, so eventually we may be able to make predictions about human concerns, such as the fluctuations of the stock market or the spread of an epidemic. In theory, life can be supposed to be predictable. We have, however, seen already from our brief encounter with chaos theory, that such an approach is not very useful; it is unlikely to be useful in these other, more general, areas either.

Chaos theory presents a universe that is deterministic, obeying the fundamental physical laws. but with a predisposition for disorder, complexity, and unpredictability. It reveals that many systems which are constantly changing are extremely sensitive to their initial state: position, velocity, and so on. As the system evolves in time, minute changes amplify rapidly through feedback. This means that systems starting off with only slightly differing conditions rapidly diverge in character at a later stage. Such behaviour imposes strict limitations on predicting a future state, since the prediction depends on how accurately you can measure the initial conditions. It is only through the widespread availability of powerful computers that it has been possible to follow the development of chaotic systems and demonstrate their behaviour. The results are often presented in computer graphics: starting with quite simple equations, one can produce breathtaking patterns of increasing complexity. Within the patterns, there are shapes that repeat themselves on smaller and smaller scales, a phenomenon which is called self-similarity and of which some examples can be found in nature. Mandelbrot (1982) coined the word "fractal" for such shapes.

1.1.4   The causes of climate change

For as long as the Earth has had an atmosphere, it has had weather and it has had a climate. Other planets too have climates. What has changed, after the emergence of life, has been the emergence of an inter-action between life forms and the climate and this interaction has become much greater following the emergence of *homo sapiens*. However, the human-induced changes are superimposed on the changes that occur naturally and generally on timescales which are much longer than the human lifespan.

The variables which are commonly used in studying the climate are concerned mainly with the atmosphere. However, we cannot look at the atmosphere alone. This is because processes in the atmosphere are strongly coupled to the oceans, to the land surface, and to the parts of the Earth which are covered with ice, i.e. the cryosphere. There is also strong coupling to the biosphere, i.e. to the vegetation and other living systems on the land and in the sea. It is convenient to consider three categories of events that may affect the climate.

    i. events that occur outside the Earth
   ii. natural events on the surface of or within, the Earth, and
  iii. human activities.

1.1.4.1   Events that occur outside the earth

The source of virtually all the energy that drives the atmosphere is the Sun. Thus, any changes in the intensity of the solar radiation that is incident on the atmosphere from above will clearly have a sig-nificant effect on the weather and on the climate. There are various possible causes of changes in the intensity of this radiation. The amount of energy emitted by the Sun itself may vary, the transmission of the radiation through space from the Sun to the Earth may change since space is not absolutely empty, and the separation and the relative orientation of the Sun and the Earth may change. Let us consider these possibilities in turn.

First there is the variation of the Sun's radiation itself. There are regular periodic variations in the Sun's behaviour; these include (a) pulsation with a period of about two hours and forty minutes, (b) rotation on its axis every 27 days, and (c) sunspot activity with an 11-year cycle. Tickell (1977) perhaps seeing things as a historian, has a nice comment on sunspots. "The connection between sun-spots and the weather on Earth is part of universal folklore and cannot easily be dismissed, although the means by which one could affect the other has not been convincingly demonstrated!" He goes on to relate that between 1645 and 1715, the coldest part of the Little Ice Age, was a period of very low sunspot activity, that the estimated intensity of the solar radiation was about 1% lower than its present value and that this is consistent with an average temperature of about 1°C or so less than at present. It is only recently that accurate measurements of the radiation output of the Sun have been made by spacecraft which operate outside the Earth's atmosphere, i.e. by remote sensing. Evidence from several carefully calibrated spacecraft instruments indicates that the total solar energy has varied by about 0.1% during the last solar cycle (Wilson and Hudson 1988; Kopp 2016). Foukal and Lean (1990) studied the spacecraft measurements and by correlating them with sunspot data were able to construct the total solar irradiance from 1874 to 1988, see Figure 1.6.

In addition to the three short-term variations which we have mentioned, there are also long-term variations in the intensity of the Sun's radiation. These are, inevitably, not well understood because of lack of data. For example, the gravitational field experienced by the Sun due to the planets, including the Earth, varies because of the varying positions of the planets as they rotate in their orbits around the Sun. This gives rise to tidal effects on the surface of the Sun, leading possibly to some effects on the Sun's radiation.

Secondly, the radiation travelling from the Sun to the Earth may encounter dust clouds and thus undergo some attenuation. Or showers of micro-meteors may enter the Earth's atmosphere causing dust and ice particles in the stratosphere; these then act as a screen and prevent some of the solar radiation from reaching the Earth.

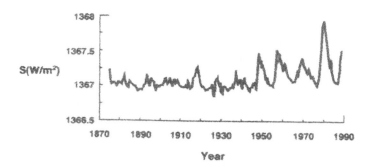

FIGURE 1.6   Changes in solar irradiance 1870–1990 (Foukal and Lean 1990).

Thirdly, there are the variations in the relative positions and orientations of the Sun and the Earth; this in turn involves three factors. First, the eccentricity of the elliptical orbit of the Earth around the Sun varies with a period of about 100,000 years. At present, the intensity of the solar radiation received at the Earth varies by about 7% during the year, but when the eccentricity of the Earth's orbit is greatest, the intensity of the solar radiation received at the Earth varies by up to 30% during the year. A second source of variation arises from the oscillation of the tilt of the Earth's axis of rotation relative to the plane of its orbit; this oscillation has a period of about 40,000 years. The greater the tilt, the greater will be the difference between summer and winter (the last maximum was about 10,000 years ago). Finally, there is the precession of the equinoxes, i.e. the variation in the time of the year when the night and day are of equal length. This arises from the wobble of the Earth around its axis and has a period of 21,000 years. These three variations clearly have an important effect on the intensity of the solar radiation that reaches the Earth, see Figure 1.7.

The first suggestion that ice ages were related to the Earth's orbit around the Sun appears to have been made by Joseph Adhémar, a mathematics teacher in Paris, in 1842; he concentrated on the 22,000 year period. The theory was extended to include the changes in the eccentricity of the Earth's orbit by James Croll (1867), the son of a Scottish crofter, who had received very little formal education. He stumbled on this idea and spent his spare time in the 1860s and 1870s working on the idea; he estimated that the last ice age ended about 80,000 years ago. There was some interest in Croll's theory at the time; however, because he was of low birth and not part of the fashionable circles of the day and because it became apparent that the last ice age ended only about 10,000 years ago rather than 80,000 years ago, his ideas were largely forgotten by the end of the 19th century. Croll's ideas were later developed by Milankovitch (1920), that the major glacial-interglacial cycles over the past few hundred thousand years might be linked with these regular variations in the distribution of solar radiation reaching the Earth. There is a quite impressive correlation between low intensity of solar radiation and large volumes of ice on the Earth's surface and this provides strong evidence to support the Milankovitch theory, see Figure 1.7.

It is important to have accurate measurements of the intensity of solar radiation (the solar-constant) over a period of years but, like all other measurements of parameters related to climatology, it has to be remembered that in climatological terms any period over which we are able to acquire measurements is infinitesimally short.

1.1.5   Natural events on the earth

1.1.5.1   Plate tectonics

It is as well to recall once more that we are, at the moment, not just concerned with short-term climate changes attributable to human activities but, rather, to the whole history of the Earth. Thus, although to us the motions of the various tectonic plates (continental drift) on the surface of the Earth may seem small and of little immediate relevance to climate change, they are – over long periods of

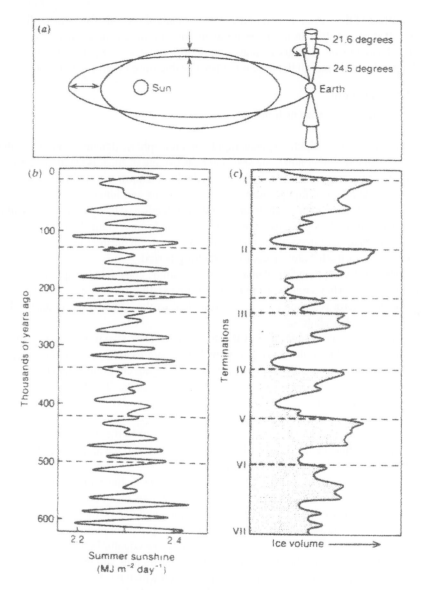

FIGURE 1.7 (a) Variations in the Earth's orbit (a) in its eccentricity, the orientation of its spin axis (between 21.6° and 24.5°) and the longitude of perihelion cause changes in the incidence of solar radiation at the poles, (b) which appear as cycles in the climate record, and (c) a mechanism for the triggering of climate change suggested by Milankovitch (1920). (After Broecker and Denton 1990).

time – very important. The transfer of energy between the surface of the Earth and the atmosphere depends on whether that surface is ocean, ice, or land and, if it is land, on the nature and elevation of the laud surface and land cover. Thus, the changes in the distribution of the land surfaces and the ocean areas of the Earth arising from the movements of the tectonic plates will have caused changes in the distribution of the input of energy into the atmosphere. Another aspect of crustal movements is associated with variations in the gravitational effect of the Sun and the Moon on the surface of the Earth. Apart from causing the familiar ocean tides, these effects give rise to stresses on the laud surface as well so that there is a greater chance of volcanic eruption when these stresses are at their maximum (see Section 5.4). Another factor that, in principle, needs to be considered in long-term studies is the effect of the variation, and indeed the reversal, of the Earth's magnetic field on incoming radiation. It is shown that in the most unfavourable, minimum field interval of the inversion process, the galactic cosmic ray flux increases by no more than a factor of three, implying that the radiation danger does not exceed the maximum permissible dose (Tsareva et al., 2018).

1.1.5.2 The cryosphere

The cryosphere can be classified as follows.

- Seasonal snow cover, which responds rapidly to atmospheric dynamics on timescales of days and longer. In a global context, the seasonal heat storage in snow is small. The primary influence of the cryosphere comes from the high albedo of a snow-covered surface.
- Sea ice, which affects climate on timescales of seasons and longer. This has a similar effect on the surface heat balance as snow on land. It also tends to decouple the ocean and the atmosphere, since it inhibits the exchange of moisture and momentum. In some regions, it influences the formation of deep water masses by salt extrusion during the freezing period and by the generation of fresh water layers in the melting period.
- Ice sheets of Greenland and the Antarctic, which can be considered as quasi-permanent topographic features. They contain 80 per cent of the existing fresh water on the globe, thereby acting as long-term reservoirs in the hydrological cycle. Any change in size will therefore influence the global sea level.
- Mountain glaciers are a small part of the cryosphere. They also represent a freshwater reservoir and can therefore influence the sea level. They are used as an important diagnostic tool for climate change since they respond rapidly to changing environmental conditions.
- Permafrost affects surface ecosystems and river discharges. It influences the thermohaline circulation of the ocean.

Variations in the nature and extent of the ice cover clearly alter the energy fluxes at the Earth's surface and thereby influence the climate. The variation in the surface area covered by ice provides examples of feedback mechanisms in the climate.

The Arctic and Antarctic ice caps play an important role in the global weather system, profoundly affecting ocean currents and winds and the transfer of heat between them and the atmosphere. Perennial ice now covers 11% of the land surface and 7% of the oceans. The drift of the Antarctic continent to its present isolated position and the subsequent partial enclosure of the Arctic Ocean from most of the warm currents moving north from the equator helped the slow accretion of ice sheets. It is the existence of these ice sheets in the polar regions and their expansion in the various ice ages and their contraction in the periods in between which has been the dominant feature of climate over the last million years or so. Over the last three and a half decades, Arctic sea ice extent has declined substantially, at an average rate of around 4% per decade annually and more than 10% per decade during the summer. In contrast, Antarctic sea ice extent has increased slightly, at a rate of about 1.5% per decade, but is projected to decrease in future (Smith et al. 2017). The Greenland and East Antarctic ice sheets, being formed over land above sea level, are relatively stable and form anchors of the systems, which from time to time have expanded to cover enormous areas. Once an ice sheet has become established, then positive feedback occurs. The reflectivity of snow and ice is very high and so the bulk of the solar radiation incident on the ice sheet is reflected and passes through the atmosphere back into space. This leads to colder weather and the expansion of the ice sheet. This expansion of the ice sheet can eventually be halted or reversed when lack of evaporation from the ice and snow over a large area reduces the amount of precipitation of fresh snow. The expansion of the ice sheet may also be halted or reversed as a result of wind-blown dust settling on the ice; this reduces the reflectivity (or albedo) of the surface and leads to warmer weather. The transfer of heat around the Earth by ocean currents also plays an important role, see Figure 1.8.

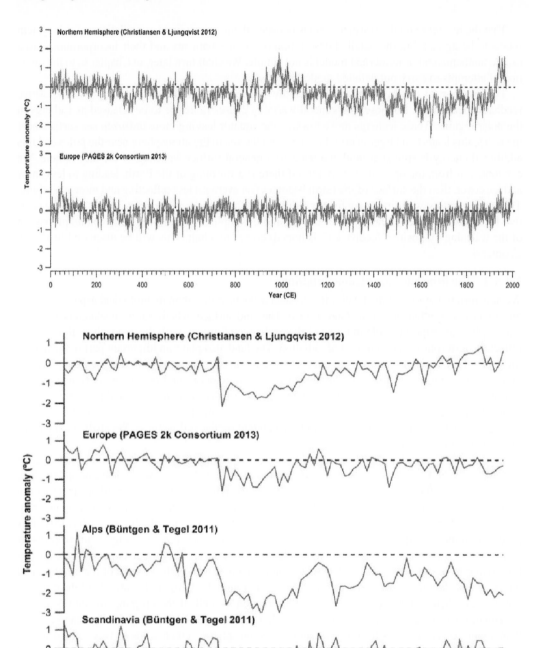

FIGURE 1.8 Summer temperature anomalies (Newfield 2018).

That the ice ages and the warmer periods between them correspond to oscillations of this system seems to be agreed, but the detailed description of the mechanisms and their incorporation into a usable mathematical or numerical model is not simple. We shall turn later, in Chapter 6, to the question of attempts to carry out detailed modelling.

The transport of ice, by floating icebergs, towards the equator also provides a possible positive feedback mechanism. Suppose that general warming causes more ice to be released as icebergs in the polar regions. These icebergs move towards the equator leaving more unfrozen sea surface near the poles; this leads to a larger energy transfer from the sea to the atmosphere near the poles and this additional energy is spread around and leads to a general further warming of the atmosphere. We can look at it from the opposite viewpoint – if there is a warming of the Earth, leading to less snow and ice cover, then the surface of the Earth becomes (on average) less reflecting and more absorbing. This causes further warming and a further reduction in the snow and ice cover, i.e. this is also a positive feedback. The real situation is, of course, more complicated and eventually the amplification of the warming by positive feedback (it is hoped) comes to a halt. This will be discussed further in Chapter 4.

1.1.5.3   The lithosphere, volcanic eruptions

We now turn to the role of the land in its physical aspects, rather than its biological aspects, and the question of energy balance at the land surface. The land surface, whether it is vegetated or not vegetated, plays an important role in terms of the exchange of energy with the atmosphere through the reflection, absorption, and emission of radiation. The land also plays an important part in the hydrological cycle. The processes involved concern the amount of fresh water stored in the ground as soil moisture (thereby interacting with the biosphere) and in underground reservoirs, or transported as run-off to different locations where it can influence the ocean circulation. The soil interacts with the atmosphere by exchanges of gases, aerosols, and moisture, and these are influenced by the soil type and the vegetation, which again are strongly dependent on the soil wetness. We shall consider this in more detail in Chapter 4.

In discussing extra-terrestrial causes of changes in the intensity of solar radiation that reaches the surface of the Earth, we mentioned dust clouds and meteor showers. Volcanic eruptions produce similar effects to those of showers of micro-meteors, though the origin of the particles is terrestrial rather than extra-terrestrial. A volcano blasts large quantities of matter into the sky during an eruption. Some of this reaches the stratosphere as dust particles around which ice can form. Dust is produced in the atmosphere by many causes other than volcanoes, e.g. by strong winds over desert areas. But the quantities arising from major volcanic eruptions are very much larger than the quantities arising from other causes. The fate of volcanic dust depends on the latitude of the eruption and on the winds. A belt of dust particles forms a screen around the Earth at the latitude of the erupting volcano and then widens out in a north-south direction as well. If the erupting volcano is near to the equator, the screen spreads out over the whole Earth towards both poles: if the volcano is more than about 20 °N or 20 °S from the equator, the stratospheric screen tends to be confined to one hemisphere. This screen reduces the intensity of solar radiation reaching the surface of the Earth and also reduces the intensity of the terrestrial radiation escaping to outer space: however, the former effect is dominant. Thus, the general effect of a volcanic eruption is to lead to a cooling of the Earth. Figure 1.7 shows the summer temperature anomalies from A.D. 500 to 600 showing a substantial drop in temperature in around A.D. 536 and continuing until nearly the end of that century. This cooling is widely attributed to volcanic activity (Newfield 2018); we shall return to the question of volcanic activity in Chapter 5.

1.1.5.4   Ocean circulation

A study of the large-scale atmospheric processes responsible for weather and climate changes requires a consideration of the interaction of the atmosphere not only with the land but also with the ocean, which is a gigantic heat reservoir. Over half of the solar radiation reaching the Earth's surface is first absorbed by the oceans, where it is stored and redistributed by ocean currents before escaping to the atmosphere, largely as latent heat through evaporation but also as longwave radiation. These currents have a complicated horizontal and vertical structure determined by the pattern of winds blowing over the sea and the distribution of continents and submerged mountain ranges.

The vertical structure of the oceans comprises three layers:

- the seasonal boundary layer, mixed annually from the surface, which is less than 100 m deep in the tropics and reaches hundreds of metres in the sub-polar seas (other than the North Pacific) and several kilometres in very small regions of the polar seas in most years;
- the warm water sphere (permanent thermocline), ventilated (i.e. exchanging heat and gases) from the seasonal boundary layer, which is pushed down to depths of many hundreds of metres in gyres by the convergence of surface (Ekman) currents driven by the wind;
- the cold water sphere (deep ocean), which fills the bottom 80 per cent of the oceans' volume, ventilated from the seasonal boundary layer in polar seas

Ocean circulation affects the weather and therefore the climate. The atmosphere–ocean interaction, showing itself in the exchange of heat, water, and momentum, is of key importance to the solution of the problem of long-range forecasts of weather and climate changes. Its detailed investigation will be discussed in Chapter 3. More than half the solar radiation reaching the surface of the Earth is absorbed by the sea, largely in the top 100 m. This layer of water acts as a giant reservoir of heat; some heat is transferred directly back into the atmosphere by evaporation, some is moved and mixed downwards, and some remains in the surface layer and travels around the world with the ocean surface circulation. Such heat transfer by ocean circulation can have very important effects on weather and climate. The main ocean currents are indicated in Figure 1.9.

Other situations, however, are much more variable. One well-known example is the El-Niño phenomenon. In normal circumstances, a strong (cold) Pacific current runs from the south northwards along and away from the coast of Peru, see Figure 1.8. This creates an upwelling of deep cold water rich in nutrient salts, which nourishes the fish on which much of the Peruvian economy depends

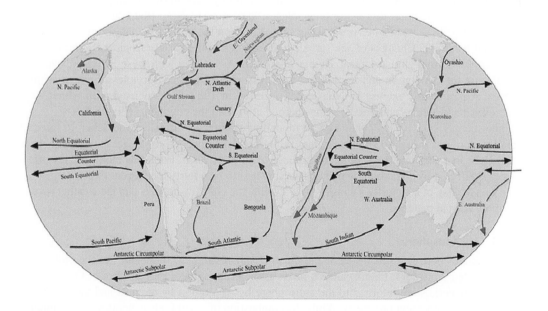

FIGURE 1.9   The main ocean currents. Source: http://www.gkplanet.in/2017/05/oceanic-currents-of-world-pdf.html.

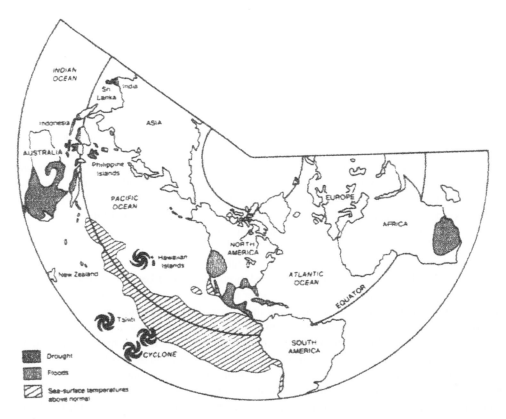

FIGURE 1.10 Regions where droughts and floods occurred associated with the 1982–3 ENSO. (After Canby 1984).

and determines the local weather. From time to time, roughly once in two to ten years, this current moves down and offshore; this is the phenomenon known as El-Niño (the Child) because of its appearance near Christmas time. Warm water moves in from the north with catastrophic results for people and fish alike. Changes of sea surface temperatures in the eastern tropical Pacific of up to 7 °C from the normal climatological average can occur. Associated with these El-Niño events are anomalies in the circulation and rainfall in all tropical regions and also to a lesser extent at mid-latitudes. A particularly intense El-Niño event occurred in 1982–83 associated with which were extreme events (droughts and floods) somewhere in almost all the continents (Figure 1.10).

Some successes have been obtained with modelling to forecast rainfall in the Sahel, to forecast rainfall in northern Brazil, to simulate winter rainfall in Australia, and to predict the consequences of the El-Niño events and, indeed, the occurrence of El-Niño itself. For a few months, the world's weather is affected by something which appeared at first to be a fairly local event (albeit somewhat larger than the flap of a butterfly's wings); after a while, the effect disappears and the weather reverts to its more normal patterns.

1.1.5.5 The biosphere

The interactions between the biosphere and the atmosphere are key components of the global carbon cycle, see Figure 1.11. Photosynthesis involves the production of biomass from CO_2 and H_2O, while these materials are returned to the atmosphere by respiration by animals and vegetation or, eventually, by combustion processes. While the combustion of fossil fuels generates similar quantities of CO_2 and H_2O, the contribution of the H_2O to the hydrological cycle is relatively unimportant because it forms a very small, and possibly negligible, contribution to the total upward flux of water

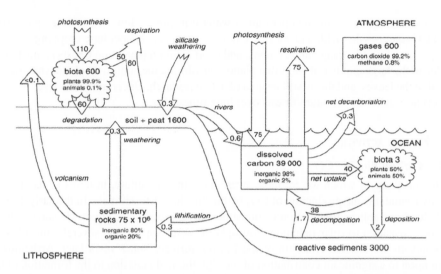

FIGURE 1.11 The preindustrial carbon cycle showing the reservoir sizes (lithosphere, atmosphere, and Ocean), annual fluxes, and processes. Reservoir sizes and annual fluxes are given in Pg carbon (10^{15} g). (Source: Lipp 2008).

vapour into the atmosphere. The contributions of the CO_2 in the combustion products form a much larger proportion of the upward flux of CO_2 into the atmosphere and are generally taken to be significant. This assumption about the relative importance of the contributions of the combustion products to the respective cycles is crucial to the activities of climate modelers looking to assess the effect on the global temperature from burning fossil fuels (see Chapter 5).

Anthropogenic impacts on the biosphere therefore play a vital role in climate-changing factors. With only small-scale anthropogenic disturbances of biospheric processes in the pre-industrial period, flows of substances due to the synthesis and decomposition in the biosphere had been mutually compensated to an accuracy of about 0.01 per cent (KC98, page 17). This had been achieved by natural selection of ecologically balanced communities of numerous types of living organisms. The effect of the significant anthropogenic changes to the biosphere in recent years will be considered in Chapter 4.

Another important aspect of the problem is that, as a rule, an increase in CO_2 concentration has a stimulating effect on photosynthesis (development of vegetation) and so increases the net primary productivity (NPP). This stimulation is determined by two mechanisms:

1. Increased quantum yield of photosynthetic reactions;
2. Increased efficiency in the use of water by vegetation (due to increasing stomatal resistance, with preserved input of CO_2 to the leaves) and decreased water loss on transpiration (Kondratyev and Cracknell 1998a, page 19).

This is sometimes referred to as the fertilisation effect of the CO_2. For the increases in CO_2 concentration which have occurred in recent years, the effect is small but not negligible, both in terms of the removal of CO_2 from the atmosphere and in terms of increased bio-productivity. The magnitude of the fertilisation effect depends on the plant species, the temperature, and the availability of water and nutrients. According to Steffen and Canadell (2005), there is a strong consensus that at the leaf level elevated CO_2 increases the instantaneous rate of photosynthesis in woody plants and in some grasses, and it decreases the amount of water lost per unit of carbon assimilated. Under most conditions and for most plants used in controlled environment experiments, these effects translate at the individual plant level to a positive growth response and an increase in water-use efficiency, that is, to an increase in carbon assimilated per unit of water transpired. However, scaling up the effects of elevated CO_2 on growth, yield, and water use to ecosystems or monoculture cropping systems (e.g. wheat), perennial pasture/rangeland systems, and short-rotation plantation forests is much more difficult, due to the large number of ways that plants can allocate the additional photosynthetic material produced in the leaves, and this is still an area of active research. Uncertainty increases further when the effects are scaled up to mature forests over long timescales.

1.1.6 HUMAN ACTIVITIES

1.1.6.1 Carbon dioxide CO_2

The human activity of which we are most aware, in common culture, in terms of its effect on the global environment is the production of CO_2 by the burning of fossil fuels. In the early 1860s, John Tyndall clearly established the notion of the atmospheric greenhouse effect. He showed that comparatively transparent to solar radiation, atmospheric water vapour and several atmospheric gases strongly absorb the thermal emission of the Earth's surface. Some years later, Arrhenius (1896) drew attention to a significant contribution of the CO_2 thermal emission to the formation of the heat balance and, consequently, of the temperature field in the atmosphere. Arrhenius estimated that a doubled CO_2 concentration should lead to a global warming of the Earth's surface by about 6 °C. Callendar (1938) was the first who suggested the possible impact of anthropogenic releases of CO_2 on climate. The concentration of CO_2 in the atmosphere has risen from about 280 ppm at the time of the Industrial Revolution to over 350 ppm at the present time; in earlier times, the concentration of CO_2 was by no means constant and was probably as low as 200 ppm during the last ice age. At present, the concentration of CO_2 in the atmosphere is rising fast because of the rapid increase in the consumption of fossil fuels and the destruction of forests leading to a reduction in the removal of CO_2 from the atmosphere by photosynthesis. This increases in the concentrations of CO_2 and the other main greenhouse gases in the atmosphere from 1970 to 2019 are shown in Figure 1.12.

The main effect of increasing the concentration of CO_2 in the atmosphere is to change the balance between incoming radiation from the Sun and outgoing radiation from the Earth. This is called the greenhouse effect because the consequence is similar to that of the glass or transparent plastic in a greenhouse although the mechanism is rather different. The bulk of the incoming solar radiation is in the visible wavelength range. After interaction with the Earth's surface, much of the energy is converted to longer wavelength (infrared) radiation which is blocked by the CO_2 (or by the glass or transparent plastic in a greenhouse). The greater the concentration of CO_2 the greater the amount of infrared radiation that is trapped near to the Earth's surface. The CO_2 thus warms the lower atmosphere and cools the upper atmosphere. This in turn increases moisture, which traps more infrared radiation. The effect of rising temperature is to reduce the areas covered by snow and ice, thereby diminishing the amount of heat reflected back into space and increasing the absorption of solar radiation. CO_2 is not the only gas contributing to the greenhouse effect in the atmosphere but it is a very important one, the most important one being water vapour (see Section 1.1.6.4).

1.1.6.2 Ozone

1.1.6.2.1 Introduction

Probably the second most well-known effect of human activity on the atmosphere is the depletion of the ozone layer. Ozone (O_3) is being created by the action of sunlight on the ordinary oxygen molecules O_2. It is unstable and reverts to ordinary (diatomic) oxygen (O_2):

$$2O_3 \rightarrow 3O_2 \qquad\qquad (1.4)$$

The importance of the ozone is that in the stratosphere it absorbs ultraviolet (UV) from the incoming solar radiation. Without the ozone layer, the UV intensity at ground level would be so high

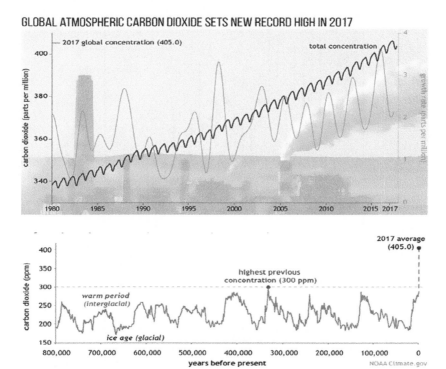

FIGURE 1.12 (top): The increase in carbon dioxide and other greenhouse gases (IPCC 2014 Figure SPM.2. Summary for Policymakers Synthesis Report). (bottom): CO_2 during ice ages and warm periods for the past 800,000 years (NOAA Climate.gov, Data: NCEI).

as to render life difficult or impossible for many species, including *homo sapiens*. Even a modest reduction in the concentration of the ozone layer would lead to significant increases in skin cancers and to various other medical problems as well. The other effect of O_3 in the stratosphere is that, by absorbing ultraviolet radiation, it increases the temperature of the stratosphere. Thus, a reduction in O_3 concentration would lead to a lowering of temperature in the stratosphere with consequential effects on weather and climate. The natural balance that leads to an equilibrium concentration of O_3 in the atmosphere has been upset by anthropogenically produced gases released into the atmosphere, principally oxides of nitrogen and CFCs (chlorofluorocarbons such as CCl_3F, CCl_2F_2, etc.). Ozone-depleting substances, including CFCs that caused holes in Earth's ozone layer in the past century, are responsible for the rapid warming of the Arctic as well. These gases accounted for up to half of the warming and sea-ice loss of the Arctic during the period 1955 and 2005 (Polvani et al., 2020).

Measurements of concentrations of O_3, as well as of the chemical species that destroy the O_3, in the stratosphere have to be made by remote instruments, either ground-based or satellite-flown systems. It is perhaps interesting to recall the story of the discovery of the notorious "ozone hole" over the Antarctic. The first thing to say about the ozone hole is that it is not a complete hole, corresponding to a zero concentration of O_3, but it does correspond to a massive reduction in the O_3 concentration. The second point to make is that the hole is not permanent. O_3 concentrations vary enormously throughout the year. What is meant by the ozone hole over Antarctica is a rather massive reduction in the O_3 concentration over a large area of Antarctica at the end of the Antarctic winter. The hole was first discovered by scientists of the British Antarctic Survey working at the Halley Bay base in Antarctica in 1982; its existence was checked in 1983 and 1984 and the discovery was eventually published in 1985 (Farman *et al.* 1985). In October 1984, Farman et al. had found a depletion of the ozone layer over Halley Bay of 30%, see Figure 1.13.

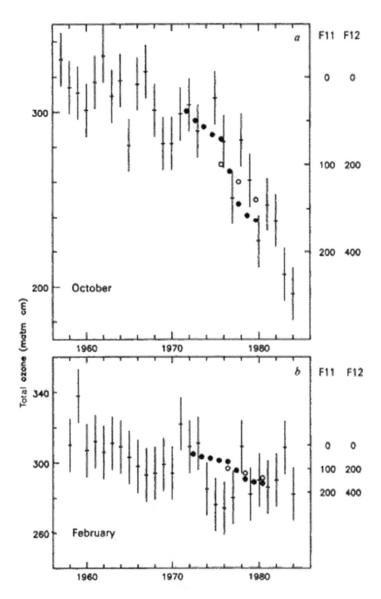

FIGURE 1.13 Monthly means of total O_3 at Halley Bay and Southern Hemisphere measurements of F-11 (CFCl$_3$) pptv and F-12 (CF$_2$Cl$_2$) pptv (Farman et al. 1985).

By October 1987, the depletion had reached 50%. The Halley Bay work was done by ground-based measurements using a Dobson spectrophotometer which is a standard instrument widely used for measuring the integrated O_3 concentration between the ground and the top of the atmosphere.

However, the Nimbus-7 satellite which was launched in 1978 carried two instruments, the Total Ozone Mapping Spectrometer (TOMS) and the Solar Backscatter Ultraviolet (SBUV) experiment, and the data from these instruments should easily have detected such a large depletion in the O_3 concentration in Antarctica. It was discovered, after the Halley Bay work was published in 1985, that at the time the computer programmes for processing the Nimbus-7 data were developed, O_3 concentrations lower than about 200 Dobson units had not been observed in ground data. (The Dobson unit

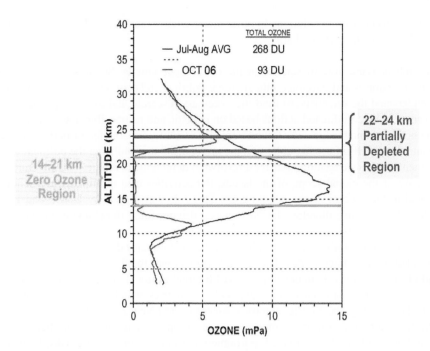

FIGURE 1.14   South pole ozone profiles in 2006 before the ozone hole developed (blue) and for the sounding that displayed the minimum ozone (red). The regions of interest at 14–21 km and 22–24 km are delineated (Hofmann et al. 2009).

is a total amount of O_3 in the atmospheric column that, when converted to 0 °C and sea-level atmospheric pressure would be 10^{-5} m thick.) The data received from Nimbus-7 at the Goddard Space Flight Center were originally processed automatically and any measurements of less than 180 Dobson units were regarded as anomalous and reset to 180 Dobson units. Fortunately, the raw Nimbus-7 data had been kept, and after the publication of the British Antarctic Survey's results, the TOMS data were re-examined. Those results were confirmed and indeed from the TOMS data it was seen that the depletion was not localised at Halley Bay but extended over the whole of Antarctica. After a few months, the hole disappears until the next Antarctic spring. The hole does not apply to the full atmospheric column, but only between heights of about 14 – 21 km, see Figure 1.14.

This is not the place to recount the whole story behind the development of the theory that CFCs, which are man-made and do not occur in nature, could destroy O_3 in the stratosphere, leading to the signing of the Montreal Protocol banning the manufacture of certain CFCs. The story can be found, for example, in Chapter 5 of Cracknell and Varotsos (2012). What is worth mentioning here is the success of the Montreal Protocol which is related to the emission of CFCs; this contrasts sharply with the relative lack of success in achieving international agreement to limit the emission of CO_2, which we shall discuss in Chapter 7 (1991).

1.1.6.2.2   Reasons for success in reaching international agreement in Montreal

Producing a document, the Montreal Protocol, which was eventually ratified by all nations was a major achievement. Then, once it was ratified, it has proved in the following 20 years to be very successful in terms of moving towards its objective, namely protecting and restoring the ozone layer.

Benedick (1991) analyses the reasons for the success in achieving the agreement in the Montreal Protocol. He cited several key factors which we summarise below and he saw in them some clues for people who are trying to find solutions to other major environmental problems, particularly global warming. The factors he identified are:

- The indispensable role of science in the ozone negotiations. Scientists were drawn out of their laboratories and into the political arena. Political and economic decision makers needed to understand the scientists, to fund the necessary research, and to be prepared to undertake internationally coordinated actions based on realistic and responsible assessments of risk.
- The power of knowledge and of public opinion was a formidable factor in the achievement at Montreal. A well-informed public was the prerequisite to mobilising the political will of governments and to weakening industry's resolve to defend the continued manufacture of CFCs.
- Because of the global scope of the issues, the activities of a multinational institution were critical to the success of the negotiations. This role was played by UNEP and Dr Mostafa Tolba, the executive director of UNEP, was a driving force in achieving the success of the negotiating process.
- An individual nation's policies and leadership made a major difference. This leadership was provided by the United States. The US government was the first to take action against CFCs and played a major role in developing the Montreal Protocol and campaigning for its acceptance by the international community. Other countries which played a leading role were Canada, Germany, the Nordic countries, Australia, and New Zealand.
- Private-sector organisations – environmental groups and industry – participated actively, though with different objectives. Environmental groups warned of the risks, lobbied governments to act, and promoted research and legislation. Industrial organisations initially sought to oppose restrictions on the manufacture and use of CFCs but later cooperated in finding substitutes that were not damaging to the ozone layer.
- The process that was adopted in reaching the ozone accord was itself a determining factor. This involved subdividing the problem into manageable components during a pre-negotiating phase. Extensive preliminary scientific and diplomatic groundwork enabled the subsequent negotiations to move forward relatively rapidly.
- Once the serious nature of ozone depletion was realised, the crucial and successful discussions of 1986–1987 were both small in attendance and short in duration by later standards. For example, the first round of one-week negotiations in 1986 was attended by only 20 nations and three nongovernmental organisations (NGOs), while at the decisive 1987 Montreal conference itself, there were only about 60 national delegations (see Benedick, 1991 for details).

Benedick (1991) makes another important point. This is that the Montreal Protocol is not like traditional international treaties which seek to cement a status quo. It was deliberately designed as a flexible and dynamic instrument which can readily be adapted to evolving conditions and developing scientific knowledge. Every year, there are Meetings of the Parties, i.e. the countries which have ratified the Protocol, and over the years, these Meetings have made substantial modifications to the Protocol based on scientific, economic, environmental, and technological assessments of changing situations. So far there have been four amendments, the London Amendment (in 1990, in force in 1992), the Copenhagen Amendment (in 1992, in force in 1994), the Montreal Amendment (in 1997, in force in 1991), and the Beijing Amendment (in 1999, in force in 2002). Up to 2011, the dedicated Multilateral Fund (MLF) had provided nearly US $3 billion in "agreed incremental costs" in three-year cycles based on an independent calculation by the Protocol's Technology and Economic Assessment Panel (TEAP) of the funding needed to ensure that developing country Parties can meet their mandatory obligations for control measures for ozone destroying substances (ODSs) (Molina et al., 2009). Facing the threat of sea level rise (Velders et al., 2007), a group of island nations led by the Federated States of Micronesia and Mauritius proposed an amendment to the Montreal Protocol that would provide jurisdiction over the production and consumption of HFCs and that would use the technical expertise and administrative structure of that treaty to start quickly phasing down

HFCs with high GWPs (39). The United States, Canada, and Mexico followed with a similar joint proposal (Molina et al., 2009). Since then, several Meetings of Parties (MOPs) have been held. At the time of writing the latest, MOP-31, was due to be held in Rome in November 2019.

Benedick quotes Dr Mostafa Tolba of UNEP as saying (rather optimistically) "The mechanisms we design(ed) for the (Montreal) Protocol will – very likely – become the blueprint for the institutional apparatus designed to control greenhouse gases and adaptation to climate change." However, such early optimism appears to have been unfounded. In some more recent papers, Benedick (2007, 2009) has pointed out that the success of the negotiators of the Montreal Protocol has not, in fact, been followed up in the more recent climate negotiations which are aimed at finding a successor to the Kyoto Protocol. Many ODSs are also greenhouse gases and it has been argued (Velders et al., 2007) that the climate protection already achieved by the Montreal Protocol alone is far larger than the reduction target of the first commitment period of the Kyoto Protocol. Moreover, additional climate benefits that are significant compared with the Kyoto Protocol reduction target could be achieved by actions under the Montreal Protocol, by managing the emissions of substitute fluorocarbon gases and/or implementing alternative gases with lower global warming potentials (Velders et al., 2007). There has been a long-running succession of United Nations global environmental megaconferences and negotiations on climate change which have been held every year since 1995 and which involve several thousand official delegates from over 190 countries, together with hundreds of NGO representatives and in the full glare of the world's media. These conferences achieve very little and Benedick (2007, 2009) has argued that there are several useful lessons which could have been learned from the history of the Montreal Protocol. In particular, he argues that small more focused meetings taking place out of the glare of media publicity could achieve more success. He has suggested that the climate negotiations should be disaggregated, or separated, into a number of components such as energy technology research and development, emission reduction policies for particular sectors such as for example transportation, agriculture, and forestry and their adaptation to changing climatic conditions, regional cooperation to combat or accommodate to climate change. It can be argued that there is no reason why every aspect of a complicated scientific and environmental problem must be addressed by every nation at the same time and in the same place. This is particularly true for the climate change negotiations. In reality, only 24 nations (half of them "developing" countries) together account for about 80% of global greenhouse gas emissions, while the remaining 170 nations each contribute less than a fraction of 1%.

1.1.6.2.3   Ratification of the Montreal Protocol

The Montreal Protocol was eventually ratified by virtually all countries; the status of ratification as at 25 November 2009 is shown in Table 1.1. It is widely held (see, e.g. Sarma and Taddonio, 2009) that it is one of the most successful international treaties for three clear reasons:

1. it is based on science, which is constantly updated and taken on board,
2. it applies the "precautionary principle" in setting technically and administratively challenging but feasible goals for phasing out ODSs before the adverse impacts of ozone become catastrophic, and
3. it is successful at motivating the development, commercialisation, and transfer of technology to developing countries, which largely have not been responsible for causing the problem of ozone depletion.

A very good example of the updating of the Protocol is that it started in 1987 with only a freeze on halon production and a 50% reduction in CFCs. "Halon" is a general term for a derivative of methane CH_4, or ethane C_2H_6, where one or more of the hydrogen atoms have been replaced by halogen atoms. CFCs are particular examples where all the H atoms have been replaced by fluorine or chlorine atoms. A number of halons have been used in agriculture, dry cleaning, fire suppression,

TABLE 1.1

Ratification of the Montreal Protocol

Instrument	Number of countries having ratified
Vienna Convention	196
Montreal Protocol	196
London Amendment	194
Copenhagen Amendment	191
Montreal Amendment	179
Beijing Amendment	161

(as at 25 November 2009, UNEP Ozone Secretariat).

TABLE 1.2

Substances originally included in the Montreal Protocol

Group	Substance	Ozone-depleting potential*
Group I		
CFC-11	$CFCl_3$	1.0
CFC-12	CF_2Cl_2	1.0
CFC-113	$C_2F_3Cl_3$	0.8
CFC-114	$C_2F_4Cl_2$	1.0
CFC-115	C_2F_5Cl	0.6
Group II		
halon-1211	CF_2ClBr	3.0
halon-1301	CF_3Br	10.0
halon-2402	$C_2F_4Br_2$	6.0

Source: (Montreal Protocol, Appendix A).

* These are the values of the ozone-depleting potentials estimated in 1987; they are subject to revision in the light of subsequent knowledge.

and other applications. The substances initially included in the Montreal Protocol in 1987 are listed in Table 1.2. A useful summary of the chemical formulae and nomenclature of the CFCs and related compounds is given in Annex V of an IPCC Special Report (IPCC CCS, 2005). New substances have been added to the Montreal Protocol's list from time to time, and at present, the Protocol aims to phase out 96 ozone depleting substances (ODSs) which are used, or have been used, in thousands of products; it provides for this to be done by changing to ozone-safe technologies throughout the world in a specified time frame. In particular, among these are the hydrochlorofluorocarbons (HCFCs), which were originally seen as substitutes for CFCs and which were later added to the Montreal Protocol's list of substances; they are being phased out more because they are powerful greenhouse gases rather than because of damage they might cause to the ozone layer. The HCFCs were scheduled to be phased out by non-Article 5 Parties (i.e. developed/industrialised countries) by 2030 and ten years later by Article 5 Parties (i.e. developing countries and countries with emerging economies, i.e. post-Soviet-Union-collapse countries). However, at the Meeting of the Parties in Montreal in September 2007, the Parties approved an adjustment to accelerate the phasing out of the HCFCs.

According to Sarma and Taddonio (2009), more than 240 sectors with products which used ozone-depleting substances had replaced most of these uses within 10 years. They said "This was a no-compromise market transformation: in addition to being ozone-safe, alternatives and substitutes are equally safe or safer for the Earth's climate, more energy-efficient, lower in toxicity, superior in safety, and more reliable and durable. Often, alternatives have reduced costs to businesses and increased employment. Corporate, military, environmental, and citizen stakeholders are proud of what they have accomplished through consensus, cooperation, and regulation.

Generally speaking, the replacement of ozone-depleting substances was accomplished with no compromise in environmental health and safety, and in a majority of cases, all aspects of environmental performance were improved."

Complying with the demands of the Montreal Protocol was a major task even in the developed countries. The governments of developed countries took various different steps to produce compliance with the Montreal Protocol; these included the adoption of policies, making regulations, conducting awareness and education campaigns, and introducing financial incentives and disincentives. Not only were the ozone-depleting substances used by large industries but they were used in numerous small industries as well. A further complication was that there were various laws and policies on safety requirements that were formulated in terms of using substances that were later found to be ozone-depleting substances.

"Fire safety laws for weapons systems, aircraft, ships, and racing cars, for example, often mandated the use of halon fire suppression systems. Quarantine standards often encouraged or required methyl bromide for certain pests and products. CFC refrigerants were required by law in some areas, and both industrial and military cleaning standards required cleaning with CFC-113. In these cases, enterprises and governments seeking change organized strategic projects to remove specific barriers." (Sarma and Taddonio, 2009).

Complying with the Montreal Protocol was rather different for developing countries. First, their production and use of ozone depletion substances was relatively small. Secondly, they were able to argue – not unreasonably – that it was the developed/industrialised nations which were largely responsible for having damaged the ozone layer. Compliance was only achieved because money was made available to developing countries to help them to comply with the Protocol. The Montreal Protocol provided for technology transfer to developing countries. According to Article 10A of the Protocol,

> "each Party shall take every practicable step, consistent with the programs supported by the financial mechanism, to ensure that the best available, environmentally safe substitutes and related technologies are expeditiously transferred to Parties operating under paragraph 1 of Article 5 (i.e. developing countries)" and that "transfers ... occur under fair and most favourable conditions".

Financial assistance to developing countries was therefore a critical part of the institutional arrangements; the Protocol created a financial mechanism to assist developing countries to implement the control measures. This was done by the Multilateral Fund (MLF) for the implementation of the Montreal Protocol on substances that deplete the ozone layer. This was established under Article 10 of the Montreal Protocol in 1990 to help developing countries to phase out the production and consumption of ozone-depleting substances. The Fund started out with US$240 million for the period 1991–1993. The Fund has been replenished every three years after needs assessments made by the Meetings of the Parties. The three-year replenishments rose to US$400.0 million by 2009–2011. By 7 November 2008, the Executive Committee of the MLF had approved the expenditure of US$2.40 billion to support about 5,789 projects and activities in 144 developing countries, planned to result in the phase out of the consumption of more than 427,000 t and the production of about 173,616 t of ozone-depleting substances.

Sarma and Taddonio (2009) cited a number of reasons for the success of the Montreal Protocol:

- One very important factor was leadership from a wide variety of organisations and people from governments, international organisations, non-governmental organisations, industry associations, scientists, engineers, and many others who took early action to tackle the problem of ozone depletion and inspired others to follow their example.

- Institutional arrangements also were important, including the Ozone Secretariat at UNEP in Nairobi and also the individual Ozone Units – the focal points for action on the Montreal Protocol in each of its Parties. The Ozone Unit focal points are a dedicated group of professionals who understand their mission and have formed both formal and informal networks with the broader ozone community to carry it out.
- Another very important factor was the acceptance of the "principle of common but differential responsibility." Thus, it was acknowledged that it was the developed countries which had produced and used most of the CFCs and therefore were largely responsible for the problem of ozone depletion
- and that the developing countries were much less responsible for the problem. It was recognised that the developed countries possessed greater resources, in terms of technological and financial resources, to tackle the problems involved in complying with the Protocol. Thus, developing countries were given longer than the developed countries to phase out the manufacture and use of ozone-depleting substances and they were provided with financial assistance via the MLF (see above).
- Article 6 of the Montreal Protocol states "… the Parties shall assess the control measures …. on the basis of available scientific, environmental, technical and economic information". Thus, the Scientific Assessment Panel, the Environmental Effects Assessment Panel, and the Technology and Economic Assessment Panel were set up. From time to time, these Panels produce reports which are made available to the Parties in readiness for their annual meetings at which they consider possible amendments to the Protocol. The latest of these at that stage was produced in 2010.
- The ability of those involved to identify and remove barriers to technology transfer, including through changes in national laws and voluntary codes.
- An extensive set of tables and graphs showing the production and consumption trends of CFCs, halons, HCFCs, etc. is given in an Ozone Secretariat report (UNEP 2005), see Figures 1.15 and 1.16. This shows that compliance with the Montreal Protocol had been excellent so far. However, compliance with the Protocol in terms of restricting the emission of ozone-depleting substances into the atmosphere is only the beginning. Ultimately the success, or otherwise, of the Montreal Protocol will have to be judged by whether (a) first of all the decline in stratospheric ozone concentration is halted and (b) secondly, if that is achieved, whether a return to the pre-CFC levels of ozone can be achieved.

1.1.6.3   Development of land areas

As a result of media coverage, it is the increasing concentration of CO_2 and the depletion of O_3 in the stratosphere that spring most immediately to mind when we think about possible effects of human activities on weather and the climate. A few decades ago, at the time of extensive above-ground nuclear testing programmes, we were concerned about the effect of those nuclear explosions on the weather and possibly, the climate. But there are many other factors which should be considered and which, in the long term, may be very important.

Human beings and domestic animals have been affecting the surface of the Earth for a long time. Tickell quotes something like 20% of the total area of the continents as having been drastically changed as a result of human activities. The cutting down of trees for settlement and agriculture, the slash-and-burn method of cultivation in primitive societies, the overgrazing of land by such animals as goats and cattle, and the overuse. Impoverishment and erosion of top soils have in the past affected the heat and water balance in various areas. The amount of solar radiation absorbed by the grass, crops, or in some cases desert land is less than that absorbed by the forests which were there before. More energy is reflected back into space, less moisture is evaporated, and so less rain is produced and of the rain which does fall more is run off the land. The inevitable increase of dust blown up from the surface can produce effects similar to those of volcanic eruptions or sandstorms.

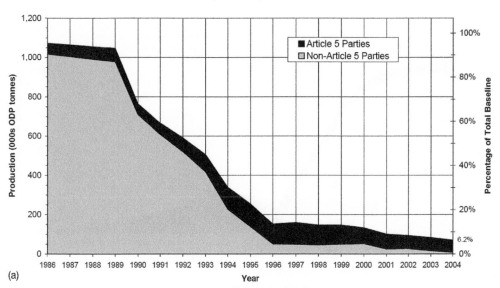

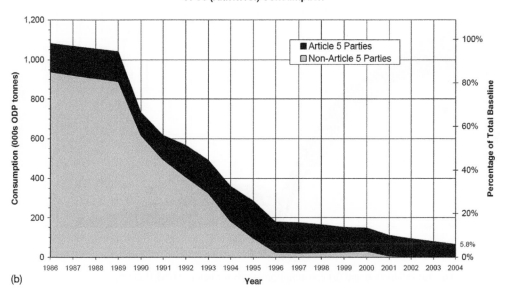

FIGURE 1.15 (a) Production and (b) consumption trends for (Annex A/I) CFCs (UNEP, 2005).

From the point of view of food production or general agricultural production, the most important effects have been a decline in rainfall in some areas. Tickell (1986), again taking the western historian's approach, cites the progressive aridity in historical times of the swath of land from the Mediterranean to northern India. This area was once covered by dense forest and later became the site of successive civilisations; its increasing aridity seems to have been caused mostly by human destruction of the natural environment. Most of the grain-growing parts of the area which once supported large populations, including those of the Roman Empire, are now largely scrub or desert. Estimates are that about 7% of the Earth's surface is man-made desert and that the process of progressive desertification is continuing. This inevitably contributes to some extent to changing weather and climate.

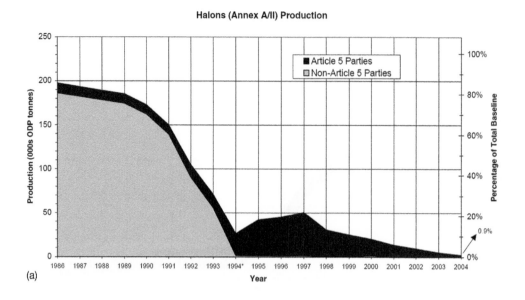

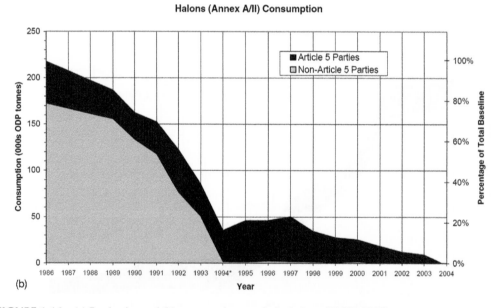

FIGURE 1.16   (a) Production and (b) consumption trends for halons (UNEP, 2005).

In more recent times, other human activities have become important. For instance, there are now situations in which human intervention occurs to modify the weather rather directly, for instance in the case of rainmaking by cloud seeding. The short-term local effects may be desired and beneficial. The effects elsewhere, or in the longer term, may be less obvious and may not necessarily be beneficial. Another human influence is the interference with the hydrology of the Earth. The creation of artificial lakes for irrigation or hydro-electric purposes, or changing the flow of rivers for irrigation purposes and consequently changing land use and land cover, will affect the local heat balance and the evaporation rate. The development of large cities in recent years is also important. Large cities are well known to form heat islands with temperatures significantly higher than their surroundings. This arises from two factors: (i) the replacement of vegetated surfaces by more highly reflecting road and roof surfaces and (ii) the heat generated by human activities within the cities. Locally the cities' effects on the heat balance will affect the weather although, in global terms, their contribution to the heat balance is relatively small (but still a good deal more significant than the flap of a butterfly's wings).

1.1.6.4 Other greenhouse gases

The other major contribution to the greenhouse effect in the atmosphere comes from water vapour, but it is quite different from CO_2. CO_2 is uniformly mixed throughout the atmosphere, that is to say it forms a (more or less) fixed proportion of the gases throughout the atmosphere. Of course, in the vicinity of an intense source of CO_2 (such as a power station or factory chimney), the concentration of CO_2 will be higher than elsewhere but the gas will disperse rather quickly. And reductions in the CO_2 concentration near CO_2 sinks, e.g. in areas of very active photosynthesis, will be compensated by movements of CO_2 into these areas. The CO_2 concentration throughout the atmosphere does not vary dramatically with time. But water vapour is quite different from the CO_2 both in terms of its spatial and temporal distribution. This is very important when it comes to constructing weather forecast models or climate models (see Chapter 6). In particular, if one is trying to model the anthropogenic effects of increasing CO_2 in the atmosphere, the CO_2 concentration can be taken as given. In the case of water vapour, its concentration is fluctuating wildly both spatially and temporally in the lower atmosphere, though its behaviour in the stratosphere may be easier to model.

The ACS (American Chemical Society) Climate Science Toolkit goes as far as to say (converted to British spelling): "It's Water Vapour, not the CO_2."

Narratives
   Remark: "The Earth has certainly been warming since we have added so much CO_2 to the atmosphere from fossil fuel burning."
   Reply: "Forget the CO_2. Water vapour is the most important greenhouse gas. It controls the Earth's temperature."

It's true that water vapour is the largest contributor to the Earth's greenhouse effect. On average, it probably accounts for about 60% of the warming effect. However, water vapo[u]r does not control the Earth's temperature but is instead controlled by the temperature. This is because the temperature of the surrounding atmosphere limits [i.e. controls] the amount of water vapour the atmosphere can contain.

Until the industrial revolution, the greenhouse effect maintained the Earth's temperature at a level suitable for human civilisation to survive. If there had been no increase in the concentrations of the non-condensable greenhouse gases (CO_2, CH_4, N_2O, and O_3), the amount of water vapour in the atmosphere would not have changed. The additional non-condensable gases created by industrial and related activities cause the temperature to increase and this leads to an increase in the concentration of water vapour that further increases the temperature, i.e. there is positive feedback. There is also a possibility that adding more water vapour to the atmosphere could produce negative feedback. Thus, the possible positive and negative feedbacks associated with increased water vapour and cloud formation act in opposite directions and complicate matters. The actual balance between them is an area of very active climate science research, see Section 6.4. While it has been realised for quite a long time that water vapour is the Earth's most abundant greenhouse gas, the extent of its contribution to global warming has been unclear.

In terms of concentration, CO_2 and water vapour are by far the most important of the greenhouse gases, see Table 1.3. However, CO_2 is not the only important product of the combustion of fossil fuels. Other gases, principally SO_2 and oxides of nitrogen (denoted generally by NO_x), are produced by the combustion of fossil fuels.

Some fuels are cleaner than others in terms of the pollution they produce. Thus, gas and oil can be treated so as to remove sulphur before they are burned and therefore to prevent SO_2 being produced when they are burned; coal, being a solid, cannot be treated in this way before it is burned and so

TABLE 1.3

Concentrations of principal atmospheric greenhouse gases

	CO_2	CH_4	CFC-11	CFC-12	N_2O
Pre-industrial concentration (1750–1800)	280 ppmv	0.8 ppbv	0	0	288 ppbv
Current concentration (2017)	405 ppmv	1.72 ppmv	235 pptv	527 pptv	325 ppbv
Approximate rate of increase	1.8 ppmv	0.015 ppmv	9.5 pptv	17 pptv	0.8 ppbv
Atmospheric lifetime (years)	(50–200)	10	65	130	150

Source: (IPCC 1990, partially updated).

Ozone and water vapour are not included.

ppmv = parts per million by volume, ppbv = parts per billion by volume, pptv = parts per trillion by volume.

For each gas in the table, except CO_2, the "lifetime" is defined here as the ratio of the atmospheric content to the total rate of removal. This time scale also characterises the rate of adjustment of the atmospheric concentrations if the emission rates are changed abruptly. CO_2 is a special case since it has no real sinks but is merely circulated between various reservoirs (atmosphere, ocean, biota). The "lifetime' of CO_2 given in the table is a rough indication of the time it would take for the CO_2 concentration to adjust to changes in the emissions.

expensive scrubbing facilities have to be used to remove the SO_2 from the gases produced by the combustion process. The production of oxides of nitrogen is more difficult to prevent because the nitrogen comes from the air used in the combustion process and not from the fuel itself. When considering the combustion of fossil fuels to produce greenhouse gases, particularly NO_x, which leads to the destruction of O_3 in the stratosphere, we should perhaps give particular attention to high-flying aircraft since they are injecting these gases directly into the stratosphere, whereas other combustion processes occur at ground level and produce gases which subsequently travel up to the stratosphere. It is interesting to note the story of the role of high-flying aircraft among the bogey men of ecologists. At the time of the development of the supersonic airliner Concorde, it was widely argued that these, and similar aircraft, would lead to a dramatic reduction in the concentration of stratospheric O_3. Subsequently it was argued that the US Shuttle programme would be responsible for producing HCl in the stratosphere which would reduce the stratospheric O_3 concentration. The story, as recounted along with the story of Concorde in Chapter 2 of the book by Gribbin (1988), is well worth reading. Basically the supersonic airliner threat came to be regarded as unreal and the Shuttle problem came to be regarded as relatively insignificant in comparison with the CFC problem. The figure given by the IPCC (IPCC 1992) for the contribution of high-flying aircraft to the atmospheric concentrations of oxides of nitrogen is relatively small, see Table 1.4.

TABLE 1.4

Estimated sources of oxides of nitrogen

Source	Nitrogen (Tg N per year)
Natural	
Soils	5–20
Lightning	2–20
Transport from stratosphere	~1
Anthropogenic	
Fossil fuel combustion	24
Biomass burning	2.5–13
Tropospheric aircraft	0.6

Source: (IPCC 1992).

TABLE 1.5
Estimated Sources of short-lived sulphur gases

Source	Emission (Tg S per year)
Anthropogenic emissions (mainly SO_2)	70–80
Biomass burning (SO_2)	0.8–2.5
Oceans (DMS)	10–50
Soils and plants (DMS and SO_2)	0.2–4
Volcanic emissions (mainly SO_2)	7–110

Source: (IPCC 1992).

The effect of NO_x as far as O_3 is concerned actually appears to depend on the height in the atmosphere. At low altitudes, the oxides of nitrogen favour the production of O_3, whereas at higher altitudes, they lead to the destruction of O_3. Apart from their role as greenhouse gases, SO_2 and NO_x are also undesirable products of combustion because through various reactions with water they produce mineral acids; these lead to acid rain (see Section 1.2.2).

As far as SO_2 is concerned, it should be pointed out that not all the gaseous sulphur compound emissions into the atmosphere arise from anthropogenic causes. SO_2 is liberated by volcanoes. H_2S and DMS (dimethyl sulphide) are liberated by plants and DMS is liberated by plankton in the ocean. The relative magnitudes of the emissions from these various sources are indicated in Table 1.5.

It has been suggested (Charlson et al. 1987) that, apart from producing acid rain, the DMS emissions by plankton lead to sulphate particles which constitute hygroscopic nuclei for cloud drops and significantly alter the cloud patterns and, thereby, the radiation balance (for an early account, see Slingo 1988 and Kreidenweis et al., 2019).

1.1.6.5   Human activities and the greenhouse effect

We have alluded to the greenhouse effect on several occasions in this section; the time has come to be a little more specific. First, from the point of view of life on Earth, the greenhouse effect is good, not bad. The average temperature of the surface of the Earth is 13°C, and in the absence of any greenhouse gases in the atmosphere, it would be reduced by 33 °C, i.e. to −18 °C. A temperature of −18 °C may not sound too cold and many of us will have experienced that temperature briefly. But as an average temperature, it would make the Earth a rather inhospitable place and any semblance of modern agriculture would only be possible in a very few locations. In past ice ages, the drop in average temperature has been less than 10 °C (not 33°C). However, what is worrying people at the moment is the possibility of a substantial global warning as a result of changes in the greenhouse effect induced by human activities, although in some parts of the world an increase in temperature would be welcome. We have touched on many of these human activities already in this section. These are the production of CO_2 and oxides of nitrogen from combustion processes and there is the release of CFCs. In addition to this there is methane, CH_4, which is produced by a whole variety of sources, some of which are natural and some of which are not, see Table 1.6.

The main greenhouse gases (except O_3), their concentrations, their rate of increase in concentration, and their lifetime in the atmosphere are given in Table 1.4. The increases in the concentrations of these gases since the Industrial Revolution are shown in Figure 1.17.

Note that O_3 is not included in this table because of lack of precise data. A discussion of the problem of obtaining global data over a long timescale for O_3 concentrations will be found in Sections 1.6 and 1.7 of the IPCC's the First Assessment (IPCC 1 1990). SO_2 is also not included in Table 1 because it does not survive very long as a gas. It does, however, react with water to produce sulphuric acid aerosols; if present in sufficient quantities in the stratosphere, where the half-life is about one year, these aerosols can significantly affect the radiation budget and thus alter the greenhouse effect.

TABLE 1.6

Estimated sources and sinks of methane (Tg CH_4 per year)

	Annual Release[1]	Range[2]
Source		
Wetlands (bogs, swamps, tundra, etc.)	115	100–200
Rice paddies	110	25–170
Enteric fermentation (animals)	80	65–100
Gas drilling, venting, transmission	45	23–50
Biomass burning	40	20–80
Termites	40	10–100
Landfills	40	20–70
Coal mining	35	19–50
Oceans	10	5–20
Freshwaters	5	1–25
CH_4 hydrate destabilisation	5	0–100
Sink		
Removal by soils	30	15–45
Reaction with OH in the atmosphere	500	400–600
Atmospheric increase	44	40–48

Source: (IPCC 1990).
[1]   (Tg CH_4 per year).
[2]   (Tg CH_4).

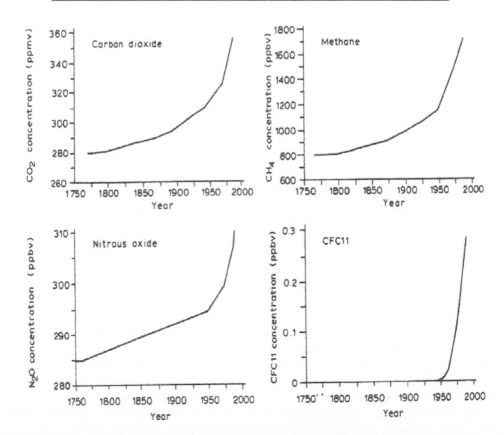

FIGURE 1.17   Increases in atmospheric gases since the industrial revolution.

The whole question of aerosol arising from SO_2 from fossil fuel combustion from SO_2 and dust from volcanic eruptions and from dust from other sources is important but very difficult to quantify. What is more important than the actual concentrations, or emissions of greenhouse gases is their actual contribution to global warning. We shall see later how the effect of all these gases is taken into account in climate modelling. The simplest thing, but one that is not particularly accurate, is to assume that for climate modelling one can represent all these other greenhouse gases by an equivalent amount of CO_2 that gives rise to an equivalent warming effect to that of all these gases.

For a while it was popular to mention above-ground nuclear explosions in the era when such tests were being undertaken. But that was only for a limited period of time and such tests are no longer carried out, so that whatever damage they might have been doing is no longer an issue; they are over and done with. The effects of those tests were probably not very different, as far as weather and climate are concerned, from those of volcanic eruptions. The question of a nuclear winter, following a supposed nuclear war, was a lively issue for a while and will be discussed in Chapter 8. The effects of such a war on the weather and climate would probably be trivial in relation to the more general damage inflicted on civilisation by such a conflict.

1.2 THE EMERGENCE OF THE MODERN ENVIRONMENTAL MOVEMENT (ALIAS THE GREEN MOVEMENT)

1.2.1 BACK TO NATURE

Our hunter-gatherer ancestors were not concerned with global climate although they would have gained some knowledge and understanding of their local climate. The population migrations that occurred over long distances at various times in the past would have been based on some knowledge of the climate in distant places. Although communications were much slower than at present, we should not assume that people were ignorant of life elsewhere. Once groups began to settle down and acquire leisure from feeding and clothing themselves, building themselves shelter and defending themselves and their homes and possessions they began to philosophise and study science. The latter included the study of the climate.

An important part of our approach to the understanding of climate change is to realise that the current public concerns about climate change have been stirred up by the green political movement and by climate modellers, the latter working through the IPCC which some people regard as having a political agenda. We now examine that claim in a little more detail and refer to an intriguing book "Green Tyranny" by Rupert Darwall (2017). Darwall's book is a highly political book as its title and subtitle "Green Tyranny: exposing the totalitarian roots of the climate industrial complex" indicates. He argues that modern environmentalism is destroying democracy. We have tried to see past the various political points that he is trying to make and to note his very important scientific points. He reminds us that Svante Arrhenius, who was awarded the 1903 Nobel prize for chemistry, produced a paper in 1896 in which he estimated that a doubling of the concentration of atmospheric CO_2 would lead to a 5 to 6°C rise in global temperature. Being from Sweden, a cold country, Arrhenius seems to have welcomed the idea of a warming of the climate as a result of fossil fuel burning. Later on, there was an element in Russian scientific thought that considered an aspect of global warming would be welcome in Russia, another cold country (see Chapter 7).

The Industrial Revolution led to enormous changes in the way of life of many people. Peasants or serfs who had lived peaceful but exploited lives in the countryside for centuries became dispossessed and flocked to the new big industrial cities to work in a new form of slavery in the mills and factories and to live in the vast urban slums which were created (see Varoufakis 2017, Chapter 2). The 19th century saw the beginnings of a "back to nature" movement in various countries. Arguably this was stronger in 19th century Germany than in other countries involved in industrialisation at that time. The Lebensreform (Life reform) movement was a reaction to the industrialisation and the urbanisation and the slums which accompanied it. According to Wikipedia, Wandervogel was the name adopted by a popular movement of German youth groups from 1896 onward. The name can be translated as rambling, hiking, or wandering bird (differing in meaning from "Zugvogel" or migratory bird) and the idea was to shake off the restrictions of industrial society and get back to nature and freedom. Wandervogel can be traced back to its predecessor the Lebensreform of earlier 19th century Germany. There were numerous groups across Germany dedicated to various concepts associated with Lebensreform, such as ecology and organic farming, vegetarianism, naturism/nudism (freie Körperkultur or FKK as it is widely known today in Germany), and abstinence from alcohol and tobacco. Dozens of magazines, books, and pamphlets were published on these topics. Different groups espoused different political philosophies and some groups were non-political. Darwall (2017) argues that an important step in the development of modern environmentalism involved the taking over of the Lebensreform/Wandervogel movement of the late 19th century and early 20th century Germany. Chapter 4 of his book is entitled "Europe's First Greens" and makes interesting reading. He quotes Biehl and Staudenmaier (1995) "Some of the themes that Nazi ideologists articulated bear an uncomfortably close resemblance to themes familiar to ecologically concerned people today." He also quotes from Hitler's *Mein Kampf* of 1925–6 "When man attempts to rebel against the iron logic of Nature, he comes into struggle with the principles to which he himself owes his existence as a man." The Nazis' involvement with nature-politics goes back to the Wandervogel which was a general back to nature and politically diverse movement. Darwall points out that "The Nazis' involvement in nature politics led them to advocate green policies half a century before any other political party." He also points out that if we leave aside the Nazis "race-hate, militarism and desire for world conquest and look at their identification with nature-politics then seen from the present day it might come as a shock, but not a surprise, to a present-day environmentalist or green politician that Hitler and the Nazis were the first to advocate renewable energy programmes." There is, understandably, a reluctance in Modern Germany to look back to 19th century culture because so much of it was taken over by the Nazis.

Darwall goes on to pursue the argument that the Nazis were the first to advocate renewable energy programmes. In 1932, the Nazis' newspaper, the *Vötkischer Beobachter*, published a long article on "National Energy Policy." The article claimed that "wind power could transform the economy and give employment to millions of Germans ("green jobs," in today's political parlance)." There would be greater fuel economy through lightweight cars and trains, and tariffs would encourage car drivers to switch to rail. Cars would also use domestically produced synthetic fuels. The program[me] would completely change the whole economy, the article concluded, improving money circulation and reducing imports and cost burdens in one of the first descriptions of green growth, " that alchemical process popular with governments in the twenty-first century whereby making the production of useful energy less efficient would help economies grow rather than do the opposite."

In 1932, Franz Lawaczeck, the Nazi party's economic spokesman and an engineer by profession, wrote a book, *Technik und Wirtschaft im Dritten Reich. Ein Arbeitsbeschaffungsprogramim* (Technology and the Economy in the Third Reich – A Programme for Work). This was both an anti-capitalist tract and a blueprint for a renewable energy future. Leaving aside the political arguments, Darwall notes that on energy, Lawaczeck considered that coal had changed the world and attained an overwhelming dominance. Because it was so cheap, it was used wastefully, and long-term priorities were not taken into consideration. Steam turbines used only one-fourth of the energy content of coal, and [steam] locomotives used only one-twelfth. Instead of being used for energy, coal would be more valuable making chemicals and other products. By contrast, hydro and wind power were much more efficient, converting up to four-fifths of their energy inputs into useable energy. Those inputs were from nature. Lawaczeck argued that the future of wind power was with very large towers like high radio masts, at least 100 metres (328 feet) high, with 100-metre diameter turbines which, unlike modern turbines, were designed to rotate in a horizontal plane. Lawaczeck conceded that hydro and wind power suffered from intermittency. It was impossible to use hydro in winter or in dry, summer months. The situation was even more difficult for wind because of changes in wind speed. The system therefore had to store energy. Lawaczeck foresaw energy from wind power being converted and stored in the form of hydrogen. Although it is technically feasible and there have been a few pilot projects, the widespread use of hydrogen to store electricity still many years later largely remains a dream. Lawaczeck questioned why coal was so dominant in capitalist societies and why only five per cent of Germany's potential hydro power was actually used. The answer lay in the upfront investment cost. Wind, like hydro, requires huge amounts of capital. As Lawaczeck and the Nazis grasped – and modern governments would discover in their turn – the market would not provide the capital if left to itself. Darwall cites proposals in 1932 for wind generators 70–90 m, 100 m, 250 m, and even 540 m high. At 540 m (1771 ft), this one would have been 218 ft higher than the Empire State Building, the world's highest building at the time. Large-scale deployment of Nazi wind power never got off the ground. Shortly after Hitler came to power, Lawaczeck lobbied the Reich Ministry of Finance to support wind power. Ministry bureaucrats were sceptical about the costs but provided funding for technical research. In 1935, the wind projects were scaled down because of technical risks. But the finance ministry did give money for a small, 80 m (262-ft.) tower with a 50-m turbine, specifying a target cost of 165 RM per kW. Although the rearmament lobby gained the upper hand in pressing for more coal-fired plants, Nazi funding of wind power research continued until 1944 under the auspices of the Reich Windpower Study Group. "Renewable energy was not the only thing that made Hitler and the Nazis Europe's first greens. As is well known, leading Nazis from Hitler down were vegetarians. In his Four Year Plan, Hermann Goring said farmers who fattened cattle on grain that could have been used to make bread were 'traitors.' Himmler blamed commercial interests,"

In the preface to his book "Green Tyranny," Rupert Darwall (2017) explains how he came to write it as a sequel to his earlier book "The Age of Global Warming – A History"(2013), He was approached by a reader with a professional knowledge of the subject going back to the mid-1980s. He recalls the gist of the conversation, that although he had written a good book, like most English and American writers, he had missed out on developments in continental Europe, especially Sweden and Germany. Darwall considered the evidence, and saw that the reader was right.

We have started with Darwall's second book (of 2017) rather than his first one (of 2013) because it goes back over the very early history of the modern environmental movement (Green politics). The earlier book was a very detailed critique of what we might call the global warming scare; we shall come back to that later (in Chapter 7). Darwall points out that were it not for its impact on industrialised societies' reliance on hydrocarbon energy, theories of man-made climate change would principally be of limited academic interest. This is an absolutely essential point if we are concerned with understanding global climate change and it is what this chapter is all about. As we have already said in the Preface, there are many threats to our way of life and, indeed, to our very survival as a species here on Earth. Climate change is only one of them and we must try and see climate change in perspective. We shall consider this further in the Section 1.3.

A landmark event in the awakening of environmental concerns was the publication of Rachel Carson's "Silent Spring" in 1962 (Carson 1962). The title "Silent Spring" arises from the account of the death of enormous numbers of birds as a result of indiscriminate spraying of pesticides and the story of the opposition the book faced from the chemical industry, the conflict of interest within the US Department of Agriculture and the eventual setting up of the Environmental Protection Agency in the USA and the worldwide restrictions on insecticides, especially DDT, is well known and we do not need to repeat it here (Davis et al., 2018).

1.2.2 Acid rain

In the later 1960s, the widespread response to *Silent Spring* was to ban DDT, but Sweden "was putting coal – the most ubiquitous source of electrical energy – in the cross hairs when it made acid rain the world's top environmental problem." We therefore need to examine the political and social situation that existed in Sweden at that time.

We return to Arrhenius. Apart from his work on CO_2 and global warming, Arrhenius also believed in technology and progress and was appointed to a Swedish government commission to investigate the potential of hydroelectricity. He also took a leading role in setting up the Nobel Prizes and was himself awarded the 1903 Nobel Prize for Chemistry. However, technology and progress had a darker side. In 1909, Arrhenius became a board member of the newly formed Swedish Society for Racial Hygiene and in 1922, following lobbying by Arrhenius, the Swedish government, established the State Institute for Racial Biology "to study ways of improving racial characteristics through selective [human] breeding." According to Björkman and Widmalm (2010), "The Swedish eugenics network may have been relatively small but it was nevertheless historically significant because of its intimate ties with that part of the German eugenics movement that would share Nazi biopolitics." The State Institute for Racial Biology was created by 1922 by Sweden's first Social Democratic government. The government wanted to develop the far north of Sweden which was inhabited only by about 50,000 Lapps who were nomads who raised reindeer and hunted moose. The Racial Biology Institute set out to find evidence that the Lapps were an inferior people to the Swedes and also gathered material to assess the racial makeup of 100,000 Swedes.

The Swedish Social Democrats were in power intermittently in the 1920s and then held power continuously from 1932 until 1976 and according to Darwall "Modern Sweden is their creation and their creature." He also quotes from Roland Huntford's book "The New Totalitarians" (1971) which "points out that of all socialist parties, only Sweden's Social Democrats could trace a direct and undefiled descent from Karl Marx himself. Of all countries, Sweden under the Social Democrats had come closest to realizing Aldous Huxley's nightmare of the *Brave New World* …. Swedes were submissive to authority and deferential to experts, which gave Sweden's technopolitical establishment a "singularly malleable population to work with, and has been able to achieve rapid and almost painless results. The system created by the Social Democrats…" Huntford wrote "has proved to be an incomparable tool for applying technology to society. They have altered the nature of government by making it a matter of economics and technology alone." The rest of this chapter (pages 25–28) of Darwall's book provides fascinating reading about Sweden's political social history under Sweden's Social Democratic governments over about 50 years up to 1976; but it is peripheral to the scientific purpose of this book.

In this section, we have dealt with the Nazis and the Swedish Social Democrats and we have relied heavily on Rupert Darwall's book because, as we noted earlier in the Section, "most English and American writers … missed out on developments in continental Europe, especially Sweden and Germany." We shall return to these developments in later sections of this chapter to provide background to the later chapters.

Darwall (2017) devotes two whole chapters (Chapters 6 and 7) to the story of acid rain from 1972 to the present day and draws a striking parallel between acid rain and global warming, both of which he regards as scares. These two chapters of Darwall's book make interesting reading from a historical view point and as general background to the issue of the case of global warming. But acid rain (like the nuclear winter to which we shall turn in Section 7.1) is no longer a live issue. Therefore, it would be of no great interest to go through all the details of the rise and fall of acid rain. Darwall's conclusion is scathing "Acid raid was not just a scare. It was a scientific scandal….. With global warming, the public is asked to put their trust in the judgment of national science academies. Acid raid provides an objective test of their trustworthiness. The national academies of five countries – those of America, Canada, Britain, Sweden and Norway – produced reports on acid rain that were biased and unscientific. Their errors remain uncorrected and unretracted. Collectively, the academies stand guilty of collusion in scientific malpractice. Their authors presumed to know too much and downplayed uncertainty and lack of knowledge in furtherance of a political agenda." (Darwall 2017, pages 81–82).

In 1970, the UN secretary-general asked member governments to present environmental case studies. The Swedish government replied with a proposal on acid rain and convened a group of experts chaired by the head of Stockholm's International Meteorological Institute, Bert Bolin. The result was, according to Darwall, that acid rain not only was a precursor of global warming, but also was the prototype. Both mobilised the same constituencies – alarmist scientists, NGOs, and credulous politicians – amplified by sensationalist media reporting. The target was the same – fossil fuels, especially coal. For both, scientists could point to a scientific pedigree stretching back into the 19th century – the term "acid rain" was first coined by the British scientist Robert Angus Smith in the 1870s. Both acid rain and global warming originated in Scandinavia and were taken up by Canada. The same scientists were active in both. The first chair of the IPCC, Bert Bolin, wrote the first governmental report on acid rain. Both had the capacity to induce hysteria – if anything, more extreme in Germany at the height of the acid rain scare in the early 1980s than anything since then. The politically favoured "solutions," pushed by activist scientists and NGOs, were the same – emissions cuts. Both produced UN conventions (the 1979 Geneva Convention on Transboundary Pollution and the 1992 UN Framework Convention on Climate Change (UNFCCC)) and protocols to implement the emissions cuts not agreed in the original conventions (the 1985 Helsinki and the 1997 Kyoto protocols, respectively).

1.2.3 THE CLUB OF ROME, LIMITS TO GROWTH, THE STOCKHOLM CONFERENCE 1972, ONLY ONE EARTH, BLUEPRINT FOR SURVIVAL

We quote from the preface to Limits to Growth (Meadows et al. 1972):

"In April 1968, a group of thirty individuals from ten countries – scientists, educators, economists, humanists, industrialists, and national and international civil servants – gathered in the Academia dei Lincei in Rome. They met at the instigation of Dr. Aurelio Peccei, an Italian industrial manager, economist, and man of vision, to discuss a subject of staggering scope – the present and future predicament of man." Out of this meeting grew The Club of Rome, an informal organization that has been aptly described as an "invisible college." "Its purposes are to foster understanding of the varied but interdependent components – economic, political, natural, and social – that make up the global system in which we all live; to bring that new understanding to the attention of policy–makers and the public worldwide; and in this way to promote new policy initiatives and action."

The Club of Rome remained an informal international association, with a membership that grew to approximately 70 persons of 25 nationalities. None of its members holds public office, nor does the group seek to express any single ideological, political, or national point of view. All are united, however, by their overriding conviction that the major problems facing mankind are of such complexity and are so interrelated that traditional institutions and policies are no longer able to cope with them, nor even to come to grips with their full content.

The members of The Club of Rome had backgrounds as varied as their nationalities. Dr. Peccei, still the prime moving force within the group, is affiliated with Fiat and Olivetti and manages a consulting firm for economic and engineering development, Italconsult, one of the largest of its kind in Europe. Other leaders of The Club of Rome include: Hugo Thiemann, head of the Battelle Institute in Geneva; Alexander King, scientific director of the Organization for Economic Cooperation and Development; Saburo Okita, head of the Japan Economic Research Center in Tokyo; Eduard Pestel of the Technical University of Hannover, Germany; and Carroll Wilson of the Massachusetts Institute of Technology. Although membership in The Club of Rome is limited, and will not exceed one hundred, it is being expanded to include representatives of an ever greater variety of cultures, nationalities, and value systems.

A series of early meetings of The Club of Rome culminated in the decision to initiate a remarkably ambitious undertaking – the Project on the Predicament of Mankind. The intent of the project is to examine the complex of problems troubling men of all nations: poverty in the midst of plenty; degradation of the environment; loss of faith in institutions; uncontrolled urban spread; insecurity of employment; alienation of youth; rejection of traditional values; and inflation and other monetary and economic disruptions. These seemingly divergent parts of the "world problematique," as The Club of Rome calls it, have three characteristics in common: they occur to some degree in all societies; they contain technical, social, economic, and political elements; and, most important of all, they interact."

The outcome of this project was the publication in 1972 of the book *Limits to Growth* (Meadows et al. 1972).

The United Nations Conference on the Human Environment (also known as the Stockholm Conference) was an international conference convened under United Nations auspices and held in Stockholm, Sweden, from 5–16 June 1972. It was the UN's first major conference on international environmental issues and marked a turning point in the development of international environmental politics. According to Darwall, "Acid rain was the reason Sweden wanted the UN conference on the environment (Bergquist and Söderholm, 2017). In 1967, Svante Odén, a Swedish soil scientist, had produced a complete theory of acid rain and wrote a sensationalist article in the Swedish daily newspaper *Dagens Nyheter* on a "chemical war" between the nations of Europe. Outside Scandinavia, acid rain was not considered important" (Darwall 2017, page 57). The formal report of the Stockholm conference was published by the UN (Stockholm 1972). The report included a declaration that is full of noble thoughts and good words (Keong, 2018). It is interesting to note that this is totally centred on *homo sapiens* as a species, i.e. the world exists to support humankind. Nowadays it is argued by some people that other species have rights and there is no apparent trace of that in this declaration. The declaration was followed by a list of 28 principles and, in turn, this was followed by a list of 108 recommendations. These recommendations, in turn, were arranged into an Action Plan, involving three components (i) Environmental assessment, (ii) Environmental management, and (iii) supporting measures. As to the Recommendations, by now they have either been adopted or assimilated into subsequent actions or are no longer relevant; there is no need to study them in detail.

The conclusions of the United Nations Conference on the Human Environment, which was held in 1972, in Stockholm and which is commonly referred to as the Stockholm Conference (see Section 1.2.3) were quite general "to inspire and guide the peoples of the world in the preservation and enhancement of the human environment" and did not specifically address the perceived problem of acid rain. Even though acid rain may no longer be a serious issue or scare, it can be argued that global warming clearly still is a scare. Acid rain was not mentioned explicitly at all in the Principles that were enunciated at the Conference but air pollution generally was (Stockholm 1972). Principles 21 and 22 stated:

Principle 21
States have, in accordance with the Charter of the United Nations and the principles of international law, the sovereign right to exploit their own resources pursuant to their own environmental policies, and the responsibility to ensure that activities within their jurisdiction or control do not cause damage to the environment of other States or of areas beyond the limits of national jurisdictions.

Principle 22
States shall co-operate to develop further the international law regarding liability and compensation for the victims of pollution and other environmental damage caused by activities within the jurisdiction or control of such States to areas beyond their jurisdiction.

The issue of transborder transport of pollutants did become important later.

Climate change was not mentioned at all in the principles enunciated by the Stockholm Conference. As far as global warming is concerned, we cannot detach ourselves from the political aspects of its history because our thesis is that global warming is only one of many threats to our way of life and, indeed, our survival as a species. The politics of environmentalism in general – and global warming in particular – in the period 1972–1988 are very complicated, but unfortunately, they are very relevant to our study of global warming.

After the Stockholm conference, Sweden used the Organisation for Economic Co-operation and Development (OECD) to push for a binding treaty to cut the emissions of sulphur dioxide, SO_2, which causes acid rain. Then, in 1975, the Soviet Union made its first strategic use of the environment as a propaganda tool, when President Leonid Brezhnev made a speech in which he said that the environment was an issue on which East and West shared a common problem. Brezhnev's intervention in the acid rain issue illustrates how the development of environmentalism became entwined in the politics of the Cold War. Brezhnev's environmental push aimed to deflect Western pressure on the Soviet Union's human rights record and focus attention to other issues covered by the 1975 Helsinki Agreement on Security and Cooperation in Europe. Scandinavian countries used Brezhnev's opening to press their claims within the framework of the United Nations Economic Commission for Europe (UNECE), which included the Soviet bloc as well as the United States and Canada. The outcome was the 1979 Geneva Convention on Long-Range Transboundary Air Pollution. Although the objective of limiting and then gradually reducing air pollution was listed as the first of the convention's objectives, like the later 1992 UN Framework Convention on Climate Change, it had no targets or timetables to cut SO_2 emissions. The Swedes and Norwegians had proposed strict standstill and rollback clauses, but these were rejected by the United States, the United Kingdom, and West Germany. Even so, a toothless agreement to monitor and evaluate long-range pollution, starting with SO_2, was a foot in the door. To mark the tenth anniversary of the Stockholm conference [i.e. in 1972], Sweden called an international conference on the acidification of the environment. It then proposed a timetable for a 30 per cent cut in SO_2 emissions. A group of ten UNECE members, including the Scandinavians, Canada, Austria, and Switzerland, then formed the 30 per cent club.

Rather than the official UN Report on the Stockholm Conference, there is a more readable account to be found in the Pelican book "Only One Earth – The Care and Maintenance of a Small Planet" by Barbara Ward and René Dubos which was first published in 1972 (Ward and Dubos 1972). This is slightly unusual in that it is described on its title page as "An unofficial report commissioned by the Secretary-General of the United Nations Conference on the Human environment, prepared with the assistance of a 152-member Committee of Corresponding Consultants in 58 Countries." The names and affiliations of these 152 individuals are listed after the Preface in the book. As the Secretary-General says in the preface "This report is the result of a unique experiment in international collaboration. A large committee of scientific and intellectual leaders from fifty-eight countries served as consultants in preparing the report ..... The names of Barbara Ward and René Dubos are listed quite properly as authors of the report. They are indeed responsible for the drafting and revision of the manuscripts to which they both contributed ... But in this case, the role of the authors is more accurately described as creative managers of a cooperative process – one which engaged many of the world's leading authorities as consultants in the multiple branches of environmental affairs." This is interesting in the sense that it was a pioneering venture and the model introduced has now become well established, perhaps most famously in the case of the IPCC in the investigations which have led to the various reports produced by that Panel.

About the same time as the Stockholm Conference, number 1 of volume 2 of *The Ecologist* published in January 1972 was republished as a Penguin Special under the title "Blueprint for Survival" (Goldsmith et al., 1972). We quote from its preface:

> This document has been drawn up by a small team of people, all of whom, in different capacities, are professionally involved in the study of global environmental problems.
>
> Four considerations have prompted us to do this:
>
> 1. An examination of the relevant information available has impressed upon us the extreme gravity of the global situation today. For, if current trends are allowed to persist, the breakdown of society and the irreversible disruption of the life-support systems on this planet, possibly by the end of the century, certainly within the lifetimes of our children, are inevitable.
> 2. Governments, and ours is no exception, are either refusing to face the relevant facts or are briefing their scientists in such a way that their seriousness is played down. Whatever the reasons, no corrective measures of any consequence are being undertaken.

3. This situation has already prompted the formation of the Club of Rome, a group of scientists and industrialists from many countries which is currently trying to persuade governments, industrial leaders, and trade unions throughout the world to face these facts and to take appropriate action while there is yet time. It must now give rise to a national movement to act at a national level, and if need be to assume political status and contest the next general election. It is hoped that such an example will be emulated in other countries, thereby giving rise to an international movement, complementing the invaluable work being done by the Club of Rome.

4. Such a movement cannot hope to succeed unless it has previously formulated a new philosophy of life, whose goals can be achieved without destroying the environment, and a precise and comprehensive programme for bringing about the sort of society in which it can be implemented.

This we have tried to do, and our Blueprint for Survival heralds the formation of the MOVEMENT FOR SURVIVAL (see p. 135) and, it is hoped, the dawn of a new age in which Man will learn to live with the rest of Nature rather than against it.

"The Blueprint" opened with the words "The principal defect of the industrial way of life with its ethos of expansion is that it is not sustainable." It identified a number of serious problems:

1. The disruption of ecosystems.
2. The depletion of non-renewable resources, both energy resources and nonenergy resources, mostly metals.
3. Population expansion and the failure of food supplies.
4. The collapse of society.

It is difficult to avoid coming to the view that people had a much more balanced view of the threats to our welfare and our survival in 1972–3 than was developed 20 years later when (some) people became obsessed with CO_2 and emissions from the use of fossil fuels. In spite of the widespread obsession with emissions of CO_2 from the burning of fossil fuels and emissions from the use of fossil fuels, there is also widespread concern that CO_2 emissions are far from being the only threat to our way of life, see e.g. Idso et al. (2015a) and Berners-Lee (2019). We shall return to this issue in Section 1.3 and in Chapter 8.

1.2.4   COLD WAR COOPERATION BETWEEN EAST AND WEST

During the Cold War, there was some interaction between Soviet and Western climate scientists on the issue of the Nuclear Winter and we shall discuss this later (see Section 8.3). In addition, there was engagement which culminated in the setting up of the International Institute for Applied Systems Analysis (IIASA) in Austria, which was driven primarily by the then two superpowers, the Soviet Union and the USA. On 4 October 1972, representatives of the Soviet Union, United States, and 10 other countries from the Eastern and Western blocs met at The Royal Society in London to sign the charter establishing the IIASA. It was the culmination of six years' effort driven forward by both the US President Lyndon Johnson and the USSR Premier Alexei Kosygin. For IIASA, it was the beginning of a remarkable project to use scientific cooperation to build bridges across the Cold War political divide and to confront growing global problems on a truly international scale. The first scientist arrived at IIASA in June, 1973. Clearly, success at bridge building and successful science would go hand in hand. But neither was a foregone conclusion. This was the 1970s, and most research organisations focused on national issues. Few encouraged researchers from different countries or disciplines to work together for the greater good. As an aside we note that when the Cold War ended (in 1991), IIASA's sponsoring countries could have said "mission accomplished" and disbanded the Institute. The IIASA had certainly helped foster mutual understanding among scientists from East and West. But it had also done more than this. The IASA had shown the scientific benefits of bringing together different nationalities and disciplines to work towards common goals. Indeed, this approach has been widely imitated, for example, in the Intergovernmental Panel on Climate Change and the International Geosphere-Biosphere Programme. Instead of closing in the 1990s, the Institute broadened its mandate from the East and West to a truly global focus. (Source: A Brief History of IIASA (webarchive.iiasa.ac.at accessed 25 November 2018)).

The road from the signing of the *Agreement on Cooperation in the field of Environmental Protection* and the setting up of IIASA in 1972 until the setting up of the IPCC in 1988 was long and complicated and, in the view of many people, highly political, regardless of the claims of the IPCC itself to be non-political.

In the 1960s, there was a general recognition that the Soviet Union and the USA shared a range of common environmental problems, in spite of their ideological differences. In addition, the prevailing rhetoric suggested that both sides anticipated benefits from the agreement beyond the general political aspects of the initiative. Thus, in addition to the establishment of IIASA, there was also an agreement on specific US-USSR interaction. The two presidents (Nixon and Brezhnev) supported the signing of the US-USSR *Agreement on Cooperation in the Field of Environmental Protection* on 23 May 1972. It was agreed to cooperate in 11 specific areas:

- air pollution;
- water pollution;
- environmental pollution associated with agricultural production;
- enhancement of the urban environment;
- preservation of nature and the organisation of reserves;
- marine pollution;
- biological and genetic consequences of environmental pollution;
- influence of environmental changes on climate;
- earthquake prediction;
- arctic and subarctic ecological systems;
- legal and administrative measures for protecting environmental quality.

1.2.5 SOME MORE CONFERENCES

The First World Climate Conference was sponsored by the World Meteorological Organization (WMO) and was held on 12–23 February 1979 in Geneva. It was one of the first major international meetings on climate change, as distinct from the Stockholm Conference which had been much more general. Essentially it was a scientific conference and it was attended by scientists from a wide range of disciplines. In addition to the main plenary sessions, the Conference organised four working groups to look into climate data, the identification of climate topics, integrated impact studies, and research on climate variability and change. The Conference emphasised that, although humans are now able to take advantage of the present favourable climate, human activity is being strongly affected by climatic variability, especially by droughts and floods, and particularly in the developing countries located in either arid or semi-arid zones or excessively humid regions. It also emphasised that the interdependence of the climates in different countries necessitates a coordinated global-scale strategy for a better understanding and rational use of climate. Whereas at the time it was realised that human activity caused inadvertent local-scale and (to some extent) regional-scale climate changes, there was deep concern that further intensification of human activity could lead to considerable regional and even global climate changes, and hence that international cooperation was urgently needed in solving this problem. The Conference was followed by the 8th World Meteorological Organization (WMO) Congress held in April–May 1979 in Geneva which made the decision to develop and accomplish a World Climate Programme (WCP). In turn the WCP established the World Climate Research Programme (WCRP) in 1980 under the joint sponsorship of the World Meteorological Organization (WMO) and the International Science Council (ISC) (previously the International Council for Science (ICSU), after its former name, the International Council of Scientific Unions until July 2018), and in 1993, the Intergovernmental Oceanographic Commission (IOC) of UNESCO also became a sponsor. The objective of the WCRP is stated as being to achieve "a better understanding of the climate system and the causes of climate variability and change" and "to determine the effect of human activities on climate "and although it has a much lower public profile than the IPCC it nevertheless has played, and continues to play, an important role in supporting and furthering climate research. The WCRP aims to foster initiatives in climate research which require or benefit from international coordination and which are unlikely to result from national efforts alone. The WCRP does not fund research. It carries out four Core Projects:

SPARC (Stratosphere-troposphere Processes And their Role in Climate)
CLIVAR (Climate and Ocean – Variability and Change)
CLiC (Climate and Cryosphere)
GEWEX (Global energy and Water Exchanges) (see WCRP 2009).

The WCRP also included TOGA (the Tropical Ocean and Global Atmosphere programme) and WOCE (the World Ocean Circulation Experiment). TOGA was a ten-year programme (1985–1995), aimed at the analysis of predictability of the interactive system "tropical ocean-atmosphere" and its impact on the global atmospheric climate on timescales from several months to several years. WOCE was aimed at the diagnosis of oceanic circulation and heat transport in the ocean with the use of new (especially satellite-based) observational means.

At a later stage, a very important programme, the Global Energy and Water Cycle Experiment (GEWEX), was initiated and entered its first phase in 1990–2002.

The Second International Conference on the assessment of the role of CO_2 and other greenhouse gases, together with aerosols and associated impacts on the biosphere and climate, was held in Villach, Austria, on 9–15 October 1985 and was sponsored by UNEP/WMO/ICSU. There is the inevitable bulky conference report (Villach 1986) but it is not worth considering this report in great detail now, except perhaps glancing at (1) the opening statement by Donald Smith, the Deputy Secretary General of WMO, and (2) the Executive Summary. The Conference was part of a process lasting from about 1972 up to the establishment of the IPCC in 1988. It was a milestone towards further consideration of climate changes. The participating countries totalled 29. The participants of the Villach Conference developed a consensus on the following: the atmospheric greenhouse gas concentration has increased (the CO_2 concentration has grown from 315 ppmv (parts per million by volume) in 1958 to 413.4 ppmv in January 2020; estimates from data on CO_2 in the bubbles of ice cores gave a preindustrial revolution concentration of 275 ± 10 ppmv. During the last 150 000 yr, the CO_2 concentration has varied between 200 ppmv and 270–280 ppmv and has been correlated with climate changes. They noted that the concentrations of other greenhouse gases (nitrous oxide, N_2O, methane CH_4, ozone O_3, and various chlorofluorocarbons, CFCs) were also increasing. It was also estimated that the combined global warming effect of these other greenhouse gases was "about as important" as that of CO_2 (World Meteorological Organization 1986).

The Statement to the conference participants included the following sentence: "As a result of increasing concentrations of greenhouse gases, it is now believed that in the first half of the next century a rise of global mean temperature can occur which is greater than any in man's history." The possibility of global climate warming requires an analysis of measures to provide successful socio-economic development in the future; this is all the more important because there is a close relation between climate changes and associated problems of sea level rise, the threat to the ozone layer, and acid rain. It was clearly expressed that there was a problem arising from climate warming and that the speed and amplitude of climate warming could be substantially reduced by more economical energy and fossil fuel consumption, which would reduce the level of greenhouse gas emissions to the atmosphere. Apart from that bland statement, the Conference recommended nothing except the setting up of a six-person Advisory Group on Greenhouse Gases (AGGG), with WMO, UNEP, and the International Council for Science (ICSU) nominating two members apiece. It was chaired by Kenneth Hare, the Canadian meteorologist. Hare had already chaired the Canadian review panel on acid rain that in 1983 had opined that the facts on acid rain were "actually much clearer" than other environmental causes célèbres and asserted that the existence of a severe environmental problem caused by acid rain was "not in doubt." If one scare wasn't enough, Hare was also chairing a Royal Society of Canada study into the nuclear winter. The by-now-inevitable Prof Bert Bolin was on the AGGG and joined by Gordon Goodman from the Stockholm-based Beijer Institute, which was an important funding channel for the effort – research money means power. At the same time, Goodman was writing the energy chapter of the Brundtland report (1987). Bolin had also been asked by ICSU to chair a committee scoping a research programme on the global dimension of chemical and biological processes. By the time it came to launch what became known as the International Geosphere Biosphere Program (IGBP) later in the year, Bolin was in an even better position to influence developments.

1.2.6 The Intergovernmental Panel on Climate Change

The Intergovernmental Panel on Climate Change (IPCC) is the leading international body for the assessment of climate change and it was created in 1988. It was set up by the World Meteorological Organization (WMO) and the United Nations Environment Programme (UNEP) to provide the world with a clear scientific view on the current state of knowledge in climate change and its potential environmental and socio-economic impacts. It was set up under the chairmanship of Professor B. Bolin. The Panel was charged with

1. Assessing the scientific information that is related to the various components of the climate change issue, such as emissions of major greenhouse gases and modification of the Earth's radiation balance resulting therefrom, and that needed to enable the environmental and socio-economic consequences of climate change to be evaluated.
2. Formulating realistic response strategies for the management of the climate change issue.

The Panel established three Working Groups:

1. (Chairman: Dr J. T. Houghton) to assess available scientific information on climate change;
2. (Chairman: Prof. Y. Izrael) to assess environmental and socio-economic impacts of climate change;
3. (Chairman: Dr F. Bernthal) to formulate response strategies.

In addition, it also established a Special Committee on the Participation of Developing Countries to promote, as quickly as possible, the full participation of such countries in the IPCC's activities.

The IPCC set out to prepare, based on available scientific information, assessments on all aspects of climate change and its impacts, with a view of formulating realistic response strategies. The initial task for the IPCC, as outlined in UN General Assembly Resolution 43/53 of 6 December 1988, was to prepare a comprehensive review and recommendations with respect to the state of knowledge of the science of climate change; the social and economic impact of climate change, and possible response strategies and elements for inclusion in a possible future international convention on climate. Today the IPCC's role is defined in the Principles Governing IPCC Work, as "to assess on a comprehensive, objective, open and transparent basis the scientific, technical and socio-economic information relevant to understanding the scientific basis of risk of human-induced climate change, its potential impacts and options for adaptation and mitigation. IPCC reports should be neutral with respect to policy, although they may need to deal objectively with scientific, technical and socio-economic factors relevant to the application of particular policies."

The IPCC reviews and assesses the most recent scientific, technical, and socio-economic information produced worldwide relevant to the understanding of climate change. It does not conduct any research nor does it monitor climate-related data or parameters. The IPCC produced its first report in 1990 and has produced several reports since then, making five in all so far:

First Report	1990
Second Report	1995
Third Report	2001
Fourth Report	2007
Fifth Report	2013–2014

The Sixth Report is expected to be finalised in 2020 or thereabouts.

In 2007, the Nobel Peace Prize was awarded to the IPCC and Al Gore (1992, 2006) "for their efforts to build up and disseminate greater knowledge about man-made climate change, and to lay the foundations for the measures that are needed to counteract such change." Winning the Nobel Peace Prize is obviously a great achievement, but there is a downside to this undoubted success. This is that it leads people to think that the emission of greenhouse gases and the consequent global warming is the major threat, or even the only threat, to the continuation of human life, or at least of our highly sophisticated society, on the planet. But this is not the case and it is our purpose in this chapter to try to put global warming into perspective among the various threats to our well-being and survival.

We shall turn to a detailed consideration of the activities of the IPCC over the last 30 years and discuss these reports in some detail in Chapter 6. In the meantime, we refer to a lecture that was given 30 years after the setting up of the IPCC on 15 October 2018 by Prof Hoesung Lee, the "Chair" of the IPCC entitled "Addressing Climate Change and Pursuing Economic Development: Reflections from the IPCC" at the Royal Society of Edinburgh. This was a useful lecture, coming as it did 30 years after the setting up of the IPCC in 1988, in the sense that it presumably reflects the current view of how the IPCC sees its role in the present world. The lecture was delivered only a week after the IPCC *Special Report on Global Warming of 1.5°C*, was published on 8 October 2018. This report essentially shows that limiting global warming to 1.5°C, rather than 2°C, would have clear benefits to people, countries, natural ecosystems, and the world. But Prof Lee indicated that achieving this, however, will require rapid and significant action and change across all parts of society.

It is important to try to separate the political use that is made of the IPCC reports from the role of the IPCC itself. Prof Lee's opening remark was to say that the role of the IPCC is to give governments, and others, information about climate change, so that policies can be based on objective analysis rather than prejudice or lobbying. It is all about "knowledge made useful," he smiled, quoting the Royal Society of Edinburgh's motto. In connection with this particular report, according to Prof Lee "some 91 authors from more than 40 countries were involved in drawing up the report and more than 6,000 pieces of evidence from the literature were cited. More than 42,000 comments from more than 1,000 experts were considered as part of the process, and five governments' hosted meetings. The hope is that the resulting documents, including a summary for policy makers, will help governments formulate appropriate policies." Of course, this begs the question of who chooses the 91 authors, the 6,000 pieces of evidence, and the handling of the 42,000 comments. Prof Lee was, indeed, questioned about this at the end of the lecture. He replied that "the IPCC has an extremely elaborate review system that involves three layers of review, including opening it up to expert review and comment. He said that the latest report (in its draft stages) attracted more than 40,000 comments, all of which were answered or addressed by the lead authors. He added that the comments and responses were made public – ensuring the openness of the process." We can only give the IPCC the benefit of the doubt and assume it tries to be "comprehensive, objective, open and transparent." Whether that succeeds is something that we individually should attempt to assess for ourselves. We shall return to the question of the IPCC and politics in Chapter 7.

1.2.7 WHY WE DISAGREE ABOUT CLIMATE CHANGE (HULME 2009)

We refer extensively to the Preface to Mike Hulme's book *Why We Disagree About Climate Change* (Hulme 2009) which we mentioned in the Preface. Hulme's book explores the idea of climate change: where the term originated, what it means to different people in different parts of the world, and why it is such a topic of disagreement. It is a book which also takes a different approach to the idea of climate change and how to work with it. He presents climate change as an idea as much as a physical phenomenon that can be observed, quantified, and measured. He observes that the latter is how climate change is mostly understood by scientists, and how science has presented climate change over recent decades. However, as people become increasingly confronted with the realities of climate change and hear of the dangers that scientists claim lie ahead, Hulme suggests that climate change has moved from being predominantly a physical phenomenon to being simultaneously a social phenomenon. These two phenomena are very different. "Far from simply being a change in physical climates – a change in the sequences of weather experienced in given places – climate change has become an idea that now travels well beyond its origins in the natural sciences. And as this idea meets new cultures on its travels and encounters the worlds of politics, economics, popular culture, commerce and religion – often through the interposing role of the media – climate change takes on new meanings and serves new purposes."

One can get a good idea of Hulme's view from a summary given in the Preface. We refer extensively to the Preface, reformatting slightly. The first two chapters set the scene for the rest of his book.

Chapter 1 – *The Social Meanings of Climate* – Hulme explores the different ways in which societies, over time, have constructed the idea of climate, and how they have related to its physical attributes. He offers a brief cultural reading of the history of climate. Some ideological purposes to which the idea of climate has previously been put are suggested and he also touches on the different ways in which changes in climate have been invoked to tell the story of the rise and fall of human civilisations.

Chapter 2 – *The Discovery of Climate Change* – Hulme explains how humans have come to believe that they are active agents in changing the physical properties of climate. He contrasts the earlier 19th-century reading of climate change as purely natural and occurring over geological timescales with the more recent appreciation that climates change on the timescales of human generations and can partly be moulded by human actions.

The following seven chapters then examine the idea of climate change from seven different standpoints, offering in each case reasons why there are disagreements about climate change. The seven facets cover: science, economics, religion, psychology, media, development, and governance.

Chapter 3 – *The Performance of Science* – In this chapter, Hulme claims that one of the reasons people disagree about climate change is because science and scientific knowledge is understood in different ways. Science thrives on disagreement; indeed, science can only progress through disagreement and challenge. But, he claims, disagreements presented as disputes about scientific evidence, theory, or prediction may often be rooted in more fundamental differences between the protagonists. These may be differences about epistemology, about values, or about the role of science in policy making. This chapter examines the changing nature of science and what significance this has for disagreements within scientific discourse about the existence, causes, and consequences of human-induced climate change.

Chapter 4 – *The Endowment of Value* – here Hulme turns his attention to economics and argues that one of the reasons people disagree about climate change is because everyone values things differently. We shall consider economics, particularly when discussing (computer) models in Chapter 5 (Section 5.6).

Chapter 5 – *The Things We Believe* suggests that one of the reasons we disagree about climate change is because people believe different things about themselves, the universe, and their place in the universe. These beliefs have a profound influence on attitudes, behaviour, and policies.

This chapter explores the different ways in which the major world religions have engaged with climate change, and also how other large-scale collective movements have recognised spiritual or non-material dimensions of the phenomenon.

People also disagree about climate change because they worry about different things; an idea explored in Chapter 6 – *The Things We Fear*. Hulme notes that there is a long history of humans relating to climate in pathological terms. Climate-related risks continue to surprise and shock. A prospective, and not fully predictable, change in climate therefore offers fertile territory for the heightening of these fears. This chapter examines the construction of risk around climate change, drawing upon insights from social and behavioural psychology, risk perception, and cultural theory.

Chapter 7 – *The Communication of Risk* – Hulme considers the ways in which knowledge is communicated and shaped by the media and argues that one of the reasons for disagreements about climate change is because of multiple and conflicting messages about climate change that are then interpreted in different ways. This chapter examines the ways in which climate change has been represented in the media, by campaigning organisations, and by advertisers. It contrasts these representations with the construction and identification of climate change by the formal knowledge community. We shall consider this briefly in Section 7.12.

Chapter 8 – *The Challenges of Development* – Hulme suggests that one of the reasons people disagree about climate change is because they prioritise development goals differently. The definition of poverty and how to understand inequalities in the world are also important. This chapter outlines some of these different views and approaches and explains why an understanding of climate change cannot be separated from an understanding of development.

People with differing political ideologies and views of appropriate forms of governance also will have differing views about climate change and this is the subject of Chapter 9 – *The Way We Govern*. Climate change has been a live public policy issue since the late 1980s. Hulme makes the point that the Kyoto Protocol, negotiated in 1997, has been the benchmark international agreement for shaping the goals of (and disputes around) mitigation policies. He goes on to comment that the last few years have seen a growing attention as to whether or not a new genre of adaptation policies is necessary or desirable. This chapter explores the various ways in which governments have approached the design and implementation of climate policy – at local, national, and international scales – and also looks at the rise of non-state actors in climate governance. There have been significant developments in this area since 2009 and we shall discuss this later in Chapter 7.

In the final chapter of the book, Chapter 10 – *Beyond Climate Change* – Hulme offers a perspective on climate change which transcends the categories and disagreements explored in earlier chapters. It looks beyond climate change. The chapter argues that climate change is not a problem that can be solved in the way that, for example, technical and political resources were mobilised to "solve" the problem of stratospheric ozone depletion. We shall discuss again in Section 8.4 the problem of ozone depletion and the relative success of the Montreal Protocol in at least stemming the expansion of the ozone hole. For a time, it was hoped that the Kyoto Protocol would be similarly successful in dealing with the perceived human-induced global warming, see Section 1.1.6.2, but this does not appear to have been the case. As Hulme states, "we need to approach the idea of climate change from a different vantage point. We need to reveal the creative psychological, spiritual and ethical work that climate change can do and is doing for us. By understanding the ways in which climate change connects with these foundational human attributes we open up a way of re-situating culture and the human spirit at the heart of our understanding of climate. Human beings are more than material objects, and climate is more than a physical entity. Rather than catalysing disagreements about how, when and where to tackle climate change, the idea of climate change is an imaginative resource around which our collective and personal identities and projects can, and should, take shape."

While this does not answer "the problem," it is a viewpoint that is complementary to the viewpoint that we expressed (also in 2009) that climate change is only one of about a dozen serious environmental threats to our way of life and, possibly, our very survival (Cracknell 2009); we shall recall this in Section 1.3.10.

The concept of the resources needed for our way of life, expressed as a percentage of the known resources of the planet Earth, was developed by the Limits to Growth (see Section 1.3). In 1972, it was about 90%, of the estimated available resources, by 1992 it had increased to about 115%, and by 2005 it was over 120% (Meadows et al. 2005). This idea has been developed by Mike Berners-Lee in his excellent popular book "There is no Planet B: A Handbook for the Make or Break Years" (Berners-Lee 2019). In the text of that book, climate occupies only eight pages, out of 251 pages of text, while for example food occupies 40 pages and energy also occupies 40 pages, and so on; that perhaps better than anything else puts the whole global warming scare into perspective. We shall return to this in Section 1.3.

1.3 SUSTAINABILITY, SURVIVAL

1.3.1 INTRODUCTION

In Section 1.2.3, we discussed the Stockholm Conference, the Blueprint for Survival published by *The Ecologist*, and the Limits to Growth published by the Club of Rome. These were all produced in the early 1970s and they were all deeply concerned with the various threats to our well-being and, indeed our survival as a species. At that stage, global warming was not perceived as being a particularly serious matter in relation to the problems facing humanity. The First IPCC Scientific Assessment on Climate Change (IPCC 1990) was dominated by human-induced generation of greenhouse gases, estimating the consequent global warming and identifying various consequences of such global warming. As a consequence, in the years that followed 1990, the consumption of fossil fuels and the emission of greenhouse gases came to be seen as the major cause for concern for our future.

1.3.2 LIMITS TO GROWTH, THE CLUB OF ROME

The Club of Rome attempted to model some of the perceived threats other than global warming, but this was less successful. This was essentially because the input data were not so reliable and the processes to be modelled are not described simply by the laws of physics. CO_2 is uniformly mixed in the atmosphere. By this we mean that although the CO_2 reaches the atmosphere from various sources and is removed by various sinks, these local variations are rapidly evened out. The molecules move around due to their own kinetic energy and to the bulk movements of the air due to the winds to even out the local variations. Thus, by measuring the atmospheric concentration of CO_2 at one location (Mauna Loa) over a period of time, one can reliably estimate the global concentration and its changes. For other factors, e.g. population, it is not like this. The rate of increase of local or global population is the result of numerous individual events (births and deaths) and the actual value to put into a computer model will have a large error associated with it. Many countries make no attempt to control this and those that have tried to control the birth rate have by and large failed while the death rate may be affected by unpredictable wars, natural disasters, disease, and famine.

In the preface to "The Limits to Growth" (Meadows et al. 1972) U Thant, the then Secretary General of the UN wrote:

> I do not wish to seem overdramatic, but I can only conclude from the Information that is available to me as Secretary-General, that the Members of the United Nations have perhaps ten years left in which to subordinate their ancient quarrels and launch a global partnership to curb the arms race, to improve the human environment, to defuse the population explosion, and to supply the required momentum to development efforts. If such a global partnership is not forged within the next decade, then I very much fear that the problems I have mentioned will have reached such staggering proportions that they will be beyond our capacity to control.

The problems U Thant mentions – the arms race, environmental deterioration, the population explosion, and economic stagnation – are often cited as the central, long-term problems of modern man. Many people believe that the future course of human society, perhaps even the survival of human society, depends on the speed and effectiveness with which the world responds to these issues. In this section, we try to redress the balance between on the one hand global warming and on the other hand a number of serious threats several of which were perceived quite clearly in the early 1970s. The strong emphasis on global warming leads to complacency; people tend to think that if global warming could be reduced by the control of CO_2 emissions, then we can ignore all the other threats.

There was obviously a time when the anthropogenic effect on the climate was negligible though it is not quite straightforward to decide when that was. When our forefathers were hunter gatherers, we can define their effect on the climate as being sufficiently small that we do not need to consider it. But at the present time, it is clear that there are various human activities which are seriously affecting the climate, although this did not begin at the time of the industrial revolution as is apparent from a study of palaeoclimatology. Ultimately our species is doomed to extinction as a result of natural causes or of our own activities. One of many ways in which our species might meet extinction would be for the climate to become sufficiently hostile for human life to become unsustainable. But it is only one of many threats to the sustainability of human life. One could regard climatology as an academic subject which could be regarded as relevant to studying the survival of human civilisation.

In 1992, the Club of Rome published *Beyond the Limits* (Meadows et al. 1992); this was a 20-year update of the original study *Limits to Growth* (see Section 1.2.3). *Beyond the Limits* repeated the original message and also concluded that the intervening two decades of history mainly supported the conclusions that had been advanced 20 years earlier, in 1972. But the 1992 book did offer one major new finding; it was suggested that by 1992 humanity had already overshot the limits of the Earth's support capacity. This was reflected in the title of the book and the view was expressed that by the early 1990s overshoot could no longer be avoided through wise policy; it was already a reality. The main task had become to move the world back "down" into sustainable territory. This was followed by further developments after a further 10 years or so by a 30-year update "Limits to Growth: The 30-year update" (Meadows *et al.* 2005). Meadows *et al.* considered the physical resources, i.e. energy and raw materials, population, economic theory, and social policy. They argued that unrestrained growth was leading the Earth towards ecological overshoot and impending disaster. The success of the work of Meadows *et al.* stems from the fact that not only did they consider the factors and mechanisms of the threats associated with "development," but they also put the various factors into computer models that enabled them to vary the parameters and examine the consequences of changing the values of these parameters. Meadows *et al.* considered the driving force and especially the difference between exponential growth and linear growth, which to a considerable extent underlies the classic work of Malthus on population (see Section 1.3.8). Meadows *et al.* then considered the limits to growth and studied the twin problems of (a) reaching a sustainable way of life and (b) reaching a fair distribution of resources to provide a common standard of living for everyone on the planet. This common level in the standard of living is almost certainly considerably below that of the present "advanced" industrial societies. According to the publicity material for the 30-year update (i.e. Meadows *et al.* 2005), the first "book went on to sell millions of copies and ignited a firestorm of controversy that burns hotter than ever in these days of soaring oil prices, wars for resources and human-induced climatic change. This substantially revised, expanded and updated edition … marshalling a vast array of new, hard data and more powerful computer modelling, and incorporating the latest thinking on sustainability, ecological footprinting and limits, presents future overshoot scenarios and makes an even more urgent case for a rapid readjustment of the global economy towards a sustainable path." It would be hard to claim, about 15 years later, that the world has taken much notice of those 2005 warnings. One can, however, begin to see a shift in the idea of how to solve the problem which is essentially the problem of human survival. Following the establishment of the IPCC and the reception of its first reports, the idea seemed to be that the problem would be able to be solved from top down by an international convention, ignoring the problems of how such a convention can be enforced. This has not been spectacularly successful and has led to the idea that individuals can come to recognise the problem and to do their own little bits to help the situation. This approach is illustrated by two books that we discuss briefly by Prof David Mackay (2009) "Sustainable Energy without the Hot Air" (see Section 1.3.6) and Prof Mike Berners-Lee (2019) "There is no Planet B." Both books focus on energy which is so fundamental. It is not a conflict between the top-down and the bottom-up approaches; both are needed.

It is useful to look back after nearly 50 years at the "Limits to Growth" from a recent standpoint as was done by Pindyck (2017), an economist, when he describes the "Limits to Growth" debate. Pindyck argues that "Limits to Growth" was based on a simple sequence of ideas that appeared quite reasonable to some environmentalists at the time: (1) The earth contains finite amounts of oil, coal, copper, iron, and other nonrenewable resources. (2) These resources are important inputs for the production of a large fraction of GDP. (3) Because they are finite, we will eventually run out of these resources. In fact, because of the growth of population and GDP, we are likely to run out very soon. (4) When we run out, the world's developed economies will contract dramatically, greatly reducing our standard of living and even causing widespread poverty. (5) Therefore, we should immediately and substantially reduce our use of natural resources (and slow or stop population growth). Although this would reduce our standard of living then, it would give us time to adapt and will push back (or even avoid) that day of reckoning when our resources run out and we are reduced to abject poverty.

Pindyck argues that while points (1) and (2) are indisputable, points (3), (4), and (5) ignore basic economics. As the various non-renewable reserves are depleted, the costs of extraction and therefore the prices of these resources will rise, causing their use to decline and create the incentive to find substitutes. Thus, we may never actually run out of these resources, although we will eventually stop using them. Most importantly, given the incentives created by rising prices and the likelihood of finding substitutes, there is no reason to expect the gradual depletion of natural resources to result in economic decline. Indeed, due partly to technological change and partly to the discovery of new reserves, the real prices of most resources have gone down over the past 40 years, and there is no evidence that reserve depletion has been or is likely to be a drag on economic growth.

Although it made little economic sense, the "Limits to Growth" argument gained considerable traction in the press and in public discourse over environmental policy. This was due in part to a lack of understanding of basic economics on the part of the public (and many politicians). But it was also due to the publication and promotion of some simulation models that gave the "Limits" argument a veneer of scientific legitimacy. The most widely promoted and cited models were those of Forrester (1973) and Meadows et al. (1974). These models were actually quite simple; as Nordhaus (1973, 1992) and others explained, they essentially boiled down to an elaboration of points (1) to (5) above, with some growth rates and other numbers attached (Pindyck, 2017; Barrage, 2019). What seemed to matter, however, was that these models required a computer for their solution and simulation. The fact that some of the underlying relationships in the models were completely ad hoc and made little sense didn't matter – the fact that they were computer models made them "scientific" and inspired a certain degree of trust.

By 1990, global warming had overtaken all the other threats in its perceived importance; the reasons for this are complicated. First, as indicated in Section 1.2, it is the result of political manoeuvring by the Environmental Movement. Secondly, it arises from the fact that, relatively speaking, it is easy – or at least feasible – to model and implement on computers the relation between atmospheric CO_2 concentration and atmospheric temperature. While the Club of Rome attempted to model some of the other threats, it was less successful because this is intrinsically much more difficult. The success of the computer models relating CO_2 and global temperature and the consequent enormous attempts that are being made to control CO_2 emissions have led to complacency. It is imagined that because global warming could be controlled by control of CO_2 emissions, we can afford to ignore all the other threats.

1.3.3   DEFINING SUSTAINABILITY

The situation regarding sustainability by Cracknell (2009), was an article written at about the time that Mike Hulme's book was written and some things have changed in the intervening ten years. The question of climate change has been brilliantly put into the context of sustainability by Prof Mike Berners-Lee (2019) in his book "There is no Planet B." He considers the resources (in terms of space, food, energy, etc.) that are needed to provide everyone on Earth (all 7.7 billion plus of us) with the standard of living currently enjoyed by the general population of a small number of "Annex I" countries (in Kyoto Protocol, i.e. UNFCCC terminology countries)... We are concerned with the long-term future of humanity; let us say at least in 500 or 1,000 years time or even later when, almost certainly, the main fossil fuels are likely to be severely depleted. Uranium, which of course is not a fossil fuel but which is nevertheless a non-renewable fuel, may or may not be exhausted by that time, but eventually it too will be exhausted. The world, of course contains pessimists and optimists. The pessimist sees that in, say, 500 years' time, our present lifestyle will have vanished, while the optimist believes that some sort of tolerable, if not luxurious, lifestyle could be possible for our descendants in 500 years' time.

There is a very extensive literature on sustainability, but we begin by going back to the simple definition of sustainability to be found in the Oxford dictionaries from which we extract the following:

Sustainable (adjective): 1: able to be sustained. 2: (of industry, development, or agriculture) avoiding depletion of natural resources.

and

Sustain (verb) ..... keep (something) going over time or continuously.

We should consider the words "over time" in this definition. How long is the time period envisaged? Harold Wilson, a mid-20th-century British prime minister, once said that a week was a long time in politics. Most elected politicians cannot see any farther than their next election after, say, four or five years. Non-elected or "permanent" rulers may have slightly longer timescales but are quite likely not to be particularly benevolent. The rest of us think in terms of the remainder of our lifetimes, our children's lifetimes and possibly our grandchildren's lifetimes; say a few decades and probably less than a century. On these timescales, the fuel minerals and the important non-fuel minerals are not (all) going to run out. However, we propose to consider a longer timescale and one on which these fuel and non-fuel minerals will be exhausted or it will become uneconomic to extract them. We cannot estimate precisely when that will be; who knows.

Daly (1990) enunciated three useful principles of sustainability as it relates to resources and pollutants:

1. For a *renewable resource*, the sustainable rate of consumption/use can be no greater than the rate of regeneration of the resource.
2. For a *non-renewable resource*, the sustainable rate of use can be no greater than the rate at which a renewable resource, used sustainably, can be substituted for it.
3. For a *pollutant,* the sustainable rate of emission can be no greater than the rate at which the pollutant can be recycled, absorbed, or rendered harmless in its sink.

Another useful definition is that of the ecological footprint of humanity. As defined by Wackernagel *et al.* (2002), this is "the area of biologically productive land and water required to produce the resources consumed and to assimilate the wastes generated by humanity, under the predominant management and production practices in any given year." According to Wackernagel *et al.,* this corresponded to 70% of the capacity of the global biosphere in 1961 but had grown to 120% (i.e. overshoot) in 1999, in some measure of agreement with the Club of Rome's timing of overshooting (see Section 1.3.1).

TABLE 1.7

Annual consumption and proven reserves of fuel minerals

Fuel mineral	Consumption in 2002 (10^6 tonnes)	Proven reserves (2002) (10^6 tonnes)
Coal	4800	909000
Oil	3400	
		148000
Gas	110	
Uranium	0.036	3.2[1]
		9.8[2]

Source: (World Energy Council http://www.worldenergy.org/wec-geis/publications/reports/foreword.asp (accessed 21 May 2007)).

[1] At a price of up to US$130 kg^{-1}.

[2] Estimated additional resources at US$130 kg^{-1}.

If we look at our way of life, we are consuming fuel resources at an annual rate of 4.8 billion tonnes of coal, 3.4 billion tonnes of oil, and 110 million tonnes of gas (these figures are for 2002, see Table 1.7). There is no stretch of the imagination by which we can regard as sustainable this use of energy which was stored up from the energy of the Sun in these minerals over hundreds of millions of years and is being consumed over a mere few centuries. It will "over time" (see the above definition) be completely used up. In discussing fuel resources, we should perhaps include uranium, as being a mineral and the (non-replaceable) nuclear fuel, see Table 1.7. This Table also includes estimates of the world's proven reserves of coal, oil, gas, and uranium. In terms of Daly's second principle, we are (slowly) attempting to achieve sustainability but we are a very long way from achieving that target.

We should perhaps not include as fuel all the oil that we consume, since some of it is used as feedstock for the petrochemicals industry; indeed, there are those who would argue that oil is too valuable to burn as a fuel for electrical power generation and that we should keep it for special uses, primarily as feedstock for the petrochemicals industry and for transportation, for which it is not easy to find a substitute. This was already argued by the Nazis in 1935 (see Section 1.2.1).

As well as the question of fossil fuel energy resources, we should also consider our consumption of non-fuel minerals, mainly metals or their ores, see Table 1.8 where figures are given for our current consumption rates and the estimated world reserves of aluminium, cobalt, gold, iron, nickel, silver, and tin are given. These minerals were formed long ago and once they are gone then, for practical purposes, they are gone forever. Recycling can, of course, help but there comes a point beyond which a material is so scattered that recycling becomes unfeasible. In terms of Daly's second principle, we are a very long way from achieving sustainability by replacing metals with renewable resources and it is very difficult to envisage how we could ever completely satisfy this principle.

There is a vast literature on sustainability, but it is mostly what it would be fairer to describe as "reducing unsustainability"; it does not take as its starting point the position that we have just described, i.e. that our present lifestyle is unsustainable in the long term and therefore how much of it can be salvaged once the fuel and non-fuel mineral resources have run out? We should not denigrate all the work which has been done on reducing unsustainability, increasing efficiency, and avoiding waste; this is important and postpones the "evil day."

People who study these problems, including the several authors of the *Limits to Growth* are divided into two camps, one of the optimists and the other of the pessimists. The optimists come in two general classes. There are what we might call the "woolly" optimists, who are usually economists and the like, who suppose that someone or something will make it turn out alright for us in the end: *Deus providebit* (God will provide), or Technology will come to our rescue, etc. They believe that human beings can make the transition to a fair and just society at a common standard of living

TABLE 1.8

Annual production, reserves, and reserve base of some important non-fuel minerals

	Production in 2005 (10^6 tonnes)	Reserves (10^6 tonnes)	Reserve base (10^6 tonnes)
Iron ore	1500	160000	370000
Bauxite (Al ore)	169	25000	32000
Nickel	1.5	64	140
Cobalt	0.058	7	13
Tin	0.290	6.1	11
Silver	0.0193	0.270	0.570
Gold	0.0025	0.042	0.090

Source: (U.S. Geological Survey, Mineral Commodity Summaries http://minerals.usgs.gov/minerals/pubs/mcs/ (accessed 21 May 2007)).

Note: As defined by the U.S. Geological Survey, the *reserves* are that part of the *reserve base* which could be economically extracted or produced at the time of determination; it does not signify that extraction facilities are actually in place and operative. The term *reserve base* refers to that part of an identified resource that meets specified minimum physical and chemical criteria related to current mining and production practices, including those for grade, quality, thickness, and depth. It includes reserves that are currently economic, reserves that are marginally economic, and reserves that are currently subeconomic.

in a peaceful and ordered manner, despite very considerable evidence throughout human history to the contrary. The others are the cautious optimists, usually scientists rather than economists, who acknowledge that there is a problem and realise that it needs political and social willpower to implement the technology necessary to solve it (e.g. Mobiot 2006, Stern 2007, Walker and King 2008, Berners-Lee 2019). The pessimists, who are usually physical or environmental scientists, fear that as resources become more scarce, we shall drift into conflict and mega-deaths and a spectacular degradation of our lifestyle. On a geological timescale, i.e. in terms of millions of years, humanity is doomed to extinction. But on the timescale, we are considering just now, say 500–1000 years, complete extinction seems unlikely. However, it seems highly likely that many parts of our present civilisation will pass away; and we can assume that those living sophisticated urban lives will be more likely to perish while those living in simple conditions much closer to nature will be more likely to survive. For a highly pessimistic view of just the next 100 years, one could read chapter 10 of the book "Peak Everything. Waking up to the century of decline in Earth's reosurces" by Heinberg (2007). This chapter is a fictitious letter written in 2107 by someone who was born in 2007 and who has seen the complete collapse of industrial civilisation as we know it and it is well worth reading. The fictitious writer says that while "attempting to pursue the career of a historian" circumstances forced him to "learn and practice the skills of farmer, forager, guerilla fighter, engineer …" He describes the energy crisis "Folks then thought it would be brief, that it was just a political or technical problem, that soon everything would get back to normal. They didn't stop to think that 'normal', in the longer-term historical sense, meant living on the energy budget of incoming sunlight and the vegetative growth of the biosphere. Perversely, they thought 'normal' meant using fossil energy like there was no tomorrow …." He describes how energy shortages led to economic recession and endless depression, the collapse of currencies, inflation, deflation, the return of barter. "We went from global casino to village flea market." Manufacturing collapsed, transportation collapsed. Supermarkets were empty. People scavenged through all our landfill sites "looking for anything that could be useful." He castigates us for taking "billions of tons of invaluable, ancient, basic resources and turn(ing) them into mountains of stinking garbage, with almost no measurable period of practical use in between!" There were purges, wars ("the generals managed to kill a few million popple … it could have been tens or hundreds of millions, even billions …."), epidemics and famine. And so on.

Another serious pessimist is Lord (Martin) Rees, whose main arguments are (a) that in many fields of scientific and technological research, notably biology and particle physics, we are now tinkering with deep fundamentals that are not completely understood and (b) in some of the new technologies there is the opportunity for small disaffected groups or individuals to cause enormous damage (Rees 2004). An example of the first was provided by the fear in 1945 that a nuclear explosion might trigger an enormously destructive chain reaction beyond the initial explosion (actually, as we now know, it did not do that) and an example of the second is provided by in the destruction of the World Trade Centre in New York on 11 September 2001.

1.3.4   OUR WAY OF LIFE

Our way of life in the so-called developed world has slowly evolved over a long period of time, but this development has recently been greatly accelerated as a result of scientific advances leading to industrialisation. There have been movements in favour of "back to nature," see Section 1.2.1 and we shall return to this question in Section 6.7. It is commonly held that global warming presents a threat to "our way of life." The thesis of this book is that, while this may well be true, it is only one of many threats to our way of life and that by attempting to address only this threat we ignore many other threats which are less able to be quantified by sophisticated computer models. We need to examine the threat of global warming a little more closely. We often think of the world as being divided into three groups of countries, (a) developed or industrialised countries, (b) newly emerging or advanced developing countries, and (c) third world countries at various stages of distance from advanced development. We know roughly what these categories mean: (a) means the USA, Canada, Japan, Western Europe, Australia, New Zealand, Singapore (if one reads Lee Kuan Yew's book "From Third World to First" (Lee 2000)) etc.; (b) includes a whole host of countries, Malaysia with its aim to be developed by 2020), China, India, some South American countries like Brazil, Eastern European countries; and (c) includes most African countries. The divisions are blurred and it does not do to define things too rigidly. Those countries in (b) and (c) aim to achieve "developed" status as quickly as they can. An alternative division which is convenient because it is well-defined is (a) Annex-1 countries as per the Kyoto Protocol, basically as defined by the UNFCCC, and (b) non-Annex-1 Countries as per the Kyoto Protocol, i.e. all the rest (see Grubb et al. (1999). There has been very little mobility between these two categories since Kytoto. However, it can be argued that "our" present lifestyle is unsustainable. Everyone wants to achieve the condition of what one might describe as luxury or extravagant luxury of the countries in the developed world. One can regard the American lifestyle as not just a luxury but as extravagant luxury which is unsustainable; it relies very heavily on extremely cheap oil. If everyone in the world, all 7.7 billion or so of us, lived in the lifestyle of people in the USA, then the planet would be wrecked very rapidly. A discussion of the economics and politics related to climate change written from the point of view of one very large developing country (India) is given by Toman et al. (2003). As Mike Berners-Lee (2019) used as the title of his recent book "There is no Planet B".

Jared Diamond (2005) wrote a very perceptive book entitled "Collapse: How societies choose to fail or survive." He recounts the stories of the collapse of various civilisations in different parts of the world. The present problem is that ours is a much more global civilisation and the threats are more global. As Diamond points out, it is not just the number of people on the planet but their impact on the environment which is important. Our numbers only cause problems insofar as each of us consumes resources and generates waste. If we only maintained world population at its present level, the average environmental footprint would increase because of economic development in various countries. People in other countries see films or watch TV about life in the developed countries, they see advertisements for First World consumer products sold in their countries, and they observe First World visitors to their countries. Not un-naturally, they want to achieve the same lifestyle. One of the problems is how to achieve this without wrecking the planet. "..... low-impact people are becoming high-impact people for two reasons: rises in living standards in Third World countries ….. and immigration, both legal and illegal, of individual Third World inhabitants in the First World, driven by political, economic, and social problems at home. Immigration from low-impact countries is now the main contributor to the increasing populations of the USA and Europe… the biggest problem is the increase in total human impact, as the result of rising Third World living standards, and of Third World individuals moving to the First world and adopting First World living standards …" (Diamond 2005 page 495) continues by pointing out that, in addition to their own aspirations, Third World countries are encouraged to follow the path of development by First World and United Nations development agencies, which hold out to them the prospect of achieving their dream if they will only adopt the right policies, like balancing their national budgets, investing in education and infrastructure, and so on.

But no one at the U.N. or in First World governments is willing to acknowledge the dream's impossibility: the unsustainability of a world in which the Third World's population were to reach and maintain current First World living standards. It is impossible for the First World to resolve that dilemma by blocking the Third World's efforts to catch up: South Korea, Malaysia, Singapore, Hong Kong, Taiwan, and Mauritius have already succeeded or are close to success; China and India are progressing rapidly by their own efforts. Diamond also refers to the European Union (EU) and the various stages of its expansion and then continues suggesting that even disregarding the human populations of the Third World, for the First World alone to maintain its present course would become impossible, as it is continuously depleting its own resources as well as those imported from the Third World. Currently, it is not politically viable for First World leaders to propose to their electorate that they should lower their living standards, as measured by lowering their resource consumption and their waste production rates. Diamond goes on to question what will happen when people in the Third World realise that First World standards are unreachable for them and that the First World refuses to abandon those standards for itself.

He comments "Life is full of agonizing choices based on trade-offs, but that's the cruelest trade-off that we shall have to resolve: encouraging and helping all people to achieve a higher standard of living, without thereby undermining that standard through overstressing global resources."

1.3.5 THE END OF FOSSIL FUELS AND OTHER MINERALS

In the 1920s, many geologists were warning that world oil supplies would be exhausted within a few years. But then huge new discoveries were made in various places, east Texas, the (Persian) Gulf, etc., apparently discrediting such predictions. Each year more oil was being discovered than was being extracted. Many people assumed that this could go on forever. However, much is made of the predictions of the geologist, Marion King Hubbert (1956), that the fossil-fuel era would prove to be very brief. That paper is quite general and refers to any mineral, whether a fossil fuel or a non-fossil fuel mineral (i.e. uranium) or to any non-fuel mineral at all. The commercial extraction of any mineral, Stone Age flint mines, Bronze Age copper or tin mines, coal mines, iron-ore mines, oil wells, etc., will follow a bell-shaped curve. If $Q(t)$ is the cumulative production from the beginning of the mining or drilling, then $P(t) = dQ(t)/dt$ is the rate of extraction at time t; the productivity at time t is sketched in Figure 1.18, but the actual curve for a real mineral extraction will differ in detail from this sketch.

Figure 1.18 shows what has come to be known as a Hubbert peak and can be applied to a particular local mining activity or on a national or global scale. For a mining activity that is finished the Hubbert curve could be drawn if accurate production records have been kept. But the question of interest for instance in oil production in the early days lies in predicting the time when the peak can be expected to occur and the expected height of the peak. Applied to oil production, for example, the theory is simple, one starts with the known quantity of oil in the ground, takes the current extraction rate, predicts future extraction rates and from then on it is just simple mathematics. But it is not that simple. First, we do not know how much oil there is in the ground with any reliable accuracy. Secondly, as time goes on, it becomes progressively more difficult to extract the oil from any given

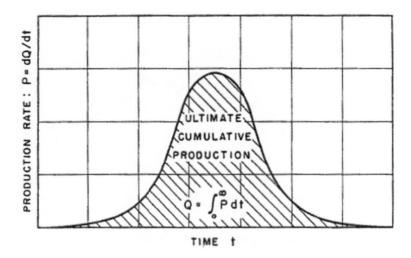

FIGURE 1.18 Sketch of the general Hubbert (1956) curve for the extraction of a resource from the ground.

well. Initially it may just gush to the surface, but later on, it has to be pumped and then, later still, water has to be pumped in at high pressure to push the oil out or in the case of fracking to break up the rocks. So the extent to which a field is exploited becomes an economic/financial decision and not just a technical matter and the future extraction rate is not easy to determine.

In 1956, Hubbert made the best estimates that he could for the USA and predicted that crude-oil production in the USA would peak between 1966 and 1972, see Figure 1.19. It is very difficult to make accurate predictions because estimates of reserves of oil in the ground are notoriously "flexible"; oil companies and some governments increase or decrease their estimates for financial or political reasons without much, if any, geological evidence. The USA crude oil production actually peaked in 1972, see Figure 1.20. The decline after 1971 led the USA to seek to assure supplies from overseas, something which lies behind a great deal of USA foreign policy, not to mention its wars, in recent decades. The increase in production in the last few years is a result of the introduction of fracking.

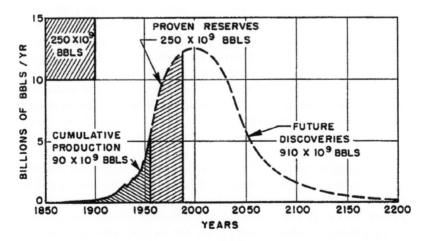

FIGURE 1.19 Hubbert's original projection of oil production in the USA based on knowledge available in 1956 (Hubbert 1956).

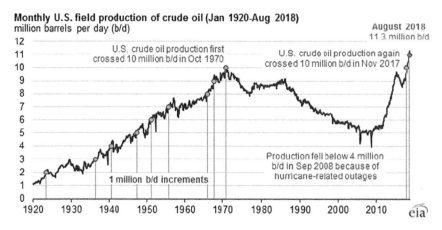

FIGURE 1.20   USA crude oil production up to August 2018 (eia.gov accessed 1 September 2019).

Hubbert's original paper, which carried the title "Nuclear Energy and the Fossil Fuels," was concerned with the need to replace coal, oil, and gas by nuclear (fission) power. As we shall see later, nuclear power fell into disfavour in certain countries and was replaced by excessive dependance on renewable energy sources (wind and solar power), see Section 5.4. Hubbert appears to have believed that society, if it is to avoid chaos during the energy decline, must give up its antiquated, debt-and-interest monetary system and adopt a system of accounts based on matter-energy – an inherently ecological system that would acknowledge the finite nature of essential resources. Hubbert is quoted as saying that we are in a "crisis in the evolution of human society. It's unique to both human and geologic history. It has never happened before and it can't possibly happen again. You can only use oil once. You can only use metals once. Soon all the oil is going to be burned and all the metals mined and scattered." We have, he believed, the necessary know-how. If society were to develop solar-energy technologies, reduce its population and its demands on resources, and develop a steady-state economy to replace the present one based on unending growth, our species' future could be rosy indeed. "We are not starting from zero," he emphasised, "we have an enormous amount of existing technical knowledge. It's just a matter of putting it all together. We still have great flexibility but our manoeuvrability will diminish with time." His optimism has not been shared by everyone.

The principles used to calculate the expected lifetimes of fuel minerals can be applied to non-fuel mineral resources too. The difficulties are similar too, namely there are the problems of estimating the resources in the ground and in estimating future rates of consumption. Meadows *et al.* (2005) made some calculations for some important metals using what they call the identified reserves and the resource base; the figures they used for the reserves are different from those we gave in Table 1.8 and they assume a growth rate of 2% per annum in consumption. Their results are shown in Table 1.9.

While the figures in this table could probably be updated, the general message is unlikely to change significantly. While there are quite a few popular books on the question of the decline in oil resources (e.g. Roberts, 2004; Leggett 2005; Heinberg 2006; Al-Fattah, 2020), there is much less written about the decline in the sources of non-fuel minerals (Tanzer 1980 and that is a bit old now, though the general ideas are still sound).

We can consider the application of the Daly principles to both the fuel and non-fuel minerals. If we apply the second of these principles to the case of fuel resources, then we see that the question is to what extent we can meet continued demands for energy by (a) economising on the use of energy and (b) using renewable energy (very nearly all of it ultimately derived from the Sun) in place of non-renewable fuels. However, when it comes to the non-fuel minerals, i.e. metal ores, it is much more difficult to envisage finding renewable substitutes for metals. Recycling can help, but of course recycling requires energy which needs to be taken into account. In some situations, metals can be

TABLE 1.9

Life expectancy of some non-fuel minerals

Mineral	Life expectancy of identified reserves (years)	Life expectancy of resource base (years)
Bauxite	81	1070
Copper	22	740
Iron	65	890
Lead	17	610
Nickel	30	530
Silver	15	730
Tin	28	760
Zinc	20	780

Source: Meadows et al. (2005).

replaced by plastics and plastics could, presumably, be made from renewable vegetable oil sources (in competition with food and biofuels). But metals have some very unique properties and there are some situations in which it is very difficult to imagine metals as ever being able to be replaced by renewable resources, for example, as conductors of electricity.

1.3.6 CAN THE PARTY CONTINUE?

Given that the end of oil will come sooner – or not much later – the problem is how to replace it and maintain our present lifestyle? There are now many general books that address this subject, but particularly worthy of mention are "The Party's Over – Oil, War and the Fate of Industrial Societies" (Heinberg 2003), "Heat – How to stop the Planet Burning" (Mobiot 2006), "The Hot Topic – How to Tackle Global Warming and Still Keep the Lights On" (Walker and King 2008), "The Uninhabitable Earth" (Wallace-Wells, 2019) and several others. Heinberg's book is focused largely on the USA and Monbiot's book is focused on the British situation, while Walker and King are particularly emphatic on the need for international agreement on reducing CO_2 emissions. There is a wealth of good quantitative material in these books, but the underlying message is that if we modify our life-style and reduce our population then we could *perhaps* manage an acceptable lifestyle after the oil (and the gas and the coal) run out. But some things would have to be given up – air travel for a start! Planes have to be run on kerosene (aviation fuel) or something very similar; they cannot be run on coal, gas (the cylinders would be too heavy), wood, or even cow dung. People are talking about fly-ing planes on batteries but the Technology has a long way to go, even for short flights. What about biofuels one might ask? Basically there is a production problem there – of serious competition with food production – and we shall discuss biofuels shortly.

The reasons why oil is so useful are simple, namely it is because oil is:

- easily transported (much more so than solids such as coal or gases such as methane)
- energy-dense (gasoline, i.e. petrol, contains approx 349.2×10^8 J m^{-3})
- capable of being refined into several fuels, gasoline, kerosene, and diesel suitable for a variety of applications
- suitable for a variety of uses including transportation, heating, and as feedstock for the petro-chemicals industry (fertilisers, plastics, etc.)

One very important concept, and something which is often neglected in the discussion of renewable energy resources, is the Energy Return on Energy Invested (EROEI). In the early days of oil, prior to 1950, it is estimated that the EROEI was in the region of 100:1; one just drilled a hole in the ground

TABLE 1.10
Energy Return On Energy Invested, EROEI; data from Heinberg (2005)

	EROEI	
Energy source	Data from Heinberg (2003)	Data from Wikipedia (accessed 27 August 2019)
Oil, pre 1950	~100	
Oil 1970	~30	
Oil present	~8–11	
Coal, US average	9.0	
Coal, western surface coal	6.0	
Coal, ditto with scrubbers	2.5	
Natural gas	6.8–10.3	28
Ethanol, sugar cane	0.8–1.7	
Ethanol, corn	1.3	
Palm oil	1.06	
Wind, aerogenerators	~2 –	16–51
Nuclear	4.5	75–106*
Hydropower	10.0	50
Geothermal	13.0	
Solar, photovoltaics	1.7–10.0	4.0–7.0

* But note that there are significant decommissioning costs.

and the oil gushed out. By the 1970s, it is estimated that the EROEI for oil production had dropped to around 30:1. Energy had to be supplied for exploration, drilling, building of rigs, transportation, housing of production workers, etc. Heinberg gives an extensive table of values of EROEI for various non-renewable and renewable sources of energy and a few of them are extracted in Table 1.10.

The values in Table 1.10 should only be regarded as indicative because so many factors need to be taken into account in determining the value of the energy invested and it is not clear whether decommissioning costs come in to the definition. Many renewable sources have (hidden) costs in terms of the energy used in setting them up (see Section 5.4). So establishing accurate values of the EROEI is therefore very important.

One can argue that most of the present work on sustainability is not concerned with real sustainability but is only tinkering at the edges and is more like "(slightly) reducing the unsustainability of our present way of life." For true sustainability, there will have to be economies no doubt but some substitution should be possible. We shall attempt to consider what happens when the oil has run out. So we shall consider what we can use in substitution:

1.3.6.1   Gas

One can dismiss gas immediately. The gas will probably run out before the oil. The present substitution in UK electricity generation, of gas for oil or coal, is driven by cost and helps to meet the country's Kyoto target on CO_2 emissions. It is not a long-term solution.

1.3.6.2   Coal

One can dismiss this too, though perhaps not quite so quickly. The world's coal reserves are enormous. But not only will they run out too, though on a much longer timescale, but also coal is very polluting (less so with scrubbers on power stations, but many power stations are not fitted with scrubbers). At present, we use oil to mine coal. Opencast mines use relatively few human miners but they use giant diesel-powered earth-moving machines so that the EROEI is not particularly good. As the oil becomes less available, the energy used to mine coal will have to come from coal or from some other sources. And as near surface coal runs out, we shall have to turn to deep mining again and there are considerable energy requirements there and so the EROEI would reduce quite severely in the future.

1.3.6.3   Nuclear

The future for nuclear energy is very unclear since it is subject to fashions of politics and public opinion, but some features can be noted:

a) it depends on uranium and this is a finite resource – (at least unless or until fusion can be controlled),
b) naively one can say it is clean as far as CO_2 emissions are concerned.
c) it has a poor public image in terms of safety and
d) the value of the EROEI is not very clear, particularly when de-commissioning costs are considered.
e) its actual safety record in terms of workers in the industry is good (Hall 2017).

However, a considerable amount of energy (coming at present from oil) is used in mining the uranium and that produces CO_2. Also the processes involved in constructing the power stations involve energy coming from oil and producing CO_2 emissions and they involve concrete, which we shall come to shortly.

As far as fusion is concerned, this may be available as a source of energy in the future but we are a long way from being there in terms of getting nuclear fusion to work as a source of energy.

1.3.6.4   Renewables

In terms of Daly's second principle, the long-term objective needs to be for our energy to be supplied entirely by renewables, but this is very long term (unknown) and by that time we may well have destroyed ourselves in some way or another. It could be argued that in the short term there is no reason for us to leave the non-renewable energy sources in the ground "for future generations"; but this is an ethical and political question, not a scientific question. On the other hand, Berners-Lee argues that we should leave all the remaining fossil fuels in the ground not for future generations but to avoid global warming. Our argument is that renewables have their role to play but that certain parties have carried the introduction of renewables to excess without considering the true costs of doing so (see Section 5.4).

1.3.6.5   Hydropower and geothermal

From Table 1.10, we see that from the point of view of the EROEI, these energy sources look particularly good. But they can only be exploited in certain situations and they come in all sorts of sizes, large, medium, and small. As far as geothermal energy is concerned, there are only a few places where it can be exploited, most notably in Iceland, but also to a modest extent elsewhere, e.g. New Zealand and Geyserville in northern California. But elsewhere in general, there is probably not much scope for the development of geothermal energy, at least not with modern technology. Nor is it truly renewable as the heat extracted from the Earth's core is not being replaced.

Hydropower has been used since ancient times to grind grain and other natural products and to perform other tasks as well. Hydroelectric power stations come in a vast range of sizes from the largest, see Table 1.11, down to the tiniest which generates an almost negligible, but nevertheless useful, quantity of electricity. The large schemes feed electricity into a national grid. At the other extreme, a small scheme is used to generate electricity for a small local consumer. The largest hydroelectric schemes are very large indeed and people generally know about them. We have in mind the Three Gorges scheme in China, the Aswan High Dam in Egypt, etc. (Figure 1.21).

The world's first hydroelectric power scheme was developed at Cragside in Northumberland, England, by William Armstrong in 1878. It was used to power a single arc lamp in his art gallery.

TABLE 1.11

The world's four largest hydroelectric power stations

Rank	Station	Country	Location	Capacity (MW)
1	Three Gorges Dam	China	30°49'15"N 111°00'08"E	22,500
2	Itaipu Dam	Brazil/Paraguay	25°24'31"S 54°35'21"W	14,000
3	Xiluodu Dam	China	28°15'35"N 103°38'58"E	13,860
4	Guri Dam	Venezuela	07°45'59"N 62°59'57"W	10.200

Source: (Wikipedia).

FIGURE 1.21   The Three Gorges Dam in Central China. The length of the dam is 2,335 m (7,660 ft) and the maximum height is 185 m (607 ft). (Creative Commons 2.0". Credit to the creators "Le Grand Portage at Flickr. com; image edited by User: Rehman Wikimedia Commons").

The old Schoelkopf Power Station No. 1, USA, near Niagara Falls, began to produce electricity in 1881. The first Edison hydroelectric power station, the Vulcan Street Plant, began operating on 30 September 1882, in Appleton, Wisconsin, with an output of about 12.5 kW. By 1886, there were 45 hydroelectric power stations in the USA and Canada, and by 1889, there were 200 in the US alone. At the beginning of the 20th century, many small hydroelectric power stations were being constructed by commercial companies in mountains near metropolitan areas. Hydroelectric power stations continued to become larger throughout the 20th century and into the 21st century, until at the present time the details of the four largest hydropower stations are shown in Table 1.11 and many new schemes are under construction. Some statistics of current world production of hydroelectricity are given in Table 1.12; the third column indicates the full installed capacity of each power station.

But no power station operates at its full power over a full year; the ratio between the annual average power and installed capacity rating is called the capacity factor and this is given in column 4. The final column shows the hydroelectricity as a fraction of total generation for each country with this percentage being greatest for Norway. There remains a large technical potential for hydropower

TABLE 1.12

Ten of the largest hydroelectric produced as at 2014

Country	Annual production (TWh)	Installed capacity (GW)	Capacity factor	Fraction of total production (%)
China	1064	311	0.37	18.7
Canada	383	76	0,59	58.3
Brazil	373	89	0.56	63.2
United States	282	102	0.42	6.5
Russia	177	51	0.42	16.7
India	132	40	0.43	10.2
Norway	129	31	0.49	96
Japan	87	50	0.37	8.4
Venezuela	87	15	0.67	68.3
France	69	25	0.46	12.2

Source: (Wikipedia).

development around the world, but much of this remains largely undeveloped due to the political realities of new reservoirs in western countries, the economic limitations in the developing world, and the lack of transmission systems in undeveloped areas.

Small hydropower generation may be developed by constructing new facilities or through re-development of existing dams whose primary purpose is flood control or irrigation. Old hydro sites may be re-developed, sometimes salvaging substantial investment in the installation such as penstock pipe and turns, or just re-using the water rights associated with an abandoned site. Either of these cost-saving advantages can make the return on investment for a small hydro site well worth the use of existing sites. One such example of this is provided by the hydroelectric scheme at Blair Castle in the Highlands of Scotland (56.7670°N, 3.8456°W). Electricity was brought to Blair Castle in 1908 by harnessing the water which had once powered an old sawmill at Blairuacher from a pond constructed in 1840 and the catchment was extended in 1907. A 4.48 km drain was dug along the contour of the hill which collects water from the hill and delivered it to a penstock, a 400 mm diameter steel pipe which is 2.06 km long which was laid in 1908 and drops 126 m. The powerhouse, about the size of a garage for a rather large car, originally contained three 25 kW Gilkes turbines. With the arrival of the National Grid in 1951, the hydroscheme was decommissioned and the turbines sold for scrap. However, subsequently the importance of local generation came to be realised. The installation was then refurbished later on when it began to be realised that (a) there were severe transmission and other losses associated with the grid and (b) there were often prohibitively high costs involved in installing long connections to the grid (e.g. in the case of the Island of Eigg in western Scotland and especially in many developing countries). Thus, the scheme was re-commissioned in 2014 with a design flow of 331,200 litres per hour and a new 84 kW Gilkes turbine and designed to produce 40 MWh per annum. In usual circumstances, the system will provide all the electricity needed by the castle. In the case of extreme loads – a big function in the winter – the power will be supplemented from the National Grid while in periods of low castle demand for electricity the surplus power is exported, i.e. sold, to the National Grid. The benefits, of course, include a source of income for the castle and a source of energy free of CO_2 emissions to the atmosphere.

Although hydroelectricity is generated from a renewable resource and does not generate CO_2, it does have some disadvantages compared with other fuel resources, including ecosystem damage and loss of agricultural land and displacement of people in the construction of a reservoir, water loss by evaporation, reduction of the methane emission (a more serious greenhouse gas than CO_2) from reservoirs, and the risk of collapse of a dam.

A tidal power station makes use of the daily rise and fall of ocean water due to tides; such sources are highly predictable, and if conditions permit construction of reservoirs, can also be dispatchable to generate power during high demand periods. Less common types of hydro schemes use water's kinetic energy or undammed sources such as undershot water wheels. Tidal power is viable in rather small amounts and in a relatively small number of locations around the world.

1.3.6.6 Wind

Wind power has been exploited for centuries but it has one fundamental weakness that the wind does not blow all the time. For centuries and in many different parts of the world, windmills were constructed, usually for grinding grain, but occasionally for pumping water. The millers would have had to adjust their working practices so as to work when the wind was blowing. However, when using the wind to generate electricity, there is the problem that electricity cannot be stored, except in minute quantities and has to be used when it is produced. We shall return to this in Chapter 7. The EROEI in Table 1.11 is worthy of detailed scrutiny; Danish studies quoted by Heinberg suggest an over optimistic value for the EROEI of 50 or more. Wind power for electrical generation has been expanding greatly and clearly has possibilities, but (a) it needs a lot of investment, including the installation of connection between the wind farms and the electricity supply grid; (b) there are limits imposed by environmental considerations etc.; (c) the wind does not blow all the time and it is necessary to have backup generation facilities that can be brought on stream quickly and automatically; and (d) it may produce electricity in very windy conditions where there is no market for it. Item (c) the question of intermittency is a very serious problem. Sometimes there is more wind than is needed and under current financial arrangements sometimes wind farms have to be switched off and the owners paid not to generate electricity. It will be argued later, in Section 5.4, that the true cost of windpower for the generation of electricity has not been assessed properly by the proponents of massive investments in wind power by the green political lobby. There is a case for the review of the subsidies paid in some countries to encourage the construction of wind farms. It is worth noting that industrial wind generation is not able to contribute significantly against the problems of global warming, pollution, nuclear waste, or dependence on imports. For example, in Denmark, with the most per-capita wind turbines in the world, the output from wind facilities is 15%-20% of their electricity consumption. The Copenhagen newspaper *Politiken* reported, however, that wind provided only 1.7% of the electricity actually used in 1999. The grid manager for western Denmark reported that in 2002, 84% of their wind-generated electricity had to be exported, i.e. dumped at extreme discount. The turbines are often shut down, because it is so rare that good wind coincides with peaking demand. A director of the western Denmark utility has stated that wind turbines do not reduce CO_2 emissions, the primary marker of fossil fuel use. But industrial wind facilities are not just problematic. They destroy the land, birds and bats, and the lives of their neighbours. Off shore, they endanger ships and boats and their low-frequency noise is likely harmful to sea mammals. They require subsidies and regulatory favours to make investment viable. They do not move us towards more sustainable energy sources and stand instead as monuments of delusion (Ribrant and Bertling, 2007).

1.3.6.7 Solar

Of course, wind and hydropower ultimately derive their energy from the Sun, but we have just dealt with them; we confine ourselves just now to direct heating and photovoltaics. Photovoltaics have their uses and the EROEI in Table 1.11 looks reasonable enough; they are very useful for small supplies of electricity or in isolated locations with a modest battery for storage. There are also many solar panel installations to produce hot water, rather than electricity, from the Sun. There are also a few installations where solar energy is concentrated to produce steam for the generation of electricity, but there are not many of them. The direct conversion of solar energy to electricity has the same big problem that we have just noted with wind power, namely intermittency. It is intermittent and there is very little capacity to store the electricity that is generated; it must be used when it is generated, backup generating facilities must be available for when the sun does not shine, and there will be costs involved in connecting large arrays of solar panels to the electricity supply grid. If one tries to connect photovoltaic solar panels to a country's electricity supply grid, there is the same problem of intermittency that we have already noted with wind power. It is in fact worse than wind because the solar power is seasonal and is subject to cloud conditions and is only available during daylight hours; the wind, however, also blows in the hours of darkness.

1.3.6.8 Biofuels

Extensive accounts of the technical, economic, and political details of biofuels are given by Pahl (2005), Worldwatch (2007), and Yin et al. (2020). But we note particularly that the use of biofuels generated from agricultural crops, rather than from waste of one form or another, is in direct competition with agricultural production of food.

We consider ethanol and biodiesel separately. Brazil provides the best-known example of the production and use of ethanol for cars starting in the 1980s, though the USA is stepping up its ethanol programme now. Following the steep rise in oil prices in the 1970s, Brazil turned to ethanol produced from sugar cane, and in 1985, 91% of cars produced in Brazil ran on ethanol. But as world oil prices fell and sugar cane prices rose, the demand for alcohol-fuelled cars subsided (Khayal and Osman 2019). Brazil could afford its ethanol programme because of its very favourable ratio of the area of cropland to the number of cars, even if topsoil was being lost and energy was being used in the process. There are also rumours of exploitation of child labour in the Brazilian sugar cane fields too. There are disputes about the EROEI of ethanol production but it is not particularly good. Heinberg (2003) does a little calculation of what would be involved if the USA tried to repeat the Brazilian experiment, but using corn oil rather than sugar cane because that is what they can grow (working in American units not SI units where the approximate conversion factors are: 1 acre = 0.405 ha, 1 pound (weight) = 0.454 kg and 1 (US) gallon − 3.79(litres)):

> The USA has 400 million acres of cropland and about 200 million cars.
> American farmers produce about 7,110 pounds of corn per acre per year and an acre of corn yields about 341 gallons of ethanol.
> A typical American driver would burn 852 gallons of ethanol per year requiring thus the production from 2.5 acres.
> Thus ethanol production from corn would need 500 million acres of cropland, or 25% more farmland than the total area available in the USA.

While this calculation was done some years ago and it would be an interesting exercise to revise the calculation using modern data, the general conclusion is still valid, namely that there is a problem of competition with agricultural resources for food production.

Then there is biodiesel. Biodiesel is a substitute for what we might call "petroleum diesel," "hydrocarbon diesel," or "mineral diesel." It is made from vegetable oil and methanol. Various oils can be used but the one with the best yield per hectare is palm oil. Malaysia and Indonesia are the world's leading producers of palm oil. Biodiesel is used in two ways. By modifying a diesel engine, it can run on biodiesel and there are some vehicles that are modified in this way. Alternatively, a small percentage of biodiesel can be added to ordinary diesel and engines need no modification to handle this. President George W. Bush's 2005 Energy Policy Act obliges fuel companies to sell 7.5 billion gallons of biodiesel and ethanol a year (see also Irwin and Good, 2017).

There is an EU (European Union) directive that 5.75% of the EU's transport fuel should come from renewable resources by 2010 (Scholz, 2019). The British government reduced the tax on biofuels by 20p a litre and the EU is paying farmers an extra €45 a hectare to grow the crops to make them. To quote Mobiot (2006) "At last, it seems a bold environmental vision is being pursued in the world's richest nations." But, then he goes on to do for the UK and road fuel what Heinberg did for the USA:

> Road transport in the UK consumes 37.8 Mtonnes of petroleum products per year.
> For oilseed rape, the most productive oil crop for the UK climate, the yield is between 3 and 3.5 tonnes per hectare.
> One tonne of rapeseed produces 415 kg of biodiesel.
> 1 hectare yields 1.45 tonnes of road fuel.
> Therefore to provide 37.8 Mt would require 25.9 Mha.
> The EU Directive quoted by Monbiot has now been replaced by a new Directive, the Fuel Quality Directive which requires a reduction of the greenhouse gas emissions of transport fuels by 6% by 2020 (Council directive (EU) 2015/652). 20% of 25.9 Mha would be very nearly 5.2 Mha
> But there is only about 5.7 Mha of arable land in the UK.
> This would leave very little for food production.

Once more, this calculation is a few years old but the essential message is still the same.

In practice therefore, in order to meet the EU target, the countries of the EU would need to import palm oil, or biodiesel, from Malaysia, Indonesia, etc. or ethanol from Brazil. This would be at the price of major deforestation. Monbiot is scathing: "The decision by governments in Europe and North America to pursue the development of biofuels is, in environmental terms, the most damaging they have ever taken. Knowing that the creation of this market will lead to a massive surge in imports of both palm oil from Malaysia and Indonesia and ethanol from rainforest land in Brazil; knowing that there is nothing meaningful they can do to prevent them"; (i.e. these imports) "and knowing that these imports will accelerate rather than ameliorate climate change; our governments have decided to go ahead anyway."

If all the developed world does the same thing as the EU, the competition between land for biofuels and land for food production could become very serious. Quoting from Monbiot again: "If the same thing is to happen throughout the rich world, the impact could be great enough to push hundreds of millions of people into starvation, as the price of food rises beyond their means. If, as some environmentalists demand, it is to happen worldwide, then much of the arable surface of the planet will be deployed to produce food for cars, not people. The market responds to money, not need. People who own cars – by definition – have more money than people at risk of starvation; their demand is 'effective', while the groans of the starving are not. In a contest between cars and people, the cars would win. Something like this is happening already. Though 800 million people are permanently malnourished, the global increase in crop production is being used mostly to feed animals: the number of livestock on earth has quintupled since 1950 (Strahan, 2011; Sharma, 2019). The reason is that those who buy meat and dairy products have more purchasing power than those who buy only subsistence crops." (Mobiot 2006).

Monbiot wrote those words in 2006 and in the following years the EU was indeed a major importer of palm oil from Malaysia and Indonesia, but in 2018 the European Commission decided to phase out imports of palm oil for transport completely by 2030, citing widespread deforestation. Not only is the EU one of the top export markets for palm oil for both Indonesia and Malaysia, but officials in Jakarta and Kuala Lumpur fear relentless campaigning against palm oil could trigger a broader global backlash that would threaten the industry's survival. With a global trade upheaval also throwing the market into flux, producers are scurrying to adjust and attempting to rehabilitate palm oil's image.

Palm oil is used in everything from margarine, chips, peanut butter and chocolate to shampoo, cosmetics, cleaning products, and biofuels. For decades, critics have claimed it is hazardous to health, while conservationists blame expanding plantations and slash-and-burn agriculture for endangering orangutan populations and the climate. Greenpeace says that over 74 million hectares of Indonesian rainforest have been "logged, burned or degraded" in the last half century.

However, Indonesia, which accounted for 56% of global palm oil supply in 2018, and Malaysia, which contributed 28%, claim the EU's real concern is protecting its own rapeseed and sunflower oils. (Nikkei Asian Review, 27 August 2019).

After all that, there is also the point that the EROEI for biodiesel (or for ethanol) is not particularly impressive, see Table 1.10, though there are arguments about the actual value. Some people even argue that the EROEI is less than 1! Instead of the term EROEI, Worldwatch (2007) in its discussion of biofuels calls it the energy balance; in this case it is the ratio of the energy in the biofuel (in joules) to the energy (in joules) used by people to plant the seeds, produce and spread agricultural chemicals and to harvest, transport and process the feedstock. Worldwatch (2007) also defines another quantity, the energy efficiency, of biofuels; this includes the energy contained in the feedstock itself in the denominator. The energy balance quoted by Worldwatch for ethanol from sugar cane is ~8 and for biodiesel from palm oil is ~9, both of which are considerably higher than the values quoted in Table 18.4 from Heinberg (2003). In addition to the question of the energy balance, or energy efficiency, another problem is that there would be a need for large supplies of methanol and a large mountain – or lake – of the main byproduct, glycerol. On a small scale, for disposing of used cooking oil from (fast food) restaurants making biodiesel is fine and on a small scale there is a good market for glycerol, but on a large scale, biofuels have limited possibilities and the market for glycerol is not unlimited.

1.3.6.9 Waves, tidal systems, etc

Wave technology is still at the research and development stage. One or two tidal systems exist, but the number of suitable sites is very small and there are costs in terms of disturbing fisheries and eco-systems.

1.3.6.10 Hydrogen and fuel cells

This is more a secondary issue, connected with how to replace oil for transportation, i.e. in cars, etc. One cannot think of the hydrogen as a primary source of energy – it would have to be made by electrolysis of water and the energy to generate the electricity would have to come from some sustainable source.

1.3.6.11 Cement/Concrete

There is a special problem of cement (and therefore of concrete) that most writers on CO_2 and global warming ignore; they concentrate on the burning of fossil fuels and ignore other aspects of cement production. Apart from the energy consumed in extracting and transporting the raw materials, in heating the kilns to about 1450 °C, and transporting the raw materials and the product, there is a special feature of cement production. This is that we are reversing the CO_2 sequestration that occurred millions of years ago when the marine micro-organisms that became chalk or limestone were formed. As far as the CO_2 is concerned, it is as if we were turning limestone or chalk into quicklime: $CaCO_3 \rightarrow CaO + CO_2$.

In fact, of course, cement is not quicklime; it consists of silicates but the effect, as far as the limestone and the CO_2 are concerned, is the same (Harrison 2019):

$$5CaCO_3 + 2SiO_2 \rightarrow \left(3CaO,SiO_2\right)\left(2CaO,SiO_2\right) + 5CO_2.$$

(The SiO_2 comes from sand) The consequence of this reaction is that the CO_2 from the energy used in the production process for 1 tonne of cement, plus the CO_2 emitted from this reaction comes to around 814 kg. Add in the quarrying and the transport and the result is something like 1 tonne of CO_2 being produced for each tonne of cement manufactured. There is some slight evidence that the construction industry is just beginning to pay attention to this question. Concrete production has quadrupled to more than four billion tonnes/year in the last three decades, but the chemical

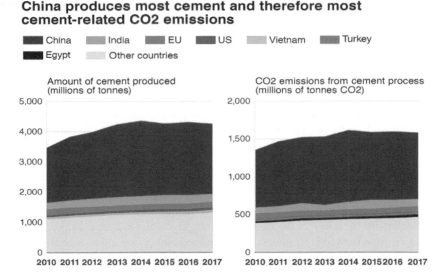

China produces most cement and therefore most cement-related CO2 emissions

FIGURE 1.22 Global cement production and consequent CO_2 emissions. (Source PBL Netherlands Environment Assessment Agency).

process required to produce cement creates enormous amounts of carbon dioxide, see Figure 1.22. The cement industry is responsible for eight per cent of global greenhouse gas emissions. A more sustainable solution is required to both provide the accommodation needed for a growing population while reducing damage to the environment. So builders are beginning to turn away from concrete. Treating timber to make it less flammable then gluing layers together, alternating the grain, can produce building blocks that are as strong as steel or concrete but lighter to produce. Figure 1.23 shows the W350 tower that is planned to be built in Tokyo. The 350 m (1,48ft) 70-storey building will be constructed of 90 per cent wooden materials with only a steel vibration-control framework to protect against earthquakes. The problem is the cost – the building is set to cost £4.lbn, almost double the cost of more conventional methods. However, the Japanese are also looking to timber for the stadium for the 2020 Tokyo Olympics. Plans for more wooden structures are branching out across the world, with large-scale wooden buildings already completed in Vancouver and Vienna and set for London and Bergen.

1.3.7 SUSTAINABLE ENERGY WITHOUT THE HOT AIR

"Sustainable Energy without the Hot Air" is the title of a truly amazing book by David MacKay (2009). It is dedicated "to those who will not have the benefit of two billion years of accumulated energy reserves" and it is a free book to download from the Internet. It was written by a Cambridge professor of physics and it is all about numbers. According to the preface:

"What's this book about?

I'm concerned about cutting UK emissions of twaddle – twaddle about sustainable energy. Everyone says getting off fossil fuels is important, and we're all encouraged to "make a difference," but many of the things that allegedly make a difference don't add up.

Twaddle emissions are high at the moment because people get emotional (for example about wind farms or nuclear power) and no one talks about numbers. Or if they do mention numbers, they select them to sound big, to make an impression, and to score points in arguments, rather than to aid thoughtful discussion.

This is a straight-talking book about the numbers. The aim is to guide the reader around the clap-trap to actions that really make a difference and to policies that add up.

FIGURE 1.23 The W350 tower in Tokyo planned to be built with 90% wooden materials. (Source *Big Issue* Scotland 2019).

This is a free book.

I didn't write this book to make money. I wrote it because sustainable energy is important. If you would like to have the book for free for your own use, please help yourself: it's on the internet at http://www.withouthotair.com. This is a free book in a second sense: you are free to use *all* the material in this book, except for the cartoons and the photos with a named photographer, under the Creative Commons Attribution-Non-Commercial-Share-Alike 2.0 UK: England & Wales Licence. (The cartoons and photos are excepted because the authors have generally given me permission only to include their work, not to share it under a Creative Commons license.) You are especially welcome to use my materials for educational purposes. My website includes separate high-quality files for each of the figures in the book."

The wonderful thing about Cambridge is that it teaches its physics students to use numbers in the way that is done by Hernberg and Monbiot that we have mentioned above and to understand the significance of their numbers. Some further examples which could serve as exercises for our readers might include the following; the beauty of these examples is that, unlike examination questions, we do not provide you with the values of all the numbers you will need – you need to find or estimate these for yourselves.

Exercise 1.3.7.1. Estimate the number of windmills (aerogenerators) needed to provide the electricity to drive one 10-car train from Edinburgh to London and also estimate the area of land that would be occupied by these windmills.

Exercise 1.3.7.2. Now estimate the number of windmills and the area of land they would occupy to provide all the electricity needed to run all the electric trains in the UK.

Exercise 1.3.7.3. An important political issue at the moment is the conversion of cars (automobiles) to run on electricity rather than on fossil fuels. Assuming that a decision were to be made to enforce the conversion of 10% of the total number of cars (automobiles) in the world to electricity, estimate the area of land that would be needed to be occupied by the required arrays of photovoltaic solar cells to provide the necessary electricity.

Exercise 1.3.7.4. Estimate the number of windmills (aerogenerators) and the area of land they would occupy to replace one 3,000 Megawatt (MW) (i.e. 3 GW) nuclear power station.

Exercise 1.3.7.5. Estimate what area of primary tropical forest would need to be cleared and converted into oil palm estates to provide enough biofuel for one flight of a modern jumbo jet from London to Sydney via Singapore.

And so on. Think of some more for yourselves.

We have not done these calculations ourselves. This is not because we are lazy (we are, of course!) but because the impact of simply presenting the results would be far more limited than if one works out the results for oneself. Moreover, the sceptics would quibble over some of the assumptions made. It is not the precise values of the answers which are important but the *indications* that the answers provide regarding the scale of the problem. If you have difficulty in performing these calculations, you can probably get help from David Mackay's book which is full of calculations like this. No one should consider themselves qualified to talk about sustainability of energy supplies until they have read this book and done some of these exercises.

1.3.8 POPULATION

Malthus (an Englishman who was born in 1766 and died in 1834) is very widely referred to in relation to population expansion and it is worthwhile giving some consideration to his work. In "An Essay on the Principle of Population," first published in 1798 and subsequently republished in various editions right up to the present time, Malthus made the famous prediction that population would outrun food supply, leading to a decrease in food per person. His Principle of Population was based on the idea that population if unchecked increases according to geometric progression (i.e. 2, 4, 8, 16, etc.), whereas the food supply only increases according to an arithmetic progression (i.e. 1, 2, 3, 4, etc.). Therefore, since food is an essential component to human life, population growth in any area or on the planet, if unchecked, would lead to starvation. He even went so far as to specifically predict that this must occur by the middle of the 19th century, a prediction which failed for several reasons. This failure has led people generally to be complacent about the problems of population growth that Malthus had highlighted. However, Malthus also argued that there are preventative checks and positive checks on population that slow its growth and keep the population from rising exponentially for too long, but still, poverty is inescapable and will continue. Positive checks are those, according to Malthus, that increase the death rate. These include disease, war, disaster, and finally, when other checks don't reduce population, famine. Malthus felt that the fear of famine or the development of famine was also a major impetus to reduce the birth rate. He indicates that potential parents are less likely to bear children when they know that their children are likely to starve:

> "The power of population is so superior to the power of the Earth to produce subsistence for man that premature death must in some shape or other visit the human race. The vices of mankind are active and able ministers of depopulation. They are the precursors in the great army of destruction, and often finish the dreadful work themselves. But should they fail in this war of extermination, sickly seasons, epidemics, pestilence, and plague advance in terrific array, and sweep off their thousands and tens of thousands. Should success be still incomplete, gigantic inevitable famine stalks in the rear and with one mighty blow levels the population with the food of the world."

The ideas that Malthus developed came before the industrial revolution and they are focused on plants, animals, and grains as the key components of diet. Therefore, for Malthus, available productive farmland was a limiting factor in population growth. With the industrial revolution and increase in agricultural production, land has become a less important factor than it was during the 18th century. His example of population growth doubling was based on the preceding 25 years of the brand-new United States of America. Malthus felt that a young country with fertile soil like the US would have one of the highest birth rates around. He liberally estimated an arithmetic increase in agricultural production of one acre at a time, acknowledging that he was overestimating but he gave agricultural development the benefit of the doubt. There was more to Malthus' work than a simple consequence of exponential growth versus linear growth. His ideas found a sequel in the "Limits to Growth" arguments of the Club of Rome (Meadows et al. 1972, 1992, 2005).

1.3.9 The collapse of former civilisations

There have been a number of former civilisations which flourished and then collapsed. It would be foolish to suggest that our present civilisation is going to collapse in the same way that any of them did; if it collapses, it will be in a different way. But, nevertheless, there are lessons to be learned. On this topic, it is worth mentioning the book "Collapse: How societies choose to fail or survive" (Diamond 2005) and in particular to concentrate briefly on his discussion of the demise of the Easter Island civilisation.

Diamond's definition of collapse is "… a drastic decrease in human population size and/or political/economic/social complexity, over a considerable area, for an extended time. The phenomenon of collapses is thus an extreme form of several milder types of decline, and it becomes arbitrary to decide how drastic the decline of a society must be before it qualifies to be labeled as a collapse." He cites examples of societies which, in his view, most people would regard as having collapsed rather than just suffering minor declines:

- The Anasazi and Cahoka in the modern USA
- The Maya cities of central America
- Moche and Tiwanaku societies in south America
- Mycenean Greece in Europe
- Minoan Crete in Europe
- Greenland Norse settlement in Europe
- The Great Zimbabwe in Africa
- Angkor Wat and the Harappan Indue Valley cities in Asia
- Easter Island in the Pacific Ocean.

The ruins left by many of these civilisations are very impressive and they testify to the existence in the past of highly populous and organised civilisations that have now simply vanished.

In studying the collapse of various former civilisations, Diamond considers a framework of five contributing factors:

- Environmental damage
- Climate change
- Hostile neighbours
- (Disappearance of) friendly trade partners
- A society's response to environmental problems.

He argues that many of these collapses were triggered, partly at least, by ecological problems, unintended ecological suicide, or "ecocide" as he calls it. That is, destroying the environmental resources on which their societies depended. This is not the only possible reason and the factors involved vary from case to case.

Among the past societies that collapsed, that of Easter Island is the one which Diamond describes as being "as close as we can get to a 'pure' ecological collapse, in this case due to total deforestation that led to war, overthrow of the elite and of the famous stone statues, and a massive population die-off." The other factors seem either not to have been relevant or not to have been particularly important; there were no hostile neighbours or trade partners – Easter Island is just so far away from anywhere else and there is no particular evidence of climate change over the period of the rise and fall of their civilisation.

1.3.10 Easter island

Easter Island is located in the Pacific Ocean at 109° 20' W and 27° 8' S and it is roughly the size and shape of Singapore or the Isle of Wight, but upside down. "It is the most remote habitable scrap of land in the world. The nearest lands are the coast of Chile 2,300 miles to the east and Polynesia's Pitcairn Islands 1,300 miles to the west." (Quotes in this section are all from Diamond 2005).

It appears that Easter Island was settled by Polynesian peoples coming from the west and arriving somewhat before A.D. 900. Recent research has shown that, for hundreds of thousands of years before human arrival and still during the early days of human settlement, Easter Island was not at all a barren wasteland, as it appeared to the early European explorers (the island was "discovered" by a Dutch explorer, Jakob Roggeveen, on Easter Day (hence the modern name of the island), 5 April 1722), but a diverse subtropical forest of tall trees and woody bushes. These included palm trees very similar to, but slightly larger than, the world's largest existing palm tree, the Chilean wine palm which grows to over 20 m tall and one metre in diameter. "Thus Easter (Island) used to support a diverse forest…. The overall picture for Easter (Island) is the most extreme example of forest destruction in the Pacific and among the most extreme in the world; the whole forest has gone and all of its tree species are extinct. The deforestation must have begun some time after human arrival by A.D. 900 and reached its peak around 1400, and been virtually complete by dates that varied locally between the early 1400s and the 1600s."

Easter Island is famous for its statues which are mostly 5 to 7 m high and the largest of which weigh about 270 tonnes. These statues must have been carved, transported, and erected by a large, well-organised, and prosperous population who had no power tools and no modern construction, lifting, or transportation machinery. Estimates of the population of Easter Island in its heyday range from 6,000 to 30,000; Diamond prefers the higher end of this range. The statues' sheer number and size suggest a population much larger than the estimated one of just a few thousand people encountered by European visitors in the 18th and early 19th centuries. What happened to the former large population? What went wrong?

In the first few centuries after the original settlers arrived, the forests were all cut down and the immediate consequences of the deforestation were losses of raw materials and losses of wild-caught foods. Crop yields also decreased because deforestation led locally to soil erosion by rain and wind, while other damage to soil that resulted from deforestation and reduced crop yields included desiccation and nutrient leaching. Farmers found themselves without most of the wild plant leaves, fruit, and twigs that they had been using as compost. These were the immediate consequences of deforestation and other human environmental impacts. Various species of fish were also fished out. The further consequences start with starvation, leading to civil war, a population crash, and a descent into cannibalism.

A small number of people survived, eking out a very meagre existence; the survivors adapted as best they could. When the European explorers arrived, they could not understand how these people could have erected all those statues.

1.3.11 Incomplete collapse: Example the end of the roman empire

There are also cases of societies which were highly organised and technologically advanced and which did not disappear completely but which collapsed and reverted to less well organised and less efficient social systems. We choose just one example, namely the Roman Empire, as being typical.

The Roman Empire was founded in the 8th century BC (in 753 B.C. according to tradition) from a small town on the Tiber. Originally a monarchy, in 509 BC, it became a republic ruled by a succession of emperors. The Empire expanded rapidly due to wars from which new territory was claimed. Initially following the Punic wars, the Empire controlled Italy, northern Africa, Sicily, and much of Spain, and following this Macedonia fell to the Empire. Eventually it was decided that the area of territory was too great to be managed by one central government and, in 27 BC, the Empire was split into the Western and Eastern Empires.

The Western Empire collapsed in AD 475 but the Eastern Empire continued until the fall of Constantinople in 1453. There are many theories for why one part of the Empire fell and the other part continued for a further thousand years. The definitive work on the subject is the six-volume "The History of the Decline and Fall of the Roman Empire" by the English historian Edward Gibbon. It traces Western civilisation (as well as the Islamic and Mongolian conquests) from the height of the Roman Empire to the fall of Constantinople. It was published between 1776 and 1789.

In the early years of the Western Empire, the army was well trained and disciplined with officers and leaders drawn from the ranks of the aristocracy. Within the army, there were central training and career structures and a central and effective legal system which served both central and local government, with effective systems and procedures in place for local government. However, the Western Empire faced many external threats from Parthians, Germans, and raids from Goths from the Aegean Sea.

In the third century, there were a series of disasters which weakened the Western Empire. Extortion and corruption became rife and career structures in the army broke down as the aristocracy began to withdraw from army careers. In addition to the loss of leaders in the army, there was also a shortage of manpower and the army was forced to employ barbarian mercenaries who proved to be unreliable. The external threats increased as barbarian groups settled on Empire territory without the army having the power to enforce the rule of law or to expel the settlers. There was also internal unrest with much political instability with separate factions trying to gain power and peasant rebellions due to social injustice. Local and central structures began to break down. The rich were able to avoid paying taxes and so social infrastructure projects were not started and roads and government buildings were not maintained. The economy was suffering and went into recession so the currency was debased leading to inflation and more unrest. Then from A.D. 249 to 262, there was a plague which led to a shortage of manpower and productivity.

The Empire was able to survive the third century crisis but in a much-weakened state. Social divisions continued and corruption became rife. The wealthiest families acquired even greater wealth but no longer supported the army nor paid taxes. The political system broke down as power was generally afforded to those who were prepared to pay for it. Public services continued to decline through lack of taxation. The greatest threat continued to be from external forces and the Empire went through a series of territorial wars with a much reduced and incompetent army, gaining in some wars but generally losing ground until the Visigoth leader Alaric had amassed sufficient support to march on Rome which, in A.D. 408, began the siege of Rome and famine and starvation, after which Rome fell.

1.3.12 Current environmental threats

As Diamond shows, the collapse of the Easter Island society followed swiftly upon the society's reaching its peak of population, monument construction, and environmental impact; he claims that the collapse was almost entirely due to environmental problems and the society's response to

---

TABLE 1.13

Twelve environmental threats

1. Destruction of natural habitats – forests, wetlands, coral reefs
2. Loss of wild food stocks – fish, shellfish
3. Loss of biodiversity
4. Soil erosion and degradation, salinisation, loss of nutrients
5. Ceiling on energy – oil, gas, coal
6. Water – shortage, pollution
7. Ceiling on photosynthesis
8. Toxic chemicals
9. Introduction of alien species
10. Generation of greenhouse gases and ozone – destroying chemicals
11. Growth of population
12. Increasing environmental impact of people

Source: (Diamond 2005).

---

environmental problems. When he comes to the end of the book and seeks to draw conclusions – or messages for us and our society – and just considering the environmental aspects he classifies the most serious environmental problems facing past and present societies into 12 groups, see Table 1.13.

"Eight of the twelve were significant already in the past, while four (numbers 5, 7, 8, and 10: energy, the photosynthetic ceiling, toxic chemicals, and atmospheric changes) became serious only recently. The first four of the 12 consist of destruction or losses of natural resources; the next three involve ceilings on natural resources; the three after that consist of harmful things that we produce or move around and the last two are population issues." (Diamond 2005).

Let's look at these briefly.

1. Destruction of natural habitats – forests, wetlands, coral reefs, and the ocean bottom. This is obvious and indisputable.
2. Wild foods – fish, shellfish. In theory, wild fish stocks could be managed sustainably – but this tends not to happen. According to Diamond, the great majority of valuable fisheries already either have collapsed or are in steep decline. Past societies that overfished included Easter Island and some other Pacific islands.
3. Loss of biodiversity. Species are becoming extinct on a daily basis. One can take a moral view and say that other species have a right to existence. Or one can take a pragmatic view and cite the benefits to human kind of the diversity of species available, for agriculture, horticulture, silviculture (forestry), medicines, etc.
4. Soil erosion and degradation (salinisation), loss of nutrients, acidification, or alkalinisation. We touched on this in discussing the consequences of global warming. It is a serious problem in many places and it has been going on for a long time. For instance, lands in Iraq and north Africa which were the bread basket of the Roman Empire are now semi-arid or full-blown desert.
5. Ceiling on energy (oil, gas, coal ….). The oil, gas and coal will run out sometime, sooner or later.
6. Water. There are well-known and serious problems involved in supplying clean unpolluted water to people in many parts of the world. We draw attention to "When the Rivers Run Dry" by Fred Pearce (2006, 2018). There is the direct interference in the natural hydrology by diverting rivers to supply water to one area at the expense of depriving another area of water.

There is the construction of dams for hydroelectric schemes with the introduction of the control of the flow of a river and eliminating the annual flooding that previously irrigated the fields downstream. There is the extraction of ground water that is not being adequately replenished. But there are also more indirect consequences such as international political disputes and even wars over water resources. But "The good news is that water is the ultimate renewable resource. We may pollute it, irrigate crops with it, and flush it down our toilets. We may even encourage it to evaporate by leaving it around in large reservoirs in the hot sun, but we never destroy it. Somewhere, sometime, it will return, purged and fresh, in rain clouds over India or Africa or the rolling hills of Europe. Each day, more than a thousand cubic kilometres (800 million acre-feet) of water rains onto the earth. There is, even today, enough to go around. The difficulty is in ensuring that water is always where we need it, when we need it, and in a suitable state – for all 7.3 billion of us." (Pearce 2019, page 264).

7. Photosynthetic ceiling. We talk about biofuels as an alternative to fossil fuels, e.g. using Malaysian palm oil to produce biodiesel or Brazilian sugar cane or US peanuts to produce ethanol for cars. But there are limits to production. And, of course, there is competition for land between biofuels and food production, see Section 1.3.5.
8. Toxic chemicals. Insecticides, pesticides, and herbicides; mercury and other metals, fire-retardant chemicals, refrigerator coolants, detergents, and components of plastics are all being discarded into the environment where many of them survive for a long time (e.g. DDT and PCBs) and the metals mostly forever. In addition, there are radioactive waste materials.
9. Introduction of alien species (animals and plants). Some alien species introduced by humans are obviously valuable to us as crops, domestic animals, and landscaping. But others have devastated populations of native species because the native species had no previous evolutionary experience of them and were unable to resist them.
10. Generation of greenhouse gases that cause global warming and of gases that damage the ozone layer.
11. The world's population is growing. More people require more food, water, shelter, space, energy, and other resources.
12. However, what really counts is not the number of people but their impact on the environment. Even without any increase in the number of people on Earth people's expectations – and therefore their demands – on the available resources increases.

The 12 sets of problems that we have just noted are not really separate from each other. They are linked and one problem exacerbates another or makes its solution more difficult. But any one of these 12 problems of non-sustainability would suffice to limit our lifestyle within the next few decades. Diamond considers that these problems are like time bombs with fuses of less than 50 years. He comments that when asked what is the single most important environmental/population problem facing the word today, his answer would be the misguided focus on identifying the single most important problem! That is because any of the dozen problems if unsolved would do severe harm, as they are all interrelated. To solve 11 of the problems, but not the 12th, would still be an issue, whichever problem remained unsolved. Every problem has to be solved.

It will be noticed that human-induced global warming as a result of burning fossil fuels is only one of these 12 threats (number 10) to our way of life. This supports our basic argument that global warming is only one of many threats to our well-being and survival and that the widespread concentration on CO_2 emissions to the exclusion of other environmental threats is misplaced. It generates complacency because by attempting to solve this problem we ignore all the other threats. The rest of this book therefore proceeds on the basis that while global warming as a result of human actions may well be a threat to our way of life, or even to our continued existence as a species, it is only one of many such threats.

2 The atmosphere

2.1 INTRODUCTION

Svante Arrhenius, see Figure 2.1, received the Nobel Prize for Chemistry in 1903 and is commonly thought of as a chemist, He was originally a physicist and can be regarded as one of the founders of the modern science of physical chemistry. He was the first to estimate the extent to which increases in atmospheric carbon dioxide are responsible for the Earth's increasing surface temperature (Arrhenius 1896) and, therefore, to suggest that CO_2 emissions, from fossil-fuel burning and other combustion processes, are large enough to cause global warming. Evidence to support this theory was produced by Guy Callendar (1938), see Figure 2.1, who was an English steam engineer and inventor. In 1938, he compiled measurements of temperatures from the 19th century on and correlated these measurements with old measurements of atmospheric CO_2 concentration. He concluded that over the previous 50 years the global land temperatures had increased and proposed that this increase could be explained as an effect of the increase in atmospheric CO_2, thereby supporting the theory which had been proposed by Arrhenius in 1896. This effect is now commonly referred to as the Callendar effect.

While the relation between CO_2 concentration and temperature is generally accepted, there have been arguments about the extent to which the observed increases in temperature and CO_2 concentration are due to human activities or natural causes. It should be possible to verify the extent to which the rate of increase of CO_2 indicated by the Keeling curve (Figure 1.12) is equal to the amount of CO_2 released by the recorded amounts of fossil fuel used.

We are confined to an ocean of air (the atmosphere), which has always been moving over space (from 10^{-6} to 10^7m) and time (from 10^{-6}s to 10^{16}s). The term *atmosphere* comes from the Greek words *atmos* (meaning *vapour*), and *sphaira* (meaning *sphere*). The spatial variability is illustrated in Figure 2.2, where tiny wisps-swirls-eddies in a cigarette smoke, cumulus clouds with bumps and wiggles kilometres across, or cloud patterns literally of the size of the planet, can be observed.

We have adapted to the air to such an extent that we often forget the importance of this tasteless, odourless, and invisible substance in our lives. Without the air, we cannot survive more than a few minutes due to breathlessness and the fatal radiation emitted by the Sun.

For this reason, the air has attracted the interest of humans since the times of the ancient Greek philosophers, who had debated on which substance was the principal element from which everything else was made (Ball, 2004). Heraclitus supported *fire*, Thales (624–546 BC) *water*, and Anaximenes (585–528 BC) *air*. More specifically, Anaximenes named air as the *arche* (a Greek word meaning the principles of knowledge). Later, Empedocles (490–430 BC) proposed four elements ("*roots*") that explain the nature and complexity of all materials: fire, earth, air, and water. Later on, Aristotle (384–322 BC) added a fifth element, *aether*, as the quintessence.

Thus, ancient Greeks used two words for air: *aer* (meant the dim lower atmosphere), and *aether* (meant the bright upper atmosphere above the clouds) (Lee, 1977; Barnes, 1987).

Nowadays, complex science has given rise to overarching theories of how our real, ordered *cosmos* actually evolved from the Big Bang through to the solar system to Earth, with its atmosphere and its biota, and even to the human brain. Chaisson, in his book *Cosmic Evolution* (Chaisson 2001) chose a precise complexity metric, the energy rate density (Φ_m), which is the typical power per unit of mass necessary to maintain a complex system in a quasi-steady state far from thermodynamic equilibrium.

FIGURE 2.1   (On the left): Svante August Arrhenius (1859–1927) who first estimated the extent to which increases in atmospheric CO_2 are responsible for global warming. (On the right): Guy Stewart Callendar (1898–1964) who correlated 19th century measurements of temperature and atmospheric CO_2 concentration and concluded that the observed increase in temperature could be explained as an effect of the increase in carbon dioxide, thereby supporting the theory which had been proposed by Arrhenius in 1896.

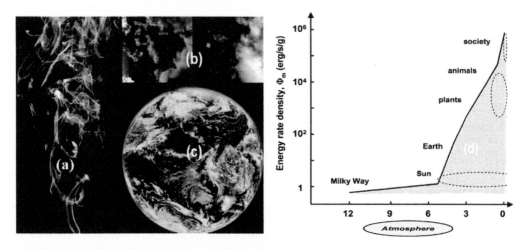

FIGURE 2.2   (a) Dimensions of cigarette smoke from mm to m. (b) Dimensions of clouds from m to a few km. (c) Clouds all over the Earth from a few km to thousands of km (*Source:* Lovejoy, 2019). (d) The evolution of complex systems since the Big Bang as quantified by the energy rate density ($x10^{-4}$ W/kg) versus the age in billions of years (*Source:* Chaisson, 2001).

According to Lovejoy (2019) placing the atmosphere's average Φ_m value into Chaisson's scheme (Figure 2.2 (d)) shows that it fits in reasonably well.

2.2   COMPOSITION OF THE ATMOSPHERE

In the early 1800s, scientists such as John Dalton realised that the atmosphere was actually made up of different chemical gases. Dalton was able to separate these gases and identify their relative quantities within the lower atmosphere. It was therefore easy for him to distinguish the main components of the atmosphere: nitrogen, oxygen, and a small amount of something un-burned, which later turned out to be argon. Joseph Black identified cabon dioxide in the atmosphere in the 1750s. In the 1920s when the spectrometer was developed, scientists were able to identify gases that were

Gas	Volume fraction (in % and ppm)	Residence time
N_2	78.084%	1.6×10^7 years
O_2	20.946%	$3 \times 10^3 - 10^4$ years
Ar	0.934%	
H_2O	0–4% (0– 40 000 ppm)	10 days
CO_2	3.94×10^{-2}% (394 ppm)	20–150 years
Ne	1.818×10^{-3}% (18.18 ppm)	
He	5.24×10^{-4}% (5.24 ppm)	10^7 years
CH_4	1.79×10^{-4}% (1.79 ppm)	10 years
Kr	1.14×10^{-4}% (1.14 ppm)	
H_2	5.3×10^{-5}% (0.53 ppm)	2 years
N_2O	3.25×10^{-5}% (0.325 ppm)	150 years
CO	$5-25 \times 10^{-6}$% (0.05–0.25 ppm)	0.2–0.5 year
Xe	8.7×10^{-6}% (0.087 ppm)	
O_3	$1-5 \times 10^{-6}$% (0.01–0.05 ppm)	weeks - months
NO_2	$0.1-5 \times 10^{-7}$% (0.001–0.05 ppm)	8–10 days
NH_3	$0.01-1 \times 10^{-7}$% (0.0001–0.01 ppm)	~5 days
SO_2	$0.003-3 \times 10^{-7}$% $(0.03-30 \times 10^{-3}$ ppm)	~2 days
H_2S	$0.01-6 \times 10^{-8}$% $(0.01-0.6 \times 10^{-3}$ ppm)	~0.5 day

Constant	Variable	Highly variable

FIGURE 2.3 Composition of the Earth's atmosphere and classification of its gases. The mixing ratio of trace gases is given in units of parts per million volume (ppmv or simply ppm). The residence time is defined as the amount of the compound in the atmosphere divided by the rate at which this compound is removed from the atmosphere. (modified from http://elte.prompt.hu/sites/default/files/tananyagok/AtmosphericChemistry/ch01s03.html).

present in much lower concentrations in the atmosphere, such as ozone and carbon dioxide. In fact, atmospheric gases are often divided into constant, variable, and highly variable gases as shown in Figure 2.3.

As seen in Figure 2.3, the residence time of constant gases is of the order of geological times-cales, while for the variable gases, it is of the order of years and days in the case of highly variable gases.

The composition of the Earth's atmosphere varies with elevation. According to the NASA MSIE E-90 atmospheric model, the composition of the atmosphere up to an elevation of 1000 km is shown in Figure 2.4. As seen in Figure 2.4 up to around 100 km (Kármán Line), the composition is fairly "normal," as depicted in Figure 2.3. After 100 km, the percentage of N_2 and O_2 decrease sharply and there is a sharp increase in O_3.

There is also a small increase in the percentage of N while Ar disappears entirely. By 200 km, He takes over as the predominant component. At the elevation of 1000 km, He makes up 93% of the human activities that are known to affect our atmosphere. Major environmental issues of the 20th and 21st centuries include air quality degradation (fog, photochemical ozone and tropospheric ozone generation, mercury pollution, etc.), acid precipitation (from the burning of coal that leads to SO_2 and therefore sulphuric acid) due to the use of ozone-depleting substances such as chlorofluoro-carbons), etc. Some of these issues have been successfully addressed through national and regional

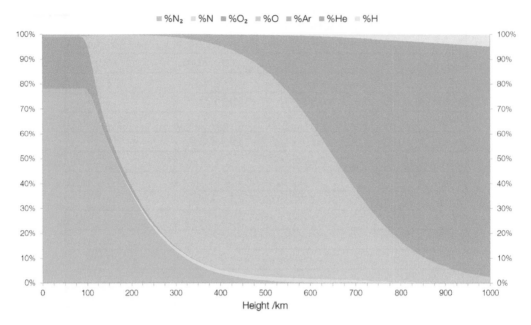

FIGURE 2.4   The composition of Earth's atmosphere as a function of height as derived from NASA's MSIE E-90 atmospheric model. Source: https://ccmc.gsfc.nasa.gov/modelweb/models/msis_vitmo.php.

legislation, international agreements, alternatives and/or changes to expectations and behaviour of people.

The role of chemistry in the Earth's climate has been examined by Ravishankara et al. (2015). According to this review, combustion of fossil fuels creates CO_2, which is considered the largest factor in predicted anthropogenic climate change. Increasing CO_2 concentration will not only affect the climate but also the acidity of the oceans. It is worth noting that CO_2 is not very chemically active in the atmosphere. Therefore, we could ask ourselves: what is the role of chemistry in the Earth's climate system, especially human-induced climate change? According to Ravishankara et al. (2015), the answer to this question is multiple:

(1) Apart from CO_2, many other chemically active species such as CH_4, halocarbons, N_2O, non-methane hydrocarbons (NMHC), and nitrogen oxides, directly or indirectly influence Earth's climate. These non-CO_2 species contribute almost as much as human-produced CO_2 to today's climate forcing. In this regard, the current CO_2 radiative forcing is about 1.68 Wm^{-2}, while the non-CO_2 emissions contribute about 1.65 Wm^{-2}. Unlike the greenhouse gases, aerosols (suspensions of liquids or solids in the air) and clouds are expected to exert global negative forcing and are now estimated to offset positive forcing by the greenhouse gases by as much as 50% of the forcing by CO_2. However, there is a great uncertainty about the effects of cooling and heating of various types of aerosols such as soot, dust, and absorbing organic molecules.

Some of the aerosols are emitted directly and some form in the atmosphere from a series of reactions starting with the oxidation of different volatile gases. Ozone is another greenhouse gas, produced by the troposphere in chemical reactions that consume emitted volatile hydrocarbons and use nitrogen oxides as a catalyst. Finally, most emissions are removed from the atmosphere by oxidants in the atmosphere such as OH^- radicals, nitrate radicals, and ozone; these determine the critical "cleansing" capacity of the atmosphere. Obviously, chemically active agents form a large part of the impact of human activities on the climate.

(2) The major impacts of climate change on Earth are an increase in sea level, changes in precipitation, drought, and extreme weather events. Many of these effects are largely involved in chemistry. For example, aerosols are at the heart of radiative forcing and the precipitation issues. Other significant impacts arise from changes in atmospheric chemical composition, such as deterioration of air quality, changes in the oxidation capacity of the atmosphere, and possible changes in the atmospheric circulation patterns.

(3) Climate change, associated with non-CO_2 gases and aerosols, is very dependent on chemical processes. The contribution of an emission that leads to greenhouse gases or aerosols, and thus alters the radiation balance of the Earth system, depends on chemical properties. The key questions about each emission include:

- the residence time, (see Figure 2.3), i.e. how long the emitted species remain in the atmosphere before being removed or transformed to another species
- where and how strongly it absorbs or scatters UV, visible, or infrared radiation, and
- how it modifies the atmospheric lifetime and the properties of other chemicals in the atmosphere?

(4) Chemistry plays an important role in any possible climate change mitigation and adaptation strategy, including intentional human intervention efforts, commonly known as "geoengineering" or "solar radiation management." For the above reasons, it is clear that chemistry plays a central role in the Earth's climate system. The essence of the chemistry role in climate is recorded on the coverage of this issue. The Earth system is highly coupled. The coupling means that the various environmental issues noted earlier are often linked. For example, the combustion of fossil fuels, which is the central issue of air quality, is clearly at the heart of man-made climate change. So, solutions to climate change are closely linked to air quality issues.

The ozone layer depletion is caused by chlorinated and brominated fluorocarbons (and related chemicals). These ozone-depleting chemicals (ODSs) are not only destructive to the ozone layer but are also powerful greenhouse gases. Therefore, ODS control has not only helped to cure the ozone layer but has also contributed greatly to the climate.

Tropospheric ozone is a greenhouse gas, and therefore its changes affect climate. Natural and anthropogenic chemical transformations of hydrocarbons account greatly for present levels and predicting future levels of tropospheric ozone. Also, climate change will change tropospheric ozone, thus playing an important role in regional and global air quality. Carbon dioxide persists for centuries in the atmosphere and therefore its impacts persist for a very long time. In contrast, the chemically active reactive species have shorter lifetimes. Therefore, there is more immediate relief for the climate system when such emissions are reduced. Consequently, there is now a focus on short-lived climate forcers in climate change mitigation approaches; this issue further highlights the importance of chemistry in the climate system today. One of the major issues that has emerged over the past decade is the large role played by aerosols in the climate system via interaction with incoming sunlight, modifying chemical composition, and influencing precipitation and clouds and, thus, Earth's radiation balance. This is particularly important since aerosols are currently thought to partially offset the positive climate forcing by greenhouse gases.

Aerosols are complex – they come in different sizes, chemical composition, phases, and properties. In addition, they are implicated in adverse health effects. They also play important roles in transforming some chemicals in the atmosphere. Their origins are diverse but are partly connected with combustion, the same source as for CO_2. However, aerosols are thought of mostly as pollutants that influence air quality. Thus, the policy instruments for dealing with aerosols are different from those for greenhouse gases. The issues related to aerosols add further layers of complexity in causes as well as in solutions. Suffice it to say, aerosols are one of the hot topics in atmospheric chemistry today. The complexity of aerosols has perplexed scientists. Yet, it is important to understand and predict the influence of aerosols on climate as well as its influences on related issues such as health, melting of snow and ice by black carbon, etc. Lastly, higher parts of the atmosphere, the ionosphere and the mesosphere, hold very little mass but still respond to climate change. There are a myriad of couplings in the climate system, some of which were noted earlier in this paragraph. Suffice to say, human actions to control one environmental issue will undoubtedly influence another. Indeed, some actions have clear impacts on multiple issues, such as climate change and air quality. It is very desirable that actions taken by society will have positive effects on climate and the environment – the so-called "win–win" strategies for the multiple issues that are involved. At least, one needs to avoid "win–lose" choices where solutions to one issue either exacerbate another issue or create a new problem. Thus, an understanding of chemical changes will continue to play a major role in better understanding and predicting climate change and providing solutions to anthropogenic climate change.

All the aforementioned issues will be described in the following sections in more detail.

2.3   THE EARTH'S RADIATION BUDGET

2.3.1   GLOBAL ENERGY FLOWS

The Earth's radiation budget refers to all the energy that enters and leaves the Earth system, which is mostly visible light and infrared radiation (IR). Figure 2.5 shows that the average energy from sunlight coming to the top of Earth's atmosphere is around 341.3 Wm^{-2}. Less than 50% of the incoming solar radiation reaches the ground and heats it. The rest is partly absorbed by the atmosphere and the residue is reflected by clouds or ice.

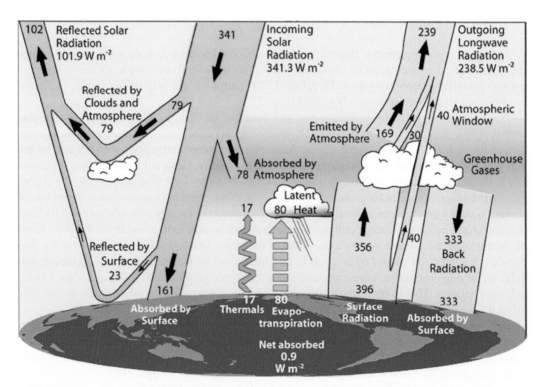

FIGURE 2.5   The global annual mean Earth's radiation budget for the Mar 2000 to May 2004 period. Arrows on the left show the solar radiation coming to Earth and arrows on the right the infrared radiation going away from Earth. The units are W m^{-2}. (Source Trenberth et al. 2009).

The sunlight that reaches the ground warms the Earth's surface. The warm ground and oceans emit IR radiation through the atmosphere where most of it is trapped by greenhouse gases - GHGs (H_2O, CO_2, CH_4), and other trace gases that are opaque to many wavelengths of thermal infrared energy.

In order to estimate the surface temperature, let us denote the solar radiative flux at the top of the atmosphere by S_o (known as the solar constant) and the albedo, i.e. the reflectivity, of the Earth by α.

We assume that the amount of the incoming radiation distributed over the Earth's sphere is equal to the amount that would be collected on the Earth's surface if it was a disk (with the same radius R as the sphere), placed perpendicular to the sunlight. Thus, the heat Q_a absorbed by Earth:

$$Q_a = (1-\alpha)\pi R^2 S_o \tag{2.1}$$

The total heat emitted from the Earth Q_e is equal to the energy flux implied by its temperature, T_e (from the Stefan-Boltzman law) times the entire surface of the Earth, or:

$$Q_e = (4\pi R^2)\sigma T^4 \tag{2.2}$$

In the case of radiative balance, we will have:

$$(4\pi R^2)\sigma T_e^4 = (1-\alpha)\pi R^2 S_o \tag{2.3}$$

and consequently:

$$T_e = \left[(1-a)S_o/4\sigma\right]^{-1/4} \tag{2.4}$$

It is worth noting that T_e (effective temperature) would be the temperature at the Earth's surface if the Earth had no atmosphere. According to Equation (2.4), T_e is about 255 K (or -18°C). With this temperature, the Earth's radiation will be centred on a wavelength of about 11 μm (in the range of IR). Because of the spectral properties of the Sun and Earth's radiation, we tend to refer to them as "shortwave" and "longwave" radiation, respectively (Kushnir 2000). In reality, T_e is much lower than the observed temperature averaged over all seasons and the entire Earth surface temperature: 288 K (or 15°C). This difference is a result of the heat absorbing components of our atmosphere.

Given that Earth neither heats up nor cools down, the energy coming to Earth as sunlight equals the energy leaving as IR. If it does not, the Earth heats up or cools down. In the case of an enhanced greenhouse effect, the energy budget is not balanced because the additional CO_2 traps more heat warming the planet.

It should be stressed that the materials of the Earth's surface (e.g. soil, rocks, water, forests, snow, and sand) behave differently in the solar radiation due to their different albedo. In general, much of the land surface and oceans are dark (i.e. low albedo), thus absorbing the major part of the solar energy reaching them. A similar behaviour is exhibited by the forests (low albedo, near 0.15). On the other hand, snow and ice have very high albedo (as high as 0.8 or 0.9), and thus absorb very little of the solar energy that reaches them. On average, the Earth's planetary albedo is about 0.31, meaning that about a third of the solar radiation reaching the Earth is reflected out to space and about two thirds is absorbed. Bearing this in mind, we realise that colder climate leads to more snow and ice, which results in more reflected solar radiation to space and the climate getting even cooler known as ice-albedo feedback.

Similarly, climate warming causes a decline in snow and ice, less reflected solar radiation, thus leading to even more warming. In addition, clouds have a high albedo and reflect a large amount of solar energy out to space. Different types of cloud reflect different amounts of solar energy. It is remarkable that if there were no clouds, the Earth's average albedo would drop by half.

2.3.2 Earth's radiation budget and climate

Because the Earth is a sphere, the Sun heats equatorial regions more than polar regions (Figure 2.6). The atmosphere and the oceans work continuously to smooth the imbalances of solar heating through surface water evaporation, transport, rainfall, winds, and ocean circulation. This coupled atmosphere with the oceans is known as the Earth's heat engine.

Solar heat is partly transferred from the equator to the poles, but also from the surface of the Earth and the lower atmosphere back into space. Otherwise, the Earth would endlessly heat up. The Earth's temperature does not increase indefinitely because, as discussed in the previous sub-section, surface and atmosphere emit heat at the same time. This net radiation flow inside and outside the Earth system is the earth's radiation budget. The amount of incoming radiation varies with latitude and season as shown in Figure 2.7 (based on data from the Earth Radiation Budget Experiment – ERBE satellite data).

When the incoming solar radiation is balanced by an equal radiation to space, the Earth is in radiative balance and the global temperature is relatively stable. Anything that disturbs the amount of incoming or outgoing energy disrupts the Earth's radiation balance and then global temperature increases or decreases.

Changes that affect how much energy enters or leaves the system alter the Earth's radiative balance and can trigger an increase or decrease in temperatures. These destabilising causes are called climate forcings and include natural phenomena, such as changes in the Sun's brightness, Milankovitch cycles (small variations in the Earth's orbit and its axis of rotation occurring over thousands of years), and large volcanic eruptions. Apart from the natural climate forcings, there are the man-made climate forcings such as particle pollution (aerosols that absorb and reflect incoming solar radiation); deforestation (which changes the surface albedo); and the rising concentration of GHGs (which decrease the heat radiated to space).

According to the current knowledge, volcanoes worldwide emit, on average, about 1.5 metric tons of CO_2 per day (only about 2% of the amount that human activity causes). This estimation may be far too low because it is based on measurements from only 33 of the world's most volcanically active peaks (only three of which are ice-covered), among the 1500 or so that have erupted in the past 10,000 years.

Ilyinskaya et al. (2018) combined high-precision airborne measurements over the period 2016–2017 with atmospheric dispersion modelling to quantify CO_2 emissions from Katla (a major subglacial volcanic caldera lying near the southernmost tip of Iceland) that had last erupted 100 years ago but has been undergoing significant unrest in recent decades. Katla's sustained CO_2 flux, 12–24 kt per day,

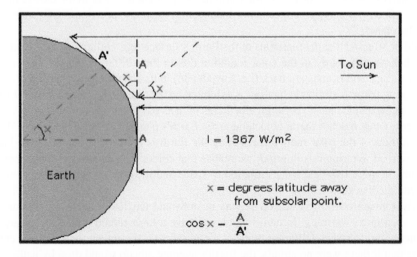

FIGURE 2.6   As the angle of incidence (x) increases, insolation decreases.

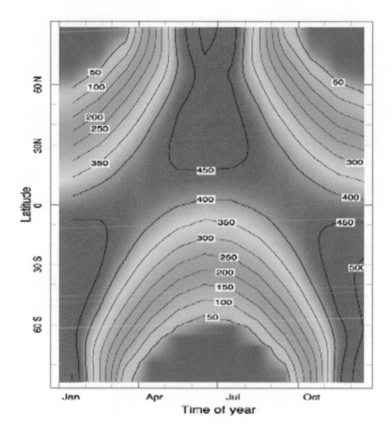

FIGURE 2.7   The temporal distribution of the incoming solar radiation at the top of the atmosphere as a function of latitude (in W m^{-2}).

is up to an order of magnitude greater than previous estimates of total CO_2 released from Iceland's natural sources. They have concluded that Katla is one of the largest volcanic sources of CO_2 on the planet, contributing up to 4% of global emissions from non-erupting volcanoes (see Figure 2.8).

2.4   OPTICALLY ACTIVE MINOR GASEOUS COMPONENTS

The composition of the atmosphere is shown in Figure 2.3 and states that nitrogen, oxygen, and argon make up the bulk of the atmosphere. These gases are said to be uniformly mixed, that is to say the percentages of these three gases are constant throughout the atmosphere, at least, up to about 100 km in height. However, over a geological timescale, the amount of oxygen is not stable because oxygen participates in life processes and other chemical interactions. Of the secondary gaseous components, the most important are carbon dioxide, water, ozone, methane, and oxides of nitrogen, and they exhibit variable concentrations in both space and time.

The sources and sinks of most minor gaseous components are found on the Earth's surface (land or sea), often by biospheric and biological activity. This applies to the case of carbon dioxide and water, as well as for most man-made and greenhouse gases, such as methane.

More details about the spatio-temporal variability of the basic minor gaseous components are given to the next few sections.

The formation of climate and its changes is determined by a complicated interaction among the components of the atmosphere-hydrosphere–lithosphere–cryosphere–biosphere system. One of the major developments in climatology over the last half century has been the opportunities provided by the exponential increase in computing power, Moore's Law, see Figure 2.9 and

FIGURE 2.8   The Icelandic Katla volcano covered by ice : (Source: NASA's Earth Observatory).

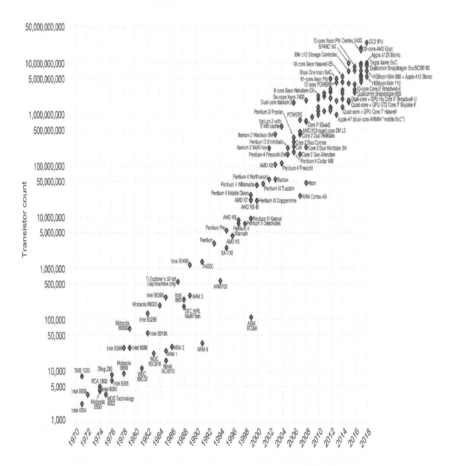

FIGURE 2.9   Moore's Law (Licensed under CC-BY-SA by the author Max Roser).

TABLE 2.1

Major impacts of atmospheric chemical components on the most important atmospheric phenomena

Components	UV radiation absorption	Greenhouse effect	Photochemical formation of oxidants	Acidification of precipitation	Drop of visibility	Corrosion of materials
O_3	+	+	+			
C (soot)		+			+	
CO_2		+				
CO						
CH_4						
C_xH_y			+		+	
NO_x			+	+	+	
N_2O		+				
NH_3/NH^+_4				+		
SO_x		+		+	+	+
H_2S						+
COS						
Organic sulphur						+
Halocarbons		+				
Other halogens						+
Minor components						+

Source: (KC98).
Note: + signifies a positive effect of a given component on the respective phenomenon.

the opportunities provided for computer modelling in meteorology and climatology which have developed into a major international scientific activity. However, the construction and operation of these models encounters great difficulties because of the complexities of the physical processes involved and the difficulties in establishing the values of the input parameters for the models.

Solar radiation is practically the only source of energy; the contributions from geothermal energy and nuclear energy are tiny. Solar radiation has an important role in the atmosphere in the formation of the greenhouse effect, due to numerous optically active minor gaseous and aérosol components. The cloudiness interaction is a key dynamical climate-forming mechanism (Marchuk *et al.* (KC98). This determines the decisive importance of chemical and photochemical processes responsible for atmospheric composition as climate forming factors.

The components of atmospheric chemical composition, directly affecting the most substantial atmospheric phenomena (Table 2.1), have been considered by Clark (KC98). The significance of these components can vary, depending on the specific problems under consideration. It is also important to take into account numerous indirect impacts (for example, the effect of halocarbon compounds on ozone) and synergism, determined by interactions among various chemical processes in the atmosphere and biosphere.

The chemical composition of the atmosphere is affected by various factors, see Tables 2.2. Table 2.3 illustrates a synoptic assessment of the impacts on the atmosphere. Much of the computer modelling of the climate has been concerned with studying the effects of increasing the concentration of CO_2 in the atmosphere (see Figure 1.12). As has been mentioned, an increasing amount of CO_2 in the atmosphere is of great concern because of its warming effect on the climate. However,

TABLE 2.2
Principal factors affecting atmospheric chemical composition

Components	Oceans and estuaries	Vegetation and soil	Wild animals	Wetlands	Biomass burning	Crop production	Domestic animals	Petrol burning	Coal burning	Industrial processes
O_3								+		
C (soot)						+		+	+	
CO_2	+	+			+	+		+	+	
CO	+	+			+	+		+	+	+
CH_4		+	+	+	+	+	+			
C_xH_y	+	+			+	+				
NO_x	+				+	+		+	+	+
N_2O	+	+			+	+		+	+	
NH_3/NH_4^+		+	+	+	+	+	+		+	
SO_x								+	+	+
H_2S	+	+		+		+				
COS	+	+		+						
Organic sulphur	+	+		+						
Halocarbons										+
Other halogens								+	+	+
Minor components					+			+	+	+

Source: (KC98).

Note: + signifies a positive effect of a given component on the respective factor.

TABLE 2.3

A synopic assessment of impacts on the atmosphere

Source	UV energy absorption	Thermal radiation budget alteration	Atmospheric property Photochemical oxidant formation	Precipitation acidification	Visibility degradation	Materials corrosion
Oceans and estuaries	II-1					II-3
Vegetation and soils		II-2	III-2	III-l		
Wild animals		III-l				
Wetlands		III-l				III-2
Biomass burning	II-2	II-2			II-2	
Crop production	1–1	II-2				
Domestic animals	1–2					
Petroleum combustion	II-2	III-3	IV-3	II-3	II-3	II-3
Coal combustion	III-2	III-3	1–3	III-3	III-3	III-3
Industry	III-3	II-2		II-3	II-3	II-3

Source: (KC98).

Note: Potential importance: I, some; II, moderate; III, major; IV, controlling.
Assessment reliability: 1, low; 2, moderate; 3, high.

the radiative regime of the atmosphere is affected by many other optically active minor gaseous constituents than CO_2 (see Figure 2.3. Therefore, there is a need for the following:

(1) a reliable parametrisation of the radiative processes of various gases in climate models, including changes in the ozone content in the atmosphere,
(2) continuous observations to reveal the trends of such components, and
(3) the continuation of laboratory studies to specify data on the rates of various photochemical reactions and to enhance databases on the spectral line parameters for various minor gaseous constituents.

Progress achieved during recent years in the development of observational techniques and in a deeper understanding of the laws of chemical and physical processes in the atmosphere has made it possible to reveal substantial changes in the concentration of numerous chemically and optically active minor gaseous constituents of the atmosphere. Theoretical calculations have led to the conclusion that the contribution of increased concentrations of such components as methane, nitrous oxide, CFCs, etc. to the formation of the atmospheric greenhouse effect (and respective climate changes) turns out to be comparable to the effect of increased CO_2 concentration, an interactive effect (non-additive effect) of various components on the greenhouse effect with climate playing the substantial role. So, for example, a temperature increase would be followed by an intensification of natural processes of the formation of methane affecting the concentration of CO_2 and O_3. In forecasting possible trends of minor gaseous constituent concentrations in the future, and associated changes in the atmospheric composition and climate, a thorough account of the complicated interaction between various processes is needed.

There is a very simple argument that has been around for decades which is based on the relationships presented in Section 2.2 and which leads to the greenhouse effect. The greenhouse effect of the atmosphere (G) may be defined by the difference between the thermal emission from the surface (E) and the outgoing longwave radiation to space (F):

$$G = E - F \qquad (2.5)$$

Assessments of global/annual average values on the basis of satellite observations (Harrison *et al.* KC98) led to the following figures: $E = 390$ Wm^{-2} and $F = 235$ Wm^{-2}, so that $G = 155$ Wm^{-2}. The clear-sky greenhouse effect (G_A) may then be defined as

$$G_A = E - F_{\text{clear}} \qquad (2.6)$$

with a global annual average value of $F_{\text{clear}} = 265$ Wm^{-2} (Harrison *et al.* KC98), so that $G_A = 125$ Wm^{-2}. Although it may seem very small, the cloud longwave radiative forcing is thus equal to only 30 Wm^{-2}.

Following Raval and Ramanathan (1989), let us introduce normalised quantities to remove the strong temperature dependence

$$G = E/F_{\text{clear}} \qquad (2.7)$$

There is a very simple possibility to assess qualitatively the influence of atmospheric temperature and humidity on the clear-sky greenhouse effect. Obviously,

$$F_{\text{clear}} = \varepsilon_A \sigma T_A^4 + \left(1 - \varepsilon_A\right)\sigma T_S^4 \qquad (2.8)$$

where ε_A is the atmospheric emissivity, σ is the Stefan-Boltzmann constant, T_A is the mean atmospheric temperature, and T_s is the mean surface temperature. Now, from Equation (4.33), we have

$$G' = T_S^4 \Big/ \left[\varepsilon_A \sigma T_A^4 + \left(1 - \varepsilon_A\right)\sigma T_S^4\right] \qquad (2.9)$$

Or, it is possible to introduce a parameter g (Raval and Ramanathan 1989) where:

$$g = \varepsilon_A \left[1 - \left(T_A/T_S\right)^4\right]. \qquad (2.10)$$

Hence:

$$G = \frac{1}{1-g} \qquad (2.11)$$

$$= \frac{1}{1 - \varepsilon_A \left[1 - \left(T_A/T_S\right)^4\right]} \qquad (2.12)$$

Here it is assumed that the emissivity ε_A is proportional to the total column moisture content, w. Although it is a clear simplification, Equations (2.11, 2.12) illustrate the essential result that there is a separation between the effects of the atmospheric humidity, as measured by w and represented by ε_A in these equations, and the temperature, as represented by T_A (Webb *et al.* KC98). As is seen, the increase of w and, consequently, of ε_A leads to the enhancement of G' or g, while the increase of T_A results in the decrease of temperature contrast between the atmosphere and the surface and, thus, to the G' decrease. It is quite obvious that the roles of w and T_A have to be geographically specific. The most obvious limitations of such a model include the restriction to one dimension and the neglect of heat flows in the oceans. Heat circulates more slowly in the oceans than in the atmosphere and so to neglect changes in the transfer of heat between the atmosphere and the oceans is more serious in climate models than in weather forecast models.

On the basis of the qualitative analysis, Webb et al. (KC98) have pointed out that, for instance, since the temperature field in the tropics is comparatively homogeneous, the dominating role in the greenhouse effect variability belongs to w, which is controlled by sea surface temperature. The opposite situation occurs in middle and high latitudes where air temperatures are highly variable and therefore exert much more powerful control over G' than at low latitudes.

The analysis of satellite data on the Earth radiation budget (Earth Radiation Budget Experiment, ERBE) and the total column moisture (Special Sensor Microwave/Imager, SSM/I) as well as the results of radiative transfer simulations made by Webb *et al.* have confirmed the conclusions of the simplified qualitative assessment. At low latitudes, the clear-sky greenhouse effect varies mainly due to w changes (which, on the other hand, are controlled by the sea surface temperature) and seasonal variations are small. In contrast, at middle and high latitudes, both G' and w exhibit strong seasonal variations. The clear-sky greenhouse effect variations are controlled by the seasonal changes in atmospheric temperatures, which are strong enough to overcome the opposing effect of the moisture variations. There are strong seasonal variations of the greenhouse effect which has its maximum in winter (when surface-atmosphere temperature difference increases, which leads to the enhancement of the greenhouse effect, in spite of small moisture content) and its minimum in summer. The combined impact of both temperature and column moisture results in the formation of the meridional profile of the greenhouse effect which is characterised by the decrease of G' towards the winter pole at a much slower rate than it does towards the summer pole.

The data in Table 2.4 characterise variations in radiative heat flux divergence with increasing concentrations of minor gaseous constituents, obtained with the use of a one-dimensional model (calculations were made for a standard atmosphere). They should only be regarded as indicative of the relative importance of these gases.

What we have described as our first edition (KC98), which was written a year or two before 1998, included a number of predictions of the situation in 2010 based on very simple models which by now have been superseded by much more sophisticated models. This material, which is concerned with predictions of mean suface temperature, and the concentrations of several minor gaseous components of the atmosphere (Tables 4.34 and 4.35 of that book KC98) would now only be of interest

TABLE 2.4

Variations in radiative heat flux divergence due to the greenhouse effect caused by different MGCs

Component	Change of concentratrion	Variations in radiative heat flux divergence (W m⁻²)			
		Stratosphere	Troposphere	Surface	Surface – troposphere system
CO_2	330 to 660 ppm	−2.64	3.72	1.16	4.88
N_2O	0.30 to 0.60 ppm	−0.20	0.55	0.52	0.07
CH_4	1.7 to 3.4 ppm	−0.06	0.42	0.29	0.71
CF_2Cl_2	0 to 2 ppb	0.17	0.00	0.87	0.87
$CFCL_3$	0 to 2 ppb				
O_3	−25%	−0.15	−0.27	−0.08	−0.35
	+ 25%	−0.21	0.12	0.17	0.29
H_2O	100%	−3.47	3.58	0.00	3.58

Source: (KC98, Table 4.33).

TABLE 2.5

CH$_4$ concentrations C, in 2100, τ, lifetime changes from 1990 to 2100 and radiative forcing ΔQ from 1900 to 2100 (direct effect of effect of CH$_4$ only) for the IPCC 1992 emissions scenarios

Scenario	C(2100)	τ (2100)/τ (1900)	ΔQ (2100–1990) (Wm^{-2})
IS92a,b	3790 (3280–4380)	1.15 (1.06–1.25)	0.62 (0.49–0.76)
IS92c	2250 (2000–2560)	1.05 (1.02–1.11)	0.19 (0.10–0.28)
IS02d	2160 (1900–2380)	0.97 (0.94–0.99)	0.15 (0.07–0.23)
IS92e	4180 (3710–4730)	1.09 (1.03–1.16)	0.72 (0.60–0.84)
IS92f	4890 (4180–5690)	1.20 (1.09–1.34)	0.88 (0.72–1.05)

Note: Figures in parentheses give lower and upper values. The CH$_4$ emissions in the six IPCC scenarios IS92a–IS92f are identified in Table 2 of IPCC 1992).

to someone who was concerned with the prehistory of the development of computer modelling of the atmosphere.

There are three particularly important minor gases beyond CO$_2$ namely methane (CH$_4$), nitrous oxide (N$_2$O), and ozone (O$_3$). We consider them in turn. The remainder are of less significance in terms of heat transfer.

The situation regarding CH$_4$ is one of the principal GHGs responsible for the radiative forcing change, due to its concentration growth since pre-industrial times, say from about 1765, of about 0.4 Wm^{-2} and is second only to CO$_2$. Unfortunately, there is still very high uncertainty concerning sources and sinks of CH$_4$ which makes it very difficult to predict future CH$_4$ concentrations. Assessments of total CH$_4$ emissions vary within the range from 330 to 880 TgCH$_4$yr^{-1} (see Table 2.5) with a best guess of 515TgCH$_4$yr^{-1} (KC98). As to the future there is very serious concern that warming of the polar regions is leading to the melting of some of the permafrost, leading in turn to the emssions of unknown additional quantities of CH$_4$.

Hao and Ward (KC98) assessed the emissions of CH$_4$ from various sources of biomass burning for tropical, temperate, and boreal regions. They showed that 85% of the total CH$_4$ is emitted in the tropical area, which is mainly the result of shifting cultivation, wood fuel use, and deforestation. Changes in land use practices and population growth in the tropics are possible causes of the increase of atmospheric CH$_4$ concentration. CH$_4$ emitted from biomass burning may have increased by at least 3Tgyr^{-1} to a total of 34Tgyr^{-1} by the late 1980s because of increases in tropical deforestation and demand for fuel-wood.

We do not include the CFCs in our discussions because, although they are much more powerful greenhouse gases (GHGs), molecule for molecule, than CO$_2$, their concentrations are much lower and their manufacture is supposedly being phased out under the Montreal Protocol on ozone-depleting substances. Ravishankara et al. (2009) discussed the role of N$_2$O as an ozone-depleting substance. By comparing the ozone depletion potential-weighted anthropogenic emissions of N$_2$O with those of other ozone-depleting substances, they showed that N$_2$O emission was the single most important ozone-depleting emission and was expected to remain the largest throughout the 21st century. N$_2$O is unregulated by the Montreal Protocol. Limiting future N$_2$O emissions would enhance the recovery of the ozone layer from its depleted state and would also reduce the anthropogenic forcing of the climate system, representing a win-win situation for both ozone and the climate.

By far the most interesting of the minor gases in the atmosphere is ozone (O$_3$). Despite its very low concentration, its presence in the stratosphere is of enormous importance in providing protection from solar ultraviolet (UV) radiation. The story of ozone depletion, a slow steady depletion of the stratospheric ozone begins with the development of the Dobson Spectrophotometer in 1924 by Prof. G.M.B. Dobson of Oxford University. Records of atmospheric concentrations of ozone began at Arosa, Switzerland in 1926 and have been made at more and more locations since then and using some other ground-based spectrometers. More recently, satellite flown instruments have contributed to the data set. Evidence of a rather slow decline in the ozone concentration was found from this data set. The discovery of the ozone hole, which is now known to be a total temporary vanishing of the ozone between heights of 14 and 21 km in each Antarctic spring, was made by scientists at research stations in the Antarctic (Chubachi 1984, 1985, Farman et al. 1985) and was dramatically verified by images from the NASA experimental satellite Nimbus-7.

A lengthy account of the discovery of ozone depletion can be found for example in Chapter 5 of Cracknell and Varotsos (2012). Ozone is an important greenhouse gas, but it is of particular relevance to climate change because ozone depletion led to the Montreal Protocol on Substances that Deplete the Ozone Layer, which was adopted by the United Nations in 1987 and subsequently ratified by enough states so that it became a part of international law. The sequence of events covering the period from the discovery in 1973 that CFCs can destroy ozone in the stratosphere through to the adoption of the Montreal Protocol in 1987 is outlined in Table 5.1 Cracknell and Varotsos (2012). The success of the Montreal Protocol is demonstrated by Figures 2.10 and Figure 2.11.

We argue later on that the success of the Montreal Protocol can be regarded as part of the reason that it was hoped to deal with the question of human-induced global warming by the adoption of a similar protocol, the Kyoto Protocol. But, at least so far it has not worked out that way. Table 2.6 shows an assessment of the role of various biogeochemical feedbacks and may be considered as a summary characterising the importance of minor gaseous constituents in climate formation.

Here the gain is $g = (w - i)/w$, where $w(i)$ is the output (input) signal defined by analogy to an electronic amplifier. The amplification (often referred to as the feedback factor) is

$$f = w / i = \frac{1}{1-g} \tag{2.13}$$

There are negative $(g < 0)$ and positive $(g > 0)$ feedbacks. The values in this Table should only be regarded as indicative.

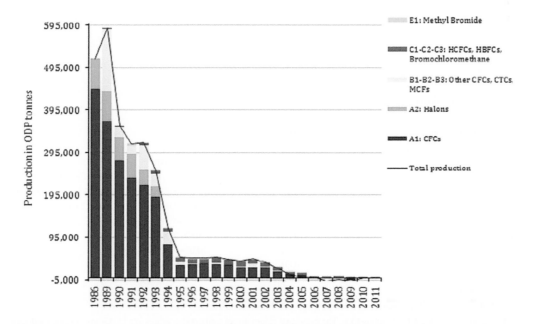

FIGURE 2.10 Production of ODS in European Environment Agency Member Countries. Production is defined under Article 1(5) of the Montreal Protocol as production minus the amount destroyed minus the amount entirely used as feedstock in the manufacture of other chemicals (Source: EEA).

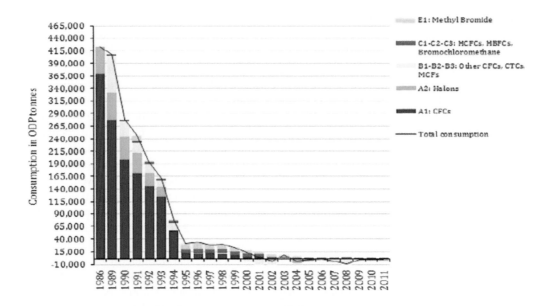

FIGURE 2.11 Consumption of ODS in European Environment Agency Member Countries. Consumption is defined as production plus imports minus exports of controlled substances under the Montreal Protocol (Source: EEA).

TABLE 2.6
Estimated gain from climate and biogeochemical feedbacks (KC98)

Feedback	Average gain	Possible range	Comments
Geophysical			
Water vapour[a]	0.39	(0.28–0.52)	
Ice and snow	0.12	(0.03–0.21)	
Clouds	0.09	−(0.12–0.29)	
Sub-total[b]	0.64	(0.17–0.77)	
Biogeochemical			
Methane hydrates	0.1	(0.01–0.2)	
Trospospheric chemistryOcean chemistry	−0.04	−(0.01–0.06)	
	0.008		
Ocean eddy-diffusion	0.02		
Ocean biology and circulation	0.06	(0.0–0.1)	
Vegetation albedo	0.05	(0.0–0.09)	
Vegetation respiration	0.01	(0.0–0.3)	Flux = 0.5 Pgyr^{-1} K^{-1}
CO_2 fertilisation	−0.02	−(0.01–0.04)	15% biomass increase
Methane from wetlands	0.01	(0.003–0.015)	for 2 × CO_2
Methane from rice	0.006	(0.0–0.01)	
Electricity demand	0.001	(0.0–0.004)	
Sub-total[c]	0.16	(0.05–0.29)[d]	
Total	0.80	(0.32–0.98)[d]	

Source: (KC98).

[a] Includes the lapse rate feedback and other geophysical climate feedbacks not included elsewhere.

[b] Based on 1.5–5.5 deg K for doubling CO_2. The individual values do not sum to these values: see Dickinson (KC98) for details.

[c] Based on selected biogeochemical feedbacks which might occur together during the next century.

[d] Ranges are combined using a least-squares approach.

2.5 AEROSOLS

2.5.1 Natural and anthropogenic aerosols in the atmosphere

As mentioned in Section 2.1, the Earth's atmosphere, apart from several different gases, contains aerosol particles. The atmosphere is enriched by aerosols every day from the burning of fossil fuels and biofuels due to human activities.

Their main constituents are mineral dust, sea salt, sulphates, nitrates, black carbon (i.e. soot), and particulate organic matter (POM). In the atmosphere, the natural aerosol species are mineral dust (39 mg m^{-2}) and sea salt (13 mg m^{-2}), whereas the main anthropogenic components, are sulphate 3.9 mg m^{-2}, POM 3.3 mg m^{-2}, and black carbon 0.4 mg m^{-2} (nitrate are semi-volatile) (Kinne et al., 2006).

Natural emission sources of primary and secondary airborne particulates include arid or semi-arid regions, oceans, vegetation, and volcanoes, among others. Anthropogenic emissions mainly stem from industrial and combustion processes. On a global scale, natural sources are responsible for ~ 98% of primary and secondary particle emissions with on average almost 12,000 Tg/year (Figure 2.12.).

The major natural emissions in terms of mass are sea spray (84%), and mineral dust (13%), with other sources being biological primary organic aerosols (POA), volcanic emissions, biogenic secondary organic aerosols (SOA), and volcanic and biogenic sulphate particles. Anthropogenic aerosols contribute only 2% to global emissions, mainly in the form of anthropogenic sulphate (49%) and industrial dust (40%), with additional emissions of anthropogenic nitrate and SOA, and fossil fuel-derived POA. On a global scale, primary aerosols are clearly dominant over secondary species (98% vs. 2%, Figure 2.9).

2.5.2 Aerosol optical properties

The direct effect of these particles (mainly in the reflectivity of solar radiation to the space) is known. In simple terms, the aerosols have a cooling effect on the Earth by blocking the incoming radiation from the Sun; this, again in simple terms, is in contrast to the effect of the CO_2 which blocks the outgoing radiation from the Earth and thereby has a warming effect. However, the possible indirect

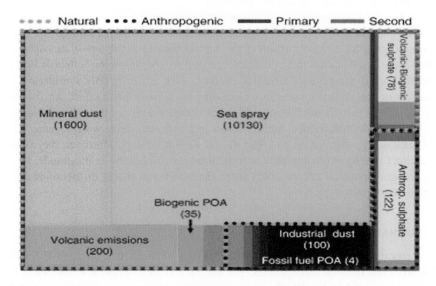

FIGURE 2.12 Atmospheric aerosol sources on the global scale. (Modified from Andreae and Rosenfeld, 2008, Durant et al., 2010 and Guieré and Querol, 2010 and Viana et al., 2014).

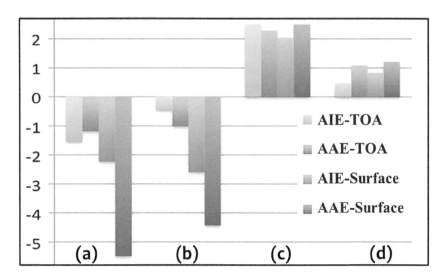

FIGURE 2.13 Comparison of aerosol radiative forcings (in Wm⁻²) between aerosol-indirect-effect-only (AIE) experiment and all-aerosol-effect (AAE) experiment for (a) East Asia, (b) South Asia, (c) Europe, and (d) United States of America (Modified from Wang et al., 2015).

effects through the modification of the cloud are uncertain. Among the indirect effects of higher aerosol concentrations in the atmosphere are: 1) an increase to the number density, 2) a decrease in the mean size of the cloud droplets, and 3) an increase in the lifetime of a cloud. Model estimations show that they could significantly increase global albedo, and so act to reduce global warming due to greenhouse gases.

The optical activity (i.e. aerosol optical depth) of the mineral dust and sea salt is less important due to their larger size, each contributing only as much as sulphate does (25%). Special attention is paid to black carbon (contributing only 3% to the optical depth), which is the main absorber of the solar radiation is leading to an air warming. The impact of this warming is the prevention of cloud formation (the so-called semi-direct effect) counteracting some of the negative aerosol forcings from scattering aerosols (e.g. sea salt and sulphate).

Wang et al. (2015) performed an experiment contrasting anthropogenic aerosol scenarios in 1970 and 2010 and concluded that the aerosol-indirect-effect-only experiment showed that aerosol-cloud interactions account for a larger portion of the aerosol forcing at the top-of-atmosphere radiation fluxes. Particularly, the aerosol indirect effect is critical for the "dimming" effect in East and South Asia, and the surface temperature reduction becomes even larger by only considering the aerosol indirect effect in East Asia (Figure 2.13.). Excluding the aerosol direct effect can also dampen the warming effect over the US and Europe, which limits the influence of the redistribution of aerosols on the global circulations. Hence, aerosol direct and indirect effects work together to modulate the meridional energy distribution and alter the circulation systems. Moreover, they found that the aerosol indirect forcing is predominant over the total aerosol forcing in magnitude, while aerosol radiative and microphysical effects jointly shape the meridional energy distributions and modulate the circulation systems.

2.5.3 SULPHATE AEROSOLS

Sulphur enters the atmosphere as SO_2, DMS (dimethyl sulphide $(CH_3)_2S$), and H_2S from various natural sources and from human activities. The natural and anthropogenic emissions of these compounds are given in Table 2.7 (KC98). There is considerable uncertainty in the size of the natural emissions of DMS by the oceans, which leads to the uncertainty of the total emissions. The emission

TABLE 2.7

Natural and anthropogenic emissions of sulphur compounds in the northern and southern hemispheres (TgS/yr)

Source of emissions	Global emissions (TgSyr^{-1})	Northern hemisphere	Southern hemisphere
Industry (SO$_2$)	70 (62–80)	64	6
Biomass burning (SO$_2$)	2.5 (2.3–2.8)	1.1	1.4
Oceans (DMS)	16 (8–51)	6.9	9.1
Volcanoes (SO$_2$)	8.5 (7.4–9.4)	5.8	2.7
Soils and vegetation (DMS, H$_2$S)	1 (0.2–4.3)	0.6	0.4

Source: (KC98).

of DMS by the ocean is characterised by strong spatial variability and annual change, especially in high and middle latitudes. The various sulphur compounds undergo various chemical reactions leading eventually to small particles of sulphates XSO$_4$ where X is some divalent cation. These particles constitute atmospheric sulphate aerosols.

2.5.4 THE SPATIAL AND TEMPORAL DISTRIBUTION OF AEROSOLS

When it comes to considering the effects of aerosols on the climate, particularly in the context of computer modelling (see Section 6.9), there is a fundamental difference from the case of the various gases in the atmosphere. The main gases N$_2$, O$_2$, and CO$_2$ and the various minor atmospheric gases are assumed to be uniformly mixed. This means that if you take a sample of air from any location at any time and measure the percentage concentration of each of these gases, you will get the same answer. Of course, there will be some local differences, for instance close to the chimney of a fossil fuel fired power station, but in the global context, such variations are soon smoothed out.

The situation for aerosols is quite different. At any given location, the concentration of aerosol particles, as well as their chemical nature and physical properties (principally their size) will vary greatly as a function of time. The processes of the generation of particles determine, first of all, their physical and chemical properties and their concentration in the atmosphere and, hence, the subsequent processes of transformation of the spectra of size distribution and the rate of removal of the particles from the atmosphere. The following types of natural aerosol particles can be identified:

1. products of the disintegration and evaporation of sea spray,
2. mineral dust, wind-driven to the atmosphere,
3. products of volcanic eruptions (both directly ejected to the atmosphere and resulting from gas-to-particle conversion),
4. particles of biogenic origin (both directly ejected to the atmosphere and resulting from the condensation of volatile organic compounds, for example, terpenes, and chemical reactions taking place between them),
5. smoke from biomass burning on land,
6. products of natural gas-to-particle conversion (for example, sulphates formed from sulphur compounds supplied by the ocean surface).

The principal types of anthropogenic aerosols are particles of industrial emissions (soot, smoke, cement and road dust, etc.), as well as products of gas-to-particle conversion. In addition, some aerosol substance results from heterogeneous chemical reactions, in particular, photocatalytic reactions.

As a result of the sufficiently long lifetime of aerosol particles, their transport, mixing, and interaction with each other and with gaseous compounds of various substances, differences in the physical and chemical properties of particles from various sources are smoothed out and the particles form what are described as background aerosols.

In different atmospheric layers, the characteristics of the background aerosols suffer substantial changes. In particular, due to the weak mixing of tropospheric and stratospheric aerosols, the particles in the stratosphere differ strongly in their composition and size distribution from the tropospheric ones. The troposphere can contain regions with particles of different types, in view of the relatively short lifetime of the particles in the troposphere, especially in the surface layer, as well as some of the sources' markedly dominating the intensity of the remainder.

The processes of transport and removal of aerosols from the atmosphere are determined by both the meteorological factors and the properties of the aerosol particles themselves. The processes which depend on meteorological conditions include organised convective and advective transports of particles as well as mixing due to the turbulent diffusion. The individual properties of particles determine, in particular, such processes of their removal from the atmosphere as sedimentation on obstacles.

In the processes of transport and removal of particles from the atmosphere, their size (mass) and composition are transformed due to coagulation and condensation growth as well as to heterogeneous reactions. These processes are responsible for the spectrum of particle sizes and determine the value of the complex refractive index and the particles' shapes, that is their optical properties.

The most substantial gas-to-particle reactions in the formation of aerosol particles are as follows:

1. reactions of sulphur dioxide with hydroxyl radicals which eventually lead to the formation of sulphuric acid molecules, which are easily condensed in the presence of water molecules;
2. reactions of hydrocarbon compounds with ozone and hydroxyl radicals with the subsequent formation of organic nitrates through the reactions of primary products with nitrogen oxides (e.g. the well-known peroxyacetyl nitrate).

All the processes of formation and evolution of aerosol systems depend on periodical variations of solar radiation in the atmosphere and at the surface. The processes of atmospheric heating near the surface, water evaporation, and chemical reactions of aerosol formation have a clear-cut diurnal change resulting in variations of the aerosol characteristics, which in some cases are stronger than the annual change (Galindo KC98). Strong daily variations of anthropogenic aerosols have been documented.

The cycles of aerosol matter are closely connected with the hydrological processes in the atmosphere. On the one hand, clouds and precipitation play an important role in the formation, transformation, and removal of aerosol particles from the atmosphere, while, on the other hand, the aerosols are condensation nuclei whose physico-chemical properties determine the microphysical processes in clouds. It is not accidental that both water vapour molecules and aerosol particles have approximately equal lifetimes. Therefore, a deeper knowledge of the interactions of aerosols and cloud elements is needed to obtain a better understanding of the processes of the formation of aerosols and of the variability of aerosol properties in time and space.

2.6 CLOUDS

2.6.1 FORMATION OF CLOUDS

The basic constituents of the clouds are tiny water droplets and ice crystals that are so small that they stay in the air and cannot be distinguished by the naked eye. Water enters the atmosphere as water vapour mainly by evaporation of the liquid water of the oceans, lakes, and rivers. When droplets become large enough, they are visible as a cloud or fog. When they become even bigger, they can fall like rain (or snow). When an air parcel rises to the atmosphere (Figure 2.14), it gradually loses its ability to retain the entire amount of its water vapour due to the gradual decrease in the atmospheric temperature and pressure.

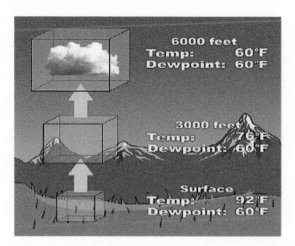

FIGURE 2.14 As air rises, it cools and decreases pressure, spreading out. Clouds form when the air cools below the dewpoint and the air cannot hold as much water vapour (Source: NOAA).

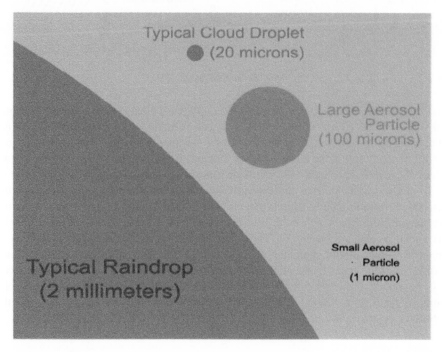

FIGURE 2.15 Comparison of the approximate sizes of large and small aerosol particles with raindrops and cloud droplets (Source: NOAA).

Consequently, a part of the water vapour condenses transforming to small droplets of water or ice crystals and then a cloud forms. The condensation is easier when water vapour coexists with particles (e.g. dust and pollen) (Figure 2.15). These condensation nuclei facilitate the cloud formation.

2.6.2 TYPES OF CLOUDS

Clouds have different heights, colours, and shapes and therefore different names. As an example, during daytime, the sunshine warms the ground and it heats the air just above it, which as it rises

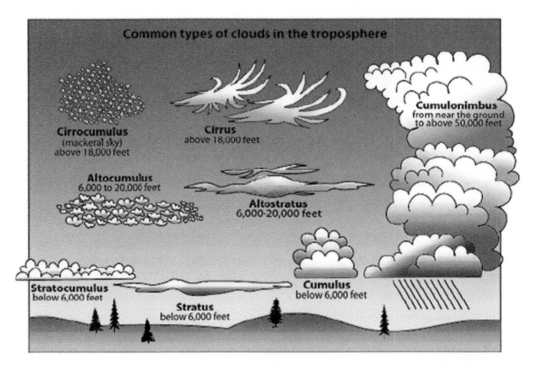

FIGURE 2.16   The different types of clouds (Source: NOAA).

cools and its water vapour condenses forming several types of clouds such as cumulus, cumulonimbus, and stratocumulus clouds (Figure 2.16). Another example is the case where the air is forced upward at areas of low pressure or at weather fronts – where two large air masses with great temperature difference collide at the Earth's surface. In the latter case, at a warm front (where a warm air mass slides above a cold air mass), the warm air is pushed upward forming many different types of clouds – from low stratus clouds to midlevel altocumulus and altostratus clouds, to high cirrus, cirrocumulus, and cirrostratus clouds. Clouds that produce rain-like nimbostratus and cumulonimbus are also common at warm fronts. At a cold front (where a heavy cold air mass pushes a warm air mass upward), cumulous clouds are common. They often grow into cumulonimbus clouds, which produce thunderstorms. Nimbostratus, stratocumulus, and stratus clouds can also form at a cold front.

The different cloud types in the atmosphere are shown in Figure 2.17. Taking into account the height of the clouds, these are categorised as follows: The highest clouds in the atmosphere are cirrocumulus, cirrus, cirrostratus, and cumulonimbus. Altocumulus and altostratus are mid-level clouds. Stratus, cumulus, and stratocumulus are the lowest clouds in the atmosphere.

2.6.3   CLOUDINESS AND RADIATION

Section 4.1 of the first edition (KC98) consists of a lengthy discussion of cloudiness and radiation. It contains some useful discussion of general principles and observations and we have tried to retain this material and augment it with some modern material. However, it also contains a considerable quantity of results of model calculations that have been superseded and are only likely to be of historical interest. We have not included these because, as we shall see in chapter 6, the problem of including clouds in computer models of the climate remains a problem of active research interest that is far from being solved.

Cloud cover is an important factor which determines the geographical distribution and the annual change of the climate regime. Analysis of the sensitivity of climate to cloud characteristics, based on numerical modelling and diagnostic studies, has now become of great concern. Such an analysis

FIGURE 2.17   Pictures of different cloud types (Source: NOAA).

is aimed at: (1) studying the features of the respective feedbacks that govern the interaction among the fields of motion, clouds, and radiation; (2) developing a technique that adequately simulates the evolution of clouds in numerical climate models; (3) studying the effect of clouds on the radiative regime; and (4) testing the results on the basis of observational data. The development of a theory of the climate-forming role of cloud cover for use in climate studies is of paramount importance (KC98). It is still probably the weakest link in computer modelling of the global climate.

Both the observational and theoretical data testify to a mutual compensation, to some extent, of the effect of clouds on the shortwave and longwave components of the radiation budget of the surface-atmosphere system and the respective components of radiative heat flux divergence. This compensation is, however, partial (even for averaged quantities), and therefore radiative heat flux divergence depends on the amount, type, and height of the upper boundary of clouds and on their spatial distribution.

Cloud anomalies can affect the natural variability of climate. Here it is important to take into account the specific influence of various types of clouds on the radiative regime in the interaction between radiative and dynamical processes which complicates the analysis of climatic sensitivity to different factors. In this connection, the development of techniques to predict the formation of clouds and their effect on radiation fluxes is of the highest priority.

As Henderson-Sellers (KC98) noted, three independent sources of information on climate show that an increasing total cloud amount would cause a climatic cooling: (1) from satellite data, an increase of the cloud amount leads to a decrease of the Earth radiation budget (i.e. the albedo effect of clouds dominates the greenhouse effect); (2) it follows from surface meteorological observations for the last hundred years that in the periods of climate warming the cloud amount grew, but in the presence of clouds, the surface air temperature was lower than it ought to have been in clear-sky conditions; and (3) results of numerical modelling agree, on the whole, with observational data, although they are controversial.

Even an adequate understanding of the climate-forming role of cloud-radiation feedback does not, however, permit a reliable forecast of the future climate, since it is unknown for certain how climate changes affect the cloud amount and it is also impossible to predict a change in the types of clouds. The climate forecast requires that three questions be answered: (1) what is the sign of the effect of clouds on climate, (2) in what direction does the cloud amount vary (increase or decrease), and (3) what is the contribution of different types of clouds to their impact on climate.

In general, one might suppose that these questions could be answered with the use of climate models. But in practice, this is impossible, in view of the inadequacy of the schemes of cloud parametrisation used in climate theory. One of the most intriguing questions facing climate modellers today is how clouds affect the climate and vice versa. Understanding these effects requires a detailed knowledge of how clouds absorb and reflect both incoming shortwave solar energy and outgoing longwave Earth radiation. Moreover, in addition, there is no complete observational database, which is needed to check the parametrisation schemes. Therefore, the existing models simulate only incomplete patterns of the effect of clouds on the formation of the climate.

Bony et al. (2015) noted that fundamental puzzles of climate science remain unsolved because of the limited understanding of how clouds, circulation, and climate interact. One example is our inability to provide robust assessments of future global and regional climate changes. However, ongoing advances in our capacity to observe, simulate, and conceptualise the climate system now make it possible to fill gaps in our knowledge. Bony et al. (2015) argued that progress can be accelerated by focusing research on a handful of important scientific questions that have become tractable as a result of recent advances. Therefore, the existing models simulate only incomplete patterns of the effect of clouds on the formation of climate. Thus, much of the early work on trying to incorporate clouds into climate models is of little relevance now. In order to give a flavour of current work on the subject we just quote one recent example.

Ming and Held (2018) introduced an idealised general circulation model (GCM) in which clouds are tracked as tracers but are not allowed to affect circulation either through latent heat release or cloud radiative effects. This model is capable of qualitatively capturing many large-scale features of cloud distributions outside of the boundary layer and deep tropics. The inclusion of cloud microphysics has a significant effect of moistening the lower troposphere in this model. A realistic perturbation, which considers the non-linearity of the Clausius-Clapeyron relation and spatial structure of CO_2-induced warming, results in a substantial reduction in the free-tropospheric cloud fraction. Ming and Held (2018) have performed several perturbation experiments with results illustrated in Figures 2.18–2.21.

In these experiments, CNTL-SA stands for: Control simulation with saturation adjustment, CNTL-C: Control simulation with the cloud scheme, NW: Narrower Width which is based on CNTL-C, but with a smaller width parameter, UN: Uniform, which is based on CNTL-C, but with a uniform increase of the saturated water vapour pressure e_s. TS: Temperature Squared, which is based on CNTL-C, but with an increase of e_s inversely proportional to T^2, TC: Temperature cubed, which is based on CNTL-C, but with an increase of e_s inversely proportional to T^3.

From Figure 2.18, we see that in the free troposphere, clouds are most prevalent in the mid- and high latitudes (especially over the storm tracks), where the cloud fraction often exceeds 20% and extends vertically through almost the entire tropospheric column. The tropical upper troposphere (100–300 hPa) is another place with large cloud fraction. Inspection of Figure 2.19 shows that both TS and TC give rise to marked reductions of similar spatial pattern. In TC, cloud fraction decreases by up to 2% in the subtropical dry zones. The entire free troposphere over 30^0–50^0 also undergoes substantial reduction of cloud fraction (2%). This trend extends to the high latitude upper troposphere. Figures 2.20 and 2.21 show that cloud liquid and ice respond largely in opposite directions.

From the viewpoint of testing the reliability of numerical global modelling results, it is important to have observational data and this has been obtained from the International Satellite Cloud Climatology Project (ISCCP), the Earth Radiation Budget Experiment (ERBE), and the Clouds and the Earth's Radiant Energy System (CERES). Only the availability of a long observational series will make it possible to study the long-term climatic trends and, in particular, to assess the anthropogenic effects on climate.

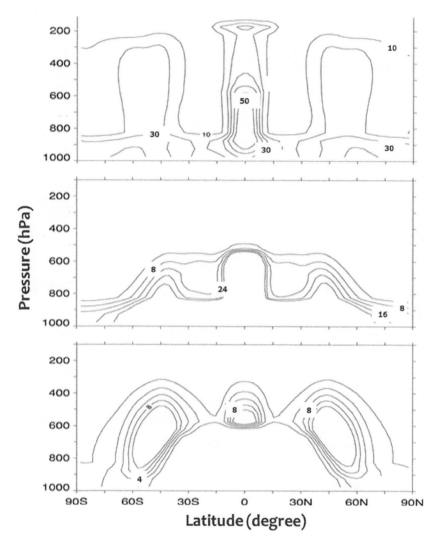

FIGURE 2.18 (top) Cloud fraction (% contours with an interval of 10%), (middle) liquid (10^{-6} kg kg^{-1}; contours with an interval of 4×10^{-6} kg kg^{-1}), and (bottom) ice (10^{-6} kg kg^{-1}; contours with an interval of 2×10^{-6} kg kg^{-1}) in CNTL-C. The vertical axis is the air-pressure in hPa (Modified from Ming and Held 2018).

The International Satellite Cloud Climatology Project (ISCCP) was established in 1982 as part of the World Climate Research Program (WCRP). It uses data from meteorological satellites to study the global distribution of clouds, their properties, and their diurnal, seasonal, and interannual variations. The resulting data sets and analysis products are used to study the role of clouds in climate, both their effects on radiative energy exchanges and their role in the global water cycle. The ISCCP cloud data sets provide our first systematic global view of cloud behaviour of the space and timescales of the weather while covering a long enough time period to encompass several El Niño – La Niña cycles.

The Earth Radiation Budget Experiment (ERBE) was designed around three Earth-orbiting satellites: the NASA Earth Radiation Budget Satellite (ERBS), and two NOAA satellites (NOAA-9 and NOAA-10). The data from these satellites is being used to study the energy exchanged between the Sun, the Earth, and space. The ERBS satellite was launched on 5 October 1984. It carried three

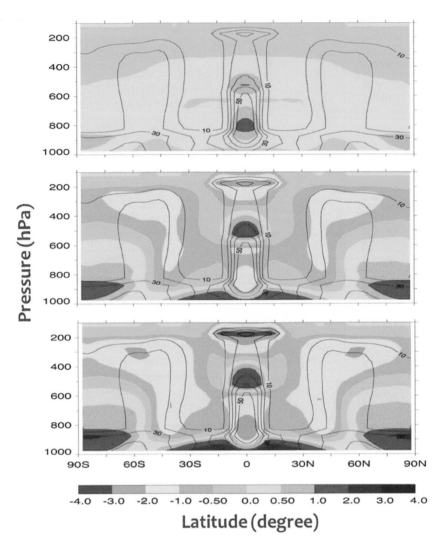

FIGURE 2.19 Cloud fraction difference (%; color shading) between (top) UN and CNTL-C, (middle) TS and CNTL-C, and (bottom) TC and CNTL-C. The contours with an interval of 10% represent the cloud fraction in CNTL-C. (Modified from: Ming and Held 2018).

instruments: the Earth Radiation Budget Experiment (ERBE) Scanner, the ERBE non-scanner, and Stratospheric Aerosol and Gas Experiment (SAGE II). ERBS had a design life of two years, with a goal of 3, but lasted 21 years suffering several minor hardware failures along the way. It was decommissioned on 14 October 2005. ERBS had the primary goals of determining, for at least one year, the Earth's average monthly energy budget and its monthly variations, the seasonal movement of energy from the tropics to the poles, and the average daily variation in the energy budget on a regional scale (data every 160 miles). All of these first-year goals were met, and the ERBE instrument continued to provide valuable data until it was decommissioned more than 20 years after its launch.

The successor to the ERBE was the Clouds and the Earth's Radiant Energy System (CERES), which was designed to extend the ERBE measurements to include the top of the atmosphere, in the atmosphere, and global surface radiation. CERES has been flown on multiple satellites starting with a launch on the Tropical Rainfall Measuring Mission (TRMM) in 1997, followed by a launch on the Earth Observing System (EOS)-AM satellite TERRA in 1998 and the EOS-PM satellite AQUA

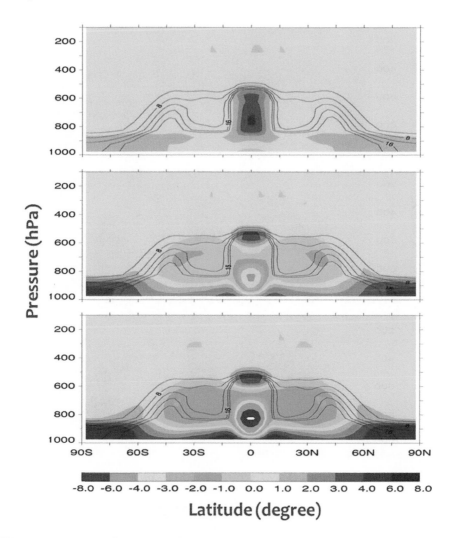

FIGURE 2.20 Cloud liquid difference (10^{-6} kg kg^{-1}; color shading) between (top) UN and CNTL-C, (middle) TS and CNTL-C, and (bottom) TC and CNTL-C. The contours with an interval of 4×10^{-6} kg kg^{-1} represent the cloud fraction in CNTL-C. (Modified from: Ming and Held 2018).

in 2000, the Suomi National Polar-orbiting Partnership (S-NPP) observatory, and the Joint Polar Satellite System, a partnership between NASA and NOAA. CERES FM6 launched on November 18, 2017 aboard NOAA-20, becoming the last in a generation of successful CERES instruments that help us to better observe and study Earth's interconnected natural systems with long-term data records.

In the early years following the launch of ERBS, a considerable amount of work was done on the solar constant. Mecherikunnel *et al.* (KC98) undertook a comparison of solar constant values obtained from the ERBE, the Stratospheric Measurement Mission (SMM), Nimbus-7, and rocket soundings. From data for the first two years of solar constant measurements on satellites ERBS and NOAA-9, the solar constant values averaged 1364.9Wm^{-2}. This value is 0.4 per cent lower than the Nimbus-7 estimate and 0.2 per cent below the solar constant value obtained with a cavity radiometer carried by SMM, and from the data of rocket soundings. The NOAA-9 and NOAA-10 measurement results averaged over the first six months gave a value of 1363.2 Wm^{-2}. Crommelynck (KC98) discussed further solar constant measurements made in 1992 and 1993 and compared them with previous observations. The solar constant values within the range 1366–1368 Wm^{-2} were obtained

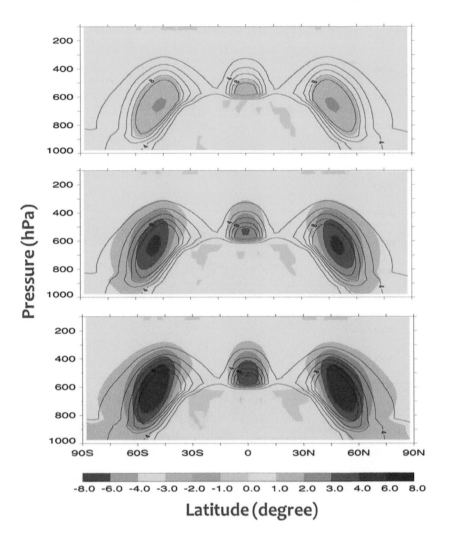

FIGURE 2.21 Cloud ice difference (10^{-6} kg kg^{-1}; color shading) between (top) UN and CNTL-C, (middle) TS and CNTL-C, and (bottom) TC and CNTL-C. The contours with an interval of 2×10^{-6} kg kg^{-1} represent the cloud fraction in CNTL-C. (Modified from: Ming and Held 2018).

with a spread of less than 1 Wm^{-2} (see also Sklyarov et al. 1994). Thus, the data of the latest solar constant measurements agree reasonably well, differing within the error of the present-day pyrheliometric observations. However, the mutual calibration of satellite-borne pyrheliometers is needed to substantiate the pyrheliometric scale for space-based solar constant measurements.

The availability of measured outgoing longwave radiation data has made it possible to assess the planetary mean greenhouse effect. Since the surface emission (at a global mean temperature of 15°C, see Section 2.3.1) constitutes 390Wm^{-2}, and the outgoing longwave radiation is 237Wm^{-2}, the atmospheric greenhouse effect reaches approximately 150 Wm^{-2}.

There was a considerable amount of work done in the early years of the ERBE and quite a lot of this is discussed in Section 4.1 of KC98, but it has now largely been superseded by more recent work the results of which are summarised by Wild et al. (2013). They discussed the global mean energy balance and its representation in recent climate models using as much as possible direct observational references from surface stations and space-born platforms. Based on this analysis, Figure 2.22 illustrates the global mean energy balance together with the uncertainty ranges.

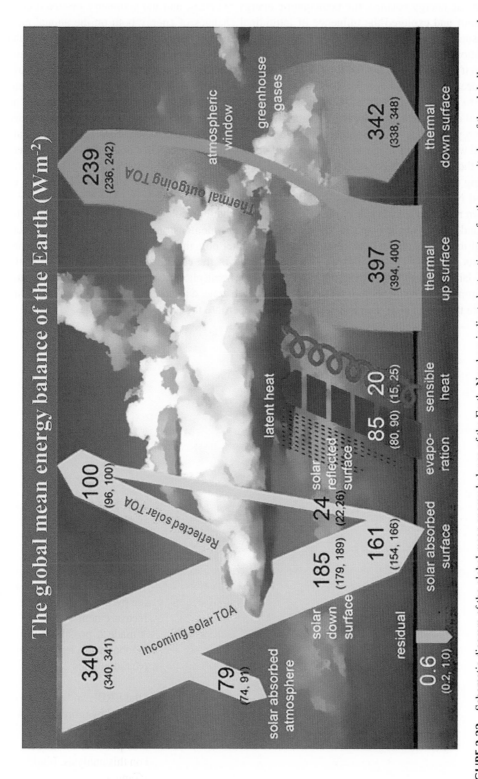

FIGURE 2.22 Schematic diagram of the global mean energy balance of the Earth. Numbers indicate best estimates for the magnitudes of the globally averaged energy balance components together with their uncertainty ranges (Modified from: Wild et al., 2013).

Stephens and L'Ecuyer (2015) reviewed the status of our understanding of the Earth's annual, global mean energy balance, the hemispheric energy balances, and the symmetry observed about the equator and explored the influence of latitudinal changes of energy both on the annual mean and seasonal transports of energy from low latitudes to higher latitudes. Based on the best available information, they showed that our planet continues to be out of balance with additional heat being added to it at the rate of 0.6 ± 0.4 Wm^{-2}. They noticed that this heat appears to be taken up primarily by the oceans of the SH and perhaps mostly equatorward of 37 °S. The nature of the adjustments applied to our best estimates of individual, annual mean fluxes of energy to produce a balance are described and the results of applying a more formal constraint for these adjustments are discussed. The energy balances of the Southern and Northern Hemispheres are then shown to be practically identical which in turn suggests the transport of energy across the equator in the net is close to zero. In fact, the hemispheres are not identically symmetrical with the SH being slightly out of balance absorbing the additional heat and transporting a small amount of net heat across the equator to the balanced NH. The symmetry in absorbed solar and the near symmetry in OLR are remarkable in their own right and are a result of the effects of clouds both on solar reflection and OLR that act to offset land–ocean interhemispheric differences. Then they demonstrate important inter-hemispheric seasonal influences on the heat transported to the winter pole that conspire to make these seasonal transports lopsided. This asymmetry is a direct result of the eccentricity of the Earth's orbit that induces larger energy losses from the southern winter hemisphere. This in turn produces a latitudinal asymmetry in the location of on the tropical trough zone, a region from which energy is always moved to the winter pole, requiring it be located deeper into the NH.

3 The hydrosphere

3.1 INTRODUCTION

As early as 800 BCE, Homer wrote in the Iliad of the ocean "from whose deeps every river and sea, every spring and well flows," suggesting the interconnectedness of all of Earth's water. About 12,000 years ago, one could walk from Alaska to Siberia. At that time, glaciers and ice sheets covered North America down to the Great Lakes and Cape Cod, although coastal areas remained largely free of ice. These extensive ice sheets appeared at a time when the sea level was very low, exposing the land where water is now filling the Bering's Strait. In fact, throughout the history of the Earth, the times of the extended glaciers are associated with the low sea level and the times when there are only minor ice sheets (as today) are associated with higher levels of the sea. These correlations are due to the fact that the amount of water on Earth is stable and is divided between reservoirs in the oceans, in the air, and on the land, so that more water stored on ice sheets means less water in the oceans. In addition, the Earth's water is constantly circulating through these reservoirs in a process known as the hydrological cycle.

We have chosen the Chinese Admiral Zheng He, see Figure 3.1(a), to feature in the introduction to this chapter on the hydrosphere, of which the oceans constitute the major component. Zheng He "led a fleet which was the greatest in the world in the fifteenth century and made in all, seven expeditions to the Western Ocean from 1405 to 1433" (Sun 1992). Figure 3.1(b) shows a model of one of his fleet along with a scale model of one of the European ships which "discovered" America in 1492. America had always been there but Zheng He got there before the Europeans and of course the indigenous people got there long before that. There are two early accounts of Zheng He's expeditions in English by Sun (1992) and Levathes (1996). In these two accounts, maps are shown of the expeditions as starting from Nanjing sailing across the South China Sea, through the Straits of Malacca and across the Indian Ocean.

However, more recently, Gavin Menzies, a retired submarine lieutenant-commander from the (British) Royal Navy, has argued that Zheng He's fleets travelled much more widely, crossing the Atlantic Ocean reaching America decades before Columbus, as well as reaching Australia and New Zealand (Menzies 2004, Menzies and Hudson 2013). Although Menzies' work has been challenged, we find it convincing and we conclude that to achieve all that the Admiral and his fleet must have had a good knowledge of the oceans, winds, and currents and they were able to navigate and to establish their latitude from observations of the stars. Hence, our choice of him to introduce this chapter. As an aside we mention that if Menzies' two books that we have just cited are controversial, his other book (Menzies 2008) is even more so. In that, he argues that the Renaissance was not sparked as a result of the rediscovery of Greek philosophy but by the visit of Zheng He's fleet to Italy in 1434. In that year, the story goes, the Chinese delegation met the influential pope Eugenius IV and presented him "with a diverse wealth of Chinese learning: art, geography (including world maps which were passed on to Columbus and Magellan), astronomy, mathematics, printing, architecture, civil engineering, military weapons and more…". The tragedy of the story is that when the ships that survived finally returned to China there had been a regime change and the ships were destroyed and the story of their voyages was largely suppressed.

Today by the term hydrosphere of the planet, we mean the total mass of water found on, under, and above its surface. A planet's hydrosphere can be liquid, vapour, or ice. In the case of the Earth,

Admiral
Zheng He
(1371-1433)

(a)

(b)

FIGURE 3.1   (a) A statue of Admiral Zheng He outside the History and Ethnography Museum at the Stadthuys in Malacca, where Zheng He is known to have called. (b) A model of one of Admiral Zheng He's treasure ships with a scale model of the Santa Maria (Christopher Columbus' flagship) beside it, also from the History and Ethnography Museum in Malacca.

the liquid water exists on the surface in the form of oceans, lakes, and rivers. It also exists below ground – as groundwater, in wells and aquifers. Water vapour is most visible as clouds and fog (Figure 3.2). The frozen part of Earth's hydrosphere, i.e. cryosphere (see Chapter 4), is made of ice: glaciers, ice caps, and icebergs. Thus, the Earth's hydrosphere consists of water in liquid and frozen forms, in groundwater (the water present beneath Earth's surface in soil pore spaces – contains the liquid and gas phases of soil and in the fractures of rock formations – the separation of soil into two or more pieces under the action of stress), oceans, lakes, and streams (a body of water with surface water flowing within the bed and banks of a channel).

The total volume of the Earth's hydrosphere is about 1,386 million cubic kilometres (97.5% of saltwater and 2.5% fresh water) and its total mass is about 1.4×10^{18} tonnes (about 0.023% of

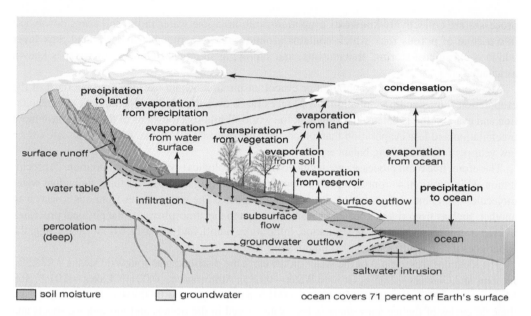

FIGURE 3.2   The components of the Earth's Hydrosphere. (Source: https://www.sciencedirect.com/topics/earth-and-planetary-sciences/hydrosphere).

Earth's total mass). 68.9% of the fresh water is ice and permanent snow cover in the Arctic, the Antarctic and mountain glaciers; 30.8% is fresh groundwater; and only 0.3% of the fresh water on Earth is in easily accessible lakes, reservoirs, and river systems. In addition, it has been estimated that about 20×10^{12} tonnes of this is in the form of water vapour in the atmosphere. The approximate coverage of the Earth's surface by ocean is 71% with an average salinity 3.5%, which is more commonly quoted as 35 parts per thousand.

3.2 THE HYDROLOGICAL CYCLE

The circulation of water through the hydrosphere is referred to as the hydrological cycle. It was not until the 17th century, however, that the notion of a finite water cycle was demonstrated in the Seine River basin by two French physicists, Edmé Mariotte and Pierre Perrault, who independently determined that the snowpack in the river's headwaters was more than sufficient to account for the river's discharge. These two studies marked the beginning of hydrology, the science of water, and also the hydrological cycle.

Water collects in clouds and then falls to Earth in the form of rain or snow. This water collects in rivers, lakes, and oceans. Then it evaporates into the atmosphere to start the cycle all over again, see Figure 3.2.

There is a continuous movement of water between the oceans, the atmosphere, and the land powered from solar radiation. The oceans hold over 97% of the planet's water, which with the power of the solar radiation evaporates rises and condenses into tiny droplets that form clouds. Water vapour usually remains in the atmosphere for a short time, from a few hours to a few days until it turns into precipitation and falls to the ground as rain, snow, sleet, or hail. Some precipitation falls onto the land and is absorbed (infiltration) or becomes surface runoff which gradually flows into gullies, streams, lakes, or rivers. Water in streams and rivers flows to the ocean, seeps into the ground, or evaporates back into the atmosphere.

Water in the soil can be absorbed by plants and is then transferred to the atmosphere by a process known as transpiration. Water from the soil is evaporated into the atmosphere. These processes are collectively known as evapotranspiration. Some water in the soil seeps downward into a zone of porous rock which contains groundwater. A permeable underground rock layer which is capable of storing, transmitting, and supplying significant amounts of water is known as an aquifer.

More precipitation than evaporation or evapotranspiration occurs over the land but most of the evaporation (86%) and precipitation (78%) takes place over the oceans. The amount of precipitation and evaporation is balanced throughout the world. While specific areas of the world have more precipitation and less evaporation than others, and the reverse is also true, on a global scale over a few year periods, everything balances out.

Therefore, five main processes are included in the hydrological cycle: 1) condensation, 2) precipitation, 3) infiltration, 4) runoff, and 5) evapotranspiration (Figure 3.3) over time, individual water molecules can come and go, in and out of the atmosphere. The water moves from one reservoir to another, such as from river to ocean, or from the ocean to the atmosphere, by the physical processes of evaporation, condensation, precipitation, infiltration, runoff, and subsurface flow. In so doing, the water goes through different phases: liquid, solid, and gas.

The locations (reservoirs) of the water on the earth are distributed as follows: Oceans – 97.25%, Ice caps and Glaciers – 2.05%, Ground Water – 0.68%, Atmosphere – 0.001%, Lakes – 0.01%, Soil Moisture – 0.005%, Streams & Rivers – 0.0001% and Biosphere – 0.00004%. Noticeably during the cold cycles of the ice ages, there is less water stored in the oceans and more in ice sheets and glaciers. In addition, it can take an individual molecule of water from a few days to thousands of years to complete the hydrological cycle from ocean to atmosphere to land to ocean again as it can be trapped in ice for a long time.

According to Bell et al. (2018) melting is pervasive along the ice surrounding Antarctica. On the surface of the grounded ice sheet and floating ice shelves, extensive networks of lakes, streams, and rivers both store and transport water. As melting increases with a warming climate, the surface hydrology of Antarctica in some regions could resemble Greenland's present-day ablation and percolation zones.

Therefore, the hydrological cycle can be thought of as a series of reservoirs, or storage areas, and a set of processes that cause water to move between those reservoirs (see Figure 3.3). The largest

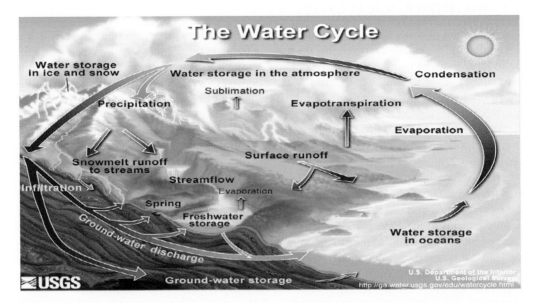

FIGURE 3.3   Scheme of the hydrological (or water) cycle. (Source: http://www.ualberta.ca/~ersc/water.pdf).

reservoir by far is the oceans, which hold about 97% of Earth's water. The remaining 3% is the freshwater so important to our survival, but about 78% of that is stored in ice in Antarctica and Greenland. About 21% of freshwater on Earth is groundwater, stored in sediments and rocks below the surface of Earth. The freshwater in rivers, streams, lakes, and rain is less than 1% of the freshwater on Earth and less than 0.1% of all the water on Earth.

Meanwhile the hydrological cycle involves the exchange of heat energy, which leads to temperature changes. For instance, in the process of evaporation, water takes up energy from the surroundings and cools the environment. Conversely, in the process of condensation, water releases energy to its surroundings, warming the environment. After evaporating, water vapour rises into the atmosphere and is carried by winds away from the tropics. Most of this vapour condenses as rain in the inter-tropical convergence zone (known as the ITCZ), releasing latent heat that warms the air. This in turn drives the atmospheric circulation.

Therefore, the water cycle figures significantly in the maintenance of life and ecosystems on Earth. Even as water in each reservoir plays an important role, the water cycle brings added significance to the presence of water on our planet. By transferring water from one reservoir to another, the water cycle purifies water, replenishes the land with freshwater, and transports minerals to different parts of the globe. It is also involved in reshaping the geological features of the Earth, through such processes as erosion and sedimentation.

The time taken for water to move from one place to another varies from seconds to thousands of years, and the amount of water stored in different parts of the hydrosphere ranges up to 1.37 billion km³, which is contained in the oceans (Table 3.1). Despite continual movement within the hydrosphere, the total amount of water at any one time remains essentially constant.

Movement of water takes place by a variety of physical and biophysical processes. The two processes responsible for moving the greatest quantities of water are precipitation and evaporation, transporting 505,000 km³ of water each year. The flow of water along rivers transports an intermediate amount of water, and sublimation of ice directly to vapour transports relatively very little.

The residence time of a reservoir within the hydrological cycle is the average time a water molecule will spend in that reservoir (Table 3.1). It is a measure of the average age of the water in that reservoir, though some water will spend much less time than average, and some much more.

In hydrology, residence times can be estimated in two ways. The more common method relies on the principle of conservation of mass and assumes the amount of water in a given reservoir is

TABLE 3.1
Average residence times of various water locations

Location	Average residence time
Oceans	3,200 years
Glaciers	20 to 100 years
Seasonal snow cover	2 to 6 months
Soil moisture	1 to 2 months
Groundwater: shallow	100 to 200 years
Groundwater: deep	10,000 years
Lakes	50 to 100 years
Rivers	2 to 6 months
Atmosphere	9 days

Source: (https://water.fandom.com/wiki/Hydrological_cycle?action=edit&section=3).

roughly constant. With this method, residence times are estimated by dividing the volume of the reservoir by the rate by which water either enters or exits the reservoir. Conceptually, this is equivalent to timing how long it would take the reservoir to become filled from empty if no water were to leave (or how long it would take the reservoir to empty from full if no water were to enter). An alternative method to estimate residence times, gaining in popularity particularly for dating groundwater, is the use of isotopic techniques.

As shown in Table 3.1 the groundwater can spend over 10,000 years beneath Earth's surface before leaving. Particularly old groundwater is called fossil water. Water stored in the soil remains there very briefly, because it is spread thinly across the Earth, and is readily lost by evaporation, transpiration, stream flow, or groundwater recharge. After evaporating, water remains in the atmosphere for about nine days before condensing and falling to the Earth as precipitation.

The effect of the ocean on climate is determined by three key factors: the role of the ocean as a source of moisture in the hydrological cycle, the gigantic thermal inertia of the ocean, and the horizontal heat transport by the ocean. Two climate models were developed in the Princeton University Geophysical Fluid Dynamics Laboratory (GFDL). One of them considered explicitly the heat transport by the ocean (interactive ocean–atmosphere model); the other used an approximation of a swamp ocean, where the ocean is considered as a wet surface with zero heat capacity, which supplies moisture but does not transport heat.

Global warming adds volume and energy to the global hydrological system resulting in an increase in precipitation, if viewed on a global basis. Unpredictable changes in the pattern of atmospheric circulation are possible, including droughts and floods which are not compatible with patterns of human settlement and infrastructure. Over the past century, the water cycle has become more intense, with the rates of evaporation and precipitation both increasing. This is an expected outcome of global warming, as higher temperatures increase the rate of evaporation due to warmer air's higher capacity for holding moisture. Human activities that impact the water cycle include: agriculture, alteration of the air chemical composition, construction of dams, deforestation and afforestation, removal of groundwater from wells, water abstraction from rivers and urbanisation. The results are: Rising Sea Levels, Decrease in Arctic Sea Ice, Abrupt Precipitation Events, and Melting Permafrost.

Human water use, climate change and land conversion have created a water crisis for billions of individuals and many ecosystems worldwide. Global water stocks and fluxes are estimated empirically and with computer models, but this information is conveyed to policymakers and researchers through water cycle diagrams. In this connection, Abbott et al. (2019) compiled a synthesis of the global water cycle, which has been compared with 464 water cycle diagrams from around the world. Although human freshwater appropriation now equals half of global river discharge, only 15% of the water cycle diagrams depicted human interaction with water. Only 2% of the diagrams showed climate change or water pollution – two of the central causes of the global water crisis – which effectively conveys a false sense of water security.

3.3   THE OCEANS

3.3.1   THE THERMOHALINE CIRCULATION

As mentioned in Chapter 1, the formation of climate and its changes is determined by a complicated interaction among the components of the atmosphere-hydrosphere–lithosphere–cryosphere–biosphere system. A study of the large-scale atmospheric processes responsible for weather and climate changes requires a consideration of the interaction of the atmosphere not only with the land but also with the ocean, which is a gigantic heat reservoir.

Over half of the solar radiation reaching the Earth's surface is first absorbed by the oceans, where it is stored and redistributed by ocean currents before escaping to the atmosphere, largely as latent heat through evaporation but also as longwave radiation. These currents have a complicated

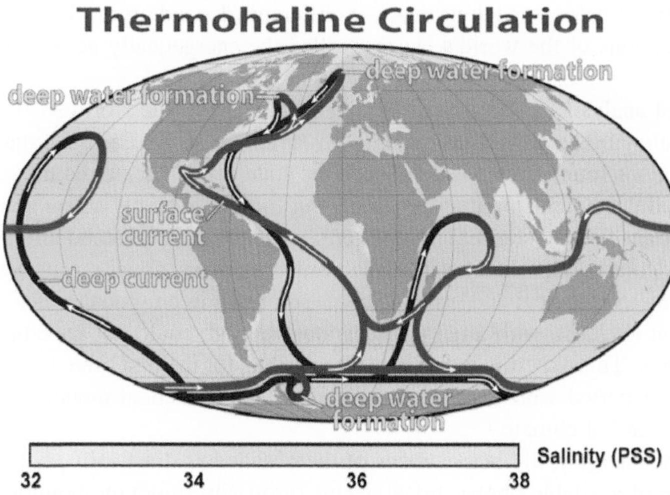

Thermohaline Circulation

FIGURE 3.4 A schematic of modern thermohaline circulation (THC). Changes in the THC are thought to have significant impacts on the Earth's radiation budget. Although poleward heat transport outside the tropics is considerably larger in the atmosphere than in the ocean, the THC plays an important role in supplying heat to the polar regions, and thus in regulating the amount of sea ice in these regions. (Source: NASA Aquarius Mission – STEM Topics).

horizontal and vertical structure determined by the pattern of winds blowing over the sea and the distribution of continents and submerged mountain ranges.

The global density gradients created by surface heat and freshwater fluxes drive a part of the large-scale ocean circulation, which is known as *thermohaline* circulation (THC), see Figure 3.4. This term comes from the two factors: *thermo-* that refers to temperature and *-haline* referring to salt content, which both determine the density of sea water.

In the Earth's polar regions, ocean water gets very cold, forming sea ice. As a consequence, the surrounding seawater gets saltier, because when sea ice forms, the salt is left behind. As the seawater gets saltier, its density increases, and it starts to sink. Surface water is pulled in to replace the sinking water, which in turn eventually becomes cold and salty enough to sink. This initiates the deep-ocean currents driving the global conveyor belt.

The vertical structure of the oceans comprises three layers:

- the seasonal boundary layer, mixed annually from the surface, which is less than 100 m deep in the tropics and reaches hundreds of metres in the sub-polar seas (other than the North Pacific) and several kilometres in very small regions of the polar seas in most years;
- the warm water sphere (permanent thermocline), ventilated (i.e. exchanging heat and gases) from the seasonal boundary layer, which is pushed down to depths of many hundreds of metres in gyres by the convergence of surface (Ekman) currents driven by the wind;
- the cold water sphere (deep ocean), which fills the bottom 80 per cent of the oceans' volume, ventilated from the seasonal boundary layer in polar seas.
- The atmosphere–ocean interaction, showing itself in the exchange of heat, water, and momentum, is of key importance to the solution of the problem of long-range forecasts of weather and climate changes.

A system of conjugated equations of thermohydrodynamics and the specially developed theory of perturbations was used by Marchuk and co-workers (KC98) to study the processes of long-term weather anomalies and climate changes manifested as surface air temperature variations over large areas. These have shown that the short-range and long-range forecasts of temperature anomalies are determined mainly by processes taking place in different (depending on the short-range or long-range forecasts) regions of the world's oceans, called the energetically active zones of the ocean (EAZO).

This theoretical analysis has been verified by global maps of the world's oceans' heat budget, based on observational data as well as satellite observations of the Earth's radiation budget, and the empirical relations found previously between the state of the ocean and the resulting weather. This has confirmed the conclusion that the concept of the energetically active zones of the oceans is the most important factor determining long-term weather anomalies and climate changes. This conclusion is particularly important because the available observational means are still inadequate to obtain regular meteorological and oceanographic information on global scales. The First GARP Global Experiment (FGGE), with a ten-yr preparatory period, could not have been accomplished before the year 1979. This experiment has made it possible, for the first time, to obtain global information for a one-yr period which can now be used to test theoretical models of the atmospheric general circulation and of climate.

Discovering the energetically active zones of the oceans has drastically changed the situation. The concentration of available observational means (primarily ships) on monitoring the processes taking place in the energetically active zones, including migration and variability therein, and also the use of satellite data, permit a deeper insight into the processes governing the long-term weather anomalies and climate changes, and provide long-range weather forecasters with much-needed information. That is why the 'Sections' programme has been initiated, with the major objective to study short-term (several years) climate changes. Based on the energetically active zones concept, the 'Sections' programme includes both field measurement in different key climatic regions and numerical experiments. There are physical and biological mechanisms in the oceans which are important in controlling the concentration of CO_2 in the climate system. CO_2 is transferred from the atmosphere into the interior of the ocean by the physical mechanism caused by differences in the partial pressure of CO_2 in the ocean and the lowest layers of the atmosphere. The rate of gas exchange depends on the air-sea difference in partial pressure of CO_2 and a coefficient which increases with wind speed.

The ocean branch of the carbon cycle involves a flux of CO_2 from the air into the sea at locations where the surface mixed layer has a partial pressure of CO_2 lower than that of the atmosphere and vice versa. The annual ventilation of the seasonal boundary layer from the surface mixed-layer controls the efficiency of the biological mechanism by which ocean plankton convert dissolved CO_2 into particulate carbon, which sinks into deep water. These two pumps are responsible for extracting CO_2 from the global carbon cycle for periods in excess of a hundred years. The mixed-layer partial pressure of CO_2 is depressed by enhanced solubility in cold water and enhanced plankton production during the spring bloom.

3.3.2 STUDIES OF HEAT AND WATER BALANCES IN THE WORLD'S OCEANS

Proper consideration of the atmosphere–ocean interaction plays a key role in resolving the problems of diagnosis and prediction of climate (Figure 3.5). Extended periods of anomalously warm ocean temperatures (marine heatwaves, compared to that of atmospheric heatwaves) can have major impacts on marine biodiversity and ecosystems (Holbrook et al. 2019). Thus, observations of the characteristics of the oceanic upper layer form an important source of information about the components of the heat and water balances of the ocean surface. We can use this kind of information to assess the role of buoyancy as a forcing impact on the oceanic circulation due to heat and fresh water inputs. The importance of fresh water inflow is particularly great at high latitudes, where the surface temperature is close to freezing.

In such conditions, the water salinity, not its temperature, controls the stratification of the upper layer of the ocean. Palaeoclimatic data show that low-frequency salinity changes could have affected substantially the formation of the THC in the northeast North Atlantic. The "cap" of fresh water can

The Ocean Heat Fluxes

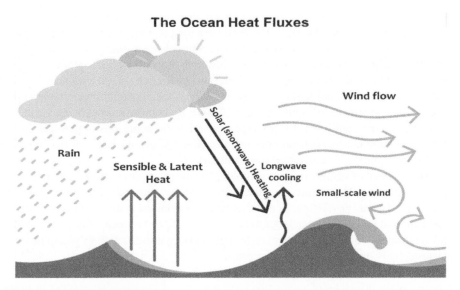

FIGURE 3.5 Schematic diagram of the main ocean heat fluxes (modified from National Oceanography Centre (Modified from https://noc.ac.uk/).

strongly limit the convection-induced cooling, isolating the deep layers from the processes taking place on the surface of the ocean. Sea ice transport from the Arctic Basin is an important source of fresh water, leading to the observed anomalies of salinity. In this respect, assessing the contributions of precipitation, river run-off, thawing, and sea ice transport to the fresh water balance in the North Atlantic is very important from the point of view of climate study.

Of the utmost interest is the retrieval from satellite data of the global field of precipitation. Browning (KC98) notes that for climatic studies, monthly means of precipitation are required over a 500 km × 500 km grid, with an error of no more than 20 per cent (especially within the inter-tropical convergence zone (ITCZ)). This corresponds to 1 mm per day, and the energy equivalent is 30 Wm^{-2}, that is, it reaches approximately 10 per cent of the average surface radiation budget. On the other hand, this error is equivalent to the heating of the 10 km layer of the atmosphere by 0.2 °C per day. For accurate weather prediction, it is important to have data on instant precipitation, with a random error of ±50 per cent. "Nowcasting" requires very high spatial (ca. 5 km) and temporal (ca. 10 min) resolution. In hydrology, data on precipitation are needed; these data should have temporal resolution from several hours (with thorough spatial details, for example, in flood predictions) up to and beyond 24 h. In oceanography, it is necessary to estimate the rain rate with an error of about 1 mm per day. The retrieval of the global distribution of rain rate from satellite data is based on three techniques, as follows.

1. The first technique is the employment of data on minimum cloud top temperature determined from given values of brightness temperature in the infra-red, and they indicate rain rate. For example, the precipitation index calculated from geostationary satellites (GOES etc.) data allows us to determine the proportion of clouds within a grid cell 2.5° × 2.5° with cloud top temperature below −40 °C. Although the infra-red threshold technique permits only a recognition of precipitating and non-precipitating clouds for each grid cell, the experience of data processing for the GARP Atlantic Tropical Experiment (GATE) has indicated that there is a high correlation between the precipitation index averaged over the area and the total precipitation for six h (though note that this only refers to precipitation from convective clouds). The combination of brightness temperature data and wind speed fields in the visible wavelength range has proved useful.

2. Techniques to retrieve rain rate from microwave data on absorption (in the frequency range 5–35 GHz) or scattering (35–183 GHz) have been used widely. In the former, precipitation is retrieved from intensified microwave emission of liquid water (under conditions of observations over the water surface, that is, with a low emissivity and a homogenous background). The minimum threshold of retrieved rain rate is about 2 mm h − 1. At a frequency of 19 GHz with a rain rate of 20 mm h − 1, a saturation occurs; the maximum measurable rain rate is much greater with measurements within 5–10 GHz. Radiometers, functioning via the scattering principle, are mainly sensitive to large ice particles near the tops of powerful convective clouds and therefore they permit only indirect estimates of rain rate near the surface. Such observations can also be conducted over land surfaces.

3. The employment of satellite radars functioning via the principles of back-scattering or attenuation is intriguing. The advantage of the former is that it allows the use of the standard technique for radar data processing; its disadvantages are manifest in the need to prescribe the size distribution of precipitation, in order to determine relations between radar echo and rain rate, and in the need to filter out the effect on background signal attenuation by precipitation, which is substantially manifest at high frequencies.

The advantage of the principles of attenuation is the presence of a linear relation between attenuation and rain rate, as well as independence of the size distribution. The possibility of very strong attenuation of the signal necessitates measurements at more than one frequency: for example, 14 GHz in the case of heavy rain and 35 GHz in the case of light rain.

Rain rate retrieval is hindered both by the complexity of the problem itself and by the strong spatial and temporal inhomogeneity of the physical nature of precipitation (size distribution, phase state); the latter complicates the estimation of representative averages. All this demands that we develop retrieval algorithms which take into account the observation conditions (convective or complex clouds, liquid or solid precipitation, etc.). To reach the vertical resolution required (ca. 0.5 km) with scanning in a plane normal to the sub-satellite trajectory (±15°), a space-borne radar at a frequency of 14 GHz should have an antenna 8 m diameter for an orbit of 600 km altitude or 4 m diameter for an orbit of 300 km altitude (Table 3.2).

TABLE 3.2

Summary of satellite methods for the retrieval of precipitation; the strengths and weaknesses of the various satellite retrieval algorithms

Observation Spectrum	Satellite Orbit	Sensor	Advantage	Disadvantage
Visible	Geostationary Orbit	GOES Imager	Cloud type	Cloud tops
	Low Earth Orbit	AVHRR	Cloud evolution	Indirect rain rate
Infrared	Geostationary Orbit	GOES Imager	Cloud temperature	Cirrus contamination
	Low Earth Orbit	AVHRR	Cloud evolution	Indirect rain rate
Passive Microwave	Low Earth Orbit	SSM/I		Poor temporal sampling
		AMSR-E	Direct measure of	Coarse spatial
		TMI	rain, especially over ocean	resolution
				Indirect rain rate (land)
Active Microwave	Low Earth Orbit	TRMM PR	Direct measure of	Narrow swath width
		CloudSat CPR	vertical structure of rain	Poor temporal sampling
				Rain rate sensitivity/ saturation

(Modified from Ferraro and Smith 2015).

Until such radars are available, we can use diverse subsidiary information to retrieve the vertical profile of rain rate:

1. retrieval of water content (to estimate evaporation) from data for a "split" transparency window,
2. cloud cover texture, which allows one to distinguish between convective and stratus clouds,
3. climatological assessment of evaporation,
4. large-scale 4-D assimilation models to assess precipitation and evaporation in the sub-cloud atmospheric layer,
5. modelling clouds with parametrised microphysics to analyse the 3-D distribution of precipitation and evaporation.

It is relatively simple to account for the horizontal inhomogeneity of precipitation in the calculation of averages with the rain rate distribution function given, while the required repeatability of observations (so as to provide the temporal representativity of averages) should be estimated as a function of the special characteristic time of autocorrelation for the rain rate field.

For example, during the GATE period, the autocorrelation time was such that, with a cloud cover area of several square kilometres, observations were repeated in less than one h; for a region 280 km × 280 km, they were repeated in about 10 h (with 12 h repeatability, the errors in the calculated monthly means constituted 10 per cent). The clear diurnal change of rain rate in many tropical regions demonstrates the uselessness of data from Sun-synchronous satellites, whereas data from geostationary satellites do not provide sufficient spatial resolution. Thus, satellites with a relatively small angle of inclination must be employed.

Recently a Global Satellite Mapping of Precipitation (GSMaP) product has been developed using data from several satellites by the Japan Aerospace Exploration Agency (JAXA). The rainfall map is provided in real time and visualised every 30 minutes on the JAXA Realtime Rainfall Watch website (sharaku.eorc.jaxa.jp accessed 2 April 2020).

An alternative approach to the real-time retrieval of precipitation has been proposed by Giannetti et al. (2017) by making use of the known attenuation of radio signals by rain which is exploited in conventional weather radars. This involves using satellite downlink signal attenuation measurements. Some preliminary results obtained in Pisa were presented and compared with the output of a conventional tipping bucket rain gauge. While this may prove useful for small local areas, it is not going to be relevant to the acquisition of global data.

3.3.3 Nonlinearities in Oceans and Climate Feedbacks

Model projections of the near-future response to anthropogenic warming show compensation between meridional heat transports by the atmosphere (AHT) and ocean (OHT) that are largely symmetric about the equator, the causes of which remain unclear. Heath et al. (2004) showed that this transient compensation – specifically during the initial stage of warming – is caused by combined changes in both atmospheric and oceanic circulations. In particular, it is caused by a southward OHT associated with a weakened Atlantic Meridional Overturning Circulation, a northward apparent OHT associated with an ocean heat storage maximum around the Southern Ocean, and a symmetric coupled response of the Hadley and Subtropical cells.

In spite of randomness or of chaos at the small-scale level, one can nevertheless make useful and meaningful calculations at the macroscopic level based on a transition from one mean (or quasi-equilibrium) state to another. Thus, we commonly assume in climate prediction that the climate system is in equilibrium with its forcing. That means, as long as its forcing is constant and the slowly varying components alter only slightly in the timescale considered, the mean state of the climate system will be stable and that if there is a change in the forcing, the mean state will change until it is again in balance with the forcing. The timescale of the transition period until an equilibrium state is re-established is determined by the adjustment time of the slowest climate system component, i.e. the ocean. The stable ("quasi-stationary") behaviour of the climate system gives us the opportunity to detect changes by taking time averages. Because the internal variability of the system is so high, the averaging interval has to be long compared with the chaotic fluctuations to detect a statistically significant signal which can be attributed to the external forcing.

It is as well to recall that we are, at the moment, not just concerned with short-term climate changes attributable to human activities but, rather, with changes arising from the whole history of the Earth. Thus, although to us the motions of the various tectonic plates (continental drift) on the surface of the Earth may seem small and of little immediate relevance to climate change, they are – over long periods of time – very important. The transfer of energy between the surface of the Earth and the atmosphere depends on whether that surface is ocean, ice, or land and, if it is land, on the nature and elevation of the land surface and land cover. Thus, the changes in the distribution of the land surfaces and ocean areas of the Earth arising from the movements of the tectonic plates will cause changes in the distribution of the input of energy into the atmosphere. Another aspect of crustal movements is associated with variations in the gravitational effect of the Sun and the Moon on the surface of the Earth. Apart from causing the familiar ocean tides, these effects give rise to stresses on the land surface as well, so that there is a greater chance of volcanic eruption when these stresses are at their maximum. Another factor that needs to be considered in long-term studies is the effect of the variation, and indeed the reversal, of the Earth's magnetic field on incoming radiation.

3.4 THE LAKES

In more recent times, other human activities have become important. For instance, there are now situations in which human intervention occurs to modify the weather rather directly, for instance in the case of rainmaking by cloud seeding. The short-term local effects may be desired and beneficial. The effects elsewhere, or in the longer term, may be less obvious and may not necessarily be beneficial. Another human influence is the interference with the hydrology of the Earth. The creation of artificial lakes for irrigation or hydro-electric purposes will affect the local heat balance and the evaporation rate. Changing the flow of rivers for irrigation purposes, and the consequent change in land use and land cover, will also affect the local heat balance and evaporation rate as well. The development of large cities in recent years is also important. Large cities are well known to form heat islands with temperatures significantly higher than their surroundings. This arises from two factors: (i) the replacement of vegetated surfaces by more highly reflecting road and roof surfaces and (ii) the heat generated by human activities within the cities. Locally the cities' effects on the heat balance will affect the weather, although in global terms their contribution to the heat balance is relatively small.

Recent advances have allowed the use of altimetry data to be extended in order to observe rivers (water extent smaller than 100 m). The scientific community may count on a number of altimetry missions (such as Geosat, ERS-1/2, TOPEX/Poseidon, Geosat Follow-on, ENVISAT, Jason-1/2/3, CryoSat-2, SARAL/AltiKa, Sentinel-3A/B/C/D, Sentinel-6) that adopt different instrumentations and provide data with various accuracies and spatial resolutions (Schumann and Domeneghetti 2016) (see Figure 3.6). Despite the wealth of measurements, in some cases made up of long time series, the satellites' low revisit time (typically ranging from 10 to 35 days), ground-track spacing (from a few kilometres, e.g. CryoSat-2, to some hundreds of kilometres, e.g. Jason), and the measurements' latency limit the use of such sources for operational procedures and for application, especially over small to medium rivers, where the time lag of the flood events is typically limited to a few days. Nevertheless, the potential of altimetry data is considerable, with applications that regard the enhancement of hydrological and hydraulic models. Undoubtedly, finer spatial resolution and climatic noise reduction techniques will improve the reliability of regional scale predictions.

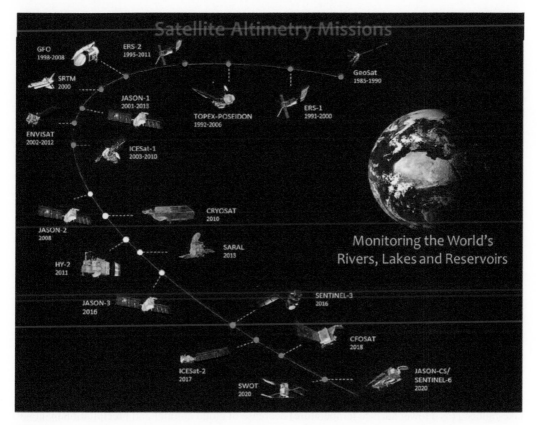

FIGURE 3.6 View of past (blue dots), current (white dots), and future (orange dots) satellite altimetry missions most used to monitor the world's rivers, lakes, and reservoirs. (Modified from Schumann and Domeneghetti 2016).

Inadequate results of using the global climate models to simulate regional climates have caused attention to be given to this problem through applying limited area models nested in a general circulation model. The combined limited area/general circulation models can be particularly effective with a complicated topography or coastline, which is typical, for example, of the western USA.

Giorgi et al. (KC98) discussed the climate numerical modelling results for this region obtained with the use of a general circulation model and a limited area model developed, respectively, at the National Center for Atmospheric Research (NCAR) and Pennsylvania University (mesoscale model MM-4). An important stage of such a numerical modelling is the validation of the results obtained. In this connection, the MM-4 model has been validated on the basis of calculations for the western USA for the period of two years, characterised, respectively, by reduced and increased humidity.

The validation has been concentrated on consideration of two parameters: surface air temperature and precipitation. A comparison with the data of observations has shown that the MM-4 model satisfactorily simulates the surface air temperature annual change (differences do not exceed several degrees). The greatest errors in predictions occur in summer in the Rocky Mountains. The annual change of precipitation is also well simulated. As a rule, the predicted precipitation corresponds better to that observed during the cold half of the year than in summer, and better in coastal regions than in inland areas.

A division of the territory considered into six sub-regions showed that the precipitation forecast was most reliable in California and the north-western Pacific, where differences in monthly mean and seasonal precipitation do not exceed ±(10–50) per cent. Over the Rocky Mountains, the calculated precipitation was close to that observed in winter, but rather overestimated in summer.

The latter suggested a revision of the mesoscale model by taking into account a less intensive horizontal diffusion along the surface, depending on the gradient of topography.

Such a revision provided an almost halved precipitation level in summer over the mountains but had almost no effect in winter. The results obtained make it possible to formulate two principal conclusions, as follows. (1) If the ECMWF data are used to prescribe initial and boundary conditions (fields of wind, temperature, humidity, and surface pressure), then the MM-4 simulated regional climate agrees satisfactorily with monthly mean and seasonal values observed at a network of 390 stations. As a rule, the differences between monthly mean and seasonal surface air temperature values do not exceed several degrees, and those of precipitation about 30–40 per cent. It is important that the reliability of the numerical modelling results does not decrease with increasing range of forecast. After a spin-up time of a few days, the performance of limited-area models is basically determined by a dynamical equilibrium between the information from the driving large-scale fields and the internal model physics. To reach equilibrium with the use of models taking into account hydrological processes, calculations are needed for a period of months to several years. (2) From the viewpoint of available computer resources, calculations are possible, with the use of a mesoscale model, for a period of many years. In this case, for models with a horizontal resolution of 60 km and a grid of $50 \times 55 \times 14$ points, the Cray Y-MP computer time constitutes about 70 hours. Thus, possibilities of numerical climate modelling using the limited area/general circulation models are quite realistic.

Based on the use of the NCAR model, Bates et al. (KC98) undertook a numerical climate modelling in the region of the Great Lakes affecting substantially the climates of adjacent territories. In particular, the thermal inertia of the lakes causes an increase of local surface air temperature and dew point values, of up to 10deg K. On synoptic scales (about 1000 km), the lakes favour the intensification of centres of low pressure in winter and high pressure in summer. The contribution of the lakes to the formation of snowfalls in winter constitutes 30–50 per cent, and in summer on the contrary a lake-induced (as a lower surface temperature) decrease of cloud amount and precipitation by 20–30 per cent occurs.

Calculations with the use of limited area models have been made with prescribed regularly renewed initial and boundary conditions from the data of objective analysis made at the ECMWF. The lake surface temperature was prescribed from the data of observations. The numerical modelling was made for the 10-day period from 22 December 1985 to 1 January 1986, with varied horizontal resolutions of 30, 60, and 90 km, bearing in mind an analysis of the role of spatial resolution. Naturally, for example, the precipitation field with the 30 km resolution is more inhomogeneous. A comparison with results of conditional calculations in the absence of lakes revealed that the surface air pressure over the lakes was higher by 2.75 hPa than with the lakes present. Preliminary calculations were made with the use of the MM-4 together with a one-dimensional model of the thermal regime, which made it possible to predict the ice-cover onset.

Hostetler et al. (KC98) analysed possibilities of the combined (in the interactive regime) use of MM-4 and the biospheric model BATS, simulating the heat and moisture exchange in the presence of vegetation cover, as well as a model of the thermal regime of lakes. The MM-4/BATS results of simulating the weather in the western USA have been used as initial data for the model of the thermal regime of lakes, used to calculate the diurnal mean surface temperature and evaporation for Pyramid Lake, Nevada, as well as a catchment-area model to calculate the stream flow discharge for the river Steamboat Creek, Oregon.

The results obtained can be considered positive. The models MM-4/BATS and of the thermal regime of lakes were used (in the interactive regime) to assess the impact of large lakes on regional climate. The MM-4 data were introduced every hour, and also every hour the data of the lake model (surface temperature, evaporation, ice cover) were put in the MM-4. Preliminary results for Pyramid Lake, Nevada, and the Great Lakes show that the interactive model considered ensures quite a realistic simulation of the lake–atmosphere interaction, including the formation of ice cover on the Great Lakes.

3.5 THE RIVERS

In the conditions of the Arctic, the principal manifestations of the climatic effects of the processes in the polar oceans include: (1) the effect of heat transport in the ocean across the Greenland–Spitsbergen passage on the variability of pack ice, (2) the possible effect of the diversion of part of the Siberian rivers run-off on the fresh water influx to the Arctic basin and subsequent transformation of sea water stratification, (3) the impact of the ocean and pack ice on the formation and variability of summertime overcast stratus clouds, (4) intensification of climate warming in high latitudes due to increasing CO_2 concentrations and its possible consequences (more intensive CO_2 release by the ocean, the effect on ice melting, etc.), (5) cyclogenesis near the edge of the polar ice cover, and (6) the influence of the atmosphere–ocean interaction in the Greenland–Norwegian Seas on the formation and variability of the North Atlantic deep waters (the state of balance between the output and input of these waters along the eastern coastline of the Atlantic; the impact of the Gulf Stream on Europe).

Estimates of the river run-off dynamics for large water basins could serve as an alternative, but in this case, an analysis of data is seriously hindered by anthropogenic impacts on the run-off as well as by the fact that the run-off is sometimes determined by the difference between precipitation and evaporation. Attempts to analyse the regional dynamics of precipitation gave no positive results; almost everywhere the interannual variability of precipitation and the trends of ten-yr means are such that it is impossible to detect a relatively weak CO_2 signal. Part of precipitation variability is determined by the effect of the El Niño/Southern Oscillation, as well as by other causes, whose combined effect cannot be filtered out reliably.

Thus, though the precipitation dynamics can be one of the most important indicators of the CO_2 signal, it is very difficult to detect this signal reliably. Broadening the database on precipitation and improving its homogeneity over the oceans, in particular, to obtain the regionally averaged estimates, must be an important goal of further studies. The emphasis must be placed on the problem of enhancing small variations in precipitation, manifested in various components of the biosphere. It was found that small variations in precipitation in the western USA cause substantial variations in the river run-off. The level and chemical composition of lakes, characteristics of vegetation canopy, etc. can also serve as respective indicators.

Recently, researchers at the University of North Carolina at Chapel Hill and Texas A&M University have charted a multitude of new rivers and streams, showing that we have 44% more of them than we ever thought (Allen and Pavelsky 2018a). According to this study, the carbon-containing pollution in our rivers and streams can also release CO_2 into the air ("outgassing") introducing a volume of it into the atmosphere roughly equivalent to one fifth of combined emissions from fossil fuel combustion and cement production.

Altogether, the same study showed that the total surface of Earth covered by streams and rivers is roughly 773,000 km² (298,457 square miles), e.g. about the same size as all of Italy or the Philippines (Figure 3.7). It is worth noting that with all these new rivers to account for, the amount of carbon dioxide actually being released is going to be even harder to mitigate.

The effect of the dynamics of permafrost on the hydrological regime is very important. Estimates show, for example, that with permafrost destroyed in the regions of the Lena and Enissey rivers, the run-off for these rivers will be halved, which will affect the salinity regime of the arctic seas and the annual change of the extent of the arctic ice cover (the process of ice formation will slow down).

Changes in the albedo of snow-ice cover caused by contamination can play a substantial role. This effect must be particularly considered in the analysis of the effect of a possible nuclear war on climate. More accurate consideration of the processes in polar regions is a fundamental aspect of further improvements on climate models.

The priority that the ocean and the atmosphere should have in studies of short-period climate variability makes land surface processes less significant, though still substantial. The urgency of the problem of parametrising the atmosphere–land-surface interaction caused the WCRP Joint Scientific

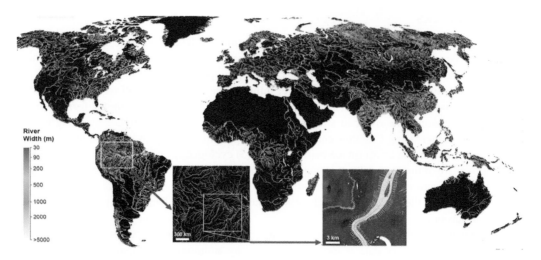

FIGURE 3.7   Long ago we thought that the rivers of the Earth cover 44 per cent less land than we believe today. Satellite maps suggest that these waterways cross some 773,000 sq km. (Modified from Allen and Pavelsky 2018a).

Commission to hold a working meeting in 1981 (Kondratyev 1985). Some aspects of this interaction are connected with the atmospheric contribution: rainfall, global radiation, and atmospheric thermal emission. In other aspects, the effects of both the atmosphere and surface are manifested (turbulent exchange of heat, moisture, and momentum), whereas the absorption and reflection of radiation by the surface, river run-off, and heat and moisture exchange in soils depend on specific topography, vegetation cover, and soil properties. The principal circumstance is the strong spatial inhomogeneity of these processes, which determines the need for parametrisation of subgrid processes and, in this connection, for a representation of input parameters on a subgrid-scale, depending on predicted quantities averaged over the scales of the applied spatial grid.

The global water cycle plays a key role in climate formation and biospheric functioning. In this cycle, water takes part in its three phases due to the unique position of the Earth in the solar system, which determines the observed thermal regime. The most important problems in hydrology are: (1) the development of new techniques for assessing soil moisture and evapotranspiration on a scale from individual fields to continents, (2) a quantitative estimation of precipitation and its distribution over the catchment area, (3) global monitoring of snow cover and its water equivalent, (4) the development of testing of reliability of hydrological models, coordinated with possibilities of obtaining information by satellite remote sensing (spatial and temporal resolution, repeatability), for events of different scales (from floods on small rivers to general circulation models), and (5) further studies of the laws of the global water cycle (Figure 3.8).

Another important parametrisation is the transfer of heat and water within the soil, for instance the balance between evaporation and precipitation, snow melt, storage of water in the ground, and river run-off. This parametrisation is of extreme relevance for climate change predictions, since it shows how local climates may change from humid to arid and vice versa depending on global circulation changes. It furthermore reflects, in some of the more sophisticated schemes, the changes that could occur through alterations in surface vegetation and land-use.

A negative feedback between the CO_2 content in the atmosphere and surface temperature observed on the Earth is largely determined by the geochemical cycle of carbonates and silicates, whose contribution to the CO_2 exchange between the Earth's crust and the atmosphere constitutes about 80 per cent (on a timescale of more than 0.5 million years). The initial phase of this cycle is connected with the solution of atmospheric CO_2 in rain water with the subsequent formation of carbonic acid (H_2CO_3). Precipitation causes the processes of water erosion of rocks containing calcium silicate

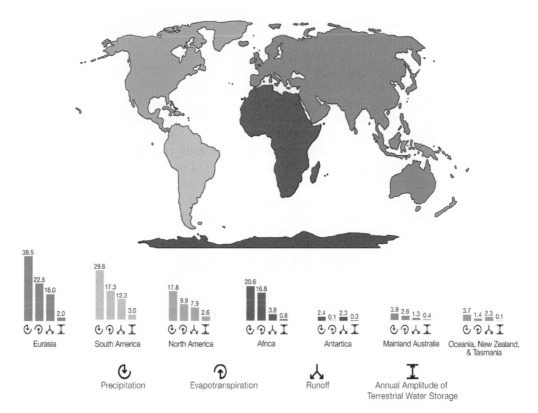

FIGURE 3.8 The amount of water per year that precipitates, evaporates, runs off into streams and rivers, or soaks into groundwater storage for each of seven main land masses. The amounts listed are in units of thousand cubic kilometers. For reference, all yearly human water use is 9.1 thousand cubic km on this scale. (Source: NASA Goddard/Conceptual Image Lab).

minerals (compounds of calcium, silicon, and oxygen), during which the carbonic acid reacts with rocks. As a result, calcium and bicarbonate ions $\left(Ca^{2+} \text{ and } HCO_3^-\right)$ are released to surface waters. These waters eventually (mainly due to river run-off) reach the ocean, where plankton and other organisms provide an inclusion of ions to the composition of shells of calcium carbonate ($CaCO_3$).

On timescales of decades and centuries, the transport of heat by the sea current systems, which is an important part of the high-latitude heat budget and, hence, determines the formation of the ocean–pole temperature gradient, becomes of fundamental importance. In this connection, the exchange between surface and deep waters and a slow reverse flow to the equator below the thermocline is crucial. The effect of sea surface temperature variations in the tropics on the moisture input to the atmosphere, which affects the cloud distribution and the atmospheric greenhouse effect, is weaker. The river freshwater run-off to the arctic seas can be a strong climatic regulator. It maintains the buoyancy of surface waters and, therefore, more favourable conditions for the formation of sea ice, which, in its turn, intensifies the surface temperature variability, limiting the heat exchange with water masses and intensifying the solar radiation reflection.

In the period January to May 1983, the atmospheric general circulation in the northern hemisphere turned out to be much more intensive than usual and was followed by numerous strong wintertime storms in the western USA. These storms were connected with a powerful polar jet stream shifted southward (compared with its usual location) by several degrees in latitude. In early winter, several storms with precipitation took place at the western coastline, exceeding the norm by 200%. In late winter, the storm-tracks continued spreading over the western USA, followed by some wintertime storms which caused intensive snowfalls in the central region of the Rocky Mountains.

This precipitation (together with an unusually cold spring followed by a sudden warming in the region of the Rocky Mountains) resulted in local heavy showers in July that caused a maximum high stream flow discharge of the Colorado river followed by floods and damage of constructions.

Based on the use of an advanced MM-4 model, Matthews et al. (KC98) undertook a numerical modelling of the events mentioned (the model was advanced by using a more reliable parametrisation of hydrological processes in soils, the atmospheric boundary layer, and radiation processes). Initial and boundary conditions were prescribed from the ECMWF data of objective analysis. The study area of the Colorado river basin is a square with a side of 3000 km centred at 40° N, 116° W. The 13-layer model has a horizontal resolution of 60 km. Analysis of the numerical modelling results has shown that the model provides a good simulation of the situation in question, including its principal feature, a longwave depression with the polar jet stream south-west of it, which is responsible for the formation of intensive precipitation. It was this feature that was responsible for a continuous moist air flow from the Pacific Ocean and frequent storms crossing the regions of the Colorado river. In conditions of a late-spring warming, the snow cover melted and a maximum intensity stream flow discharge took place.

Thus, a simulation of mesoscale processes using the atmospheric general circulation model and nested mesoscale model could be considered reliable. The ability of the atmospheric GCM to simulate reliably the structure of general circulation and the location of the longwave depression will determine the MM-4 potential from the viewpoint of simulating the mesoscale climatic regularities in the future. Further testing of the CCM-1 with the nested mesoscale unit is planned, bearing in mind an analysis of adequate simulation of climate both at present and in the future, including the doubling of the CO_2 concentration case.

Medina (KC98) discussed the prospects for the analysis of a possible impact of climate changes in the catchment area of the Delaware River (USA) based on the use of a two-dimensional stationary mesoscale model which took into account only physical processes responsible for the formation of orographic clouds. In order to simulate the present regime of precipitation, the model has been realised over a grid with a step of 10 km within the territory 340 km by 440 km, completely including the watershed.

A comparison of numerous early estimates of the mean global warming with a doubled CO_2 content point to the value of 3 ± 1.5 °C is the most probable, which is in accordance with the IPCC assessments (IPCC 1990, 1992). Gates et al. (KC98), Mitchell (KC98), and Gates and MacCracken (KC98) gave a much lower level of warming (0.2–0.3 °C), but this is probably explained by a prescribed sea surface temperature which strongly limits surface air temperature variations. Stratospheric cooling in the layer 30–45 km reaches 7–11 °C. A doubled CO_2 concentration causes a marked intensification of the hydrological cycle. Calculations of zonal mean temperature values give a high-latitudinal warming two to three times stronger than in the tropics, where it is limited by the effect of moist convection, smearing the warming through the whole atmosphere.

Other effects on the zonal mean of climate parameters, which can only be judged qualitatively (quantitatively are not quite reliable), are as follows: (1) a marked increase of the annual mean river run-off which takes place in high latitudes, caused by intensification of precipitation due to an intrusion of moist warm air masses north of 60°N, (2) earlier snow melting and later snowfalls due to the high-latitude warming, (3) a reduction in the summer northern hemisphere of soil moisture in middle and high latitudes (north of 35°N), explained by an earlier start of relatively strong evaporation (following the phase of snow melting) and by weakening of precipitation in summer and (4) a reduction in the area of polar ice cover.

3.6   CASE-STUDIES OF HYDROLOGICAL APPLICATIONS

In this section, we consider a number of seas and lakes which have been seriously affected by human activities but where these activities are not principally the combustion of fossil fuels and the emission of CO_2. These include the Okhotsk Sea, the Aral Sea, Lake Chad, the Dead Sea, and

Lake Eyre. We should perhaps clarify the use of the terms sea and lake, which are not used very logically. The Okhotsk Sea is a part of the open oceans, the Aral Sea and the Dead Sea are enclosed seas which by strict definition ought to be called lakes since they have no physical connection to the open oceans, while Lake Chad and Lake Eyre which also have no physical connection to the open oceans are logically called lakes. Enclosed water bodies with no outlet to the oceans are sometimes referred to as seas rather than lakes but the question of when does a large lake become a sea is not really answered, as illustrated by the fact that we refer to Lake Chad but the Aral Sea although they are both large.

3.6.1 Modelling the state of the Okhotsk Sea ecosystems

The need to understand and predict changes in marine ecosystems is constantly growing due to the desire of a more encompassing approach to marine fisheries and environmental management that considers a wide range of the interacting groups within an ecosystem. Because organisms at higher trophic levels have a longer life span, with significant abundant variability and complex historical events with respect to microorganisms (further complicating their coupling to lower trophic levels and the natural system), the available biogeochemical and physical oceanographic models need to be expanded (Heath et al., 2004; Chattopadhyay et al., 2012). Changing climate is one of many factors affecting marine ecosystems.

The Okhotsk Sea ecosystem is particularly interesting because it is a unique natural system which is relatively isolated from major human activities, see Figure 3.9. The Okhotsk Sea is a marginal sea of the western Pacific Ocean, between Russia and Japan. It is surrounded by the Kamchatka Peninsula on the east, the island of Hokkaido (Japan) to the south, the island of Sakhalin along the west, and a long stretch of eastern Siberian coast along the west and north. It has an area of approximately 1.583 million km^2, mean depth 859 m, and maximum depth 3372 m. However, over recent decades, global anthropogenic environmental effects have begun to impact this region with increasing intensity. Thus, the current situation of pollution in the Okhotsk Sea is progressively characterised by an increase in heavy metals flows, as well as various hazardous chemicals and organic compounds produced by man-made activities of neighbouring regions. Significant potential sources of pollutants include seven ports and the Amur River. It is expected that the level of pollution

FIGURE 3.9   The Okhotsk Sea. (Modified from https://www.worldatlas.com/aatlas/infopage/okhotsk.htm).

in the Okhotsk Sea will increase in the coming decades due to population growth in the Far-East regions of Russia and oil extraction on the shelf of the Kamchatka Peninsula. The survivability of the system (i.e. its quantified ability to continue to function during and after a natural or man-made disturbance) is of crucial importance.

The geophysical and climatic conditions of the Okhotsk Sea present many limiting factors for its ecosystem, including air temperatures, where in August it is about 14 °C and in February −24 °C. Consequently, the horizontal distribution of water temperature depends on the depth and vertical mixing that is the prerequisite for the productivity of the ecosystem. Average surface water temperature ranges from −1.6 °C to −1.8 °C in winter and from 6 °C to 14 °C in the summer. The natural evolution of the OSE creates a unique geoecosystem with high productivity but low survivability. Therefore, the search for the criterion for assessing OSE survivability is critical.

According to Krapivin et al. (2015), one of the indicators that determines the survivability level of the complex system is its biocomplexity. Michener et al. (2001) identified biocomplexity "… as property emerging from the interplay of behavioural, biological, chemical, physical, and social interactions that affect, sustain, or are modified by living organisms, including humans." This type of problem contains data sets that are too large or complex for traditional application software which are often termed as "big data" (Dedić and Stanier, 2017). It should be clarified that "cloud computing" refers to the platform for accessing large data sets. Hence, 'big data" is information and "big data cloud" is the means of getting information.

The ecological problems of the Okhotsk Sea are the focus of the annual International Symposium on Okhotsk Sea and Polar Oceans, organised by the Okhotsk Sea and Polar Oceans Research Association (Japan). This interest is explained by the concerns of Japan and Russia regarding the role of this region in climate change in relation to the further development of gas/oil resources in the Arctic region and in Sakhalin and Kamchatka, in particular. Figure 3.10 shows the block structure of the Okhotsk Sea Ecosystem Model (OESM) which has been studied in great detail. This Figure shows that climate, and therefore climate change, is one of several inputs into this model. Moreover, the IPCC's First Report (IPCC 1990) noted that while there was some measure of agreement among the predictions of global climate change by different models, there was very little agreement when it came to the prediction of local or regional climate change. Now, several decades on, while the performance of the various models has improved, that observation is still valid. Since 1990, global surface temperatures have warmed at a rate of about 0.15 °C per decade, within the range of model projections of about 0.10 °C to 0.35 °C per decade. As the IPCC (2014) notes, "global climate

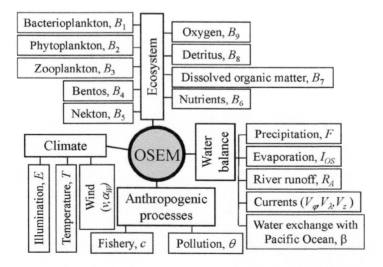

FIGURE 3.10   A block structure of OSEM and big data subjects. (Source: Varotsos and Krapivin, 2019).

models generally simulate global temperatures that compare well with observations over climate timescales … The 1990–2012 data have been shown to be consistent with the [1990 IPCC report] projections, and not consistent with zero trend from 1990 … the trend in globally-averaged surface temperatures falls within the range of the previous IPCC projections."

The functioning of the Okhotsk Sea is largely determined by severe climatic conditions. It is covered with 80–100 cm ice for six to seven months a year. The extent of the ice cover is characterised by large inter-annual variations (Ohshima and Martin, 2004; Ohshima et al., 2006; Matoba et al., 2011). In this context, the structure of the surface environment of the Okhotsk Sea exhibits a high degree of variability that evolves seasonally. The thermal regime that determines this structure is formed by large-scale cyclonic and anti-cyclonic processes whose activity changes during winter and summer according to climate trends in the area and in the neighbouring zones. The Okhotsk Sea ecosystem model OSEM has two mechanisms for modelling the ice fields (Krapivin and Varotsos, 2019):

- viewing satellite images at discrete time intervals and interpolation; and
- using models such as the Discrete-Element Model (Herman, 2016).

A simulation of the Okhotsk Sea ecosystems was conducted by Varotsos and Krapivin (2019) with initial conditions for 1 January 2015, with a horizontal spacing of 1/6 ° in latitude and longitude and with simulation time-steps of one day. Projections of these variables and of the survivability of the ecosystem over 25 years, 50 years and 100 years were made; for details, see Varotsos and Krapivin (2019). OSEM validation was based on a comparison of monitoring data and modelling of relative sea ice cover and net primary production. The OSEM validation features are listed in Table 3.3. It is known that the extent of the sea ice in the Okhotsk Sea over the last 30 years has decreased by about 10% and the ice cover formation process is a function of a global temperature increase. Projections of the sea ice extent from the OSEM up to 2100 are shown in Figure 3.11.

Although there are numerous publications on changes in marine ecosystems due to global climate change and the increasing anthropogenic impacts, this issue is still poorly understood. This is because there is a need for combined analysis of big data clouds that are characterised by heterogeneity, non-structure, instability, and diversity (Lovejoy and Varotsos, 2016; Varotsos et al., 2019a, 2019b, 2019c).

The proposed model in this case study allows the ability to organise the processing of these data for a complex geo-ecological system and to predict its evolution. The results obtained from this model demonstrate these capabilities and show the effectiveness of this model to address the different scenarios of the marine ecosystem's interaction with its surrounding environment. This example of the Okhotsk Sea was included to illustrate the use of a model and to indicate the limits associated with such models at the present time.

TABLE 3.3
The OSEM validation data

Season	Sea ice extent			Average net primary production, gCm^{-2} $yr.^{-1}$		
	Monitoring data	The OSEM result	The OSEM precision, %	Observational data	The OSEM result	The OSEM precision, %
2015	39.9%	44.2%	89.1%	449	539	79.9%
2016	69.4%	63.3%	91.2%	391	464	81.4%
2017	64.6%	72.6%	87.6%	436	509	83.2%

Source: (Varotsos and Krapivin 2019).

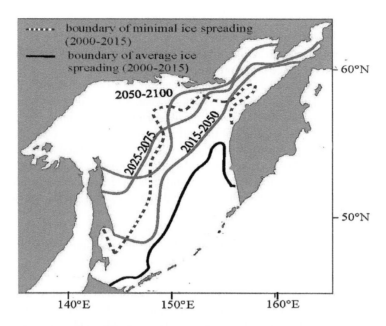

FIGURE 3.11 The results of the OSEM calculations for sea-ice extent with a forecast of up to 2100. The red lines correspond to the average ice spread over a limited period of time. (Source: Varotsos and Krapivin, 2019).

3.6.1.1 The Aral Sea

We include the case of the Aral Sea as a major example of human-induced climate change that is fairly obviously not caused by fossil fuel combustion and the emission of CO_2 but by quite different human activities. In the case of some historical examples of climate change, particularly desertification (the Fertile Crescent, some areas of north Africa, the Sahara Desert), there may be arguments about the extent of human activities in causing the process of desertification. In this case, however, there can be no reasonable doubt that a prosperous inhabited area has been turned into a desert as a very direct result of human activities, see Figure 3.12 showing a map produced in 1848–9 by Commander A. Butakoff of the Imperial Russian Navy. In those years, the Aral Sea was an enclosed sea with an area of 68,000 km^2, the fourth largest in the world.

In modern terminology, the Aral Sea lies between Kazakhstan in the north and Uzbekistan in the south. The name roughly translates as "Sea of Islands," referring to over 1100 islands that had dotted its waters; in the Turkic languages and Mongolic languages *aral* means "island, archipelago." The shrinking Aral Sea constitutes an example of dramatic events that took place in recent years; it has been described as "one of the planet's worst environmental disasters." The Aral Sea region is also heavily polluted, with consequential serious public health problems.

Undoubtedly, human-induced negative processes in Central Asia brought not only economic, ecological, and social insecurity to the resident population, but also created negative habitats with unfavourable human health conditions. The Aral Sea has been shrinking since the 1960s after the rivers that fed it were diverted by Soviet irrigation projects. By 1997, it had declined to 10% of its original size, splitting into four lakes: the North Aral Sea, the eastern and western basins of the once far larger South Aral Sea, and one smaller intermediate lake. By 2009, the south eastern lake had disappeared and the south western lake had retreated to a thin strip at the western edge of the former southern sea; in subsequent years, occasional water flows have led to the south eastern lake sometimes being replenished to a small degree. Satellite images taken by NASA in August 2014 revealed that for the first time in modern history the eastern basin of the Aral Sea had completely dried up (Figure 3.11). The eastern basin is now called the Aralkum Desert. UNESCO added the historical

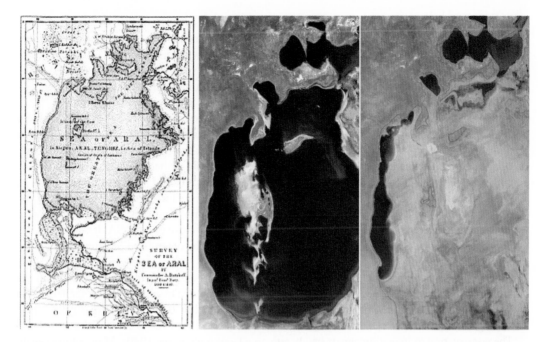

FIGURE 3.12 A map of the Aral Sea in 1848–9 (left), and satellite images in 1989 (middle) and 2014 (right). (Source: *J. Royal Geograph. Soc.*, London and "Images by NASA. Collage by Producercunningham at Wikimedia Commons).

documents concerning the development of the Aral Sea to its Memory of the World Register as a unique resource to study this environmental tragedy.

How did this disaster come about? The answer lies in the death of the two great rivers that once drained a huge swathe of Central Asia into the Aral Sea. The larger is the Amu Darya. Once named the Oxus, it was as big as the Nile. In the 4th century, Alexander the Great fought battles on its waters as he headed for Samarkand and the creation of the world's largest military empire. It still crashes out of the Hindu Kush in Afghanistan but, like its smaller twin, the Syr Darya from the Tian Shan mountains, it is largely lost in the desert lands between the mountains and the sea. During the 20th century, these two rivers were part of the Soviet Union and Soviet engineers contrived to divert almost all their flow – around 110 km³ a year – to irrigate cotton fields that they planted in the desert. Cotton is a crop that is very intensive in its demands for water. This was one of the greatest ever assaults on major rivers of the world. Perhaps nowhere else on Earth shows so vividly what can happen when rivers run dry (Pearce 2018).

The detailed story is told by several writers, but it is perhaps most poignantly described in great detail by Fred Pearce. He travelled widely in the area in 1995 and 2004 and described what he discovered in his book "When the Rivers Run Dry" (Pearce 2018, Chapter 28). His description applies to the situation around 2004 and things have only improved very slightly, if at all, since then. The following is an account of what he found:

Pearce notes that about three miles out to sea, he sees a fox running through the undergrowth on the bed of what was once the world's fourth-largest inland body of water. The sea, as marked on the map, is no longer a sea. Over the past 50 years, most of the Aral Sea in Central Asia has turned into a huge desert, most of which no human has ever set foot on. The scale of what has happened here has prompted the UN to call the disappearance of the Aral Sea "the greatest environmental disaster of the twentieth century."

Until the 1960s, the Aral Sea covered an area the size of Belgium and the Netherlands combined and contained more than a thousand km³ of water. Pearce comments that it was well loved in the

Soviet Union for its blue waters, plentiful fish, stunning beaches, and bustling fishing ports. Many maps still show a single chunk of blue, although the reality is very different. The sea is now broken into two hypersaline pools, containing less than a tenth as much water as previously. Pearce reflects: "The beach resorts and promenades where Moscow's elite once spent their summers now lie abandoned. The fish disappeared long ago. As the fox and I peered north from near the former southern port of Muynak, there was no sea for 100 miles [160 km]. It felt like the end of the world."

Anyone seriously interested in climate change should read the whole of this chapter of Pearce's book. We only summarise it very briefly here. Central Asia has a long tradition of using its two great rivers to grow crops. In the days when Alexander the Great and Mongol conqueror Tamerlane invaded these lands, when cities like Samarkand and Bukhara flourished on the Great Silk Road, people used the land and water carefully. Much of the region was covered in orchards and vineyards and grain fields. Then the Russians came. The tsars in the 19th century first saw the potential for planting cotton in the desert. They realised that the combination of near constant summer sun and water from the great rivers could produce cotton harvests to rival those in the USA. It was following the Russian revolution that the serious damage occurred under the centralised planning adopted by the former Soviet Union.

A massive network of irrigation canals supplied water to huge areas of cotton plantations and the population was turned into a near-slave society of cotton pickers. "Dissent was not tolerated. 'You cannot eat cotton, 'the prime minister of Uzbekistan complained in 1938. He was swiftly executed for 'bourgeois nationalism '(Pearce 2018)."

The construction of irrigation canals had begun on a large scale in the 1940s. Many of the canals were poorly built, allowing water to leak or evaporate. In the early 1960s, the Soviet government decided that the two rivers that fed the Aral Sea, the Amu Darya in the south and the Syr Darya in the east, would be diverted to irrigate the desert, in an attempt to grow rice, melons, cereals, and especially cotton.

The Aral Sea fishing industry, which in its heyday employed some 40,000 people and reportedly produced one-sixth of the former Soviet Union's entire fish catch, was devastated. In the 1980s, commercial harvests were becoming unsustainable, and by 1987, commercial harvest became non-existent. Due to the declining sea levels, the salinity became too high and the native fish species died. The former fishing towns along the original shores became ship graveyards. In the north the town of Aral, originally the main fishing port, is now several kilometres from the sea and has seen its population decline dramatically since the beginning of the crisis. In the south, the town of Moynaq in Uzbekistan, which had originally had a thriving harbour and fishing industry that employed about 30,000 people, now lies many kilometres from the shore. The hulks of fishing boats lie scattered on the dry land that was once covered by water. The muskrat-trapping industry in the deltas of the Amu Darya and Syr Darya, which used to yield as many as 500,000 pelts a year has also been destroyed.

From 1960 to 1998, the sea's surface area shrank by about 60%, and its volume by 80%. In 1960, the Aral Sea had been the world's fourth-largest lake, with an area around 68,000 km^2 and a volume of 1,100 km^3; by 1998 it had dropped to 28,687 km^2 and eighth largest. The salinity of the Aral Sea also increased: by 1990, it was around 376 g/l. By comparison, the salinity of ordinary seawater is typically around 35 g/l and the Dead Sea's salinity varies between 300 and 350 g/l.

Pearce travelled through Uzbekistan, the heartland of the old Soviet cotton empire, from its capital Tashkent, in the Far East, along the old Road through Samarkand and Bukhara, and then north following the Amu Darya through desert towards its delta and the final, fateful destiny with the bed of the Aral Sea. What he found was not the Soviet single-minded determination to convert water into cotton, but the sheer chaos that had resulted. Then the Soviet Union had collapsed in 1991 and the Russians went home leaving the area in the hands of several separate States, Uzbekistan and parts of Tajikistan, Turkmenistan, Kyrgyzstan, Kazakhstan, Afghanistan, and Iran. The once fertile land, which had originally many years ago been made productive by careful irrigation, was becoming degraded and polluted by agricultural chemicals and salt, had by then become totally unproductive and turned into a dust bowl. The once productive sea had dried up, the fish were all gone, and the bed of the former sea had become a desert populated the rusting hulks of abandoned ships. The population had been massively depleted by emigration, disease, and death and the people who remained were unhealthy and dying.

After the Russians went home, the governments of the region did set up an International Fund for Saving the Aral Sea and protested their desire for the Sea to return. Soon afterwards, Pearce went to the region in 1995 and heard their hopes at a big conference. But when he returned in 2004, the situation was worse than before. Even less water was reaching the Aral Sea than in Soviet times. Pearce records his journey of discovery in trying to find out how this happened by talking to victims and to see what could be done as Moscow's rule had ended. He comments that it was a deeply depressing journey. He "found a landscape of poison, disease and death" in what was once one of the most valued and desirable areas of the Soviet empire. He "found mismanagement of water on an almost unimaginable scale – a scale that has turned a showcase for socialism into a blighted land." More disturbing still, he found that in the aftermath of the collapse of the Soviet Union, there was no plan to save the Aral Sea and no one with the vision or the inclination to rethink how this area and its rivers might serve society better.

According to Pearce since 1990, the new governments that rule the Aral Sea basin have increased the area of land under cultivation by a further 12 per cent, and water abstraction from the rivers was up by a similar amount.

Pearce reports that there had been some changes and a market economy had emerged for some crops such as wheat, rice, and sunflowers, although cotton remained the region's biggest export. Pearce comments "The amount of water used here is simply insane." He notes that currently in the league table of per capita water users, the countries around the Aral Sea, Uzbekistan, Kazakhstan, Turkmenistan, Tajikistan, and Kyrgyzstan take five of the top seven places in the world. Turkmenistan and Uzbekistan, the two countries that take their water from the Amu Darya, use more water per head of population than any other country on Earth. The Aral Sea basin is not short of water, the issue is the simply the staggering level of water use. The climate is deteriorating. Without the moderating effect of the Aral Sea to cool summers, warm winters and ensure rainfall in this harsh environment, the summers in Karakalpakstan have become shorter and three degrees hotter, the winters colder and longer.

In other words, the disappearance of the Aral Sea, apart from changing the state of the soils also has had a very direct effect on the local/regional climate.

Pearce notes that rainfall has also declined, and the region is increasingly ravaged by dust storms, with an estimated seventy million tonnes of dust from the sea bed blowing across the landscape every year. He comments that this dust contains an alarming amount of chemicals previously brought to the sea in drainage water but now carried by the winds. These chemicals include pesticides such as lindane, DDT, and phosalone, traces of which have been found in the blood of Antarctic penguins and in Norwegian forests. However, worse than this, he notes that an average year brings 50 days of dust storms, most of which fall close to the former sea and is deposited on fields, inside houses and down the lungs of children.

However, Pearce comments that most researchers believe that there is a more devastating threat than the dust storms, and that is salt. He reports that salt is everywhere in Karakalpakstan. It comes on the wind, down the irrigation canals and through pipes carrying drinking water from reservoirs; it is left behind on the soil surface by the irrigation process itself. Salt destroys the productivity of the land and uses up precious water in flushing it out of the soil. All this contributes to poverty and ultimately kills the people themselves. Pearce concludes, "Salt is the true tragedy of this land. Worse than the poverty, worse than the water shortages, worse than the pesticides, the land and its people are being poisoned by salt."

There are many details of the nature of the disaster of the destruction of the Aral Sea that are heart-breaking but it is beyond the scope of this book to go into it. The simple point that we draw from it is that the story of the Aral Sea is a story of damage being done to the climate by human activities which have nothing whatsoever to do with the burning of fossil fuels and the emission of

CO_2 and that the biggest mistake we can make is to concentrate on one particular threat and neglect all the others.

What of the future? Pearce comments that due to the existing and ongoing widespread environmental degradation, any forecasts of environmental dynamics in Central Asia are largely uncertain. Reducing the scale of water is impossible and optimal water distribution in Central Asia requires a cooperative agreement between countries, which in turn requires consolidation. Water for these countries is their most valuable and conflicting natural resource despite the fact that large quantities of water are stored in the mountain glaciers of Pamir and Tien Shan.

Many different solutions to the problems of the Aral Sea have been suggested over the years, varying in feasibility and cost, including:

- Improving the quality of irrigation canals
- Using alternative cotton species that require less water
- Promoting non-agricultural economic development in upstream countries
- Using fewer chemicals on the cotton
- Cultivating crops other than cotton
- Redirecting water from the Volga, Ob, and Irtysh rivers to restore the Aral Sea to its former size in 20–30 years at a cost of US$30–50 billion
- Pumping sea water into the Aral Sea from the Caspian Sea via a pipeline, and diluting it with fresh water from local catchment areas

In January 1994, Kazakhstan, Uzbekistan, Turkmenistan, Tajikistan, and Kyrgyzstan signed a deal to pledge 1% of their budgets to help the sea recover. In March 2000, UNESCO presented their "Water-related vision for the Aral Sea basin for the year 2025" at the second World Water Forum in The Hague. This document was criticised for setting unrealistic goals and for giving insufficient attention to the interests of the area immediately around the former lakeside, implicitly giving up on the Aral Sea and the people living on the Uzbek side of the lake. By 2006, the World Bank's restoration projects, especially in the North Aral, were giving rise to some unexpected, tentative relief in what had been an extremely pessimistic picture. Work is being done to restore in part the North Aral Sea. Irrigation works on the Syr Darya have been repaired and improved to increase its water flow, and in October 2003, the Kazakh government announced a plan to build Dike Kokaral, a concrete dam separating the two halves of the Aral Sea. Work on this dam was completed in August 2005; since then, the water level of the North Aral has risen, and its salinity has decreased. As of 2006, some recovery of sea level has been recorded, sooner than expected. The dam has caused the small Aral's sea level to rise swiftly to 38 m, from a low of less than 30 m, with 42 m considered the level of viability.

Economically significant stocks of fish have returned, and observers who had written off the North Aral Sea as an environmental disaster were surprised by unexpected reports that, in 2006, its returning waters were already partly reviving the fishing industry and producing catches for export as far as Ukraine. The improvements to the fishing industry were largely due to the drop in the average salinity of the sea from 30 grams to 8 grams per litre; this drop in salinity prompted the return of almost 24 freshwater species. The restoration also reportedly gave rise to long-absent rain clouds and possible microclimate changes, bringing tentative hope to an agricultural sector swallowed by a regional dustbowl, and some expansion of the shrunken sea. The sea, which had receded almost 100 km south of the port-city of Aralsk (Aral), is now a mere 25 km away. The Kazakh Foreign Ministry stated that the North Aral Sea's surface increased from 2,550 square km² in 2003 to 3,300 km² in 2008. The sea's depth increased from 30 m in 2003 to 42 m in 2008. Now, a second dam is to be built based on a World Bank loan to Kazakhstan, with the start of construction initially slated for 2009 and postponed to 2011, to further expand the shrunken Northern Aral Sea, eventually reducing the distance to Aralsk to only 6 km. Then, it was planned to build a canal spanning the last 6 km, to reconnect the withered former port of Aralsk to the sea. The South Aral Sea, half of which lies in Uzbekistan, was abandoned to its fate. Most of Uzbekistan's part of the Aral Sea is completely shrivelled up. Only excess water from the North Aral Sea is periodically allowed to flow into the largely dried-up South Aral Sea through a sluice in the dyke. Discussions had been held on recreating a channel between the somewhat improved North and the desiccated South, along with uncertain wetland restoration plans throughout the region, but political will is lacking. Unlike Kazakhstan, which has partially revived its part of the Aral Sea, Uzbekistan shows no signs of abandoning the Amu Darya river to irrigate their cotton and is moving towards oil exploration in the drying South Aral seabed.

3.6.1.2 The future of the Aral Sea

During the former Soviet Union, various scenarios were discussed to solve the problem of the Central Asian water supply and some of them began to be understood. In the Former Soviet Union, also discussed was the transfer of Siberian River water to Central Asia and some relevant decisions were made (Micklin, 1987, 1988, 2014). In particular, the irrigation-recording channel linked to Irtysh River with Kazakhstan was built in 1971. However, many famous Soviet scientists and the Academy of Sciences call for this project to be cancelled. The main argument was that the implementation of this project would lead to a decrease in the temperature in the Arctic waters (primarily in the Kara Sea) and to unpredictable changes in the global climate. This problem remains important for the independent countries of Central Asia. Its solution is complicated by the different economic strategies of Kazakhstan and Uzbekistan that are oriented to the use of existing water resources.

There were other proposals to solve the problem of the Aral Sea. At the 48th session of the UN General Assembly on 28 September 1993 and at 50th session of 24 October 1995, the recommendation was made to support the countries of Central Asia by appealing to international financial institutions and developed countries. Micklin (2016) raised the question "What could be the future for the Aral Sea and its surrounding environments?" and replies that "the return of the sea to its 1969s situation is possible but very unlikely in the foreseeable future." Taking into account the different scenarios that were materialising, Micklin concluded that the recovery of the average river inflow in the Aral Sea at 56 km³/yr would require more than 100 years.

If the control of water resources in Central Asia was effective in the Former Soviet Union, the use of water for irrigation and electric power generation today is more complicated due to different national interests. The scenario proposed in this case study may be acceptable for Kazakhstan, Tajikistan, Kyrgyzstan, and Uzbekistan and is useful for the wider region, including Azerbaijan and Turkmenistan. This stabilises the regional water cycles and can easily be realised. Indeed, it could lead to an agreement between Central Asia countries and a closer cooperation on the implementation of this scenario.

3.6.1.3 The water balance of the Aral Sea, a new recovery scenario

Figure 3.13 shows the relation between the Caspian Sea and the Aral Sea. The water balance of the Aral Sea has been studied by many authors. The scheme of Figure 3.14 account for various water flows, including additional flows in the atmosphere over Central Asia. The arrows indicate the various flows of water between the land and the atmosphere, along rivers, canals, and underground flows, see Table 3.4. One way to restore the Aral Sea would be to transfer water from the Caspian Sea to the Aral Sea. This could in theory be done by building a pipeline. However, an alternative would be to try to allow nature to do this, see Figure 3.14 which proposes a revised scheme for the hydrology of the area shown in Figure 3.13. In this scheme, the water is transferred by evaporation into the atmosphere from the Caspian Sea (H_5) and various other sources (H_{21}, H_{37}, H_4, and several more) and subsequent precipitation is planned to occur in places from which the water can find its way eventually into the Aral Sea (H_{33}, H_{32}, and H_{38} and several other indirect flows). It sounds slightly crazy, why not just build a pipeline and be done with it? Part of the logic of the plan relies on the levels of various parts of the area shown in Figure 3.13. There are a number of land areas near the east coast of the Caspian Sea which are below the level of the Sea so that reservoirs could be constructed there and the water could flow down under gravity into these reservoirs. These reservoirs

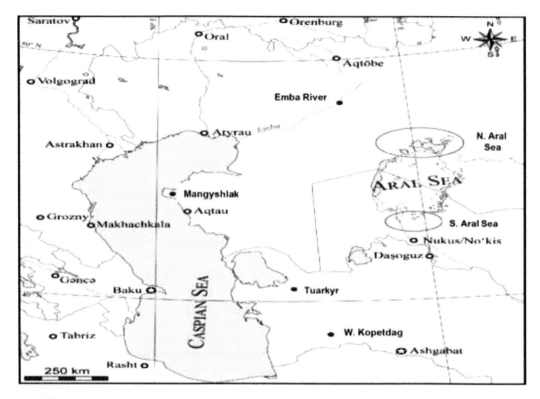

FIGURE 3.13 Map of the Caspian and Aral seas region showing localities of bryozoan material examined or mentioned (Modified from Koromyslova et al., 2018).

would be nearer than the Caspian Sea to the Aral Sea and with prevailing winds from the west the evaporated water vapour from them would tend to drift towards the Aral Sea and its catchment area.

Figure 3.14 shows that water vapour, an important minor component of the atmosphere plays the most important role in the hydrological cycle of Central Asia. Generally, the Caspian waters evaporated by Kara-Bogaz-Gol and other evaporators are an important source of water that can increase the Aral Sea level as a result of precipitation (Leroy et al., 2006). Therefore, the realisation of the WEP scenario could intensify the precipitation and thus raise the sea level.

A solution proposed by Varotsos et al. (2019c) to the problem is referred to as the watering/evaporation/precipitation (WEP) scenario and they argued that this could be the only possible and acceptable method to enable all the Central Asian countries to restore the Aral Sea to its original level. This would involve moving water from the Caspian Sea into the Aral Sea through evaporation and subsequent precipitation rather than by pipeline. To achieve this would involve moving water from the Caspian Sea by gravity to some of the natural reservoirs-evaporators on the east coast which are at levels varying between 25.7 m and 132 m below the Caspian Sea. The total area of these evaporators is approximately 90×10^3 km^2. Additional evaporation from these reservoirs could increase precipitation in the Turan plain and other parts of Central Asia. Moreover, additional water vapour could reach the mountain zone by increasing the river outflows.

An attempt to estimate the time that it would take to re-establish the 1960s level of the Aral Sea under four different scenarios using different sets of evaporators was made by Varotsos et al. (2019c). This involved the following steps:

- Classification of typical wind directions in the Aral-Caspian area.
- Some of the waters of the Caspian Sea are directed to the natural reservoirs-evaporators (saline soils and hollows) located on the East Caspian Sea coast.

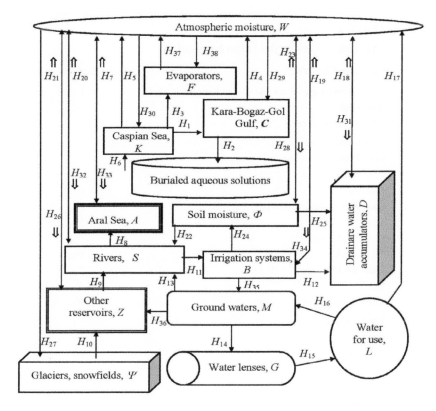

FIGURE 3.14 Block-scheme of the water-flow diagram for the Turan Lowland. The notations are explained in Table 3.4. (Source: Varotsos et al., 2019c).

- The simulation model of the Aral-Caspian hydrological area (Figure 3.14) is used to control the precipitation into water sinks that would eventually drain into the Aral Sea (possibly using rainmaking technology).

Of course, it would be necessary to consider the possible adverse effects on the Caspian Sea and the regions surrounding it. The first step involved studying the directions of the prevailing winds in the whole area of these two seas, which have been investigated by many authors and the results are shown in Figure 3.15. Data on the surface cover and the spatial distribution of the potential reservoirs were obtained by remote sensing using the IL-18 flying multi-functional laboratory from 1970 to 1990 and from ground-based studies.

Varotsos et al. (2019c) studied the WEP model for various different possible evaporators/reservoirs (scenarios 1–4) and obtained the values of the Aral Sea volume as a function of time up to 150 years from the start of the proposed project. The results are shown in Figure 3.15. This Figure provides an understanding of the Aral Sea recovery processes when proposing realistic solutions to the Central Asian water balance. The future of the Aral Sea is considered optimistic and its level fluctuations could be relatively stable after many years of WEP scenario implementation. As can be seen from Figure 3.15, there is a prospect of returning the sea to its state of the 1960s. The recovery time of the Aral Sea depends on the version of the WEP scenario, notably:

WEP scenario version 1: Use only Kara-Bogaz Gol as a Caspian water evaporator.
WEP scenario version 2: In addition to Kara-Bogas Gol, other natural Caspian water evaporators are used.

TABLE 3.4

The water flows in the Turan lowland

The water flow, mm/yr	Identifier
Runoff from Caspian Sea to the Gulf of Kara-Bogaz-Gol	H_1
Buried waters	H_2
Simulated evaporators	H_3
Evaporation from the surface:	
Gulf of Kara-Bogaz-Gol	H_4
Caspian Sea	H_5
Aral Sea	H_7
Reservoirs of drainage waters	H_{18}
Irrigation systems	H_{19}
Amu Darya River	H_{20A}
Syr Darya River	H_{20S}
Lakes and reservoirs	H_{21}
Soil	H_{23}
Artificial evaporators	H_{37}
River runoff into	
Caspian Sea	H_6
Aral Sea	H_8
Inflow due to drainage waters:	
Amu Darya	H_{9A}
Syr Darya	H_{9S}
Thawing of glaciers and snowfields	H_{10}
Water use for irrigation	H_{24}
Inflow of waters into the accumulators of drainage waters	H_{12}
Leakage from irrigation systems	H_{13}
Accumulation of waters in the lenses	H_{14}
Elimination of waters from lenses for domestic use	H_{15}
Drawoff for irrigation use	H_{11}
Surface runoff into rivers	H_{22}
Surface runoff from irrigated areas	H_{25}
Precipitation on:	
Lakes and reservoirs	H_{26}
Glaciers and snowfields	H_{27}
Soil	H_{28}
Gulf of Kara-Bogaz-Gol	H_{29}
Caspian Sea	H_{30}
Accumulators of drainage waters	H_{31}
Amu Darya River	H_{32A}
Syr Darya River	H_{32S}
Aral Sea	H_{33}
Irrigation systems	H_{34}
Artificial evaporators	H_{38}
Inflow of reservoirs at the expense of ground waters	H_{36}
Inflow of rivers at the expense of ground waters	H_{35}

Source: (Varotsos et al., 2019c).

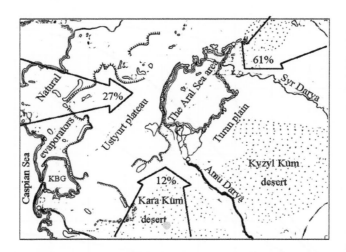

FIGURE 3.15 Prevailing wind directions in the Aral Sea zone and their recurrence. (Source: Varotsos et al. (2019c).

WEP scenario version 3: The process of evaporation of Caspian water by Kara-Bogas Gol and additional natural evaporators takes place along with the reduction of river water for irrigation by ξ percent and using rainmaking with an efficiency of μ percent.

The simulation experiments show that the implementation of the WEP scenario version 1 will restore the Aral Sea to 1960s levels over the next 500–600 years. The WEP scenario version 2 will offer a significant chance to solve the Aral Sea problem over the next 100–130 years using forced rainfalls. Using the WEP scenario version 3, the Aral Sea problem could be solved with high reliability when the irrigation strategy in Central Asia is revised and optimised. Curves 3 and 4 in Figure 3.16 indicate possible periods for increasing the Aral Sea volume under a potential reduction in water withdrawal from Amur Darya and Syr Darya by 5% (3.5 ± 0.4 km^3/yr) and 10% (7.1 ± 0.8 km^3/yr), leading to the 1960s state in 120–140 and 70–95 years, respectively. This strategy can easily be achieved by using drip irrigation.

Thus, the WEP scenario could stabilise the hydrological status of the Aral Sea under the constructive cooperation of the Central Asian governments. Irrigation systems lose about 50% of a river's water due to the irrational use and delay of technical systems. All these confirm the marketability of the WEP scenario version 3. It is envisaged that the technical implementation of the WEP scenario would not require significant financial costs and could be achieved through joint efforts by the five independent republics of Central Asia.

3.6.1.4 Other threatened seas

We have already noted that the desiccation of the Aral Sea is considered to be one of the planet's worst environmental disasters. However, there are many other water bodies experiencing significant desiccation, or are otherwise endangered because of either unsustainable anthropogenic pressures or global climate change. The negative consequences are manifold, ranging from deterioration of environmental conditions (desertification processes, increase of continental climate) to economic and social impacts (decay of fisheries, agriculture and horticulture, tourism, and other related businesses). Chapter 5 of Zavialov (2005) argues that because the Aral Sea represents an extreme case of lake degradation, insight obtained from the Aral Sea may have a broader applicability to other water bodies. That chapter discusses the cases of several other enclosed seas which are suffering severe alterations in circumstances similar to those of the Aral Sea, of which we just note a few. These include the Dead Sea (no. 6 in Figure 3.17), Lake Chad (no. 10 in Figure 3.17).

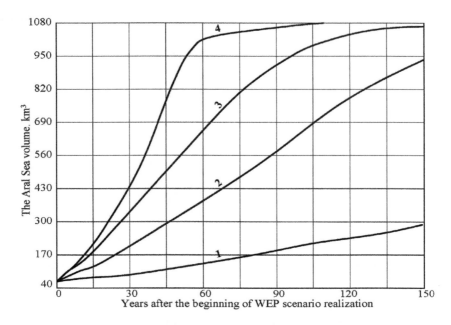

FIGURE 3.16 WEP scenario modelling results. Note: (1) – WEP scenario 1 is used; (2) – WEP scenario 2 is used; (3) – WEP scenario 3 is used for $\mu = 0\%$ and $\xi = 5\%$; (4) – WEP scenario 3 is used for $\mu = 90\%$ and $\xi = 10\%$. (Source: Varotsos et al. (2019c).

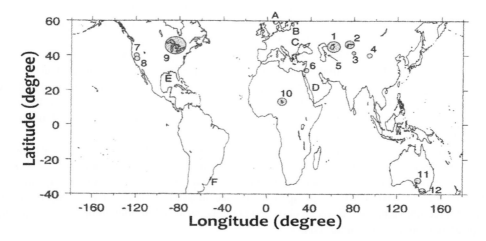

FIGURE 3.17 Geographic objects mentioned in the text: 1-Aral Sea; 2-Lake Nalkhash; 3-Lake Issyk-Kul; 4-Lake Lobnor; 5-Kara-Bogaz-Gol Bay and the Caspian Sea; 6-Dead Sea; 7-Pyramid Lake; 8-Mono Lake; 9-Great Lakes; 10-Lake Chad; 11-Lake Eyre; 12-Lake Corangamite; A-Fjords of the Norwegian Sea; B-Baltic Sea; D-Red Sea; E-Gulf of Mexico and the Mississippi delta; and F-Southern Brazilian shelf and Plata estuary. (Modified from Zavialov, 2005).

The Dead Sea is a deep terminal lake at the border between Israel and Jordan (Figure 3.17). The present Dead Sea surface is located at about 416 m below the World Ocean level, which makes it the lowest land spot on Earth. The Dead Sea whose maximum salinity is above 340 g/l and density is about 1,237 kg/m^3 (1.237 g/ml) is considered to be one of the saltiest lakes in the world. The Dead Sea desiccation continues at rates of 0.5–1 m/year.

The shallowing is believed to have been anthropogenic and resulted from major water management interventions in the drainage basin, manifested mainly through water diversions from the

TABLE 3.5
Desiccation characteristics for the Aral Sea, Dead Sea, and Lake Chad; river discharge drop is an approximate difference between the characteristic pre-desiccation and present-day inflow

Region	Time period	Sources/ Causes	Drop in River discharge (km³/year)	Percentage of the River discharge drop	Decrease in Lake level (m)	Surface Area loss (10³km²)	Percentage of Area loss
Aral Sea	1961-present	Man-made + Natural	~ 40	~ 80	23	50	75
Dead Sea	1900s-present	Mainly Man-made	~ 1	~ 90	21	0.3	35
Lake Chad	1963-present	Mainly Natural	~ 20	~ 50	4	22	90

(Modified from Zavialov 2005).

Jordan River feeding the lake. The river waters have been diverted for agricultural and industrial uses by Jordan, Syria, and Israel; the political issues and conflicts are discussed by Pearce (2018). The discharge into the Dead Sea has reduced from 1.5 km³/year in the 1950s to only 0.15 km³/year at present. It can be seen from Table 3.5 that the percentage loss in area of the Dead Sea is much less than that of the Aral Sea; it is thought to be mainly due to human activities.

The area of the surface of Lake Chad, which is in central Africa, has shrunk to nearly one-twentieth of its former extent, i.e. from approximately 25,000 km² in 1963 to 1,350 km² in 2001. It is fed by water from Nigeria, Niger, Chad, and Cameroon. Before 1960, the rivers feeding the lake supplied about 42 km³ of water per year, on average. The river discharges, however, have been highly variable at the seasonal and interannual scales. Since the early 1960s, the inflow started to decrease at a rate of about 1 km³/year until the mid-1980s and since then there has been some increase of the discharges. The corresponding changes in the area are shown in Table 3.5. There has been controversy around the relative roles of the natural climate variability and anthropogenic factors in the Lake Chad desiccation. The assessment in column 3 of Table 3.5 is that the change is mainly natural.

In the early 1960s, the absolute level of Lake Chad's surface was about 283 m a.s.l., subject to considerable spatial variability (±0.4 m) depending on the wind conditions. The present level of the remainder of the lake is below 279 m. On a long temporal scale, Lake Chad, like the Aral Sea, has undergone several regression and subsequent expansion episodes in the past. The largest of the expansions occurred between 12,000 and 6,000 years ago when the surface area of the lake is believed to have been as large as 250,000 km². More recently, a very strong regression took place in the 15th century and notable regressions occurred around 1850 and between 1904–1915.

Another notable example is Lake Eyre in Australia. Like the present Large Aral Sea, Lake Eyre consists of two separate parts, Eyre North (140 km long and 80 km wide) and Eyre South (65 km long and 25 km wide), connected through a narrow strait. Its catchment area is about 1,140,000 km². It is located in one of the driest regions of the country and the lake is episodic, i.e. it is usually dry but occasionally there is heavy rain and the lake fills up and then empties again slowly by evaporation and seepage. On such occasions, Lake Eyre temporarily becomes the largest lake in Australia, with a surface area up to 9,500 km². For example, strong flooding events occurred in 1950, 1974, and 1984 followed by desiccation periods. The process is shown rather beautifully from satellite data in Figure 3.18. During the flood events, water, sediment, and salt exchanges between the northern and the southern parts through the channel occur. These events should really be taken into account in climate models. We have argued that the drying of the Aral Sea is almost entirely anthropogenic and we have noted that there have been quite serious climatic consequences.

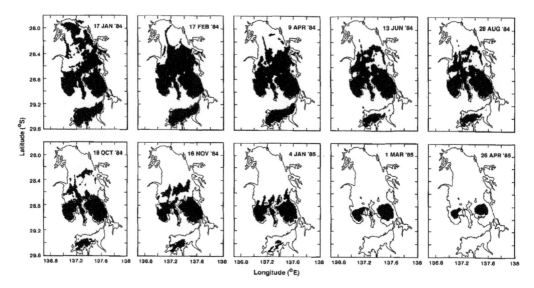

FIGURE 3.18   Silhouette maps of Lake Eyre derived from satellite data showing the area of water on the dates indicated. (Source: Prata 1990).

The summers have become hotter, the winters have become colder, and the length of growing season for crops such as cotton in irrigated areas has become shorter. When it comes to the Dead Sea and Lake Chad, the totality of anthropogenic causes appears to be less clear, see column 3 of the Table 3.5. The changes in Lake Eyre are essentially natural. In this case, there is no human interference in the water entering or leaving the lake, except possibly any changes in rainfall in the area caused by some sort of human activities. It is probably the closest example we can find to variations of a large lake that are almost entirely due to natural causes.

The general message of the whole of this Section 3.6 is that there are some climate changes that are (almost) entirely anthropogenic having nothing whatsoever to do with CO_2 emissions from the combustion of fossil fuels. Another message is that water shortage or conflicts of interest between states, or even physical conflicts, over the diversion of water resources constitute one of the many threats to our way of life(Pearce 2018). It is an area where (1) different environmental changes are caused by different combinations of human activities and natural causes and (2) there are important environmental changes – and therefore possible threats to our way of life – that have nothing whatsoever to do with CO_2 emissions from the burning of fossil fuels.

4 The biosphere, lithosphere, and cryosphere

4.1 BACKGROUND

We introduced the ideas of the biosphere, lithosphere, and cryosphere briefly in Section 1.1.5 and we now need to discuss them in a little more detail, particularly in relation to the climate and to climate change. As mentioned in Chapters 2 and 3, there are many interacting factors causing climate change, which take place in the complex subsystems (hydrosphere, atmosphere, biosphere, lithosphere, and cryosphere) of the climate system. This complexity presents the principal difficulty that occurs in formulating an adequate theory of climate and in identifying the most important factors responsible for its variations (KC98).

"The present level of parametrization of the processes in the climate system is such that climate theory can only be a first approximation to understanding the laws of formation of climate and its changes. For this reason, discussions are still continuing on the nature of climate." These words were written over 20 years ago (KC98) but they are still true today. Then, as now, the interest was in climate predictability and, in spite of all the progress made in computer modelling in the intervening period, we are not much nearer to finding a meaningful answer. Studies of the Earth's palaeoclimates, the climatic features of other planets, and the specific character of the climate of large industrial cities determined by atmospheric pollution and transformed surface characteristics are all of great importance for understanding present-day climate changes (KC98).

Climatic predictability is the key aspect of the problem under discussion. The ability to analyse and forecast climate depends on understanding that the variability of seasonal mean characteristics of climate is caused by short-term internal fluctuations of the climatic system (unpredictable climatic noise) and also by long-term fluctuations (the climatic signal). In Chapter 1, we noted that it is convenient to distinguish between weather and climate. One finds in practice that weather forecasts are reliable for a period of a few days ahead, that the temporal extent of their reliability is improving, but that the ultimate limit of their reliability may be of the order of a few weeks. This can be discussed in terms of determinism versus chaos. One is therefore entitled to ask whether there is any hope of ever being able to make meaningful climate forecast predictions over the scale of a century or two if one has no hope of being able to make reliable weather forecasts for a time such as only 6 or 12 months ahead. This question is pertinent if one takes the numerical modelling approach to climate prediction where one is using general circulation models of the atmospheric circulation which have been derived from weather forecast models and are based on the same physical mechanisms and equations used in the weather forecast models.

General circulation models are based on the physical conservation laws which describe the redistribution of momentum, heat, and water vapour by atmospheric motions. All of these processes are formulated in the "primitive" equations, which describe the behaviour of a fluid (air or water) on a rotating body (the Earth) under the influence of differential heating (the temperature contrast between equator and pole) caused by an external heat source (the Sun). These governing equations are non-linear partial differential equations, whose solution cannot be obtained except by numerical methods. If it is surprising that any success can be achieved with climate models, then it may help to appreciate this if we recall that in looking at climate we are looking at an average of the weather over a reasonably long time, e.g. 1 year or 10 years, or even longer than that. The chaotic elements, or the statistical fluctuations, in the climate are the weather systems. In looking at climate change, we are looking at long-term changes in this average. The long-term fluctuations are potentially predictable, since they are determined either by different external forcing (sea surface temperature anomalies, soil moisture variations, etc.) or by internal atmospheric dynamics. Viewed over a century or a millennium at least in the past, we find that climatic parameters (temperature, rainfall, etc.) are basically stable and vary only slowly. This is not to say that there have not been "sudden" climate changes in the past, but we need to be careful to consider what we mean by "sudden" in the historical context. It almost certainly means an event that happened over a period of several generations, not within the lifetime of an individual human being.

It is the nature of these slow, long-term variations that are of concern to us in climate studies after the local, short-term fluctuations have been smoothed out. Any hope of attempting to make climate prediction possible hinges on the assumption of the relative stability of the long-term components. In a climate model, we consider the effect of various forcing conditions on a given mean state. Analogies are always dangerous and should not be pushed too far. However, let us draw an analogy with the behaviour of the molecules in a gas. We can use very successfully an equation of state for a gas to determine the effect of a change in pressure, volume, or temperature on the gas. 22.4 litres of a gas at 0 °C and atmospheric pressure, whether it is oxygen, nitrogen, CO_2 or almost anything else, contains approximately $6.02214076 \times 10^{23}$ molecules, or 6.022×10^{23} for most practical purposes. This is strange and, as professors of physics, we have no idea why this should be so – someone probably knows. But if one were to seek to study the behaviour of the individual molecules in the gas and try to make predictions of the path of any given molecule, one would be faced with an almost impossible problem. Moreover, even if one could find the solution, it would probably not be very interesting or useful in terms of enabling one to understand the behaviour of the gas as a whole.

In spite of randomness or of chaos at the small-scale level, one can nevertheless make useful and meaningful calculations at the macroscopic level based on a transition from one mean (or quasi-equilibrium) state to another. Thus, we commonly assume in climate prediction that the climate system is in equilibrium with its forcing. That means, as long as its forcing is constant and the slowly varying components alter only slightly in the time period considered, the mean state of the climate system will be stable and that if there is a change in the forcing, the mean state will change until it is again in balance with the forcing. The timescale of the transition period until an equilibrium state is re-established is determined by the adjustment time of the slowest climate system component, i.e. the ocean. The stable ("quasi-stationary") behaviour of the climate system gives us the opportunity to detect changes by taking time averages. Because the internal variability of the system is so high, the averaging interval has to be long compared with the chaotic fluctuations to detect a statistically significant signal which can be attributed to the external forcing. Studies of the completed change from one mean state to another are called equilibrium response studies.

Studies of the time evolution of the climate change due to an altered forcing, which might also be time dependent, are called transient response experiments.

We shall now consider the principal factors and processes that determine climate changes. The variables which are commonly used in studying the climate are concerned mainly with the atmosphere. However, we cannot look at the atmosphere alone. This is because processes in the atmosphere are strongly coupled to the oceans, to the land surface, and to the parts of the Earth which are covered with ice, i.e. the cryosphere. There is also strong coupling to the biosphere, i.e. to the vegetation and other living systems on the land and in the sea.

4.2 THE BIOSPHERE

Processes in the biosphere are a driving force in the composition of the atmosphere. Anthropogenic impacts on the biosphere therefore play a vital role in climate-changing factors. In this connection, studies of the global carbon cycle are relevant. We introduced the carbon cycle in simple terms in Section 1.1.5 and it is now appropriate to consider it in more detail. In the literature, we find two slightly different approaches to quantifying the carbon cycle. One is to consider the total amount of

TABLE 4.1

Some parameters of the carbon cycle and the degree of their uncertainty

1	Amount of CO_2 due to fossil fuel burning (oil, carbon, gas) between 1861 and 1981	160 Gt	140–180 Gt
2	Average CO_2 concentration in the atmosphere	340 ppm	339–343 ppm
3	Pre-industrial CO_2 content (early 19th century)	260 ppm	240–290 ppm
4	Amount of CO_2 absorbed by the ocean by the year 1981 (calculated for equivalent carbon)	2 Gt y^{-1}	Unknown
5	Carbon in biomass at the present time (carbon content in global ecosystems and in harvests)	560 Gt	460–660 Gt
6	Annual loss of carbon by the biosphere due to soil cultivation, etc.	2Gt y^{-1}	2–5 Gt y^{-1}
7	Present-day supplies of organic carbon in soil, peat, humus	1450 Gt	1200–1700 Gt

Source: (KC98, Table 2.1).

carbon transferred from one region to another over the whole period from the Industrial Revolution to the present time. The other is to consider the annual rate of these transfers in the present time.

Table 4.1 gives an indication of some of the parameters involved in the carbon cycle and characterises their uncertainties. As can be seen from this table, there is considerable uncertainty in the quantitative characteristics, especially those which govern the exchange processes between various media. This calls into question the results of numerical modelling of the carbon cycle and explains the scatter of the estimates which have a bearing on final conclusions about possible climate changes.

In accordance with the IPCC scenario A (Business as Usual) (IPCC 1990) of the CO_2 concentration increase, a simulation has been accomplished to predict relevant biome dynamics. The results show that the largest changes occur for boreal biomes, whereas little change is seen for the Sahara and the tropical rain forests. Claussen and Esch (KC98) noted, however, that since the biome model is not capable of predicting changes in vegetation patterns caused by a rapid climate change, this simulation has to be taken as a prediction of changes in conditions favourable for the existence of certain biomes, not as a prediction of a future distribution of biomes. Because the biome model does not take into account changes in vegetation due to changes in CO_2 and soil fertility, subsequent studies need to couple a dynamic model of vegetation succession to a climate model. A further step in the study will be to use a nested approach to simulate smaller-scale dynamics. The sensitivity of a climate model to changes in global patterns of biomes will also be investigated.

4.2.1 THE CUMULATIVE CARBON TRANSFERS SINCE THE INDUSTRIAL REVOLUTION

Biospheric productivity is manifested through the synthesis and decomposition of organic matter. Stability of the chemical composition of the environment is maintained by the closed cycles of biospheric substances. As Gorshkov (KC98) showed, with only small-scale anthropogenic disturbances of biospheric processes in the pre-industrial period, flows of substances due to the synthesis and decomposition in the biosphere had been mutually compensated to an accuracy of about 0.01 per cent. This had been achieved by natural selection of ecologically balanced communities of numerous types of living organisms.

In the case of a major anthropogenic impact, i.e. following the Industrial Revolution, flows of substances due to the synthesis and decomposition of organic and industrial products became unbalanced. This determines the urgency of the following basic problems of theoretical ecology: (i) studying the laws that provide the closed nature of the cycles of substances in conditions of the undisturbed biosphere, (ii) determining the degree of disturbances that do not lead to irreversible changes in the biosphere and environment, and (iii) obtaining observational data on the present state of the biosphere and environment and, in particular, on the degree of the closing of the cycles of various substances.

Based on an analysis of biospheric energetics, Gorshkov's study estimated the distribution of energy fluxes and nutrients from the size of organisms of less than 1 cm. Highly closed geochemical cycles are possible if less than 1 per cent of the energy flux is absorbed by organisms larger than 1 cm. These conclusions agree well with available empirical data for natural communities that are undisturbed anthropogenically.

Gorshkov also studied the closed nature of the present-day carbon cycle, using data on variations in the isotopic composition of carbon in tree rings, see Section 6.8. It is now established that the redistribution of carbon can mainly take place between the four reservoirs: atmosphere, ocean, fossil fuel, and biosphere. Any leakage of carbon outside these reservoirs can be neglected.

To determine changes in the carbon content of the ocean and biosphere, it is enough to assess variations for only one of these reservoirs, since a change in the content of carbon in the second reservoir is determined from a balance equation (the law of conservation of matter). It is essential here that the absorption of carbon by the ocean is characterised by two constants. The first is determined by the physico-chemical resistance to absorption with changing total content of carbon, including bicarbonate and carbonate ions. The second is determined by the absorption resistance due to variations in the concentration of dissolved CO_2 only and corresponds to variations in biological activity.

The estimates by Gorshkov showed that with the existing poor accuracy of empirical data, emissions of carbon from land biota are 2.3 times larger than those from fossil fuel burning. A more refined value of this value will emerge in Figure 4.3. Insofar as the biosphere's emissions can be regarded as natural they can be considered to be absorbed by the ocean because of its varying biological activity. Thus, on the whole, the carbon content of the biosphere, including its near-surface and oceanic parts, remains almost stationary. Separate changes in the near-surface and oceanic parts of the biosphere and the respective flows of carbon through the atmosphere greatly exceed emissions due to fossil fuel burning.

A study by Esser (KC98) illustrates another approach to the problem of the global carbon cycle. It was argued that the human impact on the biosphere during a period of more than a century (deforestation, agricultural activity, etc.) has led to global carbon cycle perturbations which are comparable to the contribution from fossil fuel burning. The anthropogenic transformation of vegetation cover reduces its age, and agricultural activity reduces net primary productivity (NPP). Since phytomass (the total amount of living organic plant matter) depends both on NPP and on vegetation age, it follows that there is a decrease of phytomass because of human activity.

Another aspect of the problem is that an increase in the concentration of CO_2 in the atmosphere has a stimulating effect on photosynthesis (development of natural vegetation) and, apparently, the NPP. This stimulation is determined by two mechanisms:

1. Increased quantum yield of photosynthetic reactions;
2. Increased efficiency in the use of water by vegetation (due to increasing stomatal resistance, with preserved input of CO_2 to the leaves) and decreased water loss on transpiration (Idso, Idso *et al.* KC98).

The NPP can also vary under the influence of climate (temperature and humidity change). In this connection, Esser (KC98) assessed the sensitivity of the carbon cycle to the factors mentioned above, based on the use of a biospheric model (over a 2.5° by 2.5° grid on the continents except the Antarctic) developed in Osnabrück University (Germany) together with a multi-box oceanic model and a one-box model of the atmosphere. Data on regional soil use have been taken from the World Agricultural Atlas. Soil use dynamics were also taken into account. The model showed that the initial value of CO_2 concentration (for the year 1860), which is equal to 290 ppm, is close to that found in the analysis of ice cores (287 ± 3 ppm). The post-1958 growth of CO_2 concentration agrees with the data from observations of Mauna Loa, but the observed interannual variability has not been reproduced.

It was concluded that by the end of the period 1860–1981, the biospheric reservoir of carbon was estimated to be more or less balanced. Emissions due to deforestation were compensated by a stimulation of photosynthesis under the influence of increased CO_2. Before 1970, the biosphere functioned as a weak source of carbon, since CO_2 release due to forest cutting and the combustion of fossil fuels had not been compensated by a CO_2-caused increase in photosynthesis. The continuing upward trend of the Keeling curve since then indicates that this situation still continues and attempts to compensate for CO_2 emissions by planting trees has not been very successful.

If the effect of photosynthesis enhancement is not taken into account, the results of the numerical modelling of carbon reservoirs in the atmosphere and in the ocean strongly differ from those expected from the observational data. A biospheric cumulative release of carbon to the atmosphere (over the entire period mentioned above) amounted to 137 Gt (gigatonnes), whereas an emission due to fossil fuel burning reached 166.6 Gt (a total release of 303.6 Gt). As a result, CO_2 concentration in the atmosphere grew by 50 ppm (this corresponds to the increment of carbon mass by 105.2 Gt), 64 Gt of carbon being assimilated by the ocean (11.8 Gt remained in the mixed layer).

Due to photosynthesis enhancement, the NPP of the terrestrial biosphere increased from 43.1 Gt of carbon in 1860 to 48 Gt of carbon in 1981. Since the NPP of agricultural fields is much lower than that of natural vegetation, a 17 per cent increase of CO_2 has led to only a 13 per cent increase of NPP. The phytomass reservoir is the most powerful in the biosphere, reaching about 600–700 Gt of carbon, but the global phytomass decreased by 17.5 Gt of carbon (the balance of phytomass is determined by its decrease by 137 Gt of carbon because of deforestation and by its increase by 199.5 Gt of carbon because of photosynthesis enhancement). While these figures could be updated, the general message remains the same.

Different scenarios of global climate changes (annual mean temperatures and precipitation) for the period under consideration revealed their substantial effect on the balance of carbon in the biosphere and on CO_2 concentration in the atmosphere, with precipitation variations being the strongest. Of course, we return to what is more relevant to most people and that is the question of regional, rather than global, climate change and simply note that very little is known about the local effect of the biosphere. An account of regional climatic variability is even more important. A comparison of the obtained estimates of the global carbon cycle components with the data of other authors revealed considerable differences.

In this connection, an important question is whether it is possible, using observations of climate changes during the previous one hundred years, to detect (filter out) an anthropogenic signal (in particular the CO_2 signal). This is very difficult because the anthropogenic signal does not exceed the level of noise (natural climate variability). We shall turn to the consideration of this problem in Chapter 6.

For the moment we note that the questions that need to be considered in relation to the problem of the impact of CO_2 and other minor atmospheric gases on climate include:

- estimations of the possible fuel combustion in the forthcoming century, based on the consideration of population, economic development, and prospects for the use of alternative sources of energy;
- possible dynamics of the global biosphere resulting from human industrial activity;
- further study of the carbon cycle and a redistribution of carbon among the basic reservoirs (especially the monitoring of the CO_2 dynamics in the atmosphere and ocean, estimating the dynamics of the continental phytomass and analysing an interaction among the cycles of the various components);
- more reliable and adequate estimates of climatic response to increasing concentration of CO_2 and other minor atmospheric gases;
- the effect of climate on natural ecosystems and human activity;
- complex long-term observations of the climate system to monitor the dynamics of the minor atmospheric gases and to detect an anthropogenic signal.

As has been pointed out, the simulation of the biosphere as a coupled component of the climate system has become a very important task. This has also stimulated interest in making assessments of the impact of climate changes on the biosphere dynamics. It was in this context that Claussen and Esch (KC98) used the biome model developed by Prentice *et al.* (KC98) to predict global patterns of potential plant formations, or biomes, from climatologies simulated by ECHAM (European Center + HAMburg), a general circulation model created by the Max Planck Institut for Meteorology in Hamburg by modifying global forecast models developed by the European Centre for Medium-Range Weather Forecasts (ECMWF) (Cubasch *et al.* KC98). The purpose of the modelling was twofold: a qualitative test of simulated climatologies and an assessment of the effects of climate change. However, it has been suggested by Naafs et al. (2018) that models, such as ECHAM, cannot reproduce the warming at the early Paleogene, i.e. the earlier part of the tertiary period, which is characterised by an extended period of high atmospheric carbon dioxide levels in a manner to be consistent with the marine biological proxy estimates.

The first important result obtained by Claussen and Esch (KC98) was that a good overall agreement between simulation and observation of global patterns of biomes has been found. There were also discrepancies for the Kalahari Desert (southern Africa), in Australia and for the middle west of North America, which can be traced back to failures in simulated rainfall as well as summer or winter temperatures. As an example of potential vegetation changes in response to simulated changes in climate, global patterns of biomes have been computed from an ice age simulation 18 000 years ago. These calculations indicate that North America, Europe, and Siberia should have been covered largely by tundra and taiga, whereas only small differences are seen for the tropical rain forests.

4.2.2 THE CARBON CYCLE: ANNUAL DATA

An analysis of 2419 air samples made by Keeling et al. (KC98) between 1959 and 1981 at the oceanic weather station P ($50°$ N, $145°$ W) revealed three important regularities in the temporal variability of CO_2 concentration; these are a well-reproduced annual change, an interannual variability which correlates with the dynamics of the general atmospheric circulation, and a long-term increasing trend which is approximately in proportion to global industrial CO_2 emissions caused by fossil fuel burning. Figure 4.1 represents an updated version of what has become known as the Keeling Curve of which we showed an earlier version in Figure 1.12.

An increase was observed in the amplitude of the smoothed annual change (a difference between maximum and minimum values) from 13.3 ppm in 1969 to 14.5 ppm in 1981. This was apparently connected with intensified vegetation productivity in a vast region of the northern hemisphere land surface, where the vegetation processes take place mainly in summer (this assumption is suggested by data on the observed phases of the CO_2 annual change).

The interannual variability and its first derivative closely correlate with the dynamics of the Southern Oscillation. The first derivative lags 6 months behind the interannual variability, and this points to the possible existence of a remote oceanic or continental source-sink for CO_2 in the southern hemisphere tropics. Between May 1969 and June 1981, the CO_2 concentration increased from 324.9 to 340.8 ppm, which constitutes about 60 per cent of the increase which would take place if the whole amount of the CO_2 resulting from fossil fuel burning remained in the atmosphere and were uniformly distributed. The concentration of CO_2 at Station P, averaged over the period 1975–1981, turned out to be 0.8 ppm below that at Point Barrow ($71°$ N, Alaska) and 0.9 ppm above that at Mauna Loa ($19°$ N, Hawaii), which testifies to a north-to-south decrease of CO_2 in a broad latitudinal band (70–20° N).

The analysis of daily observations of atmospheric CO_2 concentration at the Mauna Loa observatory between 1952 and 1982 revealed an increase of the amplitude of the annual change of concentration (a maximum in May and a minimum in early October), reaching nearly 1 ppm, i.e. a considerable part of the average amplitude of the annual change (about 6 ppm). Borisenkov and Kondratyev (KC98) estimated that an amount of 5 Gt of carbon released as CO_2 by human activities

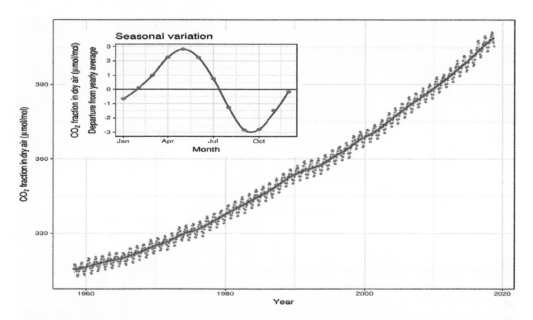

FIGURE 4.1　The monthly mean CO_2 concentration at Mauna Loa 1958–2018 and its seasonal variation. (Source Dr. P. Tans, NOAA/ESRL).

would correspond to the natural accumulation of carbon in the Earth's crust in a period of about a millennium of the Earth's prehistoric natural evolution. Thus, for the current value of 9 Gt shown in Figure 4.2, it would have taken about two millennia to lay down this quantity of carbon in the carboniferous age.

Analysis of observational data gives evidence for an increase of CO_2 concentration in the atmosphere, caused by human industrial activity, and manifested mainly through fossil fuel combustion and through impact on the biosphere (burning and cutting of forests, soil cultivation, burning of timber, etc.). This is illustrated in Figure 4.1. The negative slope of the curve in the inset of Figure 4.1,

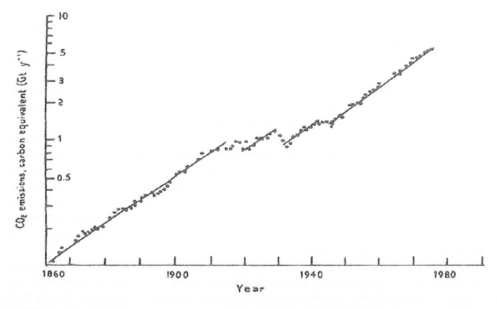

FIGURE 4.2　The CO_2 input to the atmosphere due to fossil fuel combustion (GtC y^{-1}). (Source KC98).

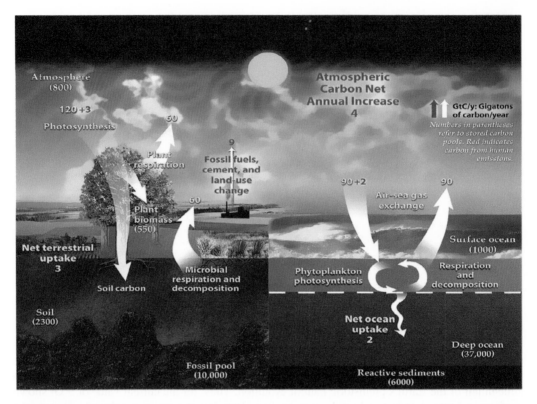

FIGURE 4.3 Movement of carbon between biosphere, land, atmosphere, and ocean in billions of tons per year. Yellow numbers are natural fluxes, red are human contributions, white are stored carbon. The effects of volcanic and tectonic activity are not included. (Source Earth Observatory. NASA. 2018 http://earthobservatory. nasa.gov/Features/CarbonCycle/).

from about May to September corresponds to the absorption of CO_2 by photosynthesis in the northern hemisphere. There is no corresponding effect in the southern hemisphere summer because of the much lower quantity of terrestrial vegetation there; during this period, the absorption of CO_2 by photosynthesis is outweighed by the emission from fossil fuel combustion.

Looking back, there were three reductions in the rate of CO_2 emissions during periods of armed conflict or economic depression around 1915–20, 1930 and 1940–45, see Figure 4.2. Viner and Hulme (KC98) note that CO_2 concentration in the atmosphere has risen from a pre-industrial level of about 280 parts per million by volume (ppmv) to the 1991 level of 355 ppmv. Figure 4.3 gives some known parameters of the carbon cycle at the present time.

Marland *et al.* (KC98) undertook an inventory of the global distribution of CO_2 emissions due to fossil fuel burning over a 5° latitude by 5° longitude grid, based on UN data on industrial development for 1980 (without taking account of the contributions by the cement industry, fossil gas burning in plumes, and some others). Data in Table 4.2 characterise total emissions for 5° latitudinal belts in both hemispheres. The total global emission (including the indicated sources left out of account) amounted to 5.284 Gt of carbon. Ninety per cent of global emissions originate within the latitudinal belt 20°–60° N (95 per cent on the northern hemisphere), with maxima in the spatial grid cells where Frankfurt, London, and Tokyo are located.

Smith *et al.* (KC98) performed numerical modelling to study changes in the carbon budget on the continents and hence, in ecosystems, associated with a doubling of the concentration of CO_2 in the atmosphere. These estimates were based on climate scenarios obtained with four climate models: Oregon State University (OSU), Geophysical Fluid Dynamics Laboratory (GFDL), Goddard Institute

TABLE 4.2
CO_2 emissions in 1980 for 5° latitudinal belts in both hemispheres

N.H. latitude (°)	CO_2 emissions (kt C)	S.H. latitude (°)	CO_2 emissions (kt C)
85–90	0	0–5	11 544
80–85	1	5–10	24 190
75–80	17	10–15	11 365
70–75	601	15–20	12 387
65–70	9 141	20–25	39 291
60–65	39 539	25–30	52 387
55–60	294 268	30–35	60 469
50–55	854 434	35–40	20 463
45–50	634 248	40–45	4434
40–45	846 680	45–50	742
35–40	726 033	50–55	336
30–35	578 493	55–60	0
20–25	140 728	65–70	0
15–20	76 712	70–75	0
10–15	45 891	75–80	0
5–10	36 094	80–85	0
0–5	25 744	85–90	0

Source: (KC98, Table 2.2).

TABLE 4.3
Changes in carbon storage (Gt) in above-ground biomass and soil for four climate change scenarios

Scenario	Above-ground biomass	Soil	Total	Change[a]
Current	737.2	1158.5	1895.7	
OSU	860.4 (16.7)	1215.8 (4.9)	2076.2	180.5 (9.5)
GFDL	782.3 (6.1)	1151.3 (−0.6)	1933.6	37.9 (2.0)
GISS	829.6 (12.5)	1213.0 (4.7)	2042.6	146.9 (7.7)
UKMO	765.2 (3.8)	1139.0 (−1.7)	1904.2	8.5 (0.4)

Source: KC98, Table 2.4).
[a] Values in parentheses are percentage change from current.

for Space Studies (GISS), and the United Kingdom Meteorological Office (UKMO). The global distribution of vegetation cover was characterised using the Holdridge Life-Zone Classification (37 life-zones were taken) over a 0.5° longitude grid. The values in Table 4.3 characterise variations in carbon reservoirs, and those in Table 4.4 are for areas covered with major biome-types.

All four scenarios suggest an increase in forest area, although details of predicted patterns vary among the scenarios. There is a significant decrease in the global extent of desert. The decline in tundra observed under all scenarios is primarily due to a shift from tundra to mesic forest. Terrestrial carbon storage increased from 0.4 per cent (8.5 Gt) to 9.5 per cent (180.5 Gt) above estimates for present conditions. These changes represent a potential reduction of 4 to 8.5 ppm on elevated atmospheric CO_2 levels.

Despite the uncertainty of the existing forecasts of CO_2 dynamics, a concern for CO_2 accumulated in the atmosphere has stimulated the development of different approaches to a solution to this problem, a key aspect of which is the dynamics of energy production, see Chapter 5.

TABLE 4.4

Changes in the areal coverage (10^3 km^2) of major biome-types under current and changed climate conditions

Major Biome-types	Current area	OSU	GFDL	GISS	UKMO
Tundra	939	−302	−515	−314	−573
Desert	3699	−619	−630	−962	−980
Grassland	1923	380	969	694	810
Dry forest	1816	4	608	487	1296
Mesic forest	5172	561	−402	120	−519

Source: KC98, Table 2.5).

Although most of the available theoretical assessments of the impact of increased CO_2 concentration on global climate testify to an unavoidable substantial climate warming, an analysis of observational data has not, so far, permitted a reliable detection of the human-induced component of the CO_2 signal. An observational database needs to include the following values: world sea level, sea surface temperature, CO_2 concentration in sea water and its alkaline property, and velocity of currents. Analysis of these and other parameters must include a study of the structure of their time series, estimation of the signal/noise ratios, and determination of the most sensitive regions. An impact on the biosphere of both increased CO_2 concentration (stimulating photosynthesis) and climate warming is not in doubt. However, detecting the CO_2 signal is very difficult in this case.

The following is a quote from KC98 (pages 87-88) which would probably be regarded as branding those authors as sceptics:

"The interpretation that the climate warming trend observed during the last century has been caused by increasing CO_2 content can be accepted only under the following four conditions:

(1) the possibility that the warming is connected with the natural process of returning to a normal, milder climate such as that during the Little Optimum (AD 900–1300) after a temperature drop during the Little Ice Age (about AD 1550–1850) is excluded;

(2) there are grounds for the supposition that CO_2 emissions reaching 100–200 Gt C due to changes in the biosphere (deforestation, agriculture) took place mainly before 1938, whereas emissions due to fossil fuel combustion (about 175 Gt C) were made after 1938;

(3) there exist other climate-forming factors (stronger than CO_2) which may have led to warmings and coolings over periods of 30–50 yr, which suppressed a persistent trend of warming due to increasing CO_2 concentration; and

(4) an assumption is made that the warming due to a doubled CO_2 concentration does not exceed 1.5 °C or that the time shift of the warming with respect to variations in CO_2 is less than 50 yr. Since the first two conditions cannot be guaranteed and the problem of the contribution by other factors remains unsolved, there are no grounds for ascribing the climate warming trend to the effect of increasing CO_2 concentrations."

However, there is some truth in some of those points. Point (1) is difficult to establish because it is still not known how much of the warming is due to anthropogenic emissions of greenhouse gases and how much is due to natural causes. Regarding point (2), the introduction of 1938 as a significant breakpoint seems somewhat artificial. Point (3) is still very valid, in particular as we note elsewhere, there are man-made emissions of some other gases which are much more powerful greenhouse gases (molecule for molecule) than CO_2 but which, at least until recently have been neglected or replaced by an estimated CO_2 equivalent in climate models. The significance of point (4) eludes us.

In connection with further development of studies within the WCRP and planning of the International Geosphere-Biosphere Programme (IGBP), a Global Energy and Water Cycle Experiment (GEWEX) were proposed by the World Meteorological Organization. Its aims are: (1) the description and analysis of the laws of the transport of water (as vapour, liquid, and solid) as well as of energy fluxes in the global atmosphere and at the surface level and (2) the development of techniques to forecast natural and anthropogenic changes in the distribution of water in all three phases in the atmosphere and at the surface level. Meeting these objectives requires advanced observational means, i.e. space-borne instruments.

The IGBP programme was reorganised in 2000 to emphasise the importance of scientific research at the interface of the major geosphere biosphere disciplines. The new structure included a new cross-disciplinary research programme, called the integrated land ecosystem-atmosphere processes study (iLEAPS) aimed at improved understanding of the processes, linkages, and feedbacks in the land-atmosphere interface (Figure 4.4).

This project was designed to build on key findings of previous IGBP projects, especially BAHC (biospheric aspects of the hydrological cycle) and IGAC (International Global Atmospheric Chemistry). The iLEAPS international project office was based at the University of Helsinki. iLEAPS activities, workshops, and scientific conferences facilitated the establishment of a community with a common goal to enhance the understanding of how interacting physical, chemical, and biological processes transport and transform energy and matter through the interface, particularly emphasising interactions and feedbacks at all scales, from past to future and from local to global. A science conference highlighting the accomplishments of the first decade of iLEAPS was held in Nanjing, China in 2014 and coincided with the transfer of the international project office to Nanjing. The current iLEAPS scientific steering committee, activities, and initiatives are described on the iLEAPS website (www.iLEAPS.org).

Of increasing concern are studies of the role of the biosphere in the formation of atmospheric gas composition and of climate. The present chemical composition of the atmosphere, including numerous MGCs, is largely a product of the biosphere, which serves as both a sink and a source of MGCs.

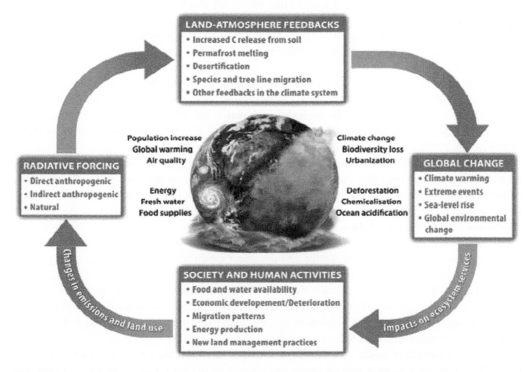

FIGURE 4.4 The land–atmosphere-society processes under global change that are the focus of iLEAPS. (Source Suni et al. 2015).

TABLE 4.5

Observed values of the mixing ratio (M), lifetime, total content in the global troposphere, dM/dT, and the contribution of anthropogenic sources for different MGCs

MGC	Mixing ratio (ppm)	Lifetime (years)	Total content (10^6 tonnes)	dM/dT (% per year)	Anthropogenic sources (%)
CH_4	1700	10	3900	1	60
N_2O	305	100	1900	0.2	30
CO	100	0.2	400	1(?)	60(?)
O_3	40	0.1	270	1(?)	50(?)
	variable (0.5)	0.002	(0.3)	1	60
CF_2Cl_2	0.35	90	6	5	100
$CFCl_3$	0.20	70	3.9	5	100

Source: (KC98, Table 3.1).

There is a very sophisticated and diverse interaction between chemical processes in the atmosphere and biosphere. As a rule, the biosphere-produced gases exhibit reducing or weakly oxidising properties, whereas gases in the atmosphere, which eventually sediment onto the Earth's surface, have clear oxidising properties due to chemical reactions in the atmosphere.

Despite a small MGC concentration, the biospheric-produced gases substantially affect the chemical composition and radiative regime of the atmosphere (Kaufman *et al.* KC98). Chemical processes and radiation transfer in the atmosphere are known to be governed by MGCs, but not by principal gases (oxygen and nitrogen, which are chemically inert and optically inactive). The MGCs contribute most to the formation of the atmospheric greenhouse effect and regulate the input of solar radiation to the troposphere. Table 4.5, compiled by Ehhalt (KC98), gives characteristics of some MGCs in the troposphere. Since the development of industry and agriculture continues, the trend of increasing MGC concentration is likely to continue. Even if anthropogenic MGC releases are stabilised, their accumulation in the atmosphere will affect the atmosphere and biosphere for a long time to come.

Since about 20 per cent of CO_2 is removed from the atmosphere by photo-synthesising plants, the role of the biosphere in the formation of the global carbon cycle is clear. It manifests itself through the effect of living organisms on the cycles of carbonates and silicates. It should be considered, however, that physical rather than biological processes play a major role in the formation of the global carbon cycle.

We note, in particular, the comment that predictions for smaller than continental scales should be treated with great caution. Only further improvements of climate models, which would consider a realistic interaction between marine and continental biota, will permit a reliable simulation of possible variabilities of various climatic parameters and, in particular, an assessment of the impact of climate changes on the biosphere.

To obtain these estimates, not only the annual mean or global mean quantities must be numerically modelled but also the annual change and the geographical distribution of calculated climatic parameters, including an estimation of probable errors as well as the transformation of changes by a large-scale model into data for characteristic scales of climate impacts on the biosphere. (One of the possible ways of solving this problem is by the use of empirical statistics.) Moreover, predicted climate changes must be analysed in terms of specific quantities which must be considered in studies of the climatic impact on the biosphere.

While the atmosphere reacts very rapidly to changes in its forcing (on a timescale of hours or days), the ocean reacts more slowly on timescales ranging from days (at the surface layer) to millennia in the greatest depths. The ice cover reacts on timescales of days for sea ice regions to millennia for ice sheets. The land processes react on timescales of days up to months, while the biosphere reacts on timescales from hours (plankton growth) to centuries (tree growth).

The Goddard Laboratory for Atmospheric Sciences AGCM model was generalised by Mintz et al. (KC98) through an account of the atmosphere–biosphere interaction manifested through evapotranspiration. The water and energy transport through the surface depends on the morphology (spatial structure) and physiology of the vegetation cover.

In the biospheric model considered, the vegetation morphology is determined by dependences of specific leaf area on height, LD(z), calculated per unit volume, and of specific length of the root system on depth RTD(z), which is determined as total length of the roots per unit volume, represented in discrete form for each of the cells of the AGCM model grid.

The simplest version of the biospheric model foresees prescribed, for each cell, values of LD(z) and RTD(z) depending on season, from data of phenological observations. In a more complicated model, an interaction is considered between these quantities and precalculated parameters of the atmosphere and soil moisture. Therefore, on timescales of phenophases, LD(z) and RTd(z) depend, for example, on winds and temperature extremes.

A more advanced model foresees a differentiation of vegetation cover (tropical forests, midlatitude forests, deserts, grass cover, etc.) and its interaction with conditions in the atmosphere and soil. Specific leaf area affects: (1) the aerodynamical resistance to latent and sensible heat transport, (2) radiation transfer through vegetation cover, and (3) evaporation from the surface and precipitation. The effect of the internal structure of roots, stems, and leaves on water and energy transport is a major physiological factor. The climate model, incorporating variables of the biosphere, is planned to be the first, using FGGE data as initial information, with a fixed sea surface temperature. At a later stage, an interactive atmosphere-ocean-biosphere model will be developed, including the parameter of heat transport in the deep layers of the ocean. The realisation of such a model is of particular importance for studying anthropogenic impacts on climate.

One of the largest challenges of surface-atmosphere exchange studies is to cope with the multitude of temporal as well as spatial scales at which land-atmosphere processes occur. Figure 4.5

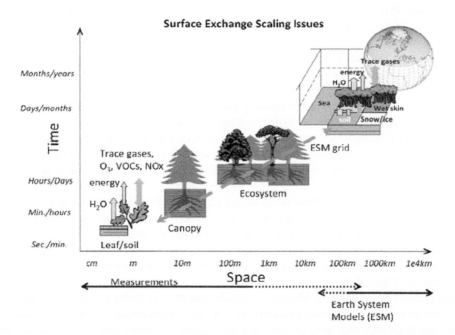

FIGURE 4.5 One of the largest challenges of land-atmosphere processes research is to cope with the multitude of temporal as well as spatial scales (Source Suni et al. 2015).

illustrates these scaling issues for surface exchange, but many more Earth system components, such as cloud processes, exhibit the same range of scales. A further complication is that measurements tend to cover a different domain compared to models.

On timescales of millenia, the oceans control the chemical composition of the atmosphere and, hence, the radiative equilibrium of the climatic system. The concentrations of CO_2, N_2O, and other minor gaseous constituents are eventually determined by the geochemical equilibrium in the ocean, modulated by interaction with the biosphere.

The rise in global CO_2 concentration since 2000 is about 20 ppm per decade, which is up to 10 times faster than any sustained rise in CO_2 during the past 800,000 years (Lüthi et al., 2008; Bereiter et al., 2015). AR5 found that the last geological epoch with similar atmospheric CO_2 concentration was the Pliocene, 3.3 to 3.0 Ma (Masson-Delmotte et al., 2013).

Since 1970, the global average temperature has been rising at a rate of 1.7°C per century, compared to a long-term decline over the past 7,000 years at a baseline rate of 0.01°C per century (Marcott et al. 2013). These global-level rates of human-driven change far exceed the rates of change driven by geophysical or biosphere forces that have altered the Earth System trajectory in the past (e.g. Summerhayes 2015; Foster et al. 2017); even abrupt geophysical events do not approach current rates of human-driven change.

To stabilise global temperature at any level, "net" CO_2 emissions would need to be reduced to zero. This means the amount of CO_2 entering the atmosphere must equal the amount that is removed. Achieving a balance between CO_2 "sources" and "sinks" is often referred to as "net zero" emissions or "carbon neutrality." The implication of net zero emissions is that the concentration of CO_2 in the atmosphere would slowly decline over time until a new equilibrium is reached, as CO_2 emissions from human activity are redistributed and taken up by the oceans and the land biosphere. This would lead to a near-constant global temperature over many centuries.

4.3   THE LITHOSPHERE

The biological involvement of the land, insofar as its surface is vegetated, in the carbon cycle has been discussed in the previous section above already. The role of the oceans in the radiation balance at the surface of the Earth has also been considered in Chapter 3. We now turn to the role of the land in its physical aspects, rather than its biological aspects, and the question of energy balance at the land surface. The land surface plays an important role in terms of its exchange of energy with the atmosphere through the reflection, absorption, and emission of radiation. The land also plays an important part in the hydrological cycle. The processes involved concern the amount of fresh water stored in the ground as soil moisture (thereby interacting with the biosphere) and in underground reservoirs, or transported as run-off to different locations where it might influence the ocean circulation, particularly in high latitudes. The soil interacts with the atmosphere by exchanges of gases, aerosols, and moisture, and these are influenced by the soil type and the vegetation, which again are strongly dependent on the soil wetness. Present knowledge about these strongly interactive processes is rather limited.

The above-mentioned multiphase processes are depicted in Figure 4.6. These deal with chemical reactions, transport processes, and transformations between gaseous, liquid, and solid matter. These processes are essential for Earth system science and climate research as well as for life and health sciences on molecular and global levels, bridging a wide range of spatial and temporal scales from below nanometres to thousands of kilometres and from less than nanoseconds to years and millennia as illustrated in Figure 4.6.

From a chemical perspective, life and the metabolism of most living organisms can be regarded as multiphase processes involving gases like oxygen and carbon dioxide; liquids like water, blood, lymph, and plant sap; and solid or semisolid substances like bone, tissue, skin, wood, and cellular membranes. On global scales, the biogeochemical cycling of chemical compounds and elements, which can be regarded as the metabolism of planet Earth, also involves chemical reactions,

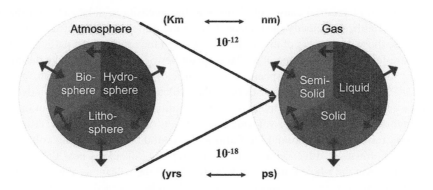

FIGURE 4.6 Multiphase-transition processes affecting the Earth's system and climate as well as life and human health on regional and global scales (Modified from: Pöschl and Shiraiwa, 2015).

mass transport, and phase transitions within and between the atmosphere, biosphere, hydrosphere, and pedosphere/lithosphere (Schlesinger and Bernhardt 2013). A Chapman Conference on the Hydrologic Aspects of Global Climate Change was held at Lake Chelan, Washington, in June 1990. The conference was motivated by interest in the description of land surfaces at scales much larger than the river basins traditionally studied by hydrologists and particularly by problems in the representation of the land surface hydrology in general circulation models.

A collection of 13 papers presented at the conference appeared in an issue of the *Journal of Geophysical Research,* volume 97, number D3 (pages 2675–2833) which was published on 28 February 1992. Of course a great deal has been published since then.

It is as well to recall that we are, at the moment, not just concerned with short-term climate changes attributable to human activities but, rather, with changes arising from the whole history of the Earth. Thus, although to us the motions of the various tectonic plates (continental drift) on the surface of the Earth may seem small and of little immediate relevance to climate change, they are – over long periods of time – very important. The transfer of energy between the surface of the Earth and the atmosphere depends on whether that surface is ocean, ice, or land and, if it is land, on the nature and elevation of the land surface and land cover. Thus, the changes in the distribution of the land surfaces and ocean areas of the Earth arising from the movements of the tectonic plates will cause changes in the distribution of the input of energy into the atmosphere. Drifting of the continents on Earth could have caused substantial climate changes on individual continents and favoured glaciations during the Pleistocene because of the movement of the continents to polar latitudes. Although no drifting of the continents has taken place on Mars, tectonic deformations of its lithosphere under certain conditions could have changed the inclination of the orbit to the plane of the ecliptic and, hence, could have caused substantial climate changes (e.g. Xu et al., 2019). In this regard, Liu et al. (2019) investigating the upper Oligocene-middle Miocene sedimentary sequence in the Xunhua Basin of the northeastern Qinghai-Tibetan Plateau found that 1) climate change at 25.1 Ma caused by tectonically forced cooling of Xunhua Basin, 2) onset of aridity during 19.2-13.9 Ma resulted from uplift of the Tibetan Plateau, and 3) intensified aridity since ~13.9 Ma due to weakening of the East-Asian summer monsoon and global cooling.

Another aspect of crustal movements is associated with variations in the gravitational effect of the Sun and the Moon on the surface of the Earth. Apart from causing the familiar ocean tides, these effects give rise to stresses on the land surface as well, so that there is a greater chance of volcanic eruption when these stresses are at their maximum. Another factor that needs to be considered in long-term studies is the effect of the variation, and indeed the reversal, of the Earth's magnetic field on incoming radiation.

In discussing extra-terrestrial causes of changes in the intensity of solar radiation that reaches the surface of the Earth, dust clouds and meteor showers are important. Volcanic eruptions produce similar effects to those of showers of micro-meteors, though the origin of the particles is terrestrial rather than extra-terrestrial. A volcano blasts large quantities of matter into the sky during an eruption. Some of this reaches the stratosphere as dust particles around which ice can form. Dust is produced in the atmosphere by many causes other than volcanoes, e.g. by strong winds over desert areas. But the quantities arising from major volcanic eruptions are much larger than the quantities arising from other causes. The fate of volcanic dust depends on the latitude of the eruption. A belt of dust particles forms a screen around the Earth at the latitude of the erupting volcano and then widens out in a north-south direction as well. If the erupting volcano is near to the equator, the screen spreads out over the whole Earth towards both poles; if the volcano is more than about 20° N or 20° S from the equator, the stratospheric screen tends to be confined to one hemisphere. This screen reduces the intensity of solar radiation reaching the surface of the Earth and also reduces the intensity of the terrestrial radiation escaping to outer space; however, the former effect is dominant. Thus, the general effect of a volcanic eruption is to lead to a cooling of the Earth. Volcanic eruptions will be discussed further in Section 6.8.

At the present stage of development of the astronomical model of paleoclimate changes, when it has been successfully tested statistically and serious ideas have appeared with respect to physical mechanisms, the time has come to investigate its reliability using numerical climate models. Berger et al. (KC98) undertook such a study with the use of a 2–3–dimensional time-dependent climate model which took into account atmospheric feedbacks, processes in the upper layer of the ocean, sea ice cover, and in the lithosphere. On the timescale of the order of a thousand years, the latitudinal variability of the seasonal insolation fields was supposed to play the most important role (from the viewpoint of the climate-forming effect of variations in the orbital parameters).

Very recently, Kutterolf et al. (2019) re-analysed four longer tephra records with the same statistical method and demonstrated that all contain the ~41,000 year and ~100,000 year Milankovitch periodicities. Most tephra records were described by event numbers, meaning they are binary records, where each point in time has a value of either 0 or 1 (e.g. Figure 4.7).

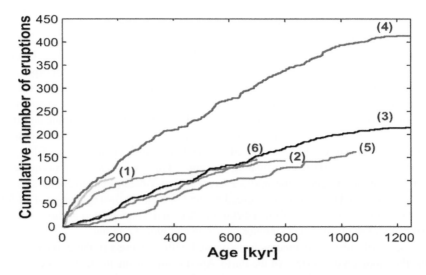

FIGURE 4.7 Cumulative number of tephras versus age for the following tephra series: (1) Central Mediterranean (Paterne et al., 1990), (2) East California (Jellinek et al., 2004; Glazner et al., 1999), (3) the northwestern Pacific (Prueher and Rea, 2001), (4) the Pacific Ring of Fire (Kutterolf et al., 2013), and (5) the Izu-Bonin-Japan region (Schindlbeck et al., 2018). (Modified from Kutterolf et al., 2019).

To investigate binary tephra records Kutterolf et al. (2019) analysed curves of cumulative event number over age (Figure 4.6). The slope of the cumulative event number curves in Figure 4.6 is proportional to the density of the eruption record and typically deviate from a straight line, which represents a homogeneous time distribution (which means no periodicities).

An increasing under-representation of volcanic events with age in the geological record has been quantitatively analysed for the Quaternary of Japan by Kiyosugi et al. (2015). This effect can be observed in Figure 4.6 where the curves of cumulative event number typically flatten to older ages. For instance, in the records of Glazner et al. (1999) and Kutterolf et al. (2013), 85% and 70%, respectively, of the eruptive events are concentrated within the first half of the time series. In the 40,000 years record of Huybers and Langmuir (2009), 80% of the events occur in the last 1000 years, i.e. in the first 1/40 of the entire time series.

The calculations made with the above-mentioned model developed by Berger et al. (1990) have made it possible to obtain information about changes in global ice volume for the last 125 000 years which, on the whole, agrees with the geological data available. This shows that an account of variations in the parameters of the orbit makes it possible to explain the principal features of the variability of the glacial–interglacial periods. In the model considered, the glacial cycle is very sensitive to changes in the heat balance of the surface of ice sheets (due to the low thermal conductivity of ice and snow, small variations in the heat input can bring about substantial changes in the heat balance).

The underestimated volume of the North-American glaciers given by the model should be partially ascribed to the ocean level variability which was not taken into account and, as a result, the propagation of glaciers over the continental plateaux and towards the Arctic Ocean turns out to be impossible. The difference between the model calculations and observations for the last six thousand years is explained, apparently, by neglecting some physical processes that affect the formation of the surface heat balance (it refers, for example, to variations in CO_2 concentrations). The prospects for further development of the climate model should be connected with taking into account the processes in the southern hemisphere and deep circulation in the ocean, as well as improving the model of the continental ice–lithosphere system and a more detailed test of the model's reliability.

4.4   THE CRYOSPHERE

The cryosphere includes all areas of our planet where water exists in its frozen state: glaciers, ice caps, ice sheets, sea ice, lake ice, and river ice, but also seasonal snow and frozen ground such as permafrost. Therefore, the cryosphere can be classified as follows (Figure 4.8).

- Seasonal snow cover, which responds rapidly to atmospheric dynamics on timescales of days and longer. In a global context, the seasonal heat storage in snow is small. The primary influence of the cryosphere comes from the high albedo of a snow-covered surface.
- Sea ice, which affects climate on timescales of seasons and longer. This has a similar effect on the surface heat balance as snow on land. It also tends to decouple the ocean and the atmosphere, since it inhibits the exchange of moisture and momentum. In some regions, it influences the formation of deep water masses by salt extrusion during the freezing period and by the generation of fresh water layers in the melting period.
- Ice sheets of Greenland and the Antarctic, which can be considered as quasipermanent topographic features. They contain 80 per cent of the existing fresh water on the globe, thereby acting as long-term reservoirs in the hydrological cycle. Any change in size will therefore influence the global sea level.
- Mountain glaciers are a small part of the cryosphere. They also represent a freshwater reservoir and can therefore influence the sea level. They are used as an important diagnostic tool for climate change since they respond rapidly to changing environmental conditions.
- Permafrost affects surface ecosystems and river discharges. It influences the thermohaline circulation of the ocean.

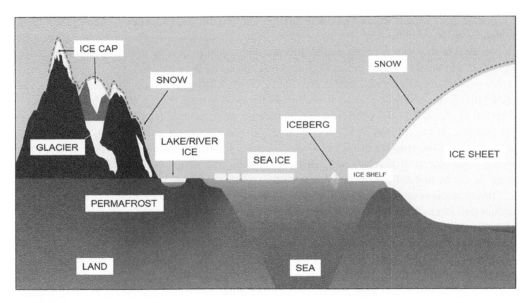

FIGURE 4.8   The components of the cryosphere. (Modified from SWIPA report 2011: Snow, Water, Ice and Permafrost in the Arctic. (Modified from Arctic Monitoring and Assessment Programme - AMAP).

TABLE 4.6
The spatial and temporal scales of the cryospheric components

Component	Area (10^6 km^2)	Time (years)
Ice sheets	16	10^3–10^5
Permafrost	25	10^4
Mid-latitude glaciers	0.35	10–10^3
Sea ice	23	10^{-1}–10
Snow cover	19	10^{-2}–10^{-1}

Source: (KC98, Table 4.13).

The cryosphere is an important component of the climate system (Henderson-Sellers and McGuffie KC98). This is testified to by the data in Table 4.6, which reflect the spatial and temporal variability of the cryospheric components. As can be seen, the sea ice, land snow, and permafrost play the leading role among the cryospheric components. A typical feature of the cryosphere is the rapid variability of its components during the course of the year. So the extent of sea ice varies from 20 × 106 km² in September to 2.5 × 105 km² in March in the southern hemisphere and within (15–8.4) × 106 km² in the northern hemisphere.

Variations in the nature and extent of the ice cover clearly alter the energy fluxes at the Earth's surface and thereby influence the climate. The variation in the surface area covered by ice provides another example of a feedback mechanism in the climate.

The transport of ice, by floating icebergs, towards the equator also provides another possible positive feedback mechanism (see Ledley KC98 for example). Suppose that a general warming causes more ice to be released as icebergs in the polar regions. These icebergs move towards the equator leaving more unfrozen sea surface near the poles; this leads to a larger energy transfer from the sea to the atmosphere near the poles and this additional energy is spread around and leads to a general further warming of the atmosphere.

We can look at it from the opposite viewpoint – if there is a warming of the Earth, leading to less snow and ice cover, then the surface of the Earth becomes (on average) less reflecting and more absorbing. This causes further warming and a further reduction in the snow and ice cover, i.e. this is an example of positive feedback. The real situation is, of course, more complicated and eventually the amplification of the warming by positive feedback (it is hoped) comes to a halt.

Global manifestations of the climatic effect on the variability of snow and ice cover can be sub-divided into two categories, to which corresponds either a fast or a slow response of the cryosphere (US Arctic Research Commission KC98). To the first category belong changes in snow and sea ice cover and the second category refers to land ice, glaciers, and ice sheets.

Continuous satellite microwave observations as well as processing of the visible and infra-red data have made it possible to accomplish a regular global mapping of sea ice and continental snow cover, to draw maps of sea ice concentration and to study the detailed regional dynamics of the cryosphere. For the 1981–1987 period, the microwave data do not reveal any marked global trend of the sea ice extent. The prospects for further development of remote sounding of ice and snow cover are connected with the use of space-borne radars, which will make it possible to obtain data with a resolution of tens of metres and better. In this connection, of great importance are the results of the Russian studies, especially those obtained with the synthetic-aperture radars carried by the satellites Kosmos-1500 and Kosmos-1870 (Almaz) as well as the ERS-1 and ERS-2 results (KC98).

An analysis of the Scanning Multi-channel Microwave Radiometer (SMMR) data for nine seasons revealed a difference in the annual changes of the sea ice in both hemispheres, determined largely by the inhomogeneous distribution of land and oceans. The most important feature of these differences is an asymmetry of the annual changes, manifesting itself in a faster decrease in the southern hemisphere ice cover extent in the summer, caused by the more unstable thermocline in the Southern Ocean. A consideration of the dynamics of the cryosphere in five sectors in high latitudes of the southern hemisphere and in eight regions of the Arctic based on data for 15 years revealed the existence of a substantial interannual variability (especially in the Antarctic) and asynchronous variations in different regions. For nine years of the period studied (1977–1987), there was no noticeable secular trend in the Arctic or the Antarctic of the variability of annual change of ice cover extent, but the global extent decreased by about 5 per cent, as a result of the change in the phase and shape of seasonal oscillations. The interpretation of the data on cryospheric dynamics, from the viewpoint of revealing the climatic variability, needs a very careful approach (see KC98). More recent work has pointed to significant reductions in the Arctic ice, while the Antarctic is less well studied, see in particular Wadhams (2017). As in other studies of remote areas of the surface of the Earth, remote sensing satellites are making very significant contributions to our knowledge of the cryosphere. As well as using passive microwave systems such as SMMR and visible and infrared systems to measure the extent of sea ice, there is Cryosat which is a European Space Agency (ESA) mission dedicated to measuring the thickness of polar sea ice and monitoring changes in the Greenland and Antarctic ice sheets. Cryosat-1 was launched in October 2005 and failed immediately. Cryosat-2 was launched successfully in April 2010 and has contributed significantly to studies of the cryosphere.

The effect of the cryosphere on climate is determined by high albedos of snow and ice cover, low heat conductivity, and large thermal inertia. In this connection, Barrie (KC98) and Barrie and Chorley (KC98) undertook a detailed overview of observational data on the variability of various components of the cryosphere. In particular, there is no clear-cut trend of snow cover extent on the northern hemisphere. The Arctic sea ice extent, having reached a minimum in the 1950s, had grown thereafter, but now is decreasing again. Though the extent of the summertime ice in the Antarctic between 1930 and 1970 has been decreasing, nevertheless it has been determined by other factors. The requirements of the data of observations of the two categories of the cryosphere components mentioned above are substantially different (Table 4.7).

In connection with the very important role of albedo feedback, of primary importance are studies of the factors determining the long-term variability of the extent and thickness of sea ice cover as a function of the sensitivity of the Arctic pack ice to variations of the climatic, oceanic, and hydrological regime. The latter determines the urgency of complex programmes of observations of the characteristics of the ice cover, ocean, and cryosphere, as well as of the development of climate models.

TABLE 4.7
Data requirements for observations of cryospheric characteristics

Characteristic	Objective	Accuracy desired / needed	Horizontal resolution (km)	Temporal resolution
Snow cover extent	T, M	3–5%	250	3 days
Water equivalent of snow cover	T	0.5–1 cm	250	1 week
Surface albedo	T	0.02–0.04	250	1 week
Ice cover boundaries	M	20–50 km	250	1 week
Ice cover concentration	T, M	3–5%	250	1 week
Sea ice drift	S	2–5km day^{-1}	500	3 weeks
Sea ice thickness	T, M	10–20 cm	250	1 month
Sea ice melting	S	?	250	1 week
Share of multi-year ice	M, S	0.1–0.1	250	1 week
Glacier topography	M	0.1–1.0m	2–250	1–10 yr
Glacier boundaries	M	1–5 km		1–10 yr
Glacier thickness	M	10–100m	250	1–10 yr
Glacier motion	S, M	(1–10)–100m	m/yr at a point	1 yr
Permafrost boundaries	M	10–50 km		1–10yr
Time-period of lake freezing	M	?	individual lakes	3 days

Source: (KC98, Table 2.10).
Note: M = monitoring; S = studying the processes; T = testing the model reliability.

An account of the interactive system atmosphere-ocean-cryosphere would be useful in pre-calculating the changes in the extent and thickness of ice cover which can be caused by increasing CO_2 concentration.

The heat balance of polar regions is one of the key aspects of the formation of climate (KC98). In this connection, Nakamura and Oort (KC98) performed calculations of four basic components of the heat balance of the atmosphere in the southern hemisphere and northern hemisphere polar regions from data of satellite and radio-sonde observations: (1) the rate of change of energy, $\Delta E/\Delta t = F_{RAD} + F_{WALL} + F_{SFC}$, (2) the radiation budget of the surface–atmosphere system, F_{RAD}, (3) poleward energy fluxes in the atmosphere at the surface level across the 70° latitudinal belt, F_{WALL}, and (4) upward energy flux in the atmosphere at the surface level F_{SFC}. The last component is calculated as a residual term, and the energy flux across the 70° S latitude circle was calculated using an atmospheric general circulation model, in view of fragmentary observational data. An analysis of the results showed that the role of heat exchange between the atmosphere and the ocean–cryosphere system is more substantial in the Arctic than in the Antarctic. However, in the case when polar caps reach the 70–60° latitudinal belt, various components of the heat balance in the Arctic and the Antarctic become closer in value. The north-polar cap is characterised by a strong annual change in F_{SFC}, with an amplitude of about 50–80 Wm^{-2}, whereas in the region of the south-polar cap, this amplitude constitutes only 20–30 Wm^{-2}. Apparently, this difference is determined by the lesser intensity of the formation processes of snow and ice cover on the south polar cap, but it needs to be followed up in more detail.

4.4.1 THE ARCTIC AND THE ANTARCTIC

The structure of the annual mean heat balance of the atmosphere in both polar regions is similar and determined by the F_{RAD}/F_{WALL} ratio, and the value of F_{SFC} is small. However, the annual change in the heat balance components is quite different. In the summertime north-polar cap zone, F_{RAD} is small and the meridional heat flux coming to this zone is largely balanced by heat losses on snow and ice melting, as well as on the ocean heating. The meridional input of heat to the south polar cap is half as much and is, apparently, equally spent on heat loss due to F_{RAD} and on snow and ice melting. In the wintertime at the north polar cap, the heat loss on emission (F_{RAD}) is very large and two thirds of it are compensated by the meridional heat flux, and one third by the heat flux through the surface. On the south polar cap, the heat loss in the winter is almost equalled by the meridional heat flux, whereas the contribution of heat exchange through the surface is negligible. A major reason for the contrast between the polar regions is the presence in the southern hemisphere of a huge ice cap, whose height reaches 3 km, as well as such a peculiar feature of the south-polar atmosphere dynamics as slope (katabatic) winds.

Bearing in mind the multi-factor dynamics of the ice cover, it can hardly be expected that monitoring the variability of its characteristics during the decades to come will make it possible to detect the CO_2 signal. Of more importance is satellite monitoring of the dynamics of freezing and ice-breaking of lakes, especially in North America and Eurasia. The corresponding data may become the most prominent cryosphere CO_2 signal.

An analysis is needed of possible regional and seasonal responses of the extent, thickness, and lifetime of snow cover to increasing CO_2 concentration. A decrease of the extent and lifetime of the northern hemisphere snow cover may become the second most important manifestation of the CO_2 signal in the cryosphere after the information on dynamics of lake ice in the decades to come. The problem of the west-Antarctic ice sheet still remains unsolved, which necessitates further theoretical studies and extended observational programmes. A more adequate observational data base is needed to assess the mass balance and changes in the volume of the two largest ice sheets as well as a regular monitoring of glacier dynamics, bearing in mind calculations of the possible rise of world sea level due to climate warming. Of great interest is the monitoring of permafrost.

Duan et al. (2018) using the National Center for Atmospheric Research Community Earth System Model investigated the contribution of sea ice and land snow to the climate sensitivity in response to increased atmospheric CO_2 content. Their main conclusion was that the existence of sea ice and land snow substantially amplifies the global temperature response to increased CO_2 with sea ice having a stronger effect than land snow. Under higher CO_2 levels, the effect of sea ice diminishes more rapidly than does the effect of land snow (Figure 4.9).

However, the validation of such models requires reliable monitoring data sets of the parameters and factors that are involved in the climate variability. In this context, Table 4.8 identifies the regions

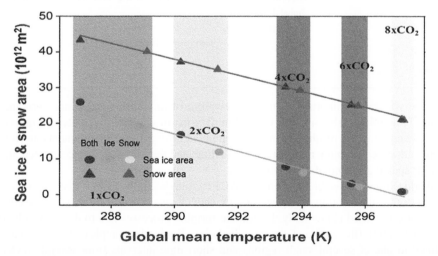

FIGURE 4.9 Model-simulated sea ice area and land snow area as a function of global mean surface temperature for "None," "Ice," "Snow," and "Both" simulations at various CO_2 levels. All results are calculated from the last 60-year simulations of 100-year slab-ocean simulations. The sea-ice and land-snow areas are calculated by multiplying the sea-ice and land-snow fraction in each grid cell with areas of those grid cells and then integrating over the globe. (Modified from Duan et al. 2018).

TABLE 4.8
Key regions with respect to monitoring the parameters and factors of climate

Climate system components and climatic parameters		Key regions, first-priority processes, and parameters
Atmosphere	1	Regions of the most frequent formation of blocking cyclones: 160–120° W; 40–70° W; East Siberia – the Chuccha Sea along 60° N
	2	Zones of jet streams
	3	Regions of teleconnections : western Pacific Ocean, western Atlantic Ocean
	4	Location of ridges and troughs of super-long waves : Greenland, Taimyr, Alaska, Aleutian Islands
	5	Subtropical high-pressure bands : eastern and western Pacific and Atlantic Oceans
	6	The Tibet maximum : from middle Asia to east Asia in summer
	7	The wind field in the tropics, including flows across the equator over Africa, the Atlantic and Pacific Oceans
	8	The South Pacific maximum, the Australian–Indonesian minimum
	9	Studies on atmospheric composition at background stations (CO_2, etc.); lidar observations at individual locations
Ocean	1	The Equatorial Eastern Pacific (El Niño): 180–90° W; 0–10° S
	2	SST along 137° E in January and July, especially in the 0–7° N band
	3	The Somali current in the Indian Ocean
	4	The Kuroshio and Oyashio currents in the north-western Pacific Ocean
	5	The central Pacific in the northern hemisphere
Cryosphere	1	The Tibet plateau
	2	Siberia
	3	North America
	4	Ice cover in the Chuccha Sea
	5	Ice cover in the Arctic Ocean
Soil surface		Soil moisture in low and middle latitudes, especially in semi-arid regions
Precipitation	1	The equatorial–tropical belt, especially from the Indian Ocean to the western Pacific Ocean
	2	The semi-desert regions : Sahel, South Africa, Australia, Mexico, northern China
Radiation	1	Global distribution
	2	Deforestated regions
	3	Regions affected by volcanic aerosols

Source: (KC98, Table 2.12).

of the globe which are important from the viewpoint of monitoring the parameters and factors of climate. In addition, Table 4.9 gives a preliminary list of climatic parameters, requirements of their spatial and temporal resolution, and measurement errors. These requirements are formulated according to the needs of numerical climate modelling, and they are therefore correlated, in particular, with the spatial grid step of climate models.

The World Climate Research Programme (WCRP) Joint Scientific Committee recommends three different classes of observational programmes, to which three classes of data on climate correspond. These data must be used (1) to study the climate-forming processes and to develop techniques for their parametrisation (these data of the so-called class P must be complex, detailed, but they can be confined to one or several small regions and short time intervals from several weeks to several years), (2) to test the adequacy of theoretical climate models (these data of the so-called class A require global-scale observations during long time periods and must contain information about numerous characteristics of the climate system) and (3) to perform long-term monitoring of slowly varying components of the climate system (the so-called class M), which is important for assessing

TABLE 4.9

Requirements for observations of climate system parameters

Type of observations	Climate system parameter	Class of data	Spatial resolution (km)	Temporal resolution (days)	Error
Basic meteorological parameters according to GARP requirements	Temperature	P, A, M	500 (100–200 hPa)	t.b.s.	1°C
	Wind speed	P, A, M	500 (100–200 hPa)	t.b.s.	1–3 ms^{-1}
	Relative humidity	P, A, M	500 (200–300 hPa)	t.b.s.	7–30%
	Surface pressure	P, A, M	500	t.b.s.	1–3 hPa
	SST	P, A, M	500	t.b.s.	1°C
Radiation budget and its components, cloudiness	Solar constant	M	–	1 month	1–5 Wm^2
	Extra-atmospheric	P, M	–	P: 1	P: 10%
	UV solar radiation			M: t.b.s.	M: t.b.s.
	Earth's radiation budget	A, M	500	15–30	2–5 W m^{-2}
	Cloudiness	A	500	5–15	5% total
				(vertically)	1°C–(top bound.temp)
	Global radiation	P, A	t.b.s.	5	1–3%
	Net longwave radiation	P, A	t.b.s.	5	1–3%
	Surface albedo	A	500	5–15	0.01–0.03
Oceanic parameters	SST	A, M	500	A: 5–10	0.5–1.5 °C
				M: 30	
	Upper-layer heat content (200 m)	A	200	5–10	1–3 kcal cm^{-2}
	Wind stress	A	200	5–10	0.1–0.4dyn cm^{-2}
	Ocean surface level (to determine the speed of currents)	A	200	5–10	2–10 cm (dynamic topography)
	Surface currents	A, M	Only critical regions	30	2–10 cm s^{-1}
	Deep-water circulation	A, M	1000	5yr	0.1–0.5 cm s^{-1}
Precipitation and hydrology	Precipitation over the ocean	A	500	5–10	4 levels (1 mm/day)
	River run-off	A	Major river basins	15–30	t.b.s
	Soil moisture	A	500	15	10% field water stress, 2 levels

(Continued)

TABLE 4.9 (*Continued*)
Requirements for observations of climate system parameters

Type of observations	Climate system parameter	Class of data	Spatial resolution (km)	Temporal resolution (days)	Error
Cryosphere	Snow cover extent	A, M	50–100	5–15	Presence/absence
	Sea ice extent	A, M	50–100	5–15	Presence/absence
	Ice thickness	A, M	200	15–30	10–20%
	Sea ice melting	P	50	5	Yes/no
	Sea ice drifting	P	400	1	5 km
	Thickness of glaciers	M	200	1 yr	0.1–1.0m
	Deformation of glaciers	M	200	1 yr	1–10 m
	Variations in the glaciers' boundaries	M	1–5	1 month	1–5 km
Atmospheric composition	CO_2	A, M	2 background and 10 regional stations	15 weeks	±0.1 ppm
	Vertical profile of ozone concentration	P, A	10 stations over the globe	1 week	± 1 ppm
	Total ozone content	M	WMO network	1	1–5%
	Global distribution of ozone	500 km; 2 km (vertically)	1	±0.5 ppm	±0.5 ppm
	Tropospheric aerosols	WMO background stations	–	–	
	Atmospheric turbidity	WMO background stations	1 week	1%	
	Stratospheric aerosols	2–4 background stations	–	–	

Source: (KC98, Table 2.13).
Note: t.b.s. = to be specified.

the present global climate, understanding natural factors of the formation and variation of climate, as well as studying the possibilities of climate forecasting.

4.4.2 Glaciers

Huss and Hock (2015) presented a new model for calculating the 21st century mass changes of all glaciers on Earth outside the ice sheets. Simulations are driven with monthly near-surface air temperature and precipitation from 14 Global Circulation Models forced by various emission scenarios, the so-called Representative Concentration Pathways: RCP2.6, RCP4.5, and RCP8.5 (IPCC, 2013). Depending on the scenario, the model predicts a global glacier volume loss of 25–48% between 2010 and 2100. For calculating glacier contribution to sea-level rise, we account for ice located below sea-level presently displacing ocean water. This effect reduces the glacier contribution by 11–14%, so that our model predicts a sea-level equivalent (multi-model mean ±1 standard deviation) of 79±24 mm (RCP2.6), 108±28 mm (RCP4.5), and 157±31 mm (RCP8.5). Mass losses by frontal ablation account for 10% of total ablation globally, and up to ~30% regionally. Regional equilibrium line altitudes (ELAs) are projected to rise by ~100–800 m until 2100, but the effect on ice wastage depends on initial glacier hypsometries.

In particular, regionally averaged glacier ELAs are projected to rise by ~100 to 500 m in the high and mid-latitudes and by ~100 to 800 m in low-latitude regions between 2010 and 2100, depending on the emission scenario (Figure 4.10). While the ELA change shows a nearly linear rise throughout the projection period for RCP8.5, the rate of ELA rise decreases substantially beyond the middle of the century for RCP4.5 and levels off/reverses for RCP2.6. At the same time, accumulation area ratios (AARs) show a tendency to increase for the latter two scenarios, indicating that many glaciers approach a new equilibrium during that period.

Concerning the observed and projected changes in the cryosphere, Bolch et al. (2019) investigated the cryosphere dynamics in the extended Hindu Kush Himalaya (HKH) region. They finally reached the following conclusions:

- There is high confidence that snow-covered areas and snow volumes will decrease in most regions over the coming decades in response to increased temperatures, and that snowline elevations will rise. The greatest changes in snow accumulations will be observed in regions

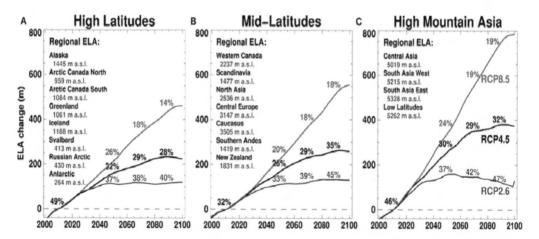

FIGURE 4.10 Changes in regional equilibrium line altitudes (ELAs) relative to 2010 for three emission scenarios. Accumulation area ratios (in %) are given as 20-year averages around 2010, 2050, 2070, and 2090. All values after 2010 are multi-model means. ELAs are averaged (weighted by glacier area) over all glaciers in the regions listed in each panel. Average ELAs are given for each region for the year 2010. (Source: Huss and Hock 2015).

with higher mean annual temperatures. Projected changes in snow volumes and snowline elevations (+400 to +900 m) will affect seasonal water storage and mountain streamflows.

- Glaciers have thinned, retreated, and lost mass since the 1970s, except for parts of the Karakoram, eastern Pamir, and western Kunlun. Trends of increased mass loss are projected to continue in most regions, with possibly large consequences for the timing and magnitude of glacier melt runoff and glacier lake expansion. Glacier volumes are projected to decline by up to 90% through the 21st century in response to decreased snowfall, increased snow-line elevations, and longer melt seasons. Lower emission pathways should, however, reduce the total volume loss (Figure 4.11). The available recent estimates of glacier area for the entire extended HKH are given in Table 4.10 (Bolch et al., 2019).
- There is high confidence that permafrost will continue to thaw and the active layer (seasonally thawed upper soil layer) thickness will increase. Projected permafrost degradation will desta-bilise some high mountain slopes and peaks, cause local changes in hydrology, and threaten the transportation infrastructure.

Based on their analysis results, Bolch et al. (2019) suggested the following policy actions:

- To reduce and slow cryospheric change, international agreements must mitigate cli-mate change through emission reductions. Lower emission pathways will reduce over-all cryospheric change and reduce secondary impacts on water resources from mountain headwaters.
- To better monitor and model cryospheric change and to assess spatial patterns and trends, researchers urgently need expanded observation networks and data-sharing agreements across the extended HKH region. This should include in situ and detailed remote sensing observa-tions on selected glaciers, rapid access to high-resolution satellite imagery, improved and expanded snow depth and snow water equivalent measurements, and ground temperatures and active layer thickness measurements in different regions, aspects, and elevations.
- Improved understanding of cryospheric change and its drivers will help reduce the risk of high-mountain hazards. Glacier lake outburst floods (GLOFs), mass movements (rockfalls, avalanches, debris flows), and glacier collapses present significant risks to mountain resi-dents. This risk can be minimised with improved observations and models of cryospheric processes.

4.4.3 OBSERVATIONS OF THE CRYOSPHERE

To solve unsolved climate problems, first of all, it is necessary to expand the programme of observa-tions in high latitudes, especially long-term observations in the key regions and special problem-orientated complex programmes (the atmosphere-ocean interaction at the polar ice edge, monitoring the properties of sea water under the ice cover and the ocean level, etc.).

For this purpose, much effort is needed to improve the observational means available and to develop new ones (especially radar altimetry and acoustic tomography). The study of the climatic effect of the cryospheric properties (sea ice) and processes taking place there is one of the WCRP's first priorities. The most important problem is to develop the coupled climate models that would take into account the processes in the cryosphere, since: (1) the strong annual change and interan-nual variability of the ice cover are followed by respective variations in surface albedo as well as heat-exchange and water-exchange between the ocean and the atmosphere, (2) the ice cover dynam-ics intensifies the variability of the temperature regime, which is the result of a substantial positive albedo feedback in the global climate system, and (3) the ice cover is an important indicator of climatic variability (in particular, that of anthropogenic origin).

It should be kept in mind that glacier surfaces are strongly irradiated due to an absence of vegeta-tive structure and snow's high photon fluency, a measure of light from all directions due to scattering

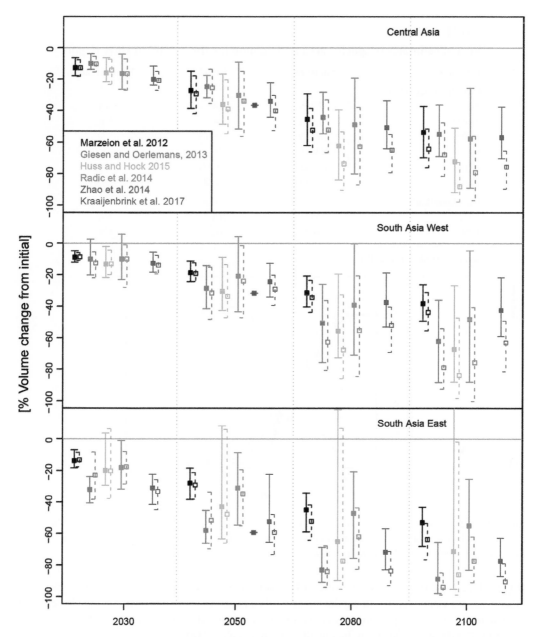

FIGURE 4.11 Regional glacier volume change projections for regions Central Asia, South Asia West, and South Asia East. Error bars show mean and range of multiple climate simulations for 2030, 2050, 2080, and 2100. Solid lines show results from RCP4.5 scenarios and dashed lines from RCP8.5 scenarios (Source Bolch et al., 2019).

in the snowpack, which is several times greater than photon irradiance. The spectral albedo of snow depends on snow grain size and wavelength of light (Figure 4.12, dashed and dotted lines).

Thus, the glacier surface is often oversaturated with photosynthetically active radiation and underprotected from UV. At the large snow grain size (~1 mm) of summer glacier surfaces occupied by snow-algae, >95% of purple through green light (350–500 nm) is reflected by snow. Snow spectral albedo drops with wavelength as a steepening curve (Figure 4.12) to <30% by the near infrared (1000 nm). In the visible spectrum, absorption by snow climbs steadily with increasing wavelength

TABLE 4.10
Recent estimates of glacier area and volume for the extended HKH region

Himalaya	Karakoram	Pamir	Tibetan-Plateau
Glacierised area (km²)			
21,973	21,205	NA	NA
22,829	17,946	NA	NA
19,991	18,563	10,403	27,622
NA	NA	NA	31,573
26,688	19,962	13,071	39,822
20,070	21,475	10,681	28,912
Ice volume (km³)			
1,212	1,683	NA	NA
1,297	1,869	NA	NA

Source: (Bolch et al. 2019).

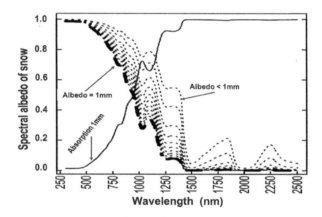

FIGURE 4.12   Snow grains of all sizes absorb little radiant energy at wavelengths below 500 nm. Theoretical spectral albedo, for clean snow by snow grain size and its complement spectral absorption, for 1 mm grain size at a given wavelength. (Modified from: Dial et al., 2018).

(Figure 4.12, solid line). Snow impurities, like black carbon glacier algae and other organics reduce albedo, absorb visible light, and conduct the absorbed radiative energy as heat into the glacier, melting ice, and snow. An impact of melt is that the presence of liquid-water enhances the rate of grain growth very efficiently, further reducing albedo (Dial et al., 2018).

The following four directions of research are of primary importance: (1) the establishment of more reliable techniques to parametrise the processes in the cryosphere (changes in albedo, heat-exchange and moisture-exchange, consideration of the contribution of polynyas and leads; cloud–radiation interaction, etc.), (2) study of interaction in the ocean–ice–atmosphere system at the ice cover edge, and the adaptation of climate theory to take account of this interaction, (3) analysis of the sensitivity of climate models to realistic changes in the extent and structure of the ice cover, and (4) continuous monitoring of the ice cover characteristics from satellites, and especially at the ice cover edge. Apparently, of key importance is an account of sea ice dynamics in the Antarctic, characterised by the strongest seasonal variability.

It should be kept in mind that large-scale singular events are components of the global Earth system that are thought to hold the risk of reaching critical tipping points under climate change, and that can result in or be associated with major shifts in the climate system. Among these components are:

- the cryosphere: West Antarctic ice sheet, Greenland ice sheet
- the thermohaline circulation: slowdown of the Atlantic Meridional Overturning Circulation
- the El Niño–Southern Oscillation as a global mode of climate variability
- role of the Southern Ocean in the global carbon cycle

Several studies assessed that the risks associated with these events become moderate between 0.6°C and 1.6°C above pre-industrial levels, based on early warning signs, and that risk was expected to become high between 1.6°C and 4.6°C based on the potential for commitment to large irreversible sea level rise from the melting of land-based ice sheets (low to medium confidence). The increase in risk between 1.6°C and 2.6°C above pre-industrial levels was assessed to be disproportionately large.

4.4.4 THERMAL RADIATION AND THE SNOW COVER: EXPLORATION TOOLS

Remote sensing surveys for snow cover are of considerable interest in several theoretical and applied areas, like:

- monitoring and modelling of the regional and global hydrologic processes;
- assessment of temporally stored snow volumes for warning about snow slides and floods as well as for agriculture;
- modelling and predicting regional and global climate change based on the database of snow-covered areas, snow depth (SD), snow water equivalent re), snow albedo, snow-soil irradiance, and other snow properties.

Knowing the possible changes in seasonal snow cover, ice fields, and the permafrost as characteristic components of the Earth's cryosphere allows for future discussions to better understand the global climate response to changing sensitive cryospheric parameters. For example, snow cover plays a special role in the ecological processes of the northern latitudes, including the Arctic Basin, through the influence on surface energy and water balances, as well as thermal regimes and trace gas fluxes. Snow cover in many latitudes is characterised by a very fast change over time. Therefore, the forecast of the snow-melt runoff is important for many areas where there are floods depending mainly on the depth of the snow cover.

The correlation between the natural microwave irradiance of the snow cover and its parameters has been studied by many authors. However, the processing of methods for determining SWE based on microwave monitoring data raises a series of difficult tasks. In particular, centimetre and millimetre wavelengths are characterised by powerful volumetric scattering in dry snow. On the one hand, this leads to an increase in the reflection coefficient of a snow layer by increasing its thickness, which mainly allows the snow layer depth to be assessed based on its radiometric observations. On the other hand, it complicates the task of modelling the microwave irradiation of the snow layer. In this respect, the determination of the depth of the snow layer using microwave observations is usually performed by semi-empirical models based on the correlations of its transmission and reflection coefficients with its geophysical characteristics, such as the depth of the layer, the crystal size, density, etc. However, the complex layered structure of the snow layer complicates the use of these models to solve the inverse task that is to estimate the characteristics of the snow layer based on its radiometric data.

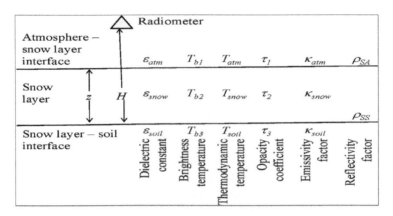

FIGURE 4.13 The profile of the layers across the atmosphere, snow, soil interface. After Varotsos et al. (2019c).

Related studies have shown that the microwave polarisation difference index (MPDI) of a snow layer can be an efficient tool for more accurately assessing the basic parameters of the snow layer. Knowledge of the polarisation indices of the snow layers leads to significant reduction in the uncertainties in the numerical estimates of the snow water equivalent, in particular. In this connection, the passive polarimetric microwave observations allow for a more accurate evaluation of the brightness temperature according to the snow structure.

The present discussion examines theoretical and empirical results of polarised measurements of the integral reflection coefficient of snow layers. Measurements were made at 6.9 and 18.7 GHz frequencies testing different types of snow, such as:

- freshly fallen snow (the granular structure is missing),
- small-grained snow (typical size of ice crystals is less than 1 mm),
- mid-grained snow (typical size of ice crystals is 1–2 mm),
- coarse-grained snow (typical size of ice crystals is 2–5 mm).

A theoretical model to study the microwave irradiation of the snow cover has been developed by Varotsos et al. (2019c). This is based on assessing the polarised microwave irradiation characteristics of the scattering layer-rough surface system based on the brightness temperature measurements. The simplified model can be synthesised using a two-level representation of the snow layer as shown in Figure 4.13.

5 Energy, the driver of the climate system

5.1 ENERGY

5.1.1 WHY DISCUSS ENERGY?

One might ask why we have chosen to devote a whole chapter to energy. The reason is twofold. First, it is energy which drives the whole climate system, but more than that it is energy which drives life itself in all its forms. Without energy there would be no life and the Earth would be a dead planet. The second reason is that the underlying reason why we now have a certain rather pleasant lifestyle, in some countries at least, is the result of the supply of energy that was unleashed a few centuries ago and which made the Industrial Revolution possible and which sustains our present lifestyle. It can be argued, e.g. by Lewis and Maslin (2018), that the Industrial Revolution was the Second Energy revolution and that the invention of agriculture was the first energy revolution. They argue that a typical resting person needs an energy input (from food) of about 120 W per day, that our hunter gatherer ancestors needed about 300 W and that a pre-industrial farmer ran at about 2,000 W by "appropriating more of the sun's energy that gets converted to plant and animal mass." What happened at the Industrial Revolution and continued afterwards was that we gained access to "buried sunlight" buried deep into the ground which had got converted into coal or oil millions of years ago and which we are consuming enormously faster than these resources were laid down.

5.1.2 RENEWABLE ENERGY

The Industrial Revolution came about as a result of various scientific, technological and social developments. And, of course, it was driven by energy. Some very big factories, spinning and weaving mills, were established beside large rivers and driven by water power. But the success of the Industrial Revolution and the powering of modern industry has been based, to a large extent, on the consumption of non-renewable resources. The definition of non-renewable is, of course, open to discussion but it is usually considered to include coal, oil, and natural gas. Historically the most important of these was coal, see Figure 5.1. This is a life-size memorial to underground coal miners in Prestonpans, a small town east of Edinburgh in Scotland which formerly was a mining town. The discovery and mining of coal by the Cistercian monks of Newbattle Abbey, the land owners of the time, in the early 13th century was arguably the first instance of coal mining in Britain. The mining of coal in Prestonpans began in the year 1210 and continued for centuries until the closure of the last deep mines in the 1980s. Coal mining was the major industry in Prestonpans but it was a dirty, back-breaking job with danger ever present. Over the years, countless miners, including women and children, were injured or killed down the mines. In the UK, no deep mining of coal survives and coal is now only obtained from domestic open cast mines or is imported from countries operating modern deep mines. Strictly speaking coal is renewable, in the sense that it is formed from decaying vegetation and the vegetation is ultimately powered by the Sun. It is, however, a question of timescale and in human terms coal is non-renewable.

We have become so seduced by our lifestyle that we face the problem of how to maintain this lifestyle without both using up these irreplaceable resources and polluting the planet in all sorts of various ways. In other words, this leads us to the question of sustainability, which to some extent is a question of sustainability of energy supplies, which we have already addressed to some extent

FIGURE 5.1   A life-size memorial to the generations of miners in Prestonpans, a former mining town in Scotland. (Photograph A.P. Cracknell).

in Chapter 1, along with other aspects of sustainability. In the long term, it would be nice to phase out coal, gas, and oil and replace them by renewables, wind, solar, and hydro power. This could be achieved by using our supplies of energy more efficiently, e.g. by better insulating our houses, by travelling less by cars, trains, and planes and investing in the supplies of renewable energy. However, the conventional Green Agenda has been less than honest in addressing the true costs of these renewable energy sources and we shall discuss this in Sections 5.3 and 5.4.

By renewable energy, in modern day terms, we mean sources of energy such as water power, which historically was turned into mechanical energy and nowadays is usually turned into electricity, or solar energy which can be turned directly into heat or into electricity in photovoltaic (PV) cells or is harnessed to produce vegetation or to dry perishable foodstuffs to preserve them. Wood, or biomass in general, is the product of photosynthesis and is renewable with its ultimate source of energy being from the Sun. However, there is a timescale involved and the land used to produce energy from biomass is in competition with land used for producing food.

One undesirable side effect of burning coal is the atmospheric pollution produced and consequent damage to the health of the citizenry. Another is the production of CO_2 and its emission to the atmosphere and the consequent human-induced global warming. A consequence of these damaging effects has been the rise of the green energy movement involving seeking to power our civilisation by renewable resources of energy rather than by consuming fossil fuels, coal, oil, and natural gas. To a large extent, the "Green" movement that seeks to replace fossil fuels by renewable energy resources has had a clear run of the political scene. One key book in countering this movement is "Green Tyranny" by Rupert Darwall (2019). The subtitle of Darwall's book is "Exposing the Totalitarian Roots of the Climate Industrial Complex" and, in the words of one reviewer, seemed "designed to appeal to readers who are already sceptical about current climate change and environmentalist policies". In one sense, this is a pity because in terms of the current standoff between alarmists and sceptics, see Section 1.3.7., it labels the author as a sceptic so that by and large his points have not been treated objectively. The book by Darwall (2017) should be read by anyone who wants to know how environmentalism could become so powerful that, in some countries, it seems like a new state religion. The author intended this book to complement his earlier work "The Age of Global Warming" (Darwall 2013) which was critical of policies and initiatives aimed at fighting climate change.

In "Green Tyranny," Darwall (2017) wanted to focus on continental Europe in general, and Sweden and Germany, in particular. In the preface, we learn how Darwall sees Germany: "German culture harbours an irrational, nihilistic reaction against industrialization, evident before and during the Nazi era. It disappeared after Hitler's defeat and only bubbled up again in the terrorism and anti-nuclear protests of the 1970s and the formation of the Green Party in 1980." By choosing Sweden as an example, he picked the ideal showcase of a Western country where the government significantly managed to shape public opinion about environmentalism over the decades. But even more inter-esting is the role of Sweden's politicians especially during the 1960s and 1970s, who were critical in setting up various UN organisations that lead, among others, to the United Nations Framework Convention on Climate – Change (UNFCCC), which we shall discuss in Chapter 7.

In the 1970s, the Swedish Social Democrats used global warming to get political support for building a string of nuclear power stations. It was the second phase of their war on coal, which began with the acid rain scare and the first big UN environment conference in Stockholm in 1969. Acid rain swept all before it. America held out for as long as Ronald Reagan was in the White House but capitulated under his successor. Like global warming, acid rain had the vocal support of the scientific establishment, but the consensus science collapsed just as Congress was passing acid rain cap-and-trade legislation. Rather than tell legislators and the nation the truth, the Environmental Protection Agency (EPA) attacked a leading scientist and suppressed the federal report showing that the scientific case for action on curbing power station emissions was baseless.

Ostensibly neutral in the Cold War, it is claimed that Sweden had a secret military alliance with Washington. A hero of the international left, Sweden's Olaf Palme used environmentalism to maintain a precarious balance between East and West. Thus, Stockholm was the conduit for the KGB-inspired nuclear winter scare, see Section 8.3. The bait was taken by Carl Sagan and leading scientists, who tried to undermine Ronald Reagan's nuclear strategy and acted as propaganda tools to end the Cold War on Moscow's terms.

Darwall explains how a powerful Green/Left network managed to occupy key political positions in Europe and the U.S. and to establish (or gain control of) institutions that gave them unquestioned authority over the subject, often supported by financing from very wealthy donors. Nuclear energy was to have been the solution to human-induced global warming. However instead, most of all thanks to Germany, a number of other countries are following Germany's lead in embracing wind and solar. German obsession with renewable energy originates deep within its culture. Few know today that the Nazis were the first political party to champion wind power, Hitler calling wind the energy of the future. Post-1945 West Germany appeared normal, but antinuclear protests in the 1970s led to the fusion of extreme left and right and the birth of the Greens in 1980. Their rise changed Germany, then Europe, and now the world. Radical environmentalism became mainstream. It demands more than the rejection of the abundant hydrocarbon energy that fuels American great-ness; Darwall argues that it is totalitarian, i.e. it requires the suppression of dissent.

He also explains how the onslaught on freedom happens openly (if unnoticed by the media and general public) by highlighting a crisis of global proportions – such as human-induced climate change – which requires solutions that "normal democracies" are not able to provide. They must be settled by a council of experts, which acts outside the democratic process, see Chapter 7.

5.2   A DIGRESSION: WHAT IS ENERGY?

A reader with a basic high school-physics background can omit this section.

We turn, once more, to the (Concise Oxford) dictionary which gives the second and third mean-ings of energy:

"2. power derived from physical or chemical resources to provide light and heat or to work machines.
3. *physics* The property of matter and radiation which is manifest as a capacity to perform work."

You cannot see energy, though you can see a source of energy by looking at a lump of coal.

Suppose you are a 70 kg person who lives on the tenth floor of a multistorey building with a height of 3 m between the floors. If you travel up from the ground to your apartment, then you gain an amount of potential energy against or in defiance of gravity amounting to

$$\text{mass} \times \text{g} \times \text{height} \tag{5.1}$$

$$\text{i.e. } 70 \times 9.81 \times 30 \, \text{J} \, (\text{J} = \text{joules}) = 20,601 \, \text{J or } 20.6 \, \text{kJ} \tag{5.2}$$

Next time you go down to ground level, this energy will be lost and it is difficult to imagine how this energy could be used profitably.

But now imagine a reservoir of water near the top of a mountain where the reservoir has been filled by precipitation, i.e. by rain or by snow which has melted. The potential energy which this water possesses, relative to some lower level, can be used to drive a mill for grinding grain for human or animal consumption, or for any number of other purposes in past times but is now most likely to be used for the generation of electricity. In this case, the ultimate source of the energy was from the sun which evaporated the water, probably from the sea into the clouds which led to the precipitation. In addition to potential energy, there is kinetic energy, the energy a body possesses by virtue of its motion, $\frac{1}{2}mv^2$, where m is the mass of the body and v is its velocity. The kinetic energy of the air in the wind can be extracted by a windmill to grind grain or pump water or it can be converted into electricity by a windmill or aerogenerator. There are other forms of energy: heat, light, electricity, and various wavelengths of electromagnetic radiation.

The sun is ultimately the major source of the energy that drives the climate and which we consume to support our way of life. This may be directly as heat or electricity derived directly from the sun, or as biomass produced through photosynthesis or by evaporation of water as part of the hydrological cycle. It may come from "buried sunlight" in the form of coal, gas, or oil that was laid down as biomass from photosynthesis millions of years ago. A rather minor contribution to our energy use comes in the form of heat from geothermal energy which is derived from the hot molten core of the Earth. And in recent times, we have started to exploit atomic energy, where the energy comes from the conversion of matter into energy according to Einstein's equation, $E = mc^2$ where E is the energy (in joules), m is the mass in kg, and c is the velocity of light. There is nothing unique about this relation in connection with atomic energy. It also applies to the combustion processes involved in burning coal, gas, or oil or in using the fuel in internal combustion or jet engines. If we add up the mass of the fuel and the oxygen before the combustion process and collect all the combustion products, we would find that there is a certain amount of missing mass, but it is very small. Basically, the c^2 factor is very large $(3 \times 10^8)^2$, or 9×10^{16}, where we are working in mks (metre kilogramme second) units. Of course, there is natural radioactive decay which also releases energy, but the amount is negligible and we do not generally put it to any useful purpose. Finally, there is energy that can be extracted from the tides, i.e. the rise and fall of the water level at the coast, or from the wave motion on the surface of the sea. If we consider the rise and fall of the water level at the coast, this comes from the gravitational force which drives the tides and this comes from the motions of the Earth and the Moon and from the rotation of the Earth, i.e. from their kinetic energy, so that all these motions will be very slightly slowed down as a result. Thus, the Earth's rotation is slowing down slightly with time so that a day is now longer than it was in the past. Atomic clocks show that a modern-day is longer by about 1.7 milliseconds than a century ago. If we take the waves as being driven by the wind, then the energy of the wind ultimately comes from the sun.

At various stages in this book, we have mentioned or will have occasion to mention that there are those, e.g. Frank Lawaczeck and Martin Ryle for example, who have argued that the buried sunlight, i.e. coal, oil, and natural gas, is too precious to burn or use in internal combustion engines or jet engines and it should be kept in the ground as feedstock for the chemical industry for future generations.

5.3 THE GREEN AGENDA AND ENERGY, ENERGIEWENDE

The effect of the Kyoto Protocol on the combustion of fossil fuels has led to an expansion of the generation of electricity from renewable non-fossil fuel resources, and in particular from wind and solar energy. The question arises as to what is the best proportion of wind and solar energy to use in an electricity supply grid network? We seek to show in this section that there is no simple answer to this question.

The Green Agenda, or sustainable development, as we have seen in Section 1.2 arose as a reaction to the evils of the industrial revolution, exploitative and sometimes dangerous conditions in the factories (the satanic mills of William Blake), and the appalling slum housing conditions in the big cities. It is sometimes portrayed as an anti-industrial movement, although climate activists and green politicians nevertheless expect to be able to enjoy the benefits brought about as a result of industrialisation and the use of fossil fuels. With the modern development of robotics and health and safety legislation, it should be possible to enjoy the benefits of industrialisation and avoid the earlier evils associated with it. While a detailed study of the modern green movement is outside the scope of this book, we do nevertheless include some discussion of the consequences of the Green Movement in relation to the provision of the energy resources that are needed to sustain our way of life.

We have already discussed the role of Sweden in the context of the acid rain scare in Section 1.2.2, and in Section 6.8, we shall discuss the role played by Sweden in the run up to the establishment of the IPCC. The politics of global warming originated in Sweden as a tool to promote nuclear power as a non-fossil fuel resource. While Sweden played a key role in the setting up of the IPCC and establishing its terms of reference, Germany played a key role in the politicising of the IPCC's Assessments after 1988. We have discussed the origins of the Green Movement in Germany in Section 1.2.1. In both Sweden and Germany, the question of nuclear power played an important role in the Green Movement's early activities.

In Sweden, the Green movement started as a campaign against coal and in favour of nuclear energy (Darwall 2017, Chapter 11). But after the Chernobyl nuclear accident of 1986, nuclear energy was rejected and thereafter nuclear energy went out of the window in Sweden. The green agenda then became a campaign for renewable energy, defined particularly in the form of solar power and wind power, though of course including hydropower, principally as sources of electricity on which our civilisation depends so heavily.

During West Germany's first three decades, environmentalism had mostly been a preserve of a tiny fringe on the Far Right (Darwall 2017, page 92). In recent years, it has emerged in the Energiewende (German for energy transition) programme in which nuclear power has had a chequered history. The Energiewende is the planned transition by Germany to a low carbon, environmentally sound, reliable, and affordable energy supply. The term *Energiewende* is regularly used in English language publications without being translated. We have included the term Energiewende in the title for his subsection not because we want to go into great detail in studying the history of Energiewende in Germany but because the issues raised by Energiewende are relevant far beyond Germany since many other countries have, in a mad rush, sought to follow Germany's lead in Energiewende. The idea behind Energiewende is to rely heavily on renewable energy (particularly wind power, photovoltaics (solar panels), and hydroelectricity), coupled with energy efficiency, and energy demand management.

The Energiewende legislation in Germany was passed in 2010 (Federal Ministry of Economics and Technology 2010). The purpose of the Energy Concept (Federal Ministry of Economics and Technology 2010) was stated as:

> Securing a reliable, economically viable and environmentally sound energy supply is one of the great challenges of the 21st century. A core element of this is the implementation of the pivotal political objectives for our future energy system: Germany is to become one of the most energy-efficient and greenest economies in the world while enjoying competitive energy prices and a high level of prosperity. At the same time a high level of energy security effective environmental and climate protection and the provision of an economically viable energy supply are necessary for Germany to remain a competitive

industrial base in the long term. We want to strengthen competition and market orientation on the energy markets, which will enable us to secure sustainable economic prosperity, jobs for the future innovation and the modernisation of our country. The challenges of sustainable energy provision derive in part from long-term global trends. The world's rising demand for energy will lead in the long term to a pronounced increase in energy prices. Our country's dependence on energy imports would also continue to increase. Energy consumption currently causes 80% of greenhouse gas emissions. For these reasons our present energy supply structures will have to be radically transformed in the medium to long term if we are to achieve energy security, value for money and the targets set by our climate protection policy. We will set the course so that the huge potential for innovation, growth and employment can be tapped as we revamp our energy system." The long-term strategy for future energy supply was to formulate guidelines "for an environmentally sound reliable and affordable energy supply and, for the first time," to map "a road to the age of renewable energy.

This is not the place to discuss the details of German politics or economics, but we do discuss Energiewende because it is often held up as an example for other countries to follow. To appreciate what is involved in Energiewende, it is necessary to appreciate one or two facts about Germany's energy supply, see Figure 5.2. Germany is the sixth largest consumer of energy in the world and has the largest national market of electricity in Europe. The country is the fifth-largest consumer of oil in the world. Oil consumption accounted for 34.3% of all energy use in 2018, and 23.7% of Germany's energy consumption came from gas. More than half of the country's energy is imported and the country largely imports its oil from Russia, Norway, and the United Kingdom and is also the world's largest importer of natural gas. Its gas imports come from the Netherlands, Norway, and Russia. Germany has a long tradition of using coal although mostly it is imported these days.

In Germany the plan, Energiewende, in 2010 was to reduce greenhouse gas (presumably referring to CO_2) by the percentages (relative to 1990) shown in Table 5.1. This would be accompanied by increasing the proportions of renewables shown in Tables 5.2 and 5.3 expressed as percentages of final energy production and electricity production, respectively. These figures were not claimed to be forecasts or targets but as rough route maps to signpost the way and it was assumed that additional investment would be required up to 2050 in order for the ambitious climate protection proposals to be achieved (Federal Ministry of Economics and Technology 2010, page 5). It was also planned to reduce energy consumption. "By 2020 primary energy consumption is to be 20% lower than in 2008 and 50% lower by 2050. This calls for an annual average gain in energy productivity of 2.1 % based on final energy consumption by around 10% by 2020 and 25% by 2050."

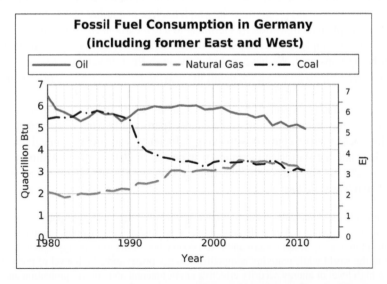

FIGURE 5.2   Fossil fuel consumption in Germany, including combined former East and West from 1980 to 2011. The use of coal declined significantly after reunification (1990). (Source *Energy Matters* based on EIA data).

TABLE 5.1
Proposed German greenhouse gas emissions cuts relative to 1990

Reduction	Scheduled date
40%	2020
55%	2030
70%	2040
≥80%–85%	2050

(Data from Federal Ministry of Economics and Technology 2010).

TABLE 5.2
Proposed German renewables* as a proportion of gross final energy production, relative to 1990

Renewables proportion	Scheduled date
18%	2020
30%	2030
45%	2040
60%	2050

*   wind, solar, hydro.
(Data from Federal Ministry of Economics and Technology 2010).

TABLE 5.3
Proposed German renewables* as a proportion of gross electricity production, relative to 1990

Renewables proportion	Scheduled date
35%	2020
50%	2030
65%	2040
80%	2050

*   wind, solar, hydro.
(Data from Federal Ministry of Economics and Technology 2010).

The Energy Concept document of (2010) goes on to discuss energy efficiency and a key component in improving this was considered to lie in renovation of buildings as a key part of a more general drive to increase the efficiency in the use of energy.

The role of nuclear power and fossil fuel power plants, which were major contributors to the electricity generating capacity in Germany, was considered in section C of this Energy Concept document. Regarding nuclear power as a bridging technology "A limited extension of the operating lives of existing nuclear power plants makes a key contribution to achieving the three energy policy goals of climate protection, economic efficiency, and supply security in Germany within a transitional period. It paves the way for the age of renewable energy. The operating lives of the 17 nuclear power plants in Germany would be extended by an average of 12 years. In the case of nuclear power plants commissioned up to and including 1980, there will be an extension of 8 years. For plants commissioned after 1980, there will be an extension of 14 years. In summary, the programme proposed the extension of the lives of the 17 operating nuclear power plants by an average of 12 years, i.e. by a total of 204 years altogether.

Energy drives our civilisation and we have already mentioned the book by David Mackay (2009) "Sustainable energy – without the hot air" in Section 1.3.7; this is outstanding for its quantitative approach to the cost per kWh of energy for each of the various forms of energy available to us and which is a factor to be taken into account when deciding (as individuals or as states) on the balance of the different sources of energy to use. That book is extremely valuable because the credibility of his science is incontrovertible. There are two other books by Mike Berners-Lee "How bad are bananas?" (2010) and "There is no Planet B" (2019) which we shall discuss in more detail and which put values in terms of energy (kWh) on many everyday items or activities. We have also referred to the two books by Rupert Darwall (2013, 2019), which do take one side in the global warming argument but which, nevertheless are very valuable in discussing the politics involved. Darwall explains how the green agenda evolved into concentration on "renewable energy" especially in the form of solar energy and wind energy and to the elimination of the use of fossil fuels and nuclear power as quickly as possible. The extreme case of this is the Energiewende which was adopted by Germany in 2010 and which some other countries have allowed themselves to emulate.

It was recognised by the Energiewende Programme that in order to achieve a high level of electrical security because of the intermittency of wind and solar power generation, sufficient balancing and reserve capacities would still need to be maintained. This would require sufficient investment in reserve and balancing capacities, in particular in more flexible coal and gas fired power stations. It was also acknowledged that the continuing development of renewables would require the ongoing development of grid infrastructure and hopefully including the connection of industrial development of renewables to the grid. The question of intermittency will be discussed in the next section.

Subsequent to the Energy Concept document of 2010, two things happened. The first was the Fukushima nuclear disaster in Japan in March 2011; this was the third major international nuclear accident, following the 1979 Three Mile Island partial nuclear meltdown in the United States and the 1986 Chernobyl disaster in the USSR. Following the Fukushima disaster, the German Government decided to shut down 8 of its 17 nuclear reactors more or less immediately and to close the rest by the end of 2022. This was a quite considerable reduction of the extension that was envisaged in 2010. In 2010, there were 17 stations with an average life expectancy of 12 years and by 2014 this had become a current 11 stations with a life expectancy of eight years (or 12 years from 2010). The efficacy of the nuclear programme to act as a bridging technology was thereby substantially reduced. The second thing was that the language changed somewhere between 2010 and 2014. Whereas the figures used in 2010 (see Tables 5.1, 5.2, and 5.3) were regarded as rough route maps to signpost the way, by 2014 they had become targets. A monitoring report on the state of the Energy Concept in 2014 presented the original figures from the Energy Concept of 2010 along with the actual figures for 2014 (see Table 5.4) as well as histograms showing the figures for the years from 2008 to 2014.

The Energy Concept of 2010 held out two attractive export opportunities for German industry, neither of which came to fruition. The first was carbon capture and storage (CCS) which involves extracting CO_2 from the atmosphere and somehow storing it, preferably in some solid form, below the ground. CCS is a fascinating issue and it is being actively explored, see for example the website of the Carbon Capture and Storage Association (CCSA) (www.ccsasociation.org accessed 28 March 2019) or there is an extensive article in Wikipedia as well as an IPCC special report on CCS (IPCC CCS 2005). However, in practical terms, nearly ten years later, the take up of CCS in the electricity generating industry is not impressive. The present state of play is typified by the UK Government's "Clean Growth – The UK Carbon Capture, Usage and Storage Deployment Pathway An Action Plan." carbon capture usage and storage deployment pathway (www.gov.ukgov.uk search "carbon capture" /beis, accessed 28 March 2019). This document sets out the next steps government and industry should take in partnership in order to achieve the government's ambition of having the option to deploy CCS at scale during the 2030s, subject to costs coming down sufficiently. The second opportunity was said to be in the manufacture and export of solar photovoltaic (PV) panels.

TABLE 5.4

Electrical generating capacity and output in Germany-2013 (preliminary data)

	Capacity		Annual output	
	GW	%	TWh	%
Solar PV	36.3	19.2	31.0	4.9
Wind	34.7	18.3	51.7	8.2
Hard coal	29.2	15.4	121.7	19.2
Natural gas	26.7	14.1	67.5	10.7
Lignite	23.1	12.2	160.9	25.4
Nuclear	12.1	6.4	97.3	15.4
Hydro	10.3	5.4	28.8	4.5
Other	17.0	9.03	74.3	11.7
Total	189.4	100.0	633.2	100.0

Source: (Federal Ministry for Economic Affairs and Energy, "Energy Data: Complete Edition" (last updated 21 October 2014), Table 22. http://www.bmwi.de/EN/Topics/Energy/Energy-data-and-forecasts/energy-data.html).

However, while there was massive investment in solar PV generating capacity in Germany, the export opportunities were seized by China.

By 2013, Germany had installed more solar capacity than any other nation, much of it imported from China. The irreversibility of Scheer's Energiewende was reflected in hard investment in Germany's electricity generating mix. Table 5.4 shows that by 2013, solar PV panels and wind turbines represented 37.5 % of total generating capacity but only provided 13.1 % of the output. This shortfall was compensated by increased output from nuclear, coal fired, and lignite (brown coal) fired power stations. In fact, windy, cloudy, Germany had more solar (19.2 % of its electrical generating capacity) than wind (18.3 p%), which was its single largest fraction of generating capacity (Table 5.4). But solar PV was even less efficient than wind. Although there was 4.6 % more solar PV capacity than wind, it generated less electricity. Overall, solar and wind accounted for three-eighths of installed generating capacity but generated only slightly more than one-eighth (13.1 %t) of Germany's electricity. Power stations burning lignite, a type of coal closer to peat than to hard coal, accounted for less than one-eighth of generating capacity but produced one-fourth of Germany's electricity. It meant that in 2013, green Germany derived over five times as much electricity from lignite, a fuel demonised by environmentalists, than it did from solar PV.

Like their forebears, Berlin's green revolutionaries knew what they wanted to destroy but had little idea how to make their Utopia work. This is probably true of all revolutionaries, though perhaps it should be modified to "had little *practical* idea how to make their Utopia work". The Energiewende was full of surprises and unintended consequences. In the plan, Germany was going to export wind turbines and solar PV panels around the globe. Instead the 2000 EEG renewable energy law spawned 100,000 profiteers from subsidies and a gigantic solar industry in China. Solar PV prices collapsed, and soaring solar take-up blew forecasts out of the water as the solar industry successfully lobbied against cutbacks to already excessively generous solar feed-in tariffs.

"Having too much wind and solar meant that on summer days when the sun shone and the wind blew, more electricity was being produced than anyone knew what to do with. Not that it greatly helped cut carbon dioxide emissions. The big fall in German power station emissions had happened earlier, when many lignite-burning power stations in East Germany were closed after reunification. Between 1990 and 1999, power station emissions fell by 80.5 million tonnes from 423.4 million tonnes, a fall of 19 %" (see Figure 5.1). "After 1999, the only two consecutive years of falling power station CO_2 emissions were in 2008 and 2009, when German industrial output slumped by more than 20 %. Second only to reunification, the subprime crisis and Lehman bankruptcy turned out to

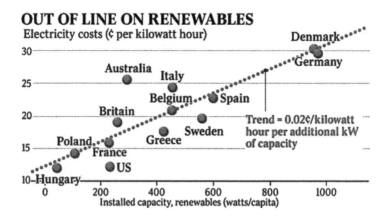

FIGURE 5.3 Electricity costs in various countries. (Source Watt's Up With That? "Germany's Energiewende program exposed as a catastrophic failure". (https://wattsupwiththat.com/2018/09/30/germanys-energiewende-program-exposed-as catastrophic-failure, accessed 8 March 2019).

be Germany's most effective decarbonisation policy, though scarcely intended as such. By 2012, CO_2 emissions from power stations were 17.2 million tonnes higher than in 1999 – an increase of 5 %. Over the same period, installed wind and solar PV capacity rose nearly 13-fold – from 5 GW in 1999 to 64.3 GW in 2012." (Darwall, 2017, pp. 143–4).

Darwall continued "'All of us under-estimated this legislation,' admitted Michaele Hustedt, a Green MP and cosponsor of the EEG law. The SPD (Sozialdemokratische Partei Deutschlands, in English: Social Democratic Party of Germany) energy minister Werner Müller, however, did not, warning that the subsidy would be far too generous. To head off the threat, Müller wanted to establish a facility in southern Europe to research solar energy. As he feared, the rush to wind and solar quickly pushed up electricity bills, see Figure 5.3. It will be noticed that the other country, apart from Germany, with very high electricity prices for customers was Denmark which had also invested heavily in wind turbines".

There has been a great deal written and published on the progress of Energiewende since 2014. This can be seen from the Wikipedia article on Energiewende which cites over 100 references. We just note three of them: one from *Energy Post* (https://energypost.eu/energiewende-enters-a-new-phase-how-is-it-performing/ accessed 26 March 2019), one from *Energy Matters* (https://euanmearns.com/germanys-energiewend-predicament/ accessed 26 March 2019), and one by Huberus Bardt from the Institut der Deutshen Wirtschaft (https://IW-Kurzbericht_2018_41_Energiewende, accessed 26 March 2019). The first of these presents some graphs showing the recorded performance (dark blue lines) and the target value (light blue lines) for 13 performance trackers value, for the first few years of the Energiewende programme. One example of these, the target for CO_2 emissions is shown in Figure 5.4, where the projected emissions have been exceeded. The second extends the first few years of data to predictions up to 2050, see Figure 5.5. The third of these, by Hubertus Bardt, shows the progress achieved by 2014 and by 2017 in achieving the objectives set out originally for Energiewende in 2010, see Table 5.5. For each of the seven quantities involved and for each year there is a rating or score. This score is determined by taking the actual value achieved and compared with the value that would have needed to be achieved on a linear progress; for instance, we see that in 2014 the investment in renewables was 20% above the running target and in 2017 this had risen to 34% above the running target. From Figs 5.4 and 5.5 and Table 5.1, we conclude that in recent years and in projections to 2050 Germany has failed to meet its target in terms of the reduction of CO_2 emissions. The investment in renewables is going as well as, or slightly better than scheduled. As to the other variables the targets are to a greater or lesser extent not being met. A more general critique of the rush to renewables is given in the Report by Mills (2019) who

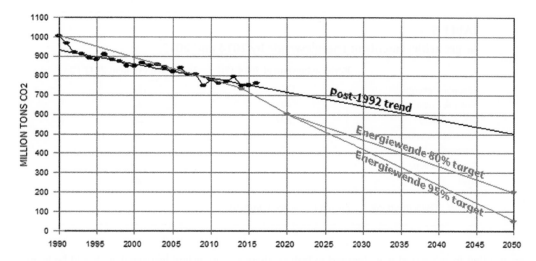

FIGURE 5.4 Annual CO_2 emissions, base year 1990, BP data. I begin the trend line in 1992 because the emissions decrease between 1990 and 1992 was largely a result of the post-reunification shutdown of CO_2-intensive plants in East Germany. The CO_2 data exclude land use changes but these appear not to be significant in Germany's case. (Source *Energy Matters*).

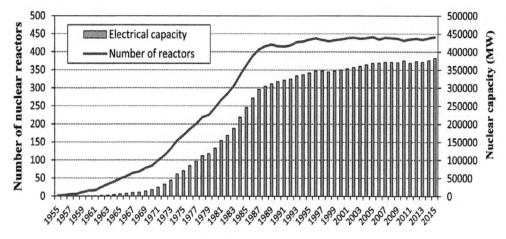

FIGURE 5.5 Temporal worldwide variation of operational nuclear reactors and of their net capacity (MW) between 1955 and 2015 (data processing from https://www.statista.com/).

commented that "The Energiewende therefore remains a fascinating large-scale energy experiment. As far as I'm concerned, the jury is still out on whether this will work or not. I will certainly continue to follow closely as Germany embarks on this next, much more complex, phase of wind/solar power expansion." There is also an article in the German magazine *Der Spiegel* of 4 May 2019 by Frank Dohmen, Alexander Jung, Stefan Schultz, and Gerald Traufetter entitled "*A botched job in Germany*" ("Murks in Germany") in which they wrote "The *Energiewende* – the biggest political project since reunification – threatens to fail." Michael Schellenberger, writing in *Technology News & Trends* on 14 January 2020 commented on the *Der Spiegel* article: "Many Germans will, like *Der Spiegel*, claim the renewables transition was merely 'botched,' but it wasn't. The transition to renewables was doomed because modern industrial people, no matter how Romantic they are, do not want to return to pre-modern life. The reason renewables can't power modern civilization is because they were never meant to. One interesting question is why anybody ever thought they could." Whether one takes the line that Energiewende was a good project that was bungled or a

TABLE 5.5

Target achievements of Energiewende by 2014 and 2017

	Target achievement 2014 (%)	Target achievement 2017 (%)
Renewable energy	120	134
Network expansion	42	41
CO_2 emissions	52	28
Power consumption	63	24
Economics	23	5
Competitiveness	−15	−53
Total	48	30

Source: (Hubertus Bardt, Institut der Deutschen Wirtshaft).

project that from its start was doomed to failure, there are important lessons that should be learned by any state, of which there are several, that are seeking to imitate it. The major lesson is the need to appreciate the implications of intermittency, which we shall address in the next section.

The term Energiewende appears to have originated in 1980 (Krause et al. 1980, Jacobs 2012). The energy transition envisaged was from nuclear power generation and the use of fossil fuels and their replacement by renewable energy sources, particularly wind energy, solar panels (photovoltaics), and hydroelectricity, echoing the Nazi programmes of the 1930s (see Section 1.2.1). It was claimed that economic growth was possible without increased energy consumption. Over the 30 years following 1980, the ideas were developed and the Energiewende legislation was passed in Germany in 2010. It included greenhouse gas reductions of 80–95% by 2050 (relative to 1990) and a renewable energy target of 60% by 2050. These targets were ambitious and the speed and scope of the Energiewende was also exceptional.

The targets for the installation of wind and solar PV generators renewable targets were achieved, but the cutback of the nuclear programme after 2011 created problems leading to the importing of electricity generated by nuclear power stations in neighbouring countries as well as the increased use of coal-fired power stations in Germany.

5.4   THE CURSE OF INTERMITTENCY

Rather than going into the details of the German Energiewende situation, we turn to some of the problems inherent in renewables technology. It can be argued that the Green movement has concealed the true costs of wind and solar energy; there are very considerable costs which arise as a result of the fact that both these sources of energy are intermittent (except perhaps for solar energy in the non-cloudy tropics).

On 22 February 2000, the final year of the 20th century, Neil Armstrong (www.greatachievements. org accessed 13 March 2019) reported to the US National Press Club on behalf of the (US) National Academy of Engineering on a poll that had been carried out by "a rather impressive consortium of professional engineering societies representing nearly every engineering discipline." They sought to "determine which engineering achievements of the twentieth century had the greatest positive effect on mankind" and they identified the 20 most important achievements. At the top of the list were:

1. Electrification
2. Automobile
3. Airplane
4. Water supply and distribution
5. Electronics

6. Radio and television
7. Agricultural mechanisation
8. Computers….

As Armstrong said, "the majority of the top 20 achievements would not have been possible without electricity."

As Darwall (2019) points out: "Wind and solar power's intermittency, unpredictability, and variability mean regressing from industrial production, where, like a factory, outputs are precisely controlled by varying the inputs, to arable farming, where output is heavily dependent on the vagaries of the weather. This is the least well understood but most damaging consequence of renewable energy. It threatens to disrupt the defining technological accomplishment of the twentieth century." Computers are needed to manage electricity flowing into and around the grid to ensure exactly the right amount of electricity is being generated and being made available to consumers. The economics of electricity generation and distribution are uniquely complex too. It can be argued that the Green Movement has been less than honest or transparent in discussing the true cost, in $ per kWh of the generation and distribution of electricity from solar panels and wind generators. This extra cost has already been shown rather clearly in Figure 5.2, from which we see that although Germany and Denmark have massive installations of wind and/or solar generation yet the price per kWh paid by consumers in these two countries is much higher than that paid by consumers elsewhere. Intermittency suffers from two problems. The first is that backup generating capacity has to be kept available for cold or calm days and this commonly involves fossil fuels; this capacity may have to be kept idle for extended periods. The second is that on very hot or very windy days the solar-generated or wind-generated capacity may exceed the immediate requirements and these generators may simply need to be switched off, when compensation is paid to the operators of those generators. The consequences of the intermittency of wind and solar power for the industrial generation and distribution of electrical power are well studied by Hirth (2014).

Wind power (aerogenerators or "Windmills") and solar energy (hot water generation and photovoltaics (solar panels)) have a role to play in isolation, i.e. where there is no attempt to connect their output to a country's electricity grid. In these situations, the users know and understand that this source of electricity is only available when the weather conditions (wind and sunshine) are suitable and life is organised accordingly by using the electricity when it is available or by using battery storage or diesel generators when the primary source of energy (the wind or the sun) is not available. We can perhaps reflect on the widespread use of windmills before the industrial revolution for grinding grain, pumping water, etc. The miller would have had to organise his or her life to operate the mill when the wind was blowing and could not have expected that the job would simply be from 9.00 to 5.00 on Mondays to Fridays only. The intermittency of the output of wind and solar power systems causes serious problems when one wants to connect these sources of electricity to a country's electricity supply grid. So much of our "developed" way of life is dependent on the availability of electricity at the flick of a switch at any time of day or night and on any day of the year.

Unlike any other commodity, electricity has to be produced the instant it is required and consumed the instant it is produced. Storing electricity therefore requires converting it into other forms of energy: as chemical energy in batteries, as potential energy in pumped-storage hydro systems to be reconverted to electricity the moment it is needed, involving energy losses on the way in and the way out. By definition, this is inefficient and expensive. Moreover, doing the sums one would find that the amount of electricity that can reasonably be stored in this way is minuscule compared to what would be needed in practice. Another option would be to use the one that Hermann Honnef and Franz Lawaczeck pressed the Nazis to adopt in the 1930s, see Section 1.2.1, namely using hydrogen. The problem here is that the technology for the electrolysis of water to produce hydrogen and for its storage and distribution facilities on an industrial scale is not sufficiently mature for the likely demand. A few hydrogen buses have now been introduced into service. Lack of storability makes the operating and economy of electricity generation and distribution entirely different from other forms of energy such as oil and gas, and from all other commodities: Supply must respond almost instantaneously to changes in demand. Just as countries cannot function safely and efficiently without buffer stocks of commodities such as food, medicines, raw materials for industry, the same is true of energy, especially electrical energy. Coal, gas, and biofuels such as wood chips can be stockpiled beside the appropriate power stations but it is not like that for solar-energy and wind-energy electricity production (Ferrey 2019, Martinez et al. 2019).

The fact that electricity cannot be stored (it has to be reconverted into other forms of energy) and has to be produced the moment it is consumed is sometimes described by saying that electricity is both homogenous and heterogeneous (e.g. Hirth 2014). By homogeneous is meant that the user cannot tell whether the electricity that is being supplied has been generated from a coal-fired power station or a wind farm or in some other way. The idea that electricity is homogeneous means that the user expects and requires that the electricity from the grid should be available at the flick of a switch any time of day or night on any day of the whole year. Moreover, the user further requires that the properties of the electricity delivered, namely the voltage, frequency, and phase, are fixed within very narrow limits. A major problem faced by the operator of a supply grid is how to deliver this electricity to the fluctuating demands of the users where the inputs come from a number of different suppliers. The situation becomes much worse when two of those sources of supply (wind energy and solar energy) can only deliver electricity intermittently and with only limited reliability of predictability.

Before wind and solar energy came on the scene, demand determined supply; yet the amount of electricity produced from wind and solar farms depends on the weather. Wind and solar thus suffer from a fundamental flaw: They are incapable of supplying on-demand electrical power. Their inability to meet such a basic requirement of electricity production means other generators, typically powered by fossil fuels, have to be kept on standby to pick up the slack. This is highly inefficient and adds to costs and to greenhouse gas emissions. Wind and solar also require more grid infrastructure. The more wind and solar energy there is on the grid, the worse it is for the electricity system taken as a whole. Even if it becomes cheaper to make and install wind turbines and solar panels, any plant-level economies of scale are more than outweighed by the system diseconomies of scale. These additional costs are often not considered or admitted by the proponents of wind and solar power.

In the past, variable renewable energy sources (VRE) for electricity generation, such as wind and solar power, have only made small contributions. However, if they are to be expected to make much larger inputs to a grid, their inherent output fluctuations create major problems, which have not been fully taken into account when political decisions have been taken. This variability has significant impacts on power system and electricity markets if VRE are deployed at large scale.

At this point, we consider the thesis of Hirth (2014) because it does two important things. First, this thesis aims at answering the question: What is the impact of wind and solar power variability on the economics of these technologies? It includes a major review of the major previous scientific papers on the impact of wind and solar power variability on the economics of electricity generation and the supply of this electricity to customers via a grid supply system. Secondly it includes the construction of a numerical Electricity Market Model (EMMA) and its implementation on a computer to address the effect of the intermittency of wind power and solar power on this problem. Hirth's thesis comprises eight chapters and, apart from Chapter 1 Introduction and Chapter 8 Findings and Conclusions, the core of the thesis is made up of six chapters, which are simply reproductions of published research papers written by Hirth and some of his collaborators (Table 5.6).

These papers are full of graphs and histograms based on both their review of previous work and on the results from their numerical Electricity Market Model EMMA. Chapter 8 introduces the numerical power market model EMMA that was developed for this dissertation. Chapter 9 presents the articles and outlines the structure of the thesis.

The economics of electricity generation are complicated. Electricity is a peculiar economic "good" (= singular of "goods") or product. Consider the different electricity factories (or power stations) that are available to supply electricity to a grid. There are coal, gas, oil, or nuclear power stations. For these, there is a significant capital cost, there are running costs, there are delivery costs relating

TABLE 5.6

Chapters of Hirth's thesis and journal papers

1: Economics of electricity: Hirth (2014).

2: Framework: Hirth (2015).

3: Market value: Hirth (2013).

4: Optimal share: Hirth (2015).

5: Redistribution: Hirth, and Ueckerdt (2013).

6: Balancing power: Hirth and Ziegenhagen (2015).

Source: (Hirth 2014).

to connection to the grid, and there are fuel costs and finally there are decommissioning costs. These are reasonably well-known and potential investors can assess these costs before deciding to invest in the construction and operation of these power stations. Another relevant consideration would be the selling price of the electricity that is produced. In this respect, these electricity factories differ from most other factories in that the selling price of the product cannot easily be estimated.

We consider the concept that the electricity generated for distribution in a supply grid system is heterogeneous. The idea of heterogeneity in relation to electricity has been around for a long time but it has been formalised and more tightly defined by Hirth (2014).

We retain the language used by Hirth which includes two peculiarities of his English. First, he uses the word "good" to describe electricity as a product; this is apparently using "good" as the singular of the word "goods" (a goods train transports things – parcels, coal, steel, grain, oil, etc.). This is not in use in British English and is not even to be found in a common dictionary (Concise Oxford). Secondly, he uses the word arbitrage, which is in the Concise Oxford Dictionary: "the buying and selling of stocks or bills of exchange to take advantage of varying prices in different markets"; however, the use of this word is not in common use and its meaning would not be known to the average native speaker of English. In the following discussion, out of respect to Lion Hirth, we shall use his language and not attempt to "purify" his English, recalling that in the modern world large numbers of people are forced into using as a second language the English language, or at least the US version of the language, making the British some of the worst linguists in the world.

We consider Hirth's definition of heterogeneity. A good (product) is classified as heterogeneous *if its marginal economic value is variable*. Apart from electricity, there are other examples. These include the case of hotel rooms where in many places the price may depend heavily on the day of the week, on special events such as sports events or concerts, or just on the season. A second example is seats on aeroplanes, where on many airlines the price may depend on the day or season or on how long in advance the seat is booked. In a free market economy or a mixed economy, the prices of many products, goods, or services may vary but usually only rather slowly depending on market conditions. The price achieved for electricity delivered to a supply grid varies much more rapidly; according to Hirth (2014, page 21) "there is not *one* electricity price per market and year, but 26,000 prices (in Germany) or three billion prices (in Texas). Hence, it is not possible to say what "the" electricity price in Germany or Texas was last year." Arbitrage, which is basically speculating against price variations of a commodity such as oil or harvest products, etc., does not apply to wind power or solar power generation of electricity because the price paid for delivery to the supply grid varies so rapidly.

For example, a good is heterogeneous in time if its marginal value differs significantly between two moments during one year. Electricity can be stored directly in inductors and capacitors, or indirectly in the form of chemical energy (battery, hydrogen), kinetic energy (flywheel), or potential energy (pumped hydro storage). In all these cases, energy losses and capital costs make storage very, often prohibitively, expensive and the amount of electricity that can be stored is tiny compared with the demand. Hence, arbitrage over time is limited. The storage constraint makes electricity heterogeneous over time; it is economically different to produce (or consume) electricity "now or then."

A good is heterogeneous in space if its marginal value differs significantly between two locations in one country. Electricity cannot be transported on ships or trucks, in the same way as tangible goods. It is transmitted on power lines which operate at various voltages meaning that there are losses in transformers as well as resistive losses. The place where it is economically efficient to produce electricity from the wind or sun may be distant from the users. The transmission constraint makes arbitrage limited between locations and electricity becomes heterogeneous across space.

There is a third way in which heterogeneity applies to the industrial generation of electricity and that is in lead-time. Lead-time might be less intuitive than the other dimensions and merits some further discussion. In an alternating power (ac) supply grid system, there has to be a balance between demand and supply at every moment in time. Imbalances cause frequency deviations, which can destroy machinery and become very costly. However, thermal power generators are limited in their ability to adjust output quickly as there are limits on temperature gradients in boilers and turbines (ramping and cycling constraints). Hence, arbitrage is limited across different lead-times and so the flexibility constraint makes electricity heterogeneous along lead-time. Thus, we come to the three dimensions of the heterogeneity of electricity. The physics of electricity imposes three arbitrage constraints, along the dimensions time, space, and lead-time (Table 5.7).

We can think of three types of generators: inflexible generators that produce according to a schedule that is specified one day in advance, like nuclear power; flexible generators that can quickly adjust, like gas-fired plants; and stochastic generators (produce synthetic time series of weather data of unlimited length for a location based on the statistical characteristics of observed weather at that location) that are subject to day-ahead forecast errors, like wind power. If demand is higher than expected, only flexible generators are able to fill the gap. In such conditions, the real-time price rises above the day-ahead price, and hence, everything else equal, flexible generators receive a higher average price than inflexible generators. Contrast this with the stochastic generators: when they generate more than expected, there tends to be oversupply in the real-time market, and hence they sell disproportionally at a lower price.

It will be laid out that the impact of intermittency can be expressed in (at least) three ways: as reduction of value, as increase of cost, or as decrease of optimal deployment. Transferring between these perspectives is not trivial, as evidenced by the confusion around the concept of "integration costs." Hence, more specifically: How does variability impact the marginal economic value of

---

TABLE 5.7
The heterogeneity of electricity along three dimensions

Dimension (differences between points in....)	Time	Space	Lead-time between contract and delivery
Arbitrage constraint	Storage*	Transmission**	Flexibility***
Differences in demand and/or supply conditions	• shifts of the demand curve (day-night pattern, temperature) • shifts of the supply curve (weather, plant availability)	• location of demand • good sites for electricity generation	• uncertainty in demand (weather) • uncertainty in supply (weather, outages)
(Hallam, 2019).	(Hallam, 2019).	(Hallam, 2019).	(Hallam, 2019).

Source: (Hirth 2014).
In this context, costly is both in the sense of losses (operational costs) and the opportunity costs of constraints.
* Storing electricity is costly.
** Transmitting electricity is costly.
*** Ramping and cycling is costly.

these power sources, their optimal deployment, and their integration costs? This is the question that Hirth's thesis addresses.

In short, the principal findings of Hirth's thesis are as follows. Electricity is a peculiar economic good, being at the same time perfectly homogenous and heterogeneous along three dimensions – time, space, and lead-time. Electricity's heterogeneity is rooted in its physics, notably the fact it cannot be stored. (Only) because of heterogeneity, the economics of wind and solar power are affected by their variability. The impact of variability, expressed in terms of marginal value, can be quite significant: for example, at 30% wind market share, electricity from wind power is worth 30-50% less than electricity from a constant source (Hirth, 2014, page 7). This value drop stems mainly from the fact that the capital embodied in thermal plants is utilised less in power systems with high VRE shares. These findings lead to seven policy conclusions:

1. Wind power will play a significant role (compared to today).
2. Wind power will play a limited role (compared to some political ambitions).
3. There are many effective options to integrate wind power into power systems, including transmission investments, flexibilising thermal generators, and advancing wind turbine design. Electricity storage, in contrast, plays a limited role (however, it can play a larger role for integrating solar).
4. For these integration measures to materialise, it is important to get both prices and policies right. Prices need to reflect marginal costs, entry barriers should be tiered down, and policy must not shield agents from incentives.
5. VRE capacity should be brought to the system at a moderate pace.
6. VRE does not go well together with nuclear power or carbon capture and storage these technologies are too capital intensive.
7. Large-scale VRE deployment is not only an efficiency issue but has also distributional consequences. Re-distribution can be large and might be an important policy driver.

Note that there is no single magic number for the "best" percentage of wind or solar power in the supply to a country's grid system. What is needed for an economical and stable electricity supply is a mix of electricity generated from a variety of sources. As a rule of thumb, it can be argued that over-dependence on wind and solar energy for a country's requirements is unwise, principally because of the problem of intermittency (Rao 2019). Identifying an actual value for this ideal percentage is not easy and it is likely to be different for different countries anyway. We have discussed Germany's Energiewende and its increasing percentages dependencies on wind and solar power and we have seen that it has not been a total success (Heyen and Wolff, 2019). What is unfortunate, not to say foolish, is the apparent enthusiasm with which various other countries seem determined to repeat the mistakes of Energiewende. One of the promises held out in 2010 was a boom for German industry in the manufacture and export of photovoltaic panels; in the event the industrial boom occurred in China.

The need for transitioning towards low-carbon energy systems, and the recent boom in available data, allows for a constant re-evaluation of global electricity sector decarbonisation progress, and its underlying theoretical assumptions. Arguably, the existing decarbonisation literature and institutional support frameworks focus on top-down supply side mechanisms, where policies, goals, access to financing, and technology innovation are suggested as the main drivers. Recently, de Leon et al., (2020) synthesised 11 global data sets that range from electricity decarbonisation progress, to quality of governance, to international fossil fuel subsidies, and environmental policies, among several others, and used methods from data mining to explore the factors that may be fostering or hindering decarbonisation progress. This exercise allows to present numerous hypotheses worth exploring in future research. Some of these hypotheses suggest that policies might be ineffective when misaligned with country-specific motivators and inherent characteristics, that even in the absence of policy there are particular inherent characteristics that foster decarbonisation progress (e.g. relatively high local energy prices, foreign energy import dependency, and the absence of a large extractive resource base), and that the interaction of country-specific enabling environments, inherent characteristics, and motivations is what determines decarbonisation progress, rather than stand-alone support mechanisms. de Leon et al., (2020) presented the hypothesis that existing support mechanisms for decarbonisation may be relying too much on blanket strategies (e.g. policies, targets), and that there is a need for support mechanisms that encompass a wider diversity of country-specific underlying conditions.

Hirth's thesis was concerned with wind generation of electricity and the results were less than totally encouraging. The results for solar are more disappointing, which is not surprising given that solar power is even more intermittent that wind power. For a start the sun only shines for only part of the day, whereas the wind may blow at any time during the day or night. Even at the optimal wind share at 60% cost reduction, the optimal solar share is below 4% in all but very few cases. This is consistent with previous findings that the marginal value of solar power drops steeply with penetration, because solar radiation is concentrated in few hours. In regions that are close to the equator, the optimal solar share could be expected to be significantly higher, both because levelled costs are lower and the generation profile is flatter. Given the large uncertainty, it is likely that realised wind shares will ex post turn out to be sub-optimal, too high, or too low.

For future research, EMMA, or comparable models, could be extended in several directions. A more thorough modelling of specific flexibility options is warranted, including a richer set of storage technologies, demand side management, long-distance interconnections, and heat storage. A special focus should be paid to the existing hydro reservoirs in Scandinavia, France, Spain, and the Alps. More generally, the integrated modelling of hydro-thermal systems and the integrated modelling of both transmission constraints and power plants investments are promising fields of model development. For certain research questions, representing existing policies and modelling the interaction of policies can be quite crucial. Developing numerically feasible approaches to incorporate internal transmission constraints into long-term power market models is another promising research direction.

Modern life depends on a stable grid providing always available electrical power at the flick of a switch, and society puts a very high value on having a reliable grid. Unlike coal, oil, gas, and nuclear, the output from wind and solar cannot be varied with demand. They can't be counted on for rapid "dispatch" to maintain grid stability but make the grid less stable. Evaluations that ignore or downplay the intermittency penalty of wind and solar systematically understate how much they cost compared to conventional dispatchable capacity.

The issues of over-reliance on solar and wind power are more matters of politics and economics rather than scientific matters and the details are outside the scope of this book. However, it could be argued that political decisions, particularly in relation to Energiewende, have been made in ignorance of the scientific, technological, and consequent economic implications. We simply acknowledge the problems and refer to other texts, particularly Darwall (2017 Chapters 13 and 14) but there are others. A major work on the economics of climate change is by the Stern Review (Stern 2007) which was commissioned by Gordon Brown (the Chancellor of the Exchequer and later Prime Minister of the UK). However, this review was produced when the IPCC's assessments held sway and before the crystallisation of the opposition to the IPCC by the publication of the NIPCC report in 2008 (Singer 2008). The Stern Review is very much concerned with the costs of curbing CO_2 emissions and does not seriously address the costs of over-dependence on wind energy and solar energy.

And what of cutting CO_2 emissions, the ostensible reason why governments intervened in the first place? A 2014 Brookings Institution report by Charles Frank compared the costs and benefits of decarbonisation for wind, solar, and combined-cycle gas turbine (CCGT) plants on the basis of a $50 per tonne cost of carbon. The analysis, which did not explicitly incorporate the extra grid and scale costs of renewables, found that wind generated annual net disbenefits of $25,333 per MW of capacity and solar generated \$188,820 of annual net disbenefits, whereas one MW of CCGT capacity generated net benefits of \$535,382 a year?

Visiting a solar company in Germany in April 2014, an exasperated Social Democratic Party SPD leader and vice chancellor, Sigmar Gabriel, exclaimed to his astonished hosts:

> Die Energiewende steht kurz vor dem Aus. Die Wahrheit ist, dass wir die Komplexität der Energiewende auf allen Feldern unterschätzt haben. Die anderen Länder in Europa halten uns sowieso für Bekloppte.
>
> (see also, Han et al., 2019)

This could be translated as:

> The Energiewende is on the brink of failure. The truth is that we completely underestimated the complexity of every aspect of the Energiewende. The other countries in Europe think we're nutters (Bekloppt is crazy, stupid crazy, literally "hit over the head") anyway.

Asked what message he would give other countries thinking about whether to adopt renewables, gave rise to the following blunt advice: "Don't follow Germany into this dead-end." (Darwall 2019, page 165).

We have already noted that the early promise of jobs in 2010 and a large export market in photovoltaics panels for Germany did not materialise. The jobs and export market opportunities materialised in China. Instead the beneficiaries were the people who reaped subsidies for the installation of imported photovoltaic panels from China. This point tends to have been overlooked, or deliberately ignored, by the Green Movement. Also promises of developing the technology of carbon capture remain largely unfulfiled.

5.5 NUCLEAR ENERGY AND THE GENERATION OF ELECTRICITY

The use of nuclear energy to generate electricity started in the early 1950s, when the first nuclear reactor (a small unit called Experimental Breeder Reactor I) became operational at the Argonne National Laboratory in Idaho, United States. In the following years, the USA, UK, Russia, France, and Germany were the first to use nuclear technology commercially, and 20 other countries followed suit over the next decades. However, even though US president Dwight D. Eisenhower, in his famous "Atoms for Peace" UN speech, urged the international community as early as 1953 to cooperate in order to develop nuclear technology, atomic energy only underwent a significant international development phase almost 20 years later, in the early 1970s (Figure 5.5).

The most important global development stage in the history of nuclear energy occurred roughly between 1970 and 1985, when the total number of nuclear reactors were from over 80 in 1970 to over 360 in 1985 (Figure 5.5). While the number of active nuclear reactors had an almost four-fold increase in this 15-year period, their installed power capacity had a much steeper 14-fold increase, i.e. from ~18000 MW in 1970 to ~250000 MW in 1985 (Figure 5.5). This corresponds to ~65% of atomic energy growth over six decades, when considering the increase in the number of nuclear reactors in 1970–1985 in relation to the entire analysed period 1955–2015.

The causes for the decrease in nuclear development after 1985 concern a series of events with global-scale effects, of which the most important are the increase of interest in oil after 1980 (as a result of price decreases) and especially the Chernobyl nuclear accident, which generated an obvious change in how countries worldwide viewed nuclear power (Albino et al., 2014). The effects of the 1986 nuclear disaster were so profound in both public and political spheres that, for instance, that same year Germany approved a resolution aimed at abandoning nuclear energy by the end of the decade, and the following year Italy completely shut down its nuclear energy programme (Albino et al., 2014). In this context, Italy became the first country to go back to a "non-nuclear energy" status. Two other states followed its lead and abandoned their nuclear reactors in the following decades – Kazakhstan (1999) and Lithuania (2009) (Schneider et al., 2011).

We turn to a UK person who should not be forgotten in connection with renewable energy, namely (Sir) Martin Ryle (1918–1984), a Cambridge radio astronomer (Cottey 2018). "In 1936 he went to Oxford University, graduating in physics in the summer of 1939. This was just before the outbreak of the Second World War and Ryle soon found himself contributing to the British military effort, working on radar counter-measures at the telecommunications Research Establishment. Radar was to play a vital role in the war – especially for Britain and Ryle's abilities flourished." In 1945, he returned to Cambridge and quickly became the world's leading pioneer of radio astronomy. He shared the 1974 Nobel Prize for Physics for the early work on radio astronomy.

For various reasons, Ryle moved away from radioastronomy in the 1970s. He became a harsh critic of both nuclear energy and nuclear weapons, and a strong advocate of renewable energy especially wind energy – as well as of energy storage and heat insulation. Ryle's ideas are still important. Back in the 1970s, our views on energy were very different from now. In particular, it was generally believed that oil and gas reserves would be exhausted within a few decades. And while a few scientists took human-induced climate change seriously, most, if they had even heard of the idea, considered it too speculative to pronounce upon. Ryle, though, entered the energy debate with characteristic urgency. He insisted that burning oil and gas for energy (heat, electricity, or motive power) must be curtailed, drastically and quickly. These precious resources, he argued, should be reserved as raw feedstock to make chemical products, especially plastics. As for coal, Ryle was not enthusiastic although he did not rule it out entirely but considered it to be a necessary stopgap as oil and gas ran out. He believed that a modern coal system, with improved pollution control of sulphur compounds and particulates, could provide electricity plus district heating with high efficiency. However, it was nuclear power to which Ryle had the strongest objections. He felt it was totally unacceptable – expensive, dangerous, subject to long lead-times, and inextricably connected with nuclear weapons. Ryle resented the vast sums of research money ploughed into this field compared with the paltry amounts targeted on efficiency improvements and alternative energy. He became an activist, writing articles in the mass media denouncing nuclear energy, the gross nuclear-weapons arsenals, and the irrational policies of the nuclear era. As a result of public opposition to nuclear weapons, see Section 8.5.2, the UK's investment in new nuclear power stations stalled for several decades.

Despite the withdrawal of Italy, Kazakhstan, and Lithuania from nuclear-energy status, several other states in Eastern Europe and Asia continued their nuclear energy projects after 1990 and went on to develop their nuclear capacity up to present day. Relevant such examples are Japan, South Korea, India, and China, which continued to build large fleets of nuclear reactors in the past two decades (Lovering et al., 2016). Nuclear energy technology therefore has kept on expanding up to the present day globally (Figure 5.5), and it is in fact going through a renaissance phase as a result of a notable rise in new power plant investments in developing economies, increases of fossil fuel prices and growing concern regarding climate change (Albino et al., 2014).

Of the entire global fleet of reactors, the United States has by far the highest number of nuclear reactors (99, 22% of 448), followed by France (58, 13%), and many other countries, see Figure 5.6(a). The amount of electricity actually generated in 2015 is shown in Figure 5.6(b).

Given this overall picture of the electrical sector, Japan's case is particularly interesting – even though it is the third country worldwide in terms of nuclear energy potential (Figure 5.6(a), its 2015 energy production was among the lowest in the world (Figure 5.6(b). This is due to the government's decision to shut down all reactors countrywide after the 2011 Fukushima accident, which caused a severe national energy shock by reducing the nuclear power share from 31% in February 2011 to 0% in May 2012 (Hayashi and Hughes, 2013). In addition to the national effects related to higher fossil fuel imports (Vivoda, 2016), this disaster had implications that went beyond the country's energy security – internationally, e.g. the event generated a significant increase of investments in other energy sources such as liquefied natural gas (Hayashi and Hughes, 2013). There are however legitimate perspectives for nuclear energy production to be resumed in the immediate future, seeing

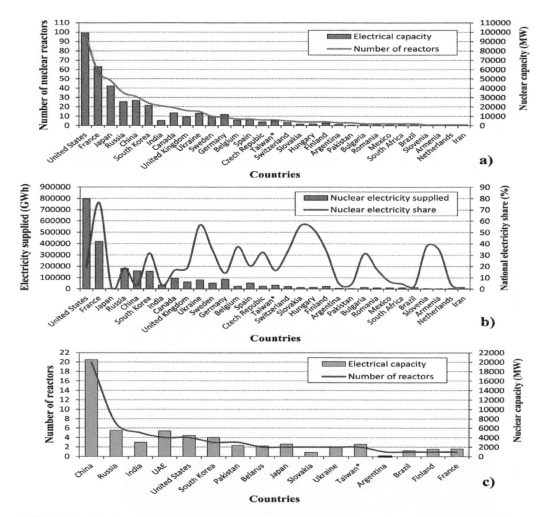

FIGURE 5.6 Total number of operational nuclear reactors worldwide and their net capacity (MW) in 2015 (a); production of nuclear energy and the corresponding share of the total national power production in 2015 (b); number of nuclear reactors under construction worldwide (in the beginning of 2017) and their net capacity (MW) in the states that are currently expanding their nuclear energy infrastructure (c) (all data processing from https://www.iaea.org/pris/).

as in 2015 two nuclear reactors were restarted under new safety standards set after the Fukushima accident.

The number of nuclear power stations under construction in 2015 was 60 reactors currently under construction in 16 countries (Figure 5.6(c)). China is by far the world leader in this respect with 20 reactors, with a projected net power capacity of 20.5 GW. A third of all reactors under construction can therefore currently be found in China. It is interesting to note that two new countries, the United Arab Emirates and Belarus (Figure 5.6(c)), will soon have operational nuclear plants, which raises the number of UN member states with nuclear energy status to 32, or 33 if the Taiwan non-member state is included. According to the International Atomic Energy Agency (IAEA), at the end of 2018 there were 450 nuclear power stations worldwide with a combined generating capacity of 396,413 MW (396.413 GW) and with a total production in that year of 2,563.0 TWh (IAEA, 2019). The IAEA (2019) predicts that the world nuclear generating capacity will rise to 496 GW by 2030 and 715 GW by 2050 in the high case but only to 371 GW by 2050 in the low case. This is, understandably, a very wide range of predictions.

5.6 ENERGY RESOURCES AND THE ENVIRONMENT

Despite the uncertainty of the existing forecasts of CO_2 dynamics, a concern for CO_2 accumulated in the atmosphere has stimulated the development of different approaches to a solution to this problem, a key aspect of which is the dynamics of energy production. In this connection, Marchetti (KC98) attempted an analytical approximation of the dynamics of the production of energy from various fuels during the last century or so based on available data, with an extrapolation of actual trends to the future. Figure 5.7 illustrates the US energy production by sources, from 1950 to 2015, while Figure 5.8 shows similar global data from 1990 to 2015 and projections from 2015 to 2040. Figure 5.9 shows the conversion of the fossil fuel consumption into tonnes of carbon liberated in the form of CO_2.

Analysis of the annual mean zonal mean profile of the warming due to a doubled CO_2 concentration has demonstrated that the warming of the surface layer intensifies from the tropics towards the mid-latitudes in both hemispheres and gradually penetrates deep into the ocean, reaching greater

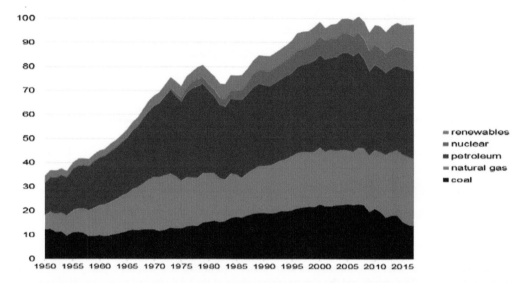

FIGURE 5.7 Relative temporal variation of energy production from various fuels. U.S. primary energy production by major sources, 1950–2017. (Source: U.S. Energy Information Administration, Monthly Energy Review April 2018).

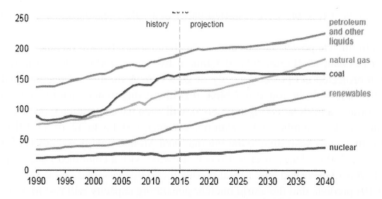

FIGURE 5.8 Global energy production from various fuels. World energy consumption by energy source (1990–2040). (Source: U.S. Energy Information Administration, International Energy Outlook 2017).

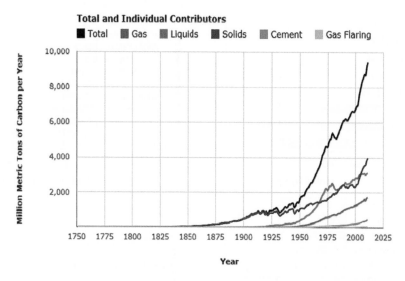

FIGURE 5.9   The evolution of global CO_2 emissions. 1 tonne coal produces 3.5 tonnes CO_2; 1 tonne CH_4 produces 3 tonnes CO_2; 10^3 m^3 CH_4 produces 2 tonnes CO_2. Source: Boden et al. 2017 Global, regional, and national fossil-fuel CO_2 emissions. Carbon dioxide information analysis center, Oak Ridge National Laboratory, U.S. Department of Energy, Oak Ridge, Tenn., USA.

depths in the subtropics and mid-latitudes than in the equatorial belt. The meridional profile of the warming turns out to be very similar to the latitudinal distribution of excess ^{14}C concentration observed (compared with its values before nuclear tests in the atmosphere).

5.7   DECARBONISATION POTENTIAL IN THE GLOBAL ENERGY SYSTEM

It can be argued that nuclear energy had considerably diminished the acceleration of global climate warming recorded in the past four decades, as its use prevented the release of over 60 billion tons CO_2 after 1970 (IAEA, 2016a, IAEA, 2016b). It is currently estimated that nuclear power is preventing the annual release of 1.2–2.4 Gt CO_2 emissions globally, assuming that, without this technology, more than 2400 TWh worth of nuclear power would be produced by natural gas combustion (which, on average, releases ~500 g CO_2/kWh) or coal combustion (~1000 g CO_2/kWh) (NEA, 2015b). In this context, nuclear power is considered to be an important contributor to the decarbonisation of the global energy system (Prăvălie and Bandoc, 2018).

Moreover, alongside renewable sources and CCS, nuclear power is also labelled a low-carbon technology (IPCC, 2013) due to the fact that the emissions of its entire life cycle are similar to those of renewable energy. Estimations show that the average quantity of CO_2 emissions per unit of electricity generated is currently 15 g CO_2/kWh, see Figure 5.10(a). This value is ~30/50/70 times smaller than the emissions generated by the combustion of gas/oil/coal and is comparable to wind power emissions.

While low-carbon energies sources currently supply ~30% of electricity worldwide, this share must grow considerably for the energy system decarbonisation to generate a concrete effect in stabilising global climate warming at the 2 °C threshold set in relation to preindustrial levels. To meet this objective and comply with the Paris Agreement, nuclear energy is being considered as a possible strategy for climate change mitigation and is included in the work schedule of the Intergovernmental Panel on Climate Change.

However, for it to become a viable option in line with the Paris Agreement 2 °C goal, a massive expansion of nuclear power capacity would be necessary in the next few decades, i.e. from ~390 GW today to 930 GW in 2050 (NEA, 2015b). This growth would require an annual increase of

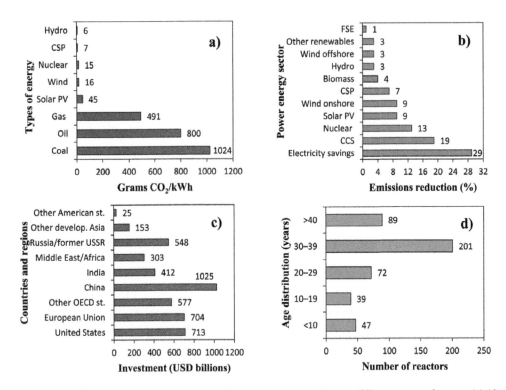

FIGURE 5.10 CO_2 emissions produced by 1 kWh electricity, according to different types of energy (a) (data processing from http://www.sfen.org/en/nuclear-for-climate); global emissions reduction required in the power sector in 2050 to keep the 2 °C warming target (b) (data processing from NEA); global investment needs in nuclear energy between 2010 and 2050 for the 2 °C climate goal achievement (c) (data processing from NEA); age distribution of operating global nuclear reactors (d) (data processing from https://www.iaea.org/pris/). Note: CSP – concentrated solar power; FSE – fuel switching and efficiency; CCS – carbon capture and storage; st. – states; develop. – developing; USSR – Union of Soviet Socialist Republics.

the total nuclear capacity of ~20 GW after 2020, which will be difficult to achieve when considering the current nuclear expansion rate of up to 5 GW/yr since 2010 (NEA, 2015a).

In terms of its contribution to reducing carbon dioxide emissions in accordance with the 2 °C threshold, nuclear energy will have to contribute with a 13%-share reduction of CO_2 annually by 2050 (NEA, 2015b) (Figure 5.10b). This percentage corresponds to an absolute value of 2.5 Gt CO_2 per year, necessary for an effective decarbonisation of the electricity sector (NEA, 2015a). In these conditions, nuclear power will provide some of the highest contributions to CO_2 reductions in the low-carbon energy spectrum (Figure 5.10).

However, considerable investments are necessary to reach this target. It was suggested that, for reaching an installed nuclear capacity of 930 GW in 2050, a total investment of over US$ 4 trillion would be necessary (NEA, 2015a), almost half of which (~US$ 2 trillion) would be required for decommissioning nuclear facilities, extending the lifetime of power plants and expanding nuclear capacity in the OECD (Organization for Economic Co-operation and Development) countries, especially in the United States and the European Union (Figure 5.10c). Outside OECD countries, China will have to invest roughly a quarter (~US$ 1 trillion) of the total global amount in new nuclear capacity, thus becoming the country with the highest contribution to nuclear development by the middle of this century (Figure 5.10(c)).

A significant part of these huge costs in OECD countries will be directed towards decommissioning power plants, seeing as, globally, about two thirds (290 units) of nuclear reactors are more than 30 years old (Figure 5.10(d)), and 60% are located in three OECD countries – United States,

Japan, and France (IAEA, 2016a). The entire decommissioning activity (which includes nuclear facilities, used fuel, and site restoration) is highly expensive – e.g. it is estimated that in the US, such a procedure costs 500 million US$ per reactor (IAEA, 2004 Chu and Majumdar, 2012). This is one of the main reasons why the country has extended the operating licence of nuclear reactors from 30 (normal lifetime) to 60 years, and there are sources that say the US is even considering further extensions (IAEA, 2016a).

In addition to costs, another challenge for meeting the Paris Agreement goal could be the uranium supply capacity. It is to be expected that the upcoming nuclear development will require an expansion of the fuel cycle around the world, more specifically an increase of uranium mining production and enrichment capacities. However, considering the stability of uranium resources, this nuclear fuel can provide support for the nuclear power increase – the most recent estimations suggest that there are currently 5.7 million tonnes of uranium in known reserves (NEA 2015a, IAEA, 2016a). Considering the current exploitation rate of almost 57000 tonnes/year, it can be deduced that uranium resources will be sufficient for at least 100 years of nuclear electricity production.

In the end, it must be noted that even though nuclear energy is widely considered to be a low-carbon solution for climate change, this reputation is questionable (Kleiner, 2008). Considering the criterion of carbon dioxide emissions, the carbon footprint of the entire nuclear cycle might currently be underestimated (Kleiner, 2008, Sovacool, 2008). There are certain studies that show that the average emissions generated by this type of energy reach 66 g CO_2/kWh (Sovacool, 2008), i.e. significantly more than the general estimations of 15 g CO_2/kWh, which we quoted above. This value, four times higher than wind power emissions (Figure 5.10(a)), is due to uranium processing (38%; e.g. mining or enrichment), decommissioning activities (18%), operation (17%), nuclear fuel processing and waste storage (15%), and nuclear power plant construction (12%) (Sovacool, 2008). However, even with these indirect emissions that are higher than the generally acknowledged estimations (but much lower compared to fossil fuel emissions), using nuclear energy in a mixed context, alongside renewable energies, could be a viable short- and medium-term solution for the global phase out of fossil fuels.

5.8 RADIOACTIVE ENVIRONMENTAL CONTAMINATION FROM NUCLEAR ENERGY

Even though nuclear energy is considered to be beneficial for human society economically and, more recently, climatically, one of its major disadvantages with global-scale implications must be noted – the risk of environmental radioactive pollution (contamination). This risk can be approached in terms of two key facets: the threat of nuclear reactor accidents and the danger associated with nuclear waste management (the risk associated with nuclear weapons testing is not included in this review since the paper explores nuclear energy for commercial/peaceful purposes). Both can cause significant environmental and societal consequences in the unfortunate event that radionuclides from inside nuclear reactors or waste storage site facilities were to be released.

5.8.1 NUCLEAR ACCIDENTS

The past decades have proved that nuclear accidents can cause critical local, regional, or global contamination. It is estimated that, after 1950, there were approximately 20 nuclear accidents worldwide in commercial (especially) and military reactors, which were linked to reactor core melting caused by cooling capacity system failures (Burns et al., 2012, Lelieveld et al., 2012). Upon analysis of the INES scale rating of these accidents (the International Nuclear and Radiological Event Scale), which groups nuclear events in seven classes based on their severity (the first three levels are called incidents, while the next four are called accidents), it can be noticed that throughout nuclear power history there have been two major accidents (maximum level – 7, Chernobyl and Fukushima), a single case of serious accident (level 6, Mayak) and several events classified as accidents with wider consequences (level 5), of which the most widely known is Three Mile Island. However, even if the most severe cases (5, 6, 7) are fewer compared to low-intensity accidents (level 4), the former are the most alarming – the difference between one level and the next amounts to a ten-fold increase in severity (IAEA, 2008). In terms of frequency, the former USSR (almost a third) and the United States (a quarter) have the highest number of accidents by far (Table 5.7).

The Chernobyl accident (Ukraine) of 26 April 1986 is known as the most severe nuclear disaster in civil nuclear power history, considering the large explosive release of radioactive material into the environment. The explosion of reactor 4 caused the uncontrolled release of a considerable amount of radioactive isotopes into the atmosphere for ~10 days, which, due to atmospheric dispersion, contaminated extensive areas in the northern hemisphere, especially in Europe and the former Soviet Union. The most important released radionuclides (in terms of radiation dose delivered to the public) were ^{137}Cs and ^{131}I, and it is estimated that ~30% of the reactor core content of ^{137}Cs (or 85 petabecquerels – PBq, 1 peta = 10^{15}), and 60% of ^{131}I (1760 PBq) (Table 5.7) was transferred into the atmosphere (Lelieveld et al., 2012). Radioactive contamination with ^{137}Cs can be considered the accident's most drastic impact form (given the prolonged environmental persistence of this radionuclide, 30-year half-life), especially in the former Soviet Union, where most of the ^{137}Cs deposition on the ground occurred (equivalent to ~40 PBq), distributed in Belarus (40%), the Russian Federation (35%), and Ukraine (24%) (UNSCEAR 2008, 2013).

At the same time, ^{131}I was a problematic radioactive isotope due to the fast transfer to the human food chain (via the transfer to pasture grass and subsequently to cows' milk), despite its much lower persistence (eight-day half-life).

The Fukushima Daiichi accident (Japan) of 11 March 2011, due to the Tohoku earthquake (magnitude 9.0) and subsequent tsunami (the waves were 14 m above sea level when they reached the nuclear power plant) that caused a power outage and the cooling system failure of reactors 1, 2, and 3, is considered the second most severe nuclear event in history after Chernobyl) and is also ranked on the top level of the INES scale (UNSCEAR-2013, 2013, Sugiyama et al., 2016) (Table 5.8). A brief comparison of the two events shows that their socio-economic impact was very large; e.g. in terms of the people who had to be evacuated/relocated immediately (over 300000 relocated at Chernobyl, ~150000 evacuated at Fukushima), costs in billion US$ (250–500 for Chernobyl, 100–500 for Fukushima) or the release of radioactive substances (Högberg, 2013). In the last instance, considering the release of ^{137}Cs into the environment (atmosphere and ocean) due to the partial melting of the reactor cores in the days following the tsunami, it is estimated the total value reached 37 PBq (Table 5.8), which represents ~44% of Chernobyl emissions (Lelieveld et al., 2012). An important particularity of the Fukushima accident is that most of the radionuclides (about 80%) were deposited in the Pacific Ocean (either directly or indirectly due to the dispersion of radioactive substances over the North Pacific Ocean, after the initial release into the atmosphere), thus becoming a possible threat for marine ecosystems (due to aquatic biota transfers), and implicitly for the human population.

Although the Mayak accident (Russia) of 29 September 1957 was a highly severe nuclear event on the INES scale (level 6) that caused a large release of radionuclides (the most important of which is ^{90}Sr) especially in the Chelyabinsk region, this disaster is not representative of the nuclear reactor accidents in question because it was due to the chemical explosion of a storage tank with liquid radioactive wastes, which was part of a weapons material production complex (UNSCEAR, 2008). However, what is important for this section is the Three Mile Island nuclear plant accident (United States) of 28 March 1979. The event, caused by the partial core melt of reactor 2 as a result of the deliberate shutdown of the cooling system (amid the occurrence of a pressure disturbance in the reactor system and a subsequent malfunction of a pressure control valve), released into the environment certain important radioactive fission products, e.g. noble gases, mainly ^{133}Xe (Högberg, 2013). However, only very small quantities of extremely dangerous radioactive substances (caesium, iodine) were released, as a result of successful efforts to maintain the nuclear reactor intact (Burns et al., 2012, Högberg, 2013).

TABLE 5.8

Nuclear accidents worldwide, their severity and radioactive atmospheric emissions – total and particular (PBq), exemplified with two representative radionuclides

No.[1]	Country	Location	Year of event	INES scale	Total radioactivity	^{137}Cs	^{131}I
1.	Canada	Chalk River	1952	5	>0.3	[2]	[2]
2.	USA	Idaho Falls	1955	4	[2]	[2]	[2]
3.	USSR	Mayak	1957	6	74–1850	[2]	[2]
4.	UK	Windscale	1957	5	1.6	0.02	0.7
5.	USA	Simi Valley	1959	5–6	>200[3]	[2]	[2]
6.	USA	Idaho Falls	1961	4	[2]	[2]	[2]
7.	USA	Monroe	1966	4	[2]	[2]	[2]
8.	Switzerland	Lucens	1969	4–5	[2]	[2]	[2]
9.	UK	Windscale	1973	4	[2]	[2]	[2]
10.	USSR	Leningrad	1974	4–5	[2]	[2]	[2]
11.	USSR	Leningrad	1974	4–5	55	[2]	[2]
12.	CSSR	Jaslovské Bohunice	1977	4	[2]	[2]	[2]
13.	USSR	Belojarsk	1977	5	[2]	[2]	[2]
14.	USA	Three Mile Island	1979	5	1.6[3]	[2]	<0.0007
15.	France	Saint-Laurent	1980	4	[2]	[2]	[2]
16.	USSR	Chernobyl	1982	5	[2]	[2]	[2]
17.	Argentina	Buenos Aires	1983	4	[2]	[2]	[2]
18.	USSR	Chernobyl	1986	7	>12 000	85	1760
19.	Japan	Tokaimura	1999	4	[2]	[2]	[2]
20.	Japan	Fukushima	2011	7	>630	12–37	190–380

[1]   the events are listed in ascending order according to the year of occurrence.

[2]   radionuclide emissions estimated indirectly by means of assumptions/no strong source of emissions/no data available.

[3]   mainly/substantial emissions of ^{85}Kr; INES level 4 – accident with local consequences; INES level 5 – accident with wider consequences; INES level 6 – serious accident; INES level 7 – major accident; CSSR – Czechoslovak Socialist Republic. (Adapted after Lelieveld et al., 2012).

There is currently a potential global risk of radioactive contamination in the hypothetical case of severe INES 7-type nuclear accidents. Based on advanced simulations, a very interesting study suggested that a major accident involving any reactor of the over 400 currently operational worldwide could affect extensive areas in Western Europe or Eastern Asia (due to the release of considerable amounts of radionuclides such as ^{137}Cs and ^{131}I), which would expose dozens of millions of people to radioactive contamination (Lelieveld et al., 2012). Considering the case of the radioactive isotope ^{137}Cs, representative due to its long persistence in the environment, it was found that, due to atmospheric dispersion, 8% would be deposited within a 50-km radius around the source, 30% within 500 km, 50% within 1000 km, and 25% within 2000 km (Lelieveld et al., 2012). Therefore, considering that over 90% of ^{137}Cs fallout will be transported within a 50-km radius around the source, as well as the fact that considerable amounts of radioactive particles will be deposited at distances of up to 2000 km, it can be concluded that any major nuclear accident could potentially cause significant environmental consequences regionally and even globally.

5.8.2   NUCLEAR WASTE MANAGEMENT

The use of nuclear energy also poses an environmental and societal threat due to the difficult management of spent/used nuclear fuel waste. Although it is estimated that after 1971 reactors have produced a huge amount of high-level waste (HLW, waste that contains 95% of the nuclear power

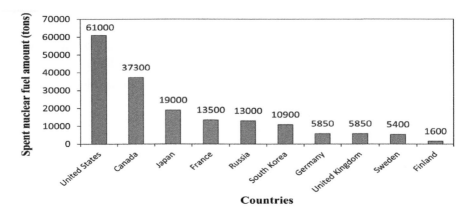

FIGURE 5.11 Amount of nuclear waste (spent nuclear fuel) in 2007, in the top ten countries with highly radioactive nuclear waste (data processing from https://www.statista.com/).

radioactivity and consists mainly of uranium spent nuclear fuel) worldwide, i.e. ~350000 tons, a part of it has been reprocessed (reprocessing aims to extract/recover the fissile elements of spent fuel, as well as the reduction of HLW volumes). Also important is low- and intermediate-level waste, as well as the waste resulting from other stages of the fuel development cycle (mining, enrichment, fabrication), which are not factored into this large amount of HLW.

The current generation rate of the most dangerous nuclear waste (HLW) is alarming, i.e. ~12000 tonnes/year, considering that each reactor (of the 448 operational worldwide) produces about 25–30 tons HLW annually (IAEA, 2009, Rosa et al., 2010). At present, it is estimated that over 250000 tons HLW need safe permanent disposal worldwide, which is all the more important seeing as their radioactivity levels can remain high for up to one million years (Rosa et al., 2010). The need for disposal is most apparent in North America, where the amount of spent nuclear fuel was getting close to 100000 tons in 2007 in the United States and Canada (Figure 5.11).

In this context, a way to partially solve the nuclear waste issue is reprocessing spent nuclear fuel. The global capacity of used nuclear fuel reprocessing is currently estimated at 5600 tons/year, and the largest reprocessing facilities are located in the United Kingdom (Sellafield, which covers 43% of global capacity), France (La Hague, 30%), Japan (Rokkasho, 14%), the Russian Federation (Mayak, 7%), India (Tarapur and Kalpakkam, 5%), and China (LanZhou, 1%) (IAEA, 2009). The United States is a particularly interesting case, as even though it is the largest producer of radioactive waste (Figure 5.11), it does not have any such facilities any more, having suspended spent fuel reprocessing back in 1977. However, although this procedure can reduce the volume of nuclear waste, the global amount of reprocessed waste is below half what is being produced annually and, additionally, it does not actually solve the urgent matter of long-term waste disposal.

Given this major challenge, the global political and scientific consensus on the most viable long-term management of highly radioactive waste is storage in deep geological repositories. The procedure entails storing HLW at a depth of at least several hundred metres and isolating the waste with especially designed containers and natural underground locations with extremely low permeability surrounding rocks, in tectonically stable areas in order to minimise the risk of radionuclide release into the environment.

While this option is being considered in several nuclear countries (Table 5.9), such underground facilities have not been completed anywhere in the world (currently there are only temporary storage facilities), although large-scale nuclear waste production dates back half a century. It is however only a matter of time before underground nuclear waste storage facilities will be built in the near future in countries such as Finland, France, Sweden, and the US (Table 5.9), which have already

TABLE 5.9

Storing and reprocessing of spent nuclear fuel and development perspectives for deep geological repositories in the top ten countries with highly radioactive nuclear waste (data processing after IEA and NEA)

No.	Country	Site storage	LIC[2]	Strategy	DGR[4] state	DGR develop. Perspective[5]
1.	US	RS[1]	LOM[3]	Direct disposal	Under construction	2021[a]
2.	Canada	RS	LOM	Direct disposal	Planned	2025[b]
3.	Japan	Rokkasho	N	Reprocessing	Planned	2035[b]
4.	France	RS, La Hague	LOM, N	Reprocessing	Under construction	2025[b]
5.	Russia	RS, Mayak, Zheleznogorsk	LOM, SW, S	Reprocessing	Considered	2025[c]
6.	South Korea	RS	LOM	Direct disposal	Considered	2028[d]
7.	Germany	RS, Ahaus, Gorleben	LOM, NW, N	Direct disposal	Considered	2031[d]
8.	UK	Sellafield	W	Reprocessing	Planned	2040[b]
9.	Sweden	Simpevarp	SE	Direct disposal	Under construction	2025[b]
10.	Finland	RS	LOM	Direct disposal	Under construction	2020[b]

Note: US – United States; UK – United Kingdom; 1 – Reactors site (the adjoining names represent central storage sites); 2 – Location in the country (N – north, S – south, W – west, NW – northwest, SW – southwest, SE – southeast; all these locations are related to the central storage sites); 3 – Location on the map; 4 – Deep geological repository; 5 – DGR development perspective; a – timeline with expected construction end/entry into operation of the underground facility (underground repository suspended in 2009); b – timeline with expected construction end/entry into operation of the underground facility; c – timeline in which a decision is expected for the repository construction; d – timeline for the underground site selection process.

made remarkable progress in implementing such projects. On a longer term, other countries (e.g. Canada, UK, Japan) might implement similar projects (Gibney, 2015).

From this table, we see that Finland is currently in the most advanced stage in terms of HLW storage in a deep underground repository, as it is the first country in the world to approve in 2015 the construction of such a facility at Olkiluoto (south-west), where approximately 6500 tonnes of uranium will be stored (Gibney, 2015) starting around the year 2020. Several other countries are planning facilities to be ready for 2025. It is clear that the great amount of highly radioactive waste is one of the major environmental risks associated with nuclear energy, especially seeing that no country in the world currently has an appropriate management system in place for this type of waste. Even though in the following decades several deep geological disposal facilities will be built, they will only partially solve the radioactive pollution issue of the nuclear sector, both in terms of amounts (not all dangerous waste will be assimilated) and safety. For the latter, it must be noted that such facilities are not absolute barriers for nuclear waste, seeing as certain possible geological perturbations can be triggered over the course of hundreds of thousands of years of operation.

6 Climate data, analysis, modelling

6.1 INTRODUCTION

It would be most useful if it were possible to identify the human activities that affect the climate and to make reliable predictions of the direct effects of these human activities on the future climate. However, this cannot be done directly because of the natural changes that are also occurring all the time. Our task is understanding global climate change. One important question is what sort of timescale we are concerned with. Discussions of climate change are very often concerned with trying to influence political decisions made by governments or by supranational organisations presumably with a view to making life better for ourselves, our contemporaries or our not too distant descendants. Given that the Earth is about 4.5 billion years old, that life evolved perhaps at least 3.7 billion years ago it is possibly an interesting intellectual exercise to consider what we can determine about the climate a long time ago. But such knowledge is not to be regarded as useful knowledge. While it may be of academic interest to consider ancient climate, in practical terms there is probably little point in going back more than about 11,000 years before the present (BP) in terms of understanding the present climate. Our hunter-gatherer forefathers are likely to have had some empirical understanding of some important features of the climate but they are not likely to have had a formal knowledge of the science of climatology.

In choosing an icon to illustrate the first section of this chapter, we chose Academician Kirill Yakovlevich Kondratyev (1920–2006), Member of the Russian Academy of Sciences, prominent scientist and outstanding geophysicist, a world-renowned expert in research on solar radiation, satellite meteorology, remote sensing of the atmosphere and the Earth's surface, and in his later years the founder of the science of global ecology, see Figure 6.1.

In constructing a model, it is necessary to take into account all the relevant natural phenomena and predict the future behaviour of the climate in the absence of any effects of human intervention. Then, it is necessary to model the effects of human activities as well as the natural phenomena and make predictions of the future climate on the basis of both the natural effects and the human activities. Then, the effect of the human activities can be determined by subtraction. Unfortunately, it is not possible to model the effect of the human activities directly on their own. The climate system is very complicated, our present knowledge is somewhat uneven, and our historical knowledge is very sparse indeed. However, in spite of all the difficulties, a great deal of effort has gone into climate modelling in recent years and some very useful results have been obtained; we shall discuss this later in this chapter as well as in Chapter 7.

A useful table of the sources of historical data on the climate is given on pages 102–107 of Lamb (1982). These include written records of climate conditions and of things related to the climate such as the price of agricultural products which reflect the success or otherwise of the harvest, records of standard meteorological measurements, lake, river, and sea bed sediments, tree rings, ice cores, ships logs, radioactive isotope measurements, pollen analysis, etc. The last 50 years have seen huge developments in the study of the Earth's climate.

(1) The advent of Earth-orbiting meteorological satellites has revolutionised the collection of meteorological data.
(2) Enormous developments in electronic computers and data storage and management systems have made it possible to run sophisticated climate computer models.

FIGURE 6.1   Academician Kirill Yakovlevich Kondratyev (1920–2006).

(3) New developments in archaeological techniques, particularly radiocarbon dating and ice core studies have provided new evidence of climate changes over tens or hundreds of thousands of years.

6.2   METEOROLOGICAL DATA

Meteorological data has been collected for several centuries at weather stations on the ground. A weather station has instruments and equipment to make observations of atmospheric conditions in order to provide information to make weather forecasts and to study the weather and climate. The measurements taken include temperature, barometric pressure, humidity, wind speed, wind direction, and precipitation amounts. Wind measurements are taken as free of other obstructions as possible, while temperature and humidity measurements are kept free from direct solar radiation, or insolation. Manual observations may be taken at least once daily, while automated observations are taken regularly and frequently and transmitted to the national meteorological service. Surface weather observations are the fundamental data used to forecast weather and issue warnings of extreme weather conditions. The International Standard Atmosphere, which is the model of the standard variation of pressure, temperature, density, and viscosity with altitude in the Earth's atmosphere, is used to correct a station pressure to sea level pressure. A 30-year average of a location's weather observations is traditionally used to determine the station's climate. However, the weather stations are sparsely and unevenly distributed. For example, see in Figure 6.2 the spatial distribution of radiosonde stations on a $5° \times 5°$ degree grid.

The atmosphere is, of course, a three-dimensional system and observations of the atmosphere as a function of height are made with radiosonde balloons. A radiosonde is a battery-powered telemetry instrument carried into the atmosphere usually by a weather balloon that measures various atmospheric parameters and transmits them by radio to a ground receiver. Modern radiosondes record and transmit the following variables: altitude, pressure, temperature, relative humidity, wind speed and wind direction, cosmic ray readings at high altitude and geographical position (latitude/longitude). Radiosondes measuring ozone concentration are known as ozonesondes. A radiosonde whose position is tracked by radar as it ascends to give wind speed and direction information is called a rawinsonde ("radar wind -sonde"). Most radiosondes have radar reflectors and are technically rawinsondes. Radiosondes are an essential source of meteorological data, and hundreds are launched all over the world daily.

A balloon filled with either helium or hydrogen lifts the device up through the atmosphere. The maximum altitude to which the balloon ascends is determined by the diameter and thickness of the balloon. As the balloon ascends through the atmosphere, the pressure decreases, causing the balloon to expand. Eventually, the balloon will burst, terminating the ascent. An 800 g balloon will

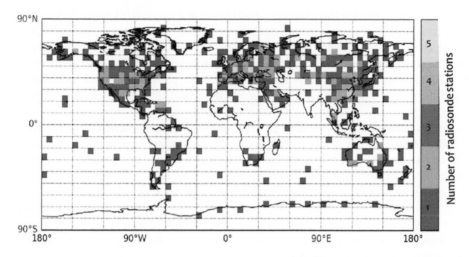

FIGURE 6.2 Map of radiosonde locations on a $5° \times 5°$ degree grid with number of stations indicated by colour. (Modified from Robert Junod, see Christy et al., (2018)).

burst at about 21 km. After bursting, a small parachute carries it back to Earth. A typical radiosonde flight lasts 60 to 90 minutes. The data are transmitted by radio and stored in a computer in real time.

Worldwide there are about 1,300 radiosonde launch sites. Most countries share data with the rest of the world through international agreements. Nearly all routine radiosonde launches occur 45 minutes before the official observation time of 0000 UTC and 1200 UTC, so as to provide an instantaneous snapshot of the atmosphere. This is especially important for numerical weather forecast modelling.

6.3   METEOROLOGICAL SATELLITES

The aim of the first weather satellites was to obtain global values of the variables that had, up to then, been obtained only at weather stations on the ground, i.e. temperature, pressure, humidity, etc. and throughout the height of the atmosphere from radiosondes. However, the use of satellites presented new opportunities to measure other quantities that play a key role in the dynamics of the atmosphere and therefore can be used to assist in studies of the climate. Faced with the opportunities provided by satellite observations and given the enormous cost of building and operating satellite systems and before an optimal global system of observations could be established, an analysis of priorities needed to be undertaken to propose the design specifications for climate monitoring satellites.

We can separate weather satellite programmes into two types: (1) operational programmes and (2) experimental programmes. Operational programmes can in turn be separated into two classes, polar-orbiting and geostationary, and they have been collecting data from 1960 so that there is now a valuable archive of data from which climates can be determined and hopefully changes in climates too. The individual satellites do not last forever but once a satellite fails it can immediately be replaced by a backup satellite which is already in orbit and so they have built up a long-term archive of climate data. Experimental programmes may involve only a single satellite or perhaps a small number of satellites but there is no guarantee of long-term continuity; they are usually designed for the study of a particularly restricted number of parameters, e.g. ozone concentration, trace gas concentrations, or ice cover.

As early as 1946, the idea of cameras in orbit to observe the weather was being developed. This was due to sparse data observation coverage and the expense of using cloud cameras on rockets. Following the launch of Sputnik in 1957, by 1958, the early prototypes for the TIROS (Television and Infrared Observation Satellite) and Vanguard (developed by the (US) Army Signal Corps) were created. The first weather satellite, Vanguard 2, was launched on 17 February 1959. It was designed to measure cloud cover for the first 19 days in orbit, but a poor axis of rotation and its elliptical orbit kept it from collecting a notable amount of useful data. It was also planned to determine the drag on the satellite by studying the change in its orbit. This was planned to provide information on the density of the upper atmosphere for the lifetime of the spacecraft which was expected to be about 300 years.

The first weather satellite to be considered a success was TIROS-1, launched by NASA on 1 April 1960. TIROS-1 operated for 78 days and proved to be much more successful than Vanguard 2. The operational meteorological satellites comprise both polar-orbiting and geostationary satellites. The first objective was to produce images of weather systems, i.e. cloud systems, to enable forecasters to visualise and predict the development of weather systems in a way that had not been possible previously and to assist in the presentation of forecasts. The weather satellites also produced material for meteorological research and to assist in the training of students of meteorology. The polar-orbiting weather satellites follow an orbit approximately 870 km above the surface of the Earth giving them a period of approximately 100 minutes.

TIROS-1 was the first in a long series of well-known polar-orbiting weather satellites operated by the US NOAA (National Oceanographic and Atmospheric Administration) which is still in operation. The spacecraft in this series operated under various names of TIROS, ESSA (Environmental Science Services Administration), ITOS (Improved TIROS Operational Satellite), and NOAA (US) (National Oceanic and Atmospheric Administration) up to NOAA-20 which was launched in 2017, see Table 6.1. There was a parallel US military programme, the DMSP (Defense Meteorological Satellite Program) with a set of similar spacecraft and instruments to the NOAA civilian programme. It had been planned to merge the NOAA programme and the DMSP programme to form the National Polar-orbiting Operational Environmental Satellite System (NPOESS) which was to be the United States' next-generation satellite system that would monitor the Earth's weather, atmosphere, oceans, land, and near-space environment. Eumetsat, the European meteorological satellite data service provider, has had a long-standing geostationary spacecraft programme (see below) and had been planning a polar-orbiting satellite series since the mid-1980s. The Eumetsat Polar System (EPS) consists of the ESA-developed MetOp (Meteorological Operational) satellites, MetOp-1, -2, and -3 which are compatible with the NOAA polar-orbiting satellites, see Table 6.1. The White House announced on 1 February 2010, that the NPOESS satellite partnership was to be dissolved, and that two separate lines of polar-orbiting satellites to serve military and civilian users would be reinstated. The NOAA/NASA portion is called the Joint Polar Satellite System (JPSS) and NOAA-20 is the first satellite in the programme. Since the early 1990s, NOAA and Eumetsat had been discussing planning future cooperation over polar-orbiting meteorological satellites and the present system involves Eumetsat using its MetOp satellites to take over the morning orbit from NOAA, while the afternoon orbit continues to be operated by the NOAA/NASA JPSS using NOAA-20.

The Joint Polar Satellite System (JPSS) is shown in Figure 6.3. For example, the orbits of the two operational NOAA/NASA JPS and Eumetsat polar-orbiting spacecraft are almost fixed relative to the instantaneous location of the centre of the Earth and the Earth rotates beneath them so that at any position on the Earth one obtains images every six hours. Each satellite is moving continuously in one of these orbits and the centres of the circles indicate some of the successive positions of the satellite. Each of these circles indicates approximately the area on the ground which can be seen from the spacecraft at a given time and is the area for which a ground station would be within range for receiving data transmitted from the satellite. The two solid curves on either side of the orbit indicate the swath that is covered by the satellite in that orbit. By the time the satellite has completed one orbit, that is about 100 minutes later, the Earth will have rotated by approximately 25°. Polar-orbiting weather satellites circle the Earth at a typical altitude of 870 km in a north to south (or vice versa) path, passing over the poles in their continuous flight. Polar-orbiting weather satellites are in

TABLE 6.1
Overview of some polar-orbiting meteorological satellite series

Satellite Series (Agency)	Launch	Major Instruments	Comments
NOAA-2 to -5 (NOAA)	21 October 1971, 29 July 1976	VHRR	2580 km swath
TIROS-N (NOAA-POES)	13 October 1978	AVHRR	>2600 km swath
NOAA-15 and NOAA-18, -19, -20	13 May 1998 to 18 November 2017	AVHRR/3	>2600 km swath
DMSP Block 5D-1 (DoD)	11 September 1976 –14 July 1980	OLS	3000 km swath
DMSP Block 5D-2 (DoD)	20 December 1982 – 4 April 1997	OLS, SSM/I	SSMIS replaces SSM/I
DMSP Block 5D-3 (DoD)	12 December 1999	OLS, SSM/1	Starting with F-16 (2001)
Meteor-3 series of Russia	24 October 1985	MR-2000M, MR-900B	3100 km, 2600 km swath
Meteor-3M series of Russia	2001 (Meteor-3M-1)		
FY-1A,-1B,-1C (CMA, China)	7 September 1988, 3 September 1990, 10 May 2000	MVISR	2800 km swath
MetOp-1 (Eumetsat)	19 October 2006	AVHRR/3, MHS, IASI	PM complement to NOAA-POES series
NPP, (NASA/IPO)	2005	VIIRS, CrIS, ATMS	NPOESS Preparatory Project
NPOESS (IPO)	2008	VIIRS, CMIS, CrIS	Successor to NOAA-POES and DMSP series

(Adapted from Kramer 2002).

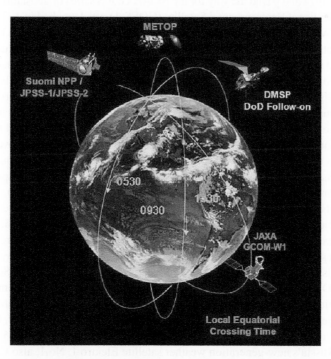

FIGURE 6.3 The Joint Polar Satellite System (JPSS) produced by NOAA (National Oceanic and Atmospheric Administration) through NASA. (https://earth.esa.int/web/eoportal/satellite-missions/j/jpss).

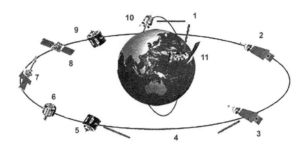

FIGURE 6.4   Main Polar Earth Orbit (PEO) (1) and Geostationary Earth Orbit (GEO) (4): GOES-E (USA) (2), GOES-W (USA) (3), GMS (Japan) (5), FY-2 (China) (6), IIISAT (India) (7), Electro-L (Russia) (8), Meteosat (EU) (9), Meteor (Russia) (10) (Modified from Ilčev, 2019).

sun-synchronous orbits, in other words their orientation relative to the line joining the centre of the Sun to the centre of the Earth stays the same, see Figure 6.4.

This means they are able to observe any place on Earth and will view every location twice each day with the same general lighting conditions due to the near-constant local solar time. Polar-orbiting weather satellites offer a much better resolution than their geostationary counterparts due their closeness to the Earth. In addition to the weather satellites operated by the United States and the MetOp satellites operated by Eumetsat, Russia has the Meteor and RESURS series of satellites, China has FY-3A, 3B, and 3C, and India has polar orbiting satellites as well.

The second class of operational weather satellites consists of geostationary satellites. The international meteorological community has committed itself to a series of operational geostationary satellites for meteorological purposes. The first geostationary weather satellite was NASA's ATS-1(Advanced Technology Satellite-1) which was launched in 1966, see Table 6.2. Six ATS satellites were launched between 1966 and 1974 and paved the way for the NOAA GOES series of geostationary satellites. The Geostationary Operational Environmental Satellite (GOES-1) was launched in 1975 and located over the Equator at 75°W and designated GOES-E. It was joined shortly afterwards by GOES-W at 135°W. Geostationary weather satellites are in an equatorial orbit above the Earth at altitudes of 35,880 km. Because the period in this particular orbit is 24 hours, these satellites remain stationary above a chosen point on the equator and can record or transmit images of the entire hemisphere below continuously with their visible-light and infrared sensors. The network of geostationary meteorological spacecraft consists of individual spacecraft which have been built, launched, and operated by a number of different countries; these spacecrafts are placed at intervals of about 60° or 70° around the equator. Given the horizon that can be seen from the geostationary height, this gives global coverage of the Earth with the exception of the polar regions. The objective is to provide the nearly continuous, repetitive observations needed to predict, detect, and track severe weather. This series of spacecraft is coordinated by the Co-ordination of Geostationary Meteorological Satellites (CGMS). These spacecrafts carry scanners that operate in the visible and infrared parts of the spectrum. The scanning is achieved by having the whole satellite spinning about its direction of motion, see Figure 6.5. They observe and measure cloud cover, surface conditions, snow and ice cover, surface temperatures, and the vertical distributions of pressure and humidity in the atmosphere. Images are transmitted by each spacecraft at 30-minute intervals, though from the very latest spacecraft, e.g. MSG (Meteosat Second Generation) and the NOAA-GOES Third Generation, this interval has been reduced to 15 minutes.

Several geostationary meteorological spacecrafts are now in operation, see Table 6.2. The United States' GOES series has three in operation: GOES-15, GOES-16, and GOES-17. GOES-16 and -17 remain stationary over the Atlantic and Pacific Oceans, respectively. GOES-15 will be retired in early July 2019. Russia's new-generation weather satellite Elektro-L No.1, also known as GOMS-2 operates at 76°E over the Indian Ocean. The Japanese have the MTSAT-2 located over the mid Pacific at 145°E and the Himawari 8 at 140°E. The Europeans have four in operation, Meteosat-8

TABLE 6.2
Overview of some geostationary meteorological satellites

Spacecraft Series (Agency)	Launch	Major Instrument	Comment
ATS-1 to ATS-6 (NASA)	6 December 1966 – 12 August 1969	SSCC (MSSCC ATS-3)	Technical demonstration
GOES-1 to –7 (NOAA)	16 October 1975 – 26 February 1987	VISSR,	1st generation
GOES-8 to –17 (NOAA)	13 April 1994 – 1 March 2018	GOES-Imager, Sounder	2nd generation
GMS-1 to -5 (JMA)	14 July 1977, 18 March 1995	VISSR (GOES heritage)	1st generation
MTSAT-1 (JMA *et al.*)	15 November 1999 (launch failure of	JAMI	2nd generation
MTSAT-1R (JMA)	H-2 vehicle) re-planned for 2003	JAMI	
Meteosat-1 to – 10 (Eumetsat)	23 November 1977 – 5 July 2012	VISSR	1st generation
MSG-1, -4 (Eumetsat)	28 August 2002 – 15 July 2015	SEVIRI, GERB	2nd generation
INSAT-1B to -1D (ISRO)	30 August 1983 – 12 June 1990	VHRR	
INSAT-2A to -2E (ISRO)	9 July 1992 – 2 April 1999	VHRR/2	Starting with -2E
INSAT-3B, -3C, 3°, 3D, 3E (ISRO)	22 March 2000 – 28 September 2003 Planned for 2001	VHRR/2	Communications only
INSAT-3A (ISRO)	Planned for Oct. 2001		Weather Satellite
MetSat-1 (ISRO)			only
GOMS-1 (Russia/Planeta)	31 October 1994	STR	1st Generation
Electro-M (Russia)	2005/6		2nd Generation
FY-2A, -2B (CMA, China)	10 June 1997, 26 July 2000	S-VISSR	
AVStar (Astro Vision Inc., Pearl River, MS, USA)	2003	Suite of 5 cameras	First commercial GEO weather satellite

(Adapted from Kramer 2002).

(3.5°W) and Meteosat-9 (0°) over the Atlantic Ocean and have Meteosat-6 (63°E) and Meteosat-7 (57.5°E) over the Indian Ocean. China currently has three Feng yun geostationary satellites (FY-2E at 86.5°E, FY-2F at 123.5°E, and FY-2G at 105°E) operated. India also operates geostationary satellites called INSAT which carry instruments for meteorological purposes. The Feng-Yun (Feng Yun = wind and cloud) meteorological satellite programme of the People's Republic of China includes both polar-orbiting and geostationary spacecraft. The Feng-Yun-1 series are polar-orbiting spacecraft; the first of which were launched in 1988, 1990, and 1999. Further information is given in Section G.3 of Kramer (2002).

A number of other countries, the USSR, Japan, India, and the European Space Agency (ESA) launched similar satellites to provide a ring of five or six well-spaced geostationary satellites to give near global coverage. The coverage by this early set of geostationary meteorological satellites is shown in Figure 6.6. Taken together, the polar-orbiting and geostationary weather satellites provide global coverage, whereas the ground-based weather stations do not provide uniform coverage but are heavily concentrated in densely populated land areas and are poorly represented in remote desert areas and do not gather data at sea. Where data are being fed as boundary conditions into a computer model, it is important to have a uniform spatial distribution of data points.

This is to be compared with the fact that the conventional weather stations cannot, by any stretch of the imagination, be described as uniformly distributed over the surface of the Earth, see Figure 6.2. Both the polar-orbiting and geostationary weather satellites carry multi-spectral scanners that produce images of weather systems in various spectral wavelength ranges. These satellites have been enormously successful in generating images of the cloud patterns in all sorts of types of weather systems.

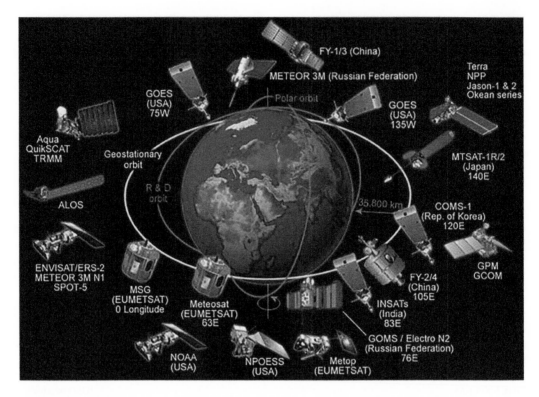

FIGURE 6.5   Global Operational Satellite Observation System Coverage. (Source from NOAA).

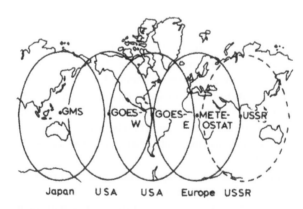

FIGURE 6.6   The coverage of geostationary meteorological satellites, *circa* 1982.

In addition to these operational systems, there have been several experimental satellite systems which have provided meteorological data for a period but with no guarantee of continuity of supply of that type of data. An overview of polar-orbiting and geostationary meteorological satellite programmes is given in Tables 6.1 and 6.2, respectively. These include the Earth Radiation Budget Experiment (ERBE), the Solar Backscatter Ultraviolet instrument (SBUV/2), and the Stratospheric Aerosol and Gas Experiment (SAGE) to name just a few. The Earth Radiation Budget Experiment (ERBE) is a NASA research instrument. ERBE data contribute to understanding the total and seasonal planetary albedo and Earth Radiation balances, zone by zone. This information is used for recognising and interpreting seasonal and annual climate variations and contributes to long-term climate monitoring, research, and prediction. The Solar Backscatter UltraViolet radiometer (SBUV) is a non-scanning, nadir viewing instrument designed to measure scene radiance in the ultraviolet spectral region from 160 to 400 nm. SBUV data are used to determine the vertical distribution of ozone and the total ozone in the atmosphere, and solar spectral irradiance.

The first task of a meteorological satellite is to provide images of cloud systems and their evolution and movements in the form of quantitative data usable for numerical weather analysis and prediction. They all have imaging instruments, originally vidicon cameras but now multispectral scanners operating at a number of different visible and infrared wavelengths. In addition to the main imaging instrument, the AVHRR (Advanced Very High Resolution Radiometer) (Cracknell 1997), each second generation (i.e. from TIROS-N (Figure 6.3) onwards) NOAA polar-orbiting satellite has an atmospheric sounding capability. This is provided by the TIROS Operational Vertical Sounder (TOVS). The TOVS is a three-instrument system consisting of

- The High-resolution Infrared Radiation Sounder (HIRS/2). The HIRS/2 is a 20-channel instrument for taking atmospheric measurements, primarily in the infrared region. The data acquired can be used to compute atmospheric profiles of pressure, temperature, and humidity.
- The Stratospheric Sounding Unit (SSU). The SSU is a three-channel instrument, provided by the United Kingdom, which uses a selective absorption technique. The pressure in a CO_2 gas cell in the optical path determines the spectral characteristics of each channel, and the mass of CO_2 in each cell determines the atmospheric level at which the weighting function of each channel peaks.
- The Microwave Sounding Unit (MSU). This four-channel Dicke radiometer makes passive microwave measurements in the 6.5 mm O_2 band. Unlike the infrared instruments of TOVS, the MSU is little influenced by clouds in the field of view.

The NOAA polar-orbiting satellites also carry sounding instruments, the TIROS Operational vertical Sounder (TOVS), which are used to generate atmospheric profiles similar to those generated by radiosondes, but there are important differences. The radiosondes generate directly the values of the atmospheric parameters, temperature, pressure, etc., but only at the locations at which weather stations have been established and not at a regular pattern of locations spread out uniformly over the surface of the Earth. Such regular patterns are important for producing data constituting boundary conditions for a weather forecast model or a climate model. The satellite sounder takes its measurements everywhere and so can provide values at all the grid points used in a model. However, it does not measure these atmospheric parameters directly but it records the intensity of infrared and microwave radiation in nearly 30 wavelength channels and involves an indirect inversion of the data to generate the values of the atmospheric parameters. The method used for inverting the data has had to be validated.

The next stage in meteorological studies by satellites involved a number of satellites aimed at studying the concentration of gases that have important radiative effects as greenhouse gases, some of them far more active, molecule for molecule, than CO_2. Three of the early satellites aimed at studying the composition of the atmosphere were SAGE (the Stratospheric Aerosol and Gas Experiment), UARS (the Upper Atmosphere Research Satellite), and AIRS (the Atmospheric Infrared Sounder). The first SAGE instrument, SAGE-1, was launched on 18 February 1979, to collect data on the various gases in the atmosphere, including ozone. The mission collected valuable data for nearly three years until the satellite's power system failed. SAGE-II was launched on board the ERBS (Earth Radiation Budget Satellite) in October 1984 and observed stratospheric O_3 from 1984 until 2006. Data from SAGE II were integral in confirming human-driven changes to O_3 concentrations in the stratosphere and thus influenced the decisions to negotiate the Montreal Protocol in 1987. Later, observations from SAGE-II showed that O_3 in the stratosphere stopped decreasing in response to the actions agreed to in the treaty.

TABLE 6.3

The SAGE-III International Space Station (ISS) legacy

Instrument	Era	Orbit/Platform	Channels	Science highlight
SAM	1975	Apollo-Soyuz	Single channel @ 850 nm	Demonstration
SAM-II	1978-1993	SSO/Nimbus-7 polar coverage	Single channel @ 1 nm	Polar stratospheric clouds
SAGE-I	1979-1981	Inclined/AEM-2 global coverage	Ozone, Aerosol, NO_2	Preceding ozone baseline
SAGE-II	1984-2005	Inclined/ERBS global coverage	Added water vapour, improved NO_2	Ozone trends, extreme aerosol variability
SAGE-III	2001-6	SSO/Meteor-3M	Added NO_3	Tropospheric measurements
• Meteor	2016 -	Inclined ISS orbit	Night-time O_3	Lunar occultation & limb
• ISS		-Int. Space Station	Mesospheric O_3	scattering
• FOO				

The SAGE family of instruments was pivotal in making accurate measurements of the amount of ozone loss in Earth's atmosphere. SAGE has also played a key role in measuring the onset of ozone recovery resulting from the internationally mandated policy changes that regulated chlorine-containing chemicals, the Montreal Protocol, which was passed in 1987. Later observations from SAGE-II showed that ozone in the stratosphere stopped decreasing in response to the implementation of that treaty. The history of this family of instruments is shown in Table 6.3.

For nearly 30 years, NASA's SAGE family of remote-sensing satellite instruments continuously measured stratospheric ozone (O_3) concentrations, aerosols, water vapour, and other trace gases. However, there has been roughly a decade-long gap in SAGE measurements since the SAGE-III Meteor-3M (launched in 2001) ended on 6 March 2006. The first SAGE mission (SAGE-I) was launched on 18 February 1979. The mission collected valuable data for nearly three years until the satellite's power system failed. SAGE-II was launched on board the ERBS (Earth Radiation Budget Satellite) in October 1984 and observed stratospheric O_3 from 1984 until 2006. Data from SAGE II were integral in confirming human-driven changes to O_3 concentrations in the stratosphere and thus influenced the decisions to negotiate the Montreal Protocol in 1987. Later, observations from SAGE-II showed that O_3 in the stratosphere stopped decreasing in response to the actions agreed to in the treaty. SAGE-III is the latest in a line of spaceborne instruments for studying the atmosphere using near UV, visible, and near IR radiation. In addition, SAGE-III provides high vertical resolution profile measurements of trace gases such as water vapour and nitrogen dioxide that play significant roles in atmospheric radiative and chemical processes.

Three copies of SAGE-III were produced. One instrument was mounted on the Meteor-3M spacecraft that was launched in 2001 and a second was stored awaiting a flight of opportunity. The third was launched on the International Space Station (ISS) and mounted looking directly down on the Earth. SAGE-III was launched from the Kennedy Space Center, Cape Canaveral SLC-39A, FL. on 19 February 2017.

The UARS satellite was launched in 1991 by the Space Shuttle Discovery and carried 10 instruments. UARS orbited at an altitude of 375 km above the surface of the Earth with an orbital inclination of 57°. It was designed to operate for three years, but six of its instruments functioned for over 14 years. UARS measured the concentration of ozone and the concentration of other chemical compounds found in the ozone layer (the stratosphere) which affect ozone chemistry and processes. UARS also measured winds and temperatures in the stratosphere as well as the energy input from the Sun. The system was designed to help define the role of the atmosphere in climate and climate variability. UARS was officially decommissioned on 14 December 2006. Most of the UARS atmospheric composition measurements were being continued with EOS AURA and all of the UARS solar irradiance measurements were being continued with SORCE (Solar Radiation and Climate Experiment) and ACRIMSAT (Active Cavity Radiometer Irradiance Monitor Satellite).

The AIRS is an instrument designed to support climate research and improve weather forecasting. It was launched into Earth orbit on 4 May 2002 on board AQUA which forms part of NASA's Earth Observing System (EOS) of satellites. AIRS is one of six instruments on board AQUA.

6.4   SATELLITES AND CLIMATE MODELLING

The first objective of the meteorological satellites that we have described above was to produce images of weather systems, i.e. cloud systems, to enable forecasters to visualise and predict the development of weather systems in a way that had not been possible previously and to assist in the presentation of forecasts. The weather satellites also produced material for meteorological research and to assist in the training of students of meteorology. Where data are being fed as boundary conditions into a weather forecasting computer model, it is important to have a uniform spatial distribution of data points. Such global coverage data can be obtained from the polar-orbiting and geostationary weather satellites taken together, whereas the ground-based weather stations do not provide uniform coverage but are heavily concentrated in densely populated land areas and are poorly represented in remote desert areas and do not gather data at sea, see Figure 6.2. These satellites have been enormously successful in generating images of the cloud patterns in all types of weather systems and in generating input data for computer models for weather forecasting.

We have noted that computer modelling of the climate developed from weather forecasting computer models. However, given the complicated nature of climate modelling, it is necessary to have far more types of global data available than are provided by the meteorological satellites that we have described so far. After a considerable amount of spadework, the WCRP Joint Scientific Committee identified the required information that climate modellers would need to obtain from satellite remote sensing from space in terms of land, sea, and cryosphere mapping and monitoring, see Table 6.4. This table represents an idealised situation and the extent to which these objectives can be achieved, or have been achieved, has been limited by technology and by financial resources; designing, building, launching, operating, and recovering the data from satellites is expensive. Naturally, in the global system of climate monitoring, apart from the decisive importance of space-based means of observation, it should be emphasised that this by no means belittles the necessity of further development of conventional techniques (ground, aircraft, and balloon observations).

Detailed sets of requirements for monitoring sea ice and snow cover parameters, as functions of time, were set out in Table 2.15 of Kondratyev and Cracknell (KC98), while requirements for measurements of atmospheric pollutants and other climate-forming factors were set out in Tables 2.17, 2.18, and 2.19 of Kondratyev and Cracknell (KC98). Some of these requirements were met by satellite programmes that already existed while others led to the development of new programmes specially designed to fulfil these requirements. While these tables have been used by individual climate modellers, the details need not concern us here because the scope for dedicated special programmes is limited by the enormous resources that need to be devoted to planning, building, and launching a dedicated satellite. The variables in Table 6.4 are envisaged as being needed as inputs to climate computer on a regular spatial grid of (x, y, z) at time t. Probably the first grid one might think of would be a regular square Cartesian grid with an identical spacing in each of three dimensions but one would soon realise that in meteorology heights are measured in pressures, i.e. hPa, not in distances, m. The next thought one might have would be to take the two horizontal coordinates in a chosen fraction of degrees of latitude and longitude. However, the problem arises that lines of longitude are closer together the nearer one approaches the poles and therefore the volume associated with each grid point decreases as one moves from the equator towards the poles. This leads to a problem if one attempts to produce a global average of e.g. temperature when cells would need to be weighted accordingly to correct for this. This can be avoided by using a projection of the Earth's surface that involves cells of equal area irrespective of latitude. This projection is used by the global high-resolution annual land cover maps (Gong et al. 2013).

In Table 6.4, it is assumed that t varies rather slowly with a temporal resolution of a few days. Cloudiness is a very important parameter because it affects the terrestrial energy flux but it varies

TABLE 6.4
WCRP requirements of surface characteristics measured using the techniques of remote sensing from space

Parameter	Spatial resolution (km)	Temporal resolution (repeatability) (days)	Accuracy
Sea surface temperature	200	5–10	0.5–1.5 °C
	500	5–10	0.5–1.5 °C
	maximum possible	30	0.5–1.5 °C
Land surface and	100	5	1.0 °C
ice cover temperature	500	5	0.1–3.0 °C
Wind stress near the ocean surface	200	5–10	0.1–0.4 dyn cm^{-2}
Sea currents and vortices	200	5–10	2–10 cm (topography)
Ice cover extent	100	5	presence/absence
	150	5–10	presence/absence
	50	5–15	presence/absence
Ice cover thickness	200	13–30	10–20%
Ice cover melting	50	5	yes/no
Ice cover drift	400	1	5 km
Snow cover extent	50	5–15	presence/absence
Soil moisture	100	5	10% field moisture content
	500	15	2 gradations
Surface albedo	100	5	0.1–0.3
	500	5–15	?

very rapidly, far too rapidly for anything other than some sort of temporal average to be used in a computer model. Cloudiness is therefore not included in Table 6.4. The problem of obtaining data on the three-dimensional cloud field remains unsolved. Information on cloud cover climatology must include the following data: global distribution of total cloud cover, types of clouds and the height of their upper boundary, interannual and seasonal variabilities of cloud cover, the diurnal change of low-level clouds and convective clouds, and statistical characteristics of cloud size. Apart from the main series of meteorological satellites that we have already discussed, there were a number of other satellites that in early days of climate computer models delivered data related to a number of the parameters in Table 6.4. Examples include particularly the Landsat and SPOT land applications satellites, some of the early active microwave (radar) programmes Seasat, Radarsat, ERS-1, and ERS-2 as well as the passive microwave scanners SMMR (Scanning Multichannel Microwave Radiometer) and SSM/I (Special Sensor Microwave/Imager).

6.5 THE NASA EARTH OBSERVING SYSTEM (EOS)

An important development was NASA's Earth Observing System (EOS). This is a coordinated series of polar-orbiting and low inclination satellites for long-term global observations of the land surface, biosphere, solid Earth, atmosphere, and oceans. By 1983, the NOAA polar orbiting meteorological satellites, the international set of geostationary meteorological satellites, and the Landsat system of (basically) land use/land cover satellites were well established and other experimental satellites, such as the Nimbus series and Seasat, had been launched. 1983 saw the first steps in the establishment of NASA's EOS; the story is told in NASA's publication The Earth Observer, volume 26, No.2, March-April 2014, pages 4-13.

About this time, a number of studies were made of the requirements to be met by any proposed satellite system for climate studies. Because of the numerous climatic parameters to be monitored and included in climate models and the difficulty of acquiring a complete system of satellites for climate monitoring, a minimum set of requirements needed to be established. Discussions took place at a meeting in NCAR (National Center for Atmospheric Research) (Boulder, Colorado, USA, July 1976) convened by the COSPAR (Committee on Space Research) Working Group-6 on the recommendation of the GARP (Global Atmospheric Research Program) Organizing Committee. The problems discussed were grouped as follows: (1) interaction of cloud cover and radiation, (2) the effect of land surface processes, (3) minor atmospheric greenhouse gases, (4) ocean–atmosphere interaction, and (5) cryosphere and hydrosphere, and processes taking place there. The main objective was to determine a set of climatic parameters and factors which can be remotely sensed from space, as well as to choose adequate observational means. A report was prepared (COSPAR 1978) containing the results of discussions on the available possibilities and prospects of satellite observations for studying the physical processes that determine the climate which is still very important and informative. Table 2.19 of Kondratyev and Cracknell (1998b pages 112–118) summarises these results. Designing, building, launching, and operating a satellite system, as well as receiving, processing, archiving, and distributing the data, is an expensive business and it is necessary to prioritise the items in a wish list such as the Table that we have just mentioned. The realistic inclusion of the effects of the oceans was seen early on as an important extension of the very early climate models. Hooper and Sherman (1986) analysed the requirements for satellite-derived oceanographic information determined by the needs of governmental interest, as well as of interest for scientific research and commercial organisations. Data in Table 6.5 characterise the

TABLE 6.5
Requirements for operational oceanographic information

Parameter	Error	Resolution Spatial (km)	Resolution Temporal	Delay in obtaining information
Winds				
Speed	2 ms^{-1}	25	12 h	3 h
Direction	10°	25	12 h	3 h
SST				
Global	1.0 °C	25	3 days	12 h
Local	0.5 °C	10	1 day	12 h
Roughness				
Substantial height of wave	0.3 m	25	12 h	3 h
Direction	10°	25	12 h	3 h
Ice cover				
Extent	15%	20	3 days	12 h
Thickness	2 m	50	3 days	12 h
Age	Fresh, 1-yr or multi-year	20	3 days	12 h
Height of glaciers	0.5 m variation	10	1 yr	1 month
Water masses				
Chlorophyll content	100%	0.4	2 days	8 h
Turbidity	Low, middle, high	0.4	1 day	10 h
Subsurface currents				
Velocity	5 cm s^{-1}	20	1 day	1 day
Direction	10°	20	1 day	12 days

(KC98 Table 2.16).

general requirements which form the basis for the development of a scientific complex for the United States national system of oceanographic satellites in order to obtain operational geophysical information. This table contains summaries of the special requirements of spatial and temporal resolution of various parameters for different fields of applications.

The period 1983 – 1988 saw the development of the concept of the EOS leading up to an Announcement of Opportunity (AO) for EOS in 1988. After 30 years, the system had matured to the stage shown in Figure 6.7.

SAGE-1 and SAGE-II, which we have already mentioned in Section 6.2, were developed pre-EOS but the SAGE system was taken over into the EOS programme. SAGE-III was developed within the EOS programme and was the latest in a line of spaceborne instruments for studying the atmosphere using near UV, visible, and near IR radiation.

In the early days, EOS concentrated on the development of three large satellites devoted to the study of the atmosphere, the land, and the seas. Each of these three satellites carried several instruments. The first of these satellites to be launched was TERRA (EOS-AM1) (18 December 1999) (Latin = Earth) and it was principally concerned with the land (Figure 6.7).

TERRA collects data about the Earth's bio-geochemical and energy systems using five instruments that observe the atmosphere, land surface, oceans, snow and ice, and energy budget. Each instrument has unique features that meet a wide range of science objectives. The five instruments on board TERRA are:

- ASTER (Advanced Spaceborne Thermal Emission and Reflection Radiometer). This is a high-spatial resolution multispectral scanner (an imager operating in a number of different wavelength ranges)
- CERES (Clouds and Earth's Radiant Energy System) – its purpose is described by its name.
- MISR (Multi-angle Imaging SpectroRadiometer). Whereas most satellite flown imagers look vertically down or to the horizon, this one has 9 cameras, one looking straight down, four looking forwards at various angles of inclination to the vertical and another four looking similarly aft.
- MODIS (Moderate-resolution Imaging Spectroradiometer). This is a multi-purpose imaging system which has found a wide range of applications.
- MOPITT (Measurements of Pollution in the Troposphere). This is concerned with monitoring the distribution, transport, sources, and sinks of carbon monoxide near the ground.

The second of these satellites to be launched was AQUA (EOS-PM1) (Latin = water) (4 May 2002). AQUA carries six instruments, two of which, MODIS and CERES, are copies of instruments already flown on TERRA. The six instruments are the

- AIRS (Atmospheric Infrared Sounder),
- AMSU-A (Advanced Microwave Sounding Unit),
- HSB Humidity Sounder for Brazil (HSB),
- AMSR-E (Advanced Microwave Scanning Radiometer for EOS),
- MODIS Moderate-Resolution Imaging Spectroradiometer.
- CERES (Clouds and the Earth's Radiant Energy) System (CERES).

Each instrument has unique characteristics and capabilities, and all six serve together to form a powerful package for Earth observations. The first three of these instruments are descendants of earlier atmospheric sounding instruments, principally TOVS, while the AMSR-E is a descendant of earlier (passive) microwave scanners, SMMR and SMM/I, and provides global sea surface temperatures which are very important for climate models.

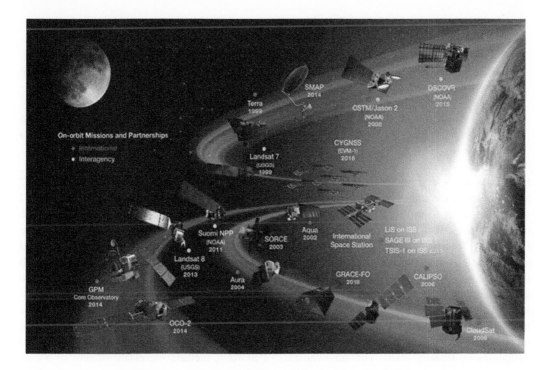

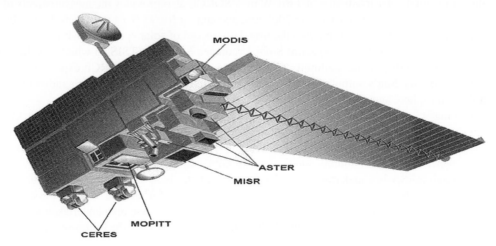

FIGURE 6.7 (Top) The Earth Observing System (EOS) comprises a series of artificial satellite missions and scientific instruments in Earth orbit designed for long-term global observations of the land surface, biosphere, atmosphere, and oceans. (Bottom) Terra spacecraft with instrument locations labelled.

The third of these satellites to be launched was a multinational satellite AURA (EOS-CH1) (Latin = air) (15 July 2004) and involving the Netherlands, Finland, and the UK. The four instruments flown on AURA were:

- HIRDLS (High-Resolution Dynamics Limb Sounder). This determines the temperature and the concentrations of O_3, H_2O, CH_4, N_2O, NO_2, HNO_3, N_2O_5, CFC11, CFC12, $ClONO_2$, and aerosols in the upper troposphere, stratosphere, and mesosphere.

- MLS (Microwave Limb Sounder). This was the first instrument to provide global measurements of OH, HO_2, and BrO, constituents that play an important role in stratospheric chemistry. MLS also provides unique measurements of the ice content of cirrus clouds.
- OMI (Ozone Monitoring Instrument). This is principally concerned with measuring ozone concentration and monitoring progress in the attempts to repair the ozone hole in the stratosphere following the adoption of the Montreal Protocol.
- TES (Tropospheric Emission Spectrometer). This was designed to monitor ozone in the lowest layers of the atmosphere (i.e. the troposphere) where ozone is regarded as a pollutant.

In addition to these three satellites, it will be seen from Figure 6.9 that EOS includes Landsat-7 and Landsat-8, the latest two satellites in the Landsat programme which began in 1972. It will also be apparent from this Figure that the EOS contains several other satellites which we have not mentioned. We mention just one more and that is SORCE. The Solar Radiation and Climate Experiment (SORCE) measures the incoming X-ray, ultraviolet, visible, near-infrared, and total solar radiation. It continues the precise measurements of total solar irradiance that began with the ERB instrument in 1979 and continued with the ACRIMSAT series of measurements (see also subsection 6.2). These measurements specifically address long-term climate change, natural variability and enhanced climate prediction, and atmospheric ozone and UV-B radiation. These measurements are critical to studies of the Sun, its effect on our Earth system, and its influence on humankind. The SORCE spacecraft was launched on 25 January 2003 and is only due to end operations in January 2020. The achievements of SORCE include the establishment of a new value of the total solar irradiance of 1361 W m^{-2}, SORCE also provides the measurements of the solar spectral irradiance from 1 nm to 2000 nm, accounting for 95 % of the spectral contribution to the total solar irradiance. It initiated the first daily record of solar spectral irradiance, and it has overlapped with the Total and Spectral Solar Irradiance Sensor-1 (TSIS-1) which was launched in 2017 and so will provide continuity of data. Data obtained by the SORCE experiment can be used to model the Sun's output and to explain and predict the effect of the Sun's radiation on the Earth's atmosphere and climate.

As a major component of the Earth Science Division of NASA's Science Mission Directorate, EOS enables an improved understanding of the Earth as an integrated system. The EOS Project Science Office (EOSPSO) is committed to bringing programme information and resources to the Earth science research community and the general public alike.

Although NASA has undertaken the development of the EOS programme and is perhaps the major player in the provision of satellite remote sensing data for climate studies, we should not overlook the fact that several other countries and space agencies have also undertaken major projects in this area too. These include China, India, Brazil, Russia, France, the European Space Agency (ESA), etc.

Remote sensing, by which in this context we mean principally observations from Earth-orbiting satellites, plays three major roles in terms of numerical modelling of atmospheric general circulation (AGC), i.e. weather and climate. They are (1) direct observation of meteorological and climatological variables on a global basis, (2) providing input data to numerical weather forecasting or climate models, and (3) validation of the outputs from these models. However, although in the global system of climate monitoring, the importance of space-based means of observation is decisive, it should be emphasised that this by no means belittles the necessity of further development of conventional techniques (ground, aircraft, and balloon observations). In this connection, it is relevant to recall the recommendations of the 1974 Stockholm Conference, see Section 1.2.3, as well as plans for satellite monitoring of climate developed in the USA and elsewhere. The progress in climate studies based on satellites in the first ten years of or so of the EOS programme, i.e. for the decade 1980–1990, is described in some considerable detail in Section 2.3 of Kondratyev and Cracknell (KC98).

The WCRP Joint Scientific Committee recommended three different classes of observational programmes related to climate modelling. These data must be used:

(1) to study the climate-forming processes and to develop techniques for their parameterisation; these must be complex and detailed, but they can be confined to one or several small regions, i.e. local, and short time intervals from several weeks to several years,
(2) to test the adequacy of global theoretical climate models that require global-scale observations over long time periods and must contain information about numerous characteristics of the climate system, and
(3) to perform long-term monitoring of slowly varying components of the climate system, which is important for assessing the present global climate, understanding natural factors of the formation and variation of climate, as well as studying the possibilities of climate forecasting.

In testing the adequacy of climate models by comparing them with observational data, there is the problem of objective analysis. Experience has shown that application of various techniques for objective analysis (with identical input data) can give substantially different results.

6.6   TRACE GASES AND POLLUTANTS

The urgency of assessing the anthropogenic impact on climate indicates numerous requirements for monitoring atmospheric pollutants (minor optically active gas and aerosol components). In this connection, an informative overview was made (Committee on Atmospheric Sciences 1980, NSF-UCAR Long-Range Planning Committee 1987) of possibilities for detecting and monitoring environmental pollution based on the technique of remote sensing from space. Table 2.17 of Kondratyev and Cracknell (1998a, b) characterises the atmospheric pollutants, objectives, and requirements in terms of measurement accuracy while Table 2.18 gives summaries of the components whose effect on the environment required further study. A very extensive table of the possibilities for studying climate-forming factors and processes using observations from space is given in Table 2.19 of Kondratyev and Cracknell (1998a, b). The main optically active minor gaseous components and particulate matter in the atmosphere are indicated in Table 2.2 and the sources of these materials are indicated in Table 2.3.

Important data have been accumulated on the basis of the application of spectroscopic techniques used on the manned orbital stations Soyuz, Salyut and Mir, as well as Nimbus-7 and SAGE and later satellites. In this connection, the development of spectroscopic techniques is very important. These techniques have been used to measure concentrations of such minor atmospheric components as O_3, CO, NH_3, H_2O, NO_2, CH_4, NO_x, SO_2, H_2S, CFCs, halogens, etc.

To study the numerous and complicated chemical processes taking place in the troposphere, it is necessary to employ space-based observational means for their investigation, principally to assess the content of optically active minor gaseous and aerosol components (MGACs). Natural and anthropogenic surface sources of MGACs are as follows: volcanoes, biogenic processes in vegetation cover, soil and oceans, burning of fossil fuel and plant residuals, functioning of internal-combustion engines, etc. Tropospheric sources of MGCs include lightning, photochemical reactions induced by ultraviolet (UV) solar radiation, and numerous chemical reactions which transform atmospheric properties. The urgency of adequate information about MGCs is determined by their important role as factors of air quality, visibility, acid rain, ozone layer depletion, and climate formation.

Studies of the chemical composition of the troposphere from satellite observations are aimed at (in priority): (1) determination of continental and oceanic sources and sinks of such optically and chemically active MGCs as CO_2, CO, CH_4, and other hydrocarbon components, N_2O, NH_3, $(CH_3)_2S$ (DMS), H_2S, OCS, and SO_2; (2) analysis of spatial and temporal variability (in particular, annual and latitudinal change, regionally and globally averaged trends) of various MGACs, in particular, H_2O, CO, O_3, NO_2, SO_2, acid and soot aerosols.

In this connection, it is important to emphasise that all first-priority components (H_2O, O_3, CO, and CH_4) participate in chemical and photochemical reactions of the formation and destruction of hydroxyl OH. Thus, OH is the key chemically active component of the troposphere, taking part in reactions with almost all the other components and governing their residence time in the troposphere (NO, whose lifetime is controlled by O_3, is the only important exception). The very short lifetime of OH (about a second) and its very low concentration (about 10^5 to 10^6 molecules cm^{-3}) exclude any possibility of reliable direct measurements of its content in the troposphere, determined by the reaction

$$O\left(^1D\right) + H_2O \rightarrow 2OH \tag{6.1}$$

with excited atoms of oxygen $O(^1D)$ resulting from ozone photolysis:

$$O_3 + \rightarrow O\left(^1D\right) + O_2 \tag{6.2}$$

The destruction of hydroxyl is controlled by two reactions:

$$OH + CO \rightarrow CO_2 + H \qquad OH + CH_4 \rightarrow CH_3 + H_2O \tag{6.3}$$

Thus, H_2O and O_3 are responsible for the photochemical formation of OH, whereas CO and CH_4 cause the chemical destruction of OH.

Table 6.6 is of some historical interest in that it contains a brief, but complex, outline of the plans for EOS in the early days of its planning is complex, but these plans have subsequently been changed and the evolution of the EOS programme over the last 40 years is a major subject (<EOSPSO. NASA.GOV> accessed 14 November 2019).

An important component of the global-scale observational system is provided by developments within the International Satellite Cloud Climatology Project (ISCCP) which have led to the choice of an operational algorithm to retrieve cloud characteristics. The International Satellite Cloud Climatology Project (ISCCP) was established in 1982 as part of the World Climate Research Program (WCRP) to collect weather satellite radiance measurements and to analyse them to infer the global distribution of clouds, their properties, and their diurnal, seasonal, and interannual variations. The resulting data sets and analysis products are being used to study the role of clouds in climate, both their effects on radiative energy exchanges and their role in the global water cycle.

The ISCCP cloud data sets provide our first systematic global view of cloud behaviour of the space and timescales of the weather yet covering a long enough time period to encompass several El Niño – La Niña cycles. Figure 6.8 is a snap-shot of clouds.

The ISCCP was planned to run from 1983 to 1988, but the necessity to obtain a more adequate database determined the extension of the project. Special attention was paid to the problem of obtaining data on precipitation (in the interests of the WCRP) with an accuracy of ± 1 cm and spatial resolution on land and over the oceans of 250 km × 250 km. Therefore, a suggestion was approved to accomplish a special Project on Global Precipitation Climatology, with the use of both conventional and satellite information.

An important stage in the development of remote sensing to assist in acquiring information about the wind field near the ocean surface involved the development of the European Earth Resources Satellite (ERS-1) which was launched in 1991 followed by ERS-2 which was launched in the same orbit in 1995. Their payloads included a synthetic aperture imaging radar, a radar altimeter, and instruments to measure ocean surface temperature and wind fields. Table 6.7 indicates the information about the sea surface conditions that can be obtained from these instruments. ERS-2 carried an additional sensor for atmospheric ozone monitoring. The two satellites acquired a combined data set extending over two decades. The ERS-1 mission ended on 10 March 2000 and ERS-2 was retired on 5 September 2011. Similar instruments were carried on the Japanese satellite JERS-1, which was launched in 1992. A French-American satellite TOPEX/POSEIDON was also launched in 1992; it

TABLE 6.6

Early plans for the measurement complex of satellites of the Earth Observation System (EOS)

Objective	Instruments	Basic characteristics	Parameters to be measured
Obtaining surface images and remote sounding	Slant-axis moderate-resolution video-spectrometer, MODIS-T	0.4–1.1 µm; 64 channels; spatial resolution 1 km	Biogeochemical cycles; ocean colour
	Nadir-viewing videospectrometer, MODIS-N	0.4–1.1 µm; 36 channels; 0.5 km	Biogeochemical cycles; temperature
	High-resolution videospectrometer, HIRIS	0.4–2.2 µm; 200 channels; 30 m	Biogeochemical cycles; geology
	Thermal video-spectrometer, TIMS	3–5, 8–15 µm; 9 channels; 30 m	Biogeochemical cycles; geology
	High-resolution multifrequency microwave radiometer	1.4–91 GHz; 6 channels; 1.1 km × 1.8 km at 91 GHz	Radiobrightness temperature of the ocean and land
	Lidar sounder and altimeter, LASA	A set of wavelengths; the receiving mirror is 1.25 m in diameter	Atmospheric correction; aerosols and clouds soundings; altimetry
	Advanced system for data collection and coordination referencing	Referencing accuracy; 2–5 km	Mainly, data from sea buoys
Active sounding in radiowave region	Synthetic-aperture radar, SAR	Bands C, L, X at different polarisations; 30m	Characteristics of land, vegetation, and ice cover
	Radio-altimeter, ALT Scatterometer	Height measuring accuracy 10 cm Wind speed retrieval accuracy ±2 ms^{-1}	Surface topography Wind shear near the ocean surface
Monitoring of physical and chemical characteristics of the atmosphere	Correlation radiometer		Retrieval of the vertical profiles of CO and NH$_3$ concentrations
	Nadir-viewing Fourier spectrometer		Retrieval of NO$_3$, CH$_4$, H$_2$S
	Lidar sounder-altimeter		Profiles of temperature, pressure, water vapour, and ozone in the troposphere
	Doppler lidar		Tropospheric wind
	IR radiometer		Limb scanning of O$_3$, N$_2$O, etc.
	Pressure-modulated radiometer		Limb scanning for CFC1$_3$, etc.
	Sub-millimetre spectrometer		Limb scanning for OH, HC1,
	Microwave limb sounder		etc.
			Sounding the gas composition of the atmosphere
	Visible and IR spectrometers Fabry-Perot interferometer		Limb sounding of CO, O$_3$, etc. Profiles of wind and temperature
	Cryogenic Fourier spectrometer		Profiles of wind and temperature
Monitors	Solar constant; particles and fields; Earth's radiation budget		

(From Kondratyev and Cracknell (1998a, b, Table 2.20).

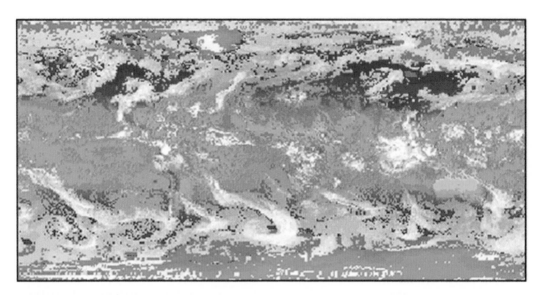

FIGURE 6.8 A single global snap-shot of clouds from the ISCCP website (https://www.isccp.giss.nasa.gov accessed 14 November 2019).

TABLE 6.7

Geophysical parameters that can be retrieved from satellite radar data

Parameter	Range	Error	Instruments
Wind speed (ms−1)	4–24	±2 ms−1 or 10%	Scatterometer and radar-altimeter
Wind direction	0–360°	±20°	Scatterometer
Significant wave- height (m)	1– 20	±0.5 m or 10%	Radar-altimeter
Wave direction	0–360°	±15°	SAR (synthetic-aperture radar)
Wavelength (m)	500–1000	±20%	SAR
Images: ocean; coastal zones; ice cover; land	Minimum viewing bandwidth 80 km	Geometric radiometric resolution: (a) 30 m/3.5 dB; (b) 100 m/1.0 dB	SAR
Satellite altitude: over the ocean; over the polar glaciers	745–845 km	2 m (absolute) ±10 cm (relative)	Radar-altimeter
SST	Viewing bandwidth 500 km	±0.5 K	Multi-channel radiometer scanning along the orbit
Water vapour content	Spatial resolution 25 km	10%	Microwave radiometer

was in a circular orbit 1320 km high (the inclination angle 63.1° providing a 10-day repeatability). The measuring complex includes an advanced radiometer (13.6 and 6.3 GHz) and a three-channel microwave radiometer (18, 21, and 37 GHz). The latter is designed to solve problems of atmospheric correction which ensures the retrieval of data on sea surface level with an accuracy of 1–3 cm.

Jason-1 and Jason-2 (launched in 2001 and 2008, respectively) continued the same sea surface temperature measurements begun by the ERS-1, ERS-2, and JERS-1 and Topex-Poseidon satellites in the Ocean Surface Topography Mission and are planned to be followed by a planned future Jason-3. The data collected by various satellites carrying optical, infrared, and microwave instruments over the years since 1978 (e.g. see Fowler et al. 2004) have made enormous contributions to the studies of the ocean (a) to retrieve, for example, sea surface temperature, wind speed and wind shear near the surface, as well as parameters of the ocean topography, etc.; (b) to provide input data for atmospheric and oceanographic computer models; and (c) to validate the predictions of these models.

As noted above, in connection with further development of studies within the WCRP and planning of the International Geosphere-Biosphere Programme (IGBP), a Global Energy and Water Cycle Experiment (GEWEX) was proposed (WMO-World Meteorological Organization 1987, 1988). The GEWEX project is dedicated to understanding Earth's water cycle and energy fluxes at the surface and in the atmosphere. GEWEX is a network of scientists gathering information on and researching the global water and energy cycles, which help to predict changes in the world's climate. The International GEWEX Project Office, or IGPO, supports these activities by planning meetings, implementing research goals, and producing a quarterly newsletter to keep the GEWEX community informed.

GEWEX coordinates science activities to facilitate research into the global water cycle and interactions between the land and the atmosphere. One of the primary influences on humans and the environments they live in, the global water cycle encompasses the continuous journey of water as it moves between the Earth's surface, the atmosphere, and beneath the Earth's surface. Clouds, precipitation, water vapour, surface radiation, aerosols, and other phenomena each play a role in the cycle. Many GEWEX scientists conduct research on those and other elements to help fine-tune our understanding of them and their impact on the climate. GEWEX also points out important gaps in knowledge and implements ways to fix those gaps, whether through new studies, reviews of data sets, gatherings of experts, or other opportunities. (www.GEWEX.ORG Accessed 16 November 2019).

6.7   EARTH RADIATION BUDGET, CLOUDS, AND AEROSOL

In all kinds of satellite observations, obtaining the observational series, the duration of which in some cases already exceeds at the present time two decades, plays the key role. This involves the necessity (when archiving the data) to ensure the reliability of calibration and to take into account the characteristics of instruments, orbital parameters, etc. For example, the needs of the problem of climate change are known to determine the need for data on the global mean ERB components with an error of not more than 1 Wm^{-2}, whereas in the case of regional averages the error can increase to 5 Wm^{-2}, and the errors in estimating the long-term trends must not exceed 1 Wm^{-2}. Hence, very strict requirements have to be met for the stability of the instruments and the reliability of their calibration.

A proposed set of requirements for earth radiation budget (ERB) data is shown in Table 6.8.

One of the efforts of NOAA has been the development of algorithms to retrieve the amount and radiative properties of clouds from the data of the five-channel AVHRR, as well as algorithms to retrieve air temperature and pressure at the cloud top level and cloud liquid water content from the data of the TOVS. Beginning from 1989, the aerosol optical thickness (AOT) of the atmosphere over the oceans has been retrieved regularly from the data of one shortwave AVHRR channel, and in the future, the results of observations at two to three wavelengths are planned to be used to estimate not only the aerosol optical thickness but also the aerosol number density. An important problem in the future will be the further development and improvement of algorithms to retrieve: (1) the ERB components, including the spectral distribution of outgoing longwave radiation (OLR) and radiative heat flux divergence from the TIROS Operational Vertical Sounder (TOVS) data, (2) surface radiation budget component, (3) cloud cover characteristics (cloud amount, temperature and pressure at cloud top level, water and ice content), and (4) total number concentration and size distribution of aerosol (not only over the ocean but also over land). Of particular importance is meeting the requirements of the Global Energy and Water Cycle Experiment (GEWEX) and an experiment on studies of the tropical ocean and global atmosphere (TOGA) for observational data.

TABLE 6.8.

Spheres of application and requirements for ERB data

Application	ERB components		Spatial resolution (km)	Temporal resolution		Error (W m⁻²)
				Period of averaging	Repeatability	
Input information	Solar constant		–	1 month	Dailly	0.2–1
for monitoring and	OLR	Global means	–	1 month	1–3 hours	0.2–1
numerical modelling of	OSR	Zonal means	–	1 month	1–3 hours	1–3
climate	SRB	Regional means	100–200	1 month to 1 day	1–3 hours	2–5
	OLR	Daily means	100–250	1 month	1–3 hours	2–5
	OSR	Synoptic			Synoptic periods	
	SRB	studies	100–250	Without averaging		10
	SRB	components	100–250	1 month	1–3 hours	10
Input data and	OLR, OSR, SRB:					
validation of numerical	Regional means		100–250	Without averaging	1–3 hours	10
weather forecasts						

Note: OLR = outgoing longwave radiation; OSR = outgoing shortwave radiation; SRB = surface radiation budget. (KC98 Table 2.24).

The Copernicus Sentinels programme developed by the European Space Agency consists of a family of missions designed for the operational monitoring of the Earth system with continuity up to 2030 and beyond (Table 6.9). The Sentinels' concept is based on two satellites per mission necessary to guarantee a good revisit and global coverage and to provide more robust data sets in support of the Copernicus Services. On-board sensors include both radar and multi-spectral imagers for land, ocean, and atmospheric monitoring: Sentinel-1 is a polar-orbiting, all-weather, day-and-night radar imaging mission for land and ocean services. Sentinel-1A was launched on 3 April 2014 and Sentinel-1B on 25 April 2016. Sentinel-2 is a polar-orbiting, multi-spectral high-resolution imaging mission for land monitoring to provide, for example, imagery of vegetation, soil and water cover, inland waterways, and coastal areas. Sentinel-2 can also deliver information for emergency services. Sentinel-2A was launched on 23 June 2015 and Sentinel-2B followed on 7 March 2017. Sentinel-3 is a polar-orbiting multi-instrument mission to measure sea-surface topography, sea- and land-surface temperature, ocean colour and land colour with high-end accuracy and reliability. The mission will support ocean forecasting systems, as well as environmental and climate

TABLE 6.9

Copernicus Sentinels launch dates and relevance in modelling application (aggregate use refers to composites in space and time, while direct use refers to single observation)

Sentinel 1	Launch Date	ESM Relevance
Sentinel 1A	3 April 2014	aggregate
Sentinel 1B	25 April 2016	aggregate
Sentinel 2A	23 June 2015	aggregate
Sentinel 2B	7 March 2017	aggregate
Sentinel 3A	16 February 2016	direct
Sentinel 3B	25 April 2018	direct
Sentinel 5P	13 October 2017	direct
Sentinel 6	2020	(expected) direct

monitoring. Sentinel-3A was launched on 16 February 2016 and Sentinel-3B has been launched on 25 April 2018. The Sentinel-5 Precursor – Sentinel-5P – polar-orbiting mission is dedicated to trace gases and aerosols with a focus on air quality and climate. It has been developed to reduce the data gaps between the Envisat satellite – in particular the Sciamachy instrument – and the launch of Sentinel-6. Sentinel-5P has been orbiting since 13 October 2017. Sentinel-4 is devoted to atmospheric monitoring that will be embarked on a Meteosat Third Generation-Sounder (MTG-S) satellite in geostationary orbit and it will provide European and North African coverage. Sentinel-5 will monitor the atmosphere from polar orbit aboard a MetOp Second Generation satellite. Sentinel-6 will be a polar-orbiting mission carrying a radar altimeter to measure global sea-surface height, primarily for operational oceanography and for climate studies. Copernicus Sentinels launch dates and relevance in modelling application (aggregate use refers to composites in space and time, while direct use refers to single observation).

6.7.1 THE STRATOSPHERE

In this area of observations, the problem lies in estimating the vertical distribution and total ozone content (TOC), air temperature, and UV solar radiation flux. The basic requirements for observational data reliability are that the errors in the total ozone concentration trends on a decade-long timescale do not exceed 1–1.5 per cent, that errors in the vertical profile of ozone concentration do not exceed 3–5 per cent, and that errors in the stratospheric temperature do not exceed 1–1.5 per cent. Since the dynamics of the ozone layer depend on the variability of extra-atmospheric UV solar radiation in the wavelength interval 170–240 nm (radiation at these wavelengths is responsible for the formation of atomic oxygen in the upper stratosphere), a particular problem is the monitoring of UV radiation variations (which can reach 20 per cent during an 11-year solar cycle). A reliable retrieval of the natural variability of stratospheric ozone content forms the basis for reliable recognition of anthropogenically induced effects.

One of the most poorly estimated stratospheric parameters is its temperature; since the top height of radiosondes is confined, as a rule, to altitudes below 30 km, rocket sondes are rarely launched and remote sounding techniques (especially lidar) have not been adequately developed. Meanwhile, information about the trend of the atmospheric temperature is very important. As demonstrated by numerical modelling of climate changes caused by the enhanced atmospheric greenhouse effect, the stratospheric response has been manifested as a cooling. It is very important here to consider the interactive variations in the ozone content and temperature.

The NOAA-9 satellite launched in March 1985 was the first to carry the SBUV-2 (Solar Backscattered Ultraviolet) instrument, which was designed to obtain total ozone content data. An improved version of this instrument was installed on NOAA-11, which was launched in January 1989. Though these observations have been planned as operational (following the observations, from 1978 onwards, with the Nimbus-7 SBUV complex), they cannot be considered climatic, because of problems of data interpretation. A new stage of ozonometric observations has been introduced with the Upper Atmosphere Research Satellite (UARS) and Meteor/TOMS instrumentation.

Initial data from remote temperature sounding of the atmosphere were obtained in 1969 via the Satellite Infra-red Spectrometer (SIRS) developed for the Nimbus series satellites. In 1978, the TIROS-N series of NOAA satellites began to be employed for operational remote temperature sounding of the atmosphere. The current TOVS complex comprises three units: the HIRS/2 (High Resolution Infrared Spectrometer), the MSU (Microwave Sounding Unit) and the SSU (Stratospheric Sounding Unit). Each instrument scans in a plane normal to the orbit, with a spatial resolution between about 17 km and 200 km. This ensures the retrieval of temperature vertical profiles in the 1000–0.4 hPa (0–54 km) atmospheric layer, with a vertical resolution of about 10 km. Of course, consideration must be given to the problems of retrieval error, and the reliable determination of trends is very important. Comparison with rocket sounding data shows that the differences between temperatures at levels 5, 2, 1, and 0.4 hPa reached, respectively, and on average 6, −3, −7, and 6°C, with considerable temporal variability; for example, errors at the 2 hPa level varied between 0 and 6 °C.

In situ surface observations are an important way of excluding the drift of the space-borne instruments' sensitivity in ozone and temperature observations. To achieve this aim, atmospheric monitoring is under way in order to provide homogeneity of space-based observations via the use of surface observations at various locations; five stations are in place, including the tropics, mid-latitudes of both hemispheres, the Arctic and the Antarctic, with the use of reliably calibrated instruments. Lidar sounding is also a major source of control information about the stratospheric temperature. Simultaneous remote soundings in the UV, visible, infra-red, and microwave intervals are planned.

Since SBUV/TOMS data make it possible to retrieve the content of SO_2 in the stratosphere, at least after volcanic eruptions when the SO_2 content is substantial, a new retrieval algorithm is being developed to facilitate the collection of both O_3 and SO_2 data.

Concordiasi was a field campaign, part of the Thorpex experiment dedicated to the study of the atmosphere above Antarctica as well as the land surface of the Antarctic continent and the surrounding sea-ice. In 2010, an innovative constellation of balloons, including 13 drift-sondes, provided a unique set of measurements covering both a large volume and time. The balloons drifted for several months on isopycnic surfaces in the lowermost stratosphere around 18 km, circling over Antarctica in the winter vortex, performing, on command, hundreds of soundings of the troposphere. Many dropsondes were released to coincide with overpasses of the Meteorological Operation (MetOp) satellite, allowing comparison with data from the Infrared Atmospheric Sounding Interferometer (IASI). IASI is an advanced infrared sounder that has a large impact on Numerical Weather Prediction systems in general. However, there are some difficulties in its use over polar areas because the extremely cold polar environment makes it more difficult to extract temperature information from infrared spectra and makes it difficult to detect cloud properties (Parsons et al., 2017; Balsamo et al., 2018).

6.7.2 THE TROPOSPHERE

From the viewpoint of monitoring climate dynamics, satellite observations of atmospheric general circulation (AGC) characteristics (including water vapour and cloud cover) are essential. In this respect, the collection of three levels of data is planned:

1. direct observation data on outgoing radiation (both shortwave and longwave), with special emphasis on the homogeneity of the observation series (removal of differences caused by special properties of different instruments is essential),
2. retrieved values of geophysical parameters, and
3. data of four-dimensional assimilation.

Such data will allow reprocessing of observational results if this should prove to be necessary as a result of the development of more reliable retrieval algorithms.

Atmospheric remote sounding accumulated (archived) data from 1969 include observational results from the following instruments: SIRS-A (Nimbus-3 satellite, 11–15 μm spectral interval, spatial resolution in nadir 250 km, 1969–1972), YTPR (NOAA-2-NOAA-5, 12–19 μm, 1972–1978), and TOVS (TIROS-N, NOAA-6, October 1978 to date).

The data base obtained from these instruments includes the following information, with a nominal horizontal resolution of 60 km: vertical profiles of temperature and humidity, total ozone content, pressure at top level and effective cloud cover, sea and land surface temperatures, outgoing longwave radiation, cloud-induced longwave radiation forcing, rain rate, microwave emissivity of snow and ice cover; and wind field (from either cloud or inhomogeneous water vapour distribution motions).

In order to interpret data from tropospheric observations, one should consider the cloud-radiation interaction in the context of the global energy balance. This requires reliable information on water vapour content and cloud cover characteristics. Available space-derived data include information on total water content of the atmosphere above the ocean (SSM/I, SSMR), and relative humidity of the upper troposphere (6.7 μm channel, GOES satellite, and equivalent TOVS data). The use of data for channel 2 of the MSU instruments is very important for intercalibrations. Since instrument sensitivity could have changed over time, retrospective data must be reprocessed to secure observation series homogeneity.

A comparison between short-range forecasts and the Concordiasi data was investigated for various Numerical Weather Prediction centres. Results show that models suffer from deficiencies in representing near-surface temperature over the Antarctic high terrain. The very strong thermal inversion observed in the data is a challenge in numerical modelling because models need both a very good representation of turbulent exchanges in the atmosphere and of snow processes to be able to simulate this extreme atmospheric behaviour. It has been shown that satellite retrievals also have problems representing the sharp inversions over the area. An example of comparison of such profiles is provided in the Figure 6.9. At the surface, particular attention has been paid to observing and modelling the interaction between snow and the atmosphere.

This research has led to an improvement of snow representation over Antarctica in the Integrated Forecasting System (IFS) model at the European Centre for Medium-range Weather Forecast (Rabier et al. 2010). Coupled snow–atmosphere simulations performed at Météo-France with the Crocus and Applications of Research to Operations at Mesoscale (AROME) models have been shown to realistically reproduce the internal and surface temperatures of the snow as well as boundary layer characteristics (Brun et al., 2011). In this example of the Concordiasi campaign, data sets obtained from the field experiment over snow and sea ice have proved invaluable to document the quality and deficiencies of both satellite retrievals and Numerical Weather Prediction model fields. This has helped to provide directions in which to explore the improvement of models, which will be beneficial for weather prediction and climate monitoring. The validation and diagnostics of numerical weather modelling over snow and sea-ice with Earth observations pose challenges because of the

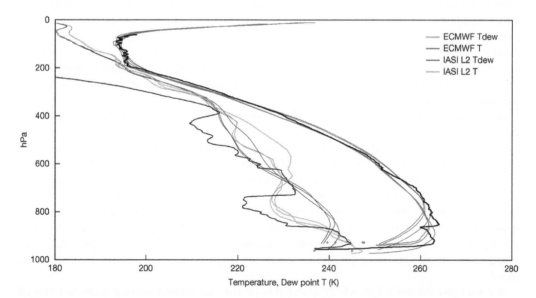

FIGURE 6.9 A dropsonde vertical profile of temperatures (in black) on 23/10/2010 together with nearest ECMWF model profiles (in red and orange) and EUMETSAT IASI retrievals (in magenta and blue).

lack of regular in situ data and because satellite observations can be difficult to interpret in those areas. Field experiments can help to provide precious information to assist scientists in this validation work (Balsamo et al., 2018).

6.8 VOLCANIC ERUPTIONS

Volcanic eruptions are some of the major unpredictable natural events that have major consequences for the climate. There are a number of written records of climatic effects that were not realised at the time to be consequences of volcanic eruptions.

6.8.1 THE ERUPTION OF CIRCA AD 535-6

Probably the most widespread collection of historical writings related to a volcanic eruption is that put together by David Keys (1999). These relate to an eruption of a volcano in A.D. 535, or thereabouts, that was thought to have been in the Sunda Straits that now separates the Indonesian islands of Java and Sumatra. This is slightly to the west of Krakatoa which erupted in 1883 (Winchester 2004). It is even argued (see Keys) that previously Java and Sumatra constituted one island that was ripped apart by this volcanic eruption and the caldera sank below sea level and the sea flooded in to form the present Straits. It is thought that this was the largest volcanic eruption in the last 50,000 years and Keys collected written evidence of its effects from a large number of countries and we quote extensively from Chapter 30 of Keys' book. These included Chinese records of having heard the explosion from Nanjing as well as records of droughts and floods. Keys even argued that the volcanic eruption was responsible for a number of social and political changes too.

Keys quotes from John of Ephesus, a 6th-century historian and prominent church leader writing in 535–536. 'There was a sign from the sun, the like of which had never been seen and reported before. The sun became dark and its darkness lasted for 18 months. Each day, it shone for about four hours, and still this light was only a feeble shadow. Everyone declared that the sun would never recover its full light again.' In the 530's the Roman Empire was resurgent. Under Justinian's rule generals led seaborne expeditions to recapture Italy, North Africa, and parts of southern Spain. In order to record the progress of reconquest, the historian Procopius was sent with the expeditionary forces. In 536 the general Belisarius sailed west to North Africa but, as Carthage and other cities were retaken, the skies began to darken and the sun was dimmed. 'During this year that a most dread portent took place. For the sun gave forth its light without brightness like the moon during this whole year, and it seemed exceedingly like the sun in eclipse, for the beams it shed were not clear, nor such as it is accustomed to shed,' said the historian – a top Palestinian-born government and military official.

Another 6th-century writer, Zacharias of Mytilene, was the author of a chronicle containing a third account of the 535/536 "Dark Sun" event. "the sun began to be darkened by day and the moon by night," he recorded. A fourth account was written by a Roman official and academic of Anatolian origin, known as John the Lydian, who reported that "the sun became dim for nearly the whole year."

All these reports were compiled by eyewitnesses in the Roman imperial capital, Constantinople. But in Italy, a very senior local civil servant also recorded the solar phenomenon. "The sun seems to have lost its wonted light, and appears of a bluish colour. We marvel to see no shadows of our bodies at noon, to feel the mighty vigour of the sun's heat wasted into feebleness, and the phenomena which accompany a transitory eclipse prolonged through almost a whole year," wrote Cassiodorus Senator in late summer 536. "The moon too, even when its orb is full, is empty of its natural splendour," he added.

It wasn't just the sun's light which appeared to be reduced. Its heat seemed weakened as well. Unseasonable frosts disrupted agriculture. "We have had a spring without mildness and a summer without heat," wrote Cassiodorus. The months which should have been maturing the crops have been chilled by north winds. Rain is denied and the reaper fears new frosts.

In normally warm Mesopotamia, the winter was 'a severe one, so much so that from the large and unwonted quantity of snow, the birds perished' and there was 'distress among men,' says the chronicle written by Zacharias of Mytilene.

John of Ephesus (reported through Michael the Syrian) said that 'the fruits did not ripen and the wine tasted like sour grapes,' while John the Lydian noted that 'the fruits were killed at an unseasonable time.'

On the other side of the planet, the abnormal weather was also being recorded. The Japanese Great King is reported in the ancient chronicle of Japan (the Nihon shoki) to have issued an edict lamenting hunger and cold: 'Food is the basis of the empire. Yellow gold and ten thousand strings of cash cannot cure hunger. What avails a thousand boxes of pearls to him who is starving of cold?' said the King.

We have already mentioned China; the disaster is chronicled in greater detail. In 535, there was a massive drought in the north of the country. The Bei shi (the north-Chinese chronicle) says in an entry for late April/early May that 'because of drought, there was an imperial edict which ordered that in the capital [Chang'An], in all provinces, commanderies, and districts, one should bury the corpses.' By the fifth month, the situation had deteriorated to such an extent that in the capital itself the government 'was forced to provide water' for the population 'at the city gates.'

Soon the drought had become so intense that hundreds of thousands of square miles of normally fertile or semi-fertile land became totally arid. The evidence suggests that huge dust storms began to rage.

Between 11 November and 9 December 535, the capital of south China, Nanjing itself, was deluged by dust falling from the sky, 'yellow dust rained down like snow'. The time of year, the colour, and the apparent quantity strongly suggests that this dust from the sky was, in fact, a yellow-coloured fine sand called loess which had been carried by the wind from the interior of China. In normal conditions, loess dust comes only from the Gobi Desert and other inland and areas – and storms only affect areas hundreds of miles north and west of Nanjing. But in extreme drought conditions, when unusually large areas become arid, much wider areas can be inundated by the dust. As the drought worsened, the Bei shi says that in 536 in the central-Chinese province of Xi'an, seven or eight out of every ten people died. Survivors were forced to eat the corpses of the dead.

As the months rolled on, the climate became increasingly bizarre. The Bei shi reports that in some areas of north China (Bias, Si, Zhuo, and Ran) hail fell in September 536 – but there was still 'a great famine'. Between 29 November and 27 December 536 and again in February 537, in the south-Chinese capital Nanjing, even greater dust storms covered the city in a saffron-coloured blanket: 'Yellow dust rained down like snow. It could be scooped up in handfuls,' said the Nan shi (History of the Southern Dynasties).

In early 537 in nine provinces of north China, the drought continued, but was increasingly interrupted by hail. Then finally, in 538, the drought ended, but the climatological chaos continued – there were now huge floods. In the summer of that year, the toads and frogs were said to be croaking from the trees' so torrential was the rain. The instability continued into the 540s with major droughts in 544, 548, 549, and 550.

In Korea, the situation was appalling. 535–542 had the worst climate recorded for the peninsula for any time in the 90-year period 510–600, and 535/536 was the worst 24 months in that nine-decade time span.

This mid-sixth-century climatic disaster also struck the Americas, the steppes of Russia, western Europe, and other regions. But many of these areas left no written records. It is a plethora of non-written sources that must, therefore, provide the evidence for the climatic situations in these regions.

What happened in 536 was recently confirmed by an expedition, led by Harald Sigurdsson and recorded by Ken Wohletz of the Los Alamos National Laboratory in the USA. On the seabed between the Indonesian islands of Java and Sumatra, Sigurdsson's bathyspheres discovered a huge caldera with deposits which could be reliably radiocarbon-dated to the 6th century. This volcanic depression was about 40-60 km across and it was close to the point where the island of Krakatoa exploded in 1883. What happened in 536 was a much larger eruption than Krakatoa.

Apart from written records, there is a great deal of evidence to be obtained from the measurement of tree rings and of ice cores. The growth rings of many species of trees preserve an indelible annual record of climatic history. Tree-ring specialists (dendro-chronologists) can attempt to reconstruct past climate by studying two sets of data. One is the width of each annual growth ring, which reveals the exact amount of growth in a given year (i.e. in a given growing season, usually spring and summer). A drought or unseasonable frosts which restrict growth will therefore produce narrow rings. The second, the density of each ring, in conifers in cool climates provides information about the temperature. Keys cites tree-ring evidence from around the world, Finland, Sweden, the British Isles, central Europe, the Aegean, Siberia, North America, Chile, Argentina, and Tasmania, showing that 535–550, or later, was a period of unusually low tree-ring growth. Newfield (2018) also gives a table of 28 dendroclimatological studies, several of them going back beyond 5,000 B.C. Keys also refers to dust from ice-cores providing evidence of drought during this same period. Keys even went further than that and claimed that the volcanic eruption and its consequences for the climate, in the form of famine for instance, also caused great social and political changes too. The eruptions have also been cited as one of the causes of the spread of plague, bringing to an end the reconquest of the western Roman Empire.

6.8.2 ICE CORE RECORDS

Ice cores yield various types of information about past climates (volcanic dust, CO_2, ^{18}O, ^{2}D (deuterium), etc. The information comes from two principal sources, cores of ice from Greenland and Antarctica and also from cores of sediments from the deep oceans. Progressively, over the last 250,000 years, nearly three kilometres of ice have accumulated to make up the Greenland ice cap, while an even longer record is available from Antarctica. Every year new snowfall collects and, as time goes on, it becomes gradually compressed into fine layers of solid ice, each one a perfect record of a year of climate, the temperature, the gases in the atmosphere, the wind-blown dust, and the amount of snow. For instance, the thickness of each layer gives an indication of the amount of snow that fell that year. The colour of the ice, measured by laser, gives an indication of the amount of dust it contains, which may have come from known historical volcanic eruptions. Sigl et al. (2015) analysed the records of atmospheric sulphate aerosol loading developed from high-resolution, multi-parameter measurements from an array of Greenland and Antarctic ice cores as well as distinctive age markers to constrain chronologies to study large volcanic eruptions in the tropics and high latitudes that were primary drivers of interannual-to-decadal temperature variability in the Northern Hemisphere during the past 2,500 years. Previous inconsistencies in the timing of atmospheric volcanic aerosol loading determined from ice cores and subsequent cooling from climate proxies such as tree rings were resolved. It is even possible to analyse the past atmosphere. When snow falls it is quite loose and has many holes filled with air. As the snow gradually compacts, much of the air is expelled, but some is permanently enclosed in the ice as small bubbles. These bubbles still contain the original atmospheric gases such as carbon dioxide and methane, which can be extracted and analysed, giving a record of their past concentrations in the atmosphere.

By counting downwards from the surface, the date on which each layer was set down can be determined. The thickness of a layer is an indication of the amount of snowfall in that year. The ice itself also provides useful information because in addition to the common isotopes of hydrogen ^{1}H and oxygen ^{16}O, there are small concentrations of heavy hydrogen, ^{2}H or ^{2}D, i.e. deuterium, and heavy oxygen ^{18}O. The proportion of deuterium and the proportion of ^{18}O in water vary slightly with the temperature and so, by measuring the proportion of ^{2}D or ^{18}O in the ice one can estimate the temperature in that season. These proportions are described as δD and $\delta^{18}O$ values, respectively

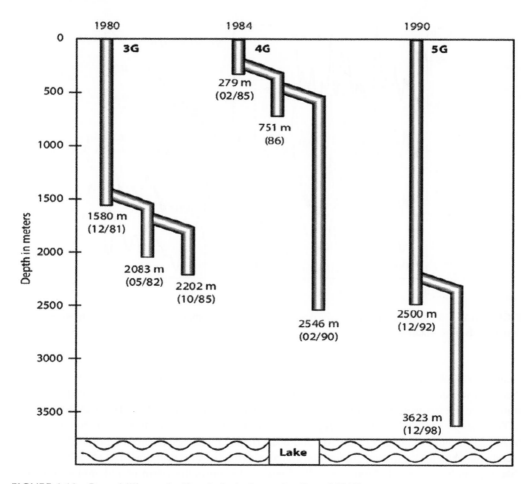

FIGURE 6.10 Deep drilling at the Vostok site in Antarctica (Jouzel 2013).

(pronounced "delta-Deuterium" and "delta-O-18 value"). The values are always given in per mille (‰), i.e. parts per thousand and are typically around -460‰ for δD and -57‰ for $\delta^{18}O$ (Figure 6.10).

For about 50 years, ice cores have provided a wealth of information about past climatic and environmental changes and we cite two reviews in particular by Langway (2008) and Jouzel (2013). The first attempt to obtain historical evidence from within an ice sheet was carried out by Sorge (1935) by studying a 15m-deep pit at station Eismitte in Greenland, while drilling of ice cores for research purposes began about 20 years later. Ice cores from Greenland, Antarctica, and other glacier-covered regions now encompass a variety of timescales. However, the longer timescales (e.g. at least back to the Last Glacial period) are covered by deep ice cores, the number of which is still very limited, according to Jouzel (2013) to seven from Greenland, with only one providing an undisturbed record of a part of the last interglacial period, and a dozen from Antarctica, with the longest record covering the last 800 000 years. Figure 6.12 shows a sketch of the deep drilling arrangement at the Vostok site in Antarctica (Jouzel 2013). Drilling continues and it is hoped to extend the record back to 1.5 million years ago. For anyone interested in reading more about ice cores work, the review by Jouzel (2013) is thoroughly recommended.

Figure 6.13 shows an example of results compiled by Jouzel (2013) from several sources; it goes back 800,000 before the present time, i.e. including the previous ice age. The top two curves show the values of $\delta^{18}O$ from two different bore holes in Antarctica; these two curves show very

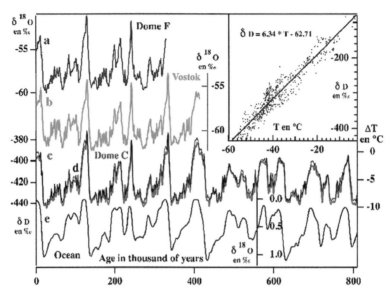

FIGURE 6.11 $\delta^{18}O$, δD records and derived temperature records compiled by Jouzel (2013) from various ice core records from Antarctica.

clear correlation. The lowest curve is for δD and the correlation with the $\delta^{18}O$ curves is also very clear. These isotope results have been converted into temperature variations which are shown in the third curve. Earlier work was reproduced in Figs. 35 and 36 of Lamb (1995). The results for the last 2,000 years are particularly interesting. Some more recent results are shown in Figure 6.11 for 0 to AD 2,000 and in Figure 6.12 for AD 500-600. These Figures, especially Figure 6.12 show very clearly the sudden drop in temperature associated with the volcanic eruption in AD 535-6 which was discussed in Section 6.8.1. More generally, Figure 6.12 shows the warm period around AD 1,000 and the Little Ice age of the late mediaeval period. However, as Lamb (1995) points out a close analysis of the last thousand years of the record from a site in far northwest Greenland indicates a great deal about the changes of temperature and the downput of snow there, but it should not be assumed that the temperature sequence in Greenland is identical with that anywhere in Europe. The medieval warmth reached its climax, and also ended, earlier there than in Europe. The Little Ice Age affected north Greenland too, but there were some differences of phasing. Indeed, the climatic history obtained by various techniques from central longitudes of Canada appears to parallel that in western and northern Europe more closely than that from Greenland. As well as extracting historical information about temperature from the ice core data, Petit et al. (1999), for example, using data from the Vostok ice core in Antarctica going back 420,000 years deduced the concentration of CO_2 and CH_4 in past times.

420,000 or 800,000 years may seem a long time, but it is important to realise what that period means in relation to the ice ages. In the history of the Earth, there have been long periods of warm climate when the Earth was warm all the way from the equator to the poles. Then at other times, during ice ages, there have been cold periods in which glaciation reached the regions which are currently temperate. These fluctuations seem to have happened periodically. There has probably always been some ice, but it may sometimes have been limited, perhaps, to mountains or polar regions. The important thing about ice ages is that they change the character of the Earth's surface. The existence of the ice ages has been known since the middle of the 19th century, but it is not really known how many ice ages have occurred in Earth's history. There have been at least five, and possibly as many as seven, major ice ages, since the formation of the Earth. There was a major period

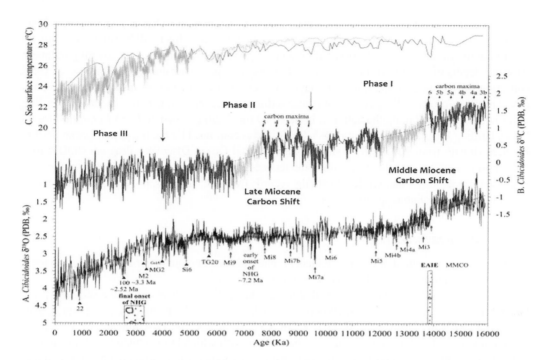

FIGURE 6.12 A. Benthic foraminiferal $\delta^{18}O$; B. Benthic foraminiferal $\delta^{13}C$; C. Derived sea surface temperature record of ODP (green) and IODP (red). The red lines in A and B denote 30% weighted mean of the isotopic records. The red arrows and blue triangles in A denote the Miocene glaciation events and the Marine Isotope Stages (MISs), respectively. The sky-blue lines in B denote MMCS (middle Miocene carbon shift) and LMCS (late Miocene carbon shift). The short black arrows in B denote carbon maxima events. The long black arrows separate the phases of the Pacific long-term ocean carbon isotope decrease. NHG, northern hemisphere glaciation. EAIE, east Antarctic ice sheet expansion. MMCO, middle Miocene carbon optimum. Phase I: Pacific long-term C decrease (9.4-15.8 Ma); Phase II: Pacific long-term C decrease (4.0-9.4 Ma); Phase III: Pacific long-term C decrease (0-4.0 Ma) (Modified from Tian et al., 2018).

of glaciation about 700 million years ago which is sometimes referred to as "Snowball Earth." The next glaciation occurred in the Ordovician period about 450 million years ago. Then there was the Permo-Carboniferous Glaciation which was a very major ice age which occurred on the southern continents, often called Gondwanaland, when they were arranged around the South Pole. Those continents were Antarctica, Australia, South America, India, and Africa, and there are glacial deposits covering extensive regions which date from that period, about 300 million years ago. The most recent glaciation is the one we are living in now, but it is not very clear when it began. It appears to have begun progressively about 50 million years ago in Antarctica. However, according to "Ice Ages – <serious-science.org> "the major Ice Age which we are living in now (the Quaternary) really started, as far as we're concerned, when glaciers in the northern hemisphere reached the sea, about 2.5 million years ago. And we're still living in that ice age." We have described this situation with regard to ice ages to make the point that even though 480,000 years seems a long time it is still quite short in terms of our present ice age.

As well as being present in ice, heavy oxygen, ^{18}O, is also present in deep sea floor sediments. When animals living in the sea make a skeleton, they use the elements, including oxygen, from the surrounding water. The calcium carbonate, $CaCO_3$, of their shells, especially of foraminifera, tiny, sea-surface living plankton, will contain the same proportion of ^{18}O as the seawater of the time. This means that as the ice volume grows during ice ages and ^{16}O becomes progressively locked up in ice, foraminifera shells will show a progressive enrichment in ^{18}O. The final amount of each isotope will also depend on seawater temperature as well, but this effect will be small. At death, the tiny foraminifera shells fall to the ocean floor and are preserved in thin layers of covering sediment. At water depths of 3000 m or more, these sediments stay quite undisturbed for millions of years, collecting slowly shell by shell, layer by layer, a fraction of an inch every millennium. There is a global programme for drilling cores from these sediments and analysing their contents. The drilling of the cores and distributing the data was initiated by the drilling vessel *JOIDES Resolution* for the Ocean Drilling Program (ODP; from 1985–2003) and is continued by the Integrated Ocean Drilling Program (IODP; from 2003–2013), and the International Ocean Discovery Program (IODP; from 2013–now). Now it is possible to combine the records from the deep-sea sediments of past ice volumes, sea temperatures, and sea-level with the records from the ice cores, providing air temperature and concentrations of atmospheric gases.

6.8.3 RADIOCARBON DATING

Another important method of dating artefacts or materials that contain material that was once living is radiocarbon dating and its role in dating evidence.

For times beyond the range of identifiable year-layers in ice-sheets and lake sediments, and beyond the longest yearly tree ring chronologies or human records in any part of the world, dates can be estimated by radiometric methods or, in the case of material in sediments, by assuming a broad constancy of sedimentation rate. For the periods with which human history and archaeology are concerned, the most important of the radiometric methods is radiocarbon dating. The common form of carbon is ^{12}C with an atomic weight of 12. However, a radioactive form ^{14}C with an atomic weight of 14 is produced in the atmosphere by the effect of cosmic ray bombardment on some of the nitrogen ^{14}N atoms in the atmosphere. The Carbon 14 is assimilated into the structure of the living vegetation with the carbon dioxide breathed in from the atmosphere.

About one per cent of the carbon in the atmosphere is Carbon 14 and this finds its way by photosynthesis into living vegetation. These atoms undergo radioactive decay, producing on average about 15 disintegrations minute per gram of carbon present. After the death of the vegetation, which means cessation of the absorption of atmospheric carbon dioxide, its store of radioactivity is no longer being renewed. The activity therefore decays. The half-life of radiocarbon (^{14}C) is 5730 years; this means that the activity falls by a half every 5730 years. In practice, as the amount of radioactivity dwindles, it ultimately becomes very difficult to measure. The errors produced by any contamination become greater, the older the material to be dated. The effective limit of radiocarbon dating is about 50 thousand years. Estimates of the error arising from experimental difficulties are always quoted. There is an additional source of error, however, established by radiocarbon dating of objects of known age and attributed to the fact that the proportion of radioactive carbon in the atmosphere has not been precisely constant throughout the ages. These variations give rise to an error amounting to 500–1000 years in middle post-glacial times and to over 100 years in another period as recent as about AD 1400-1800. These errors can, however, in most cases be corrected by using a calibration curve relating apparent radiocarbon ages to true ages obtained from specimens wood of bristlecone pine dated by its rings.

6.8.4 TREE RINGS

According to Wikipedia, The Greek botanist Theophrastus (c. 371 – c. 287 BC) first mentioned that the wood of trees has rings and Leonardo da Vinci (AD 1452–1519) was the first person to mention that trees form rings annually. This leads to the well-known fact that the age of a tree can be determined by counting the rings in a complete horizontal section. Leonardo da Vinci also observed that the thickness of the rings is determined by the conditions under which they grew. Because trees are

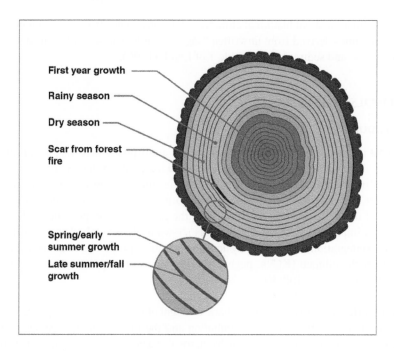

First year growth

Rainy season

Dry season

Scar from forest fire

Spring/early summer growth

Late summer/fall growth

FIGURE 6.13 Schematic showing the effect of climatic conditions on tree ring growth. (Source NAA).

sensitive to local climate conditions, such as rain and temperature, the rings provide some information about the past climate in the area where the tree was growing. For example, tree rings usually grow wider in warm, wet years and they are thinner in years when it is cold and dry. If the tree has experienced stressful conditions, such as a drought, the tree might hardly grow at all in those years, see Figure 6.13 rings.

Dendrochronology (or tree-ring dating) is the scientific method of dating tree rings to the exact year they were formed. Dendrochronology is useful for determining the precise age of samples, especially those that are too recent for radiocarbon dating, which always produces a range rather than an exact date, to be very accurate. However, for a precise date of the death of the tree, a full sample to the edge is needed, which most trimmed timber will not provide. As well as dating them, this can give data for dendroclimatology, the study of climate, and atmospheric conditions during different periods in history from wood.

Dates are often represented as estimated calendar years B.P., for before present, where "present" refers to 1 January 1950. A fully anchored and cross-matched chronology for oak and pine in central Europe extends back 12,460 years (Friedrich et al. 2004) and an oak chronology goes back 7,429 years in Ireland and 6,939 years in England (Walker 2013). Comparison of radiocarbon and dendrochronological ages supports the consistency of these two independent dendrochronological sequences (Stuiver et al. 1986). Another fully anchored chronology that extends back 8500 years exists for the bristlecone pine in the Southwest US (White Mountains of California) (Ferguson and Graybill, 1983). In 2004, a new radiocarbon calibration curve, INTCAL04, was internationally ratified to provide calibrated dates back to 26,000 P B. For the period back to 12,400 B.P., the radiocarbon dates are calibrated against dendrochronological (tree ring) dates.

One of the criticisms levelled against the IPCC's reports (e.g. by the NIPCC, see Section 7.9) is that they paid scant attention to the history of the Earth's climate. For a long time, the Earth's climate was totally controlled by nature processes and it is only relatively recently – in the history of the Earth – that human activities began to affect the climate. Of course, human behaviour has also been affected by the climate. The relation between mankind and the climate is a coupled system. Our information about historical climatic conditions comes from two sources, first the written records, sometimes derived from unwritten "saga" records, and secondly physical records from tree rings, ice cores, and sediments, see Table 1 of Lamb (1995).

6.9 AEROSOLS AND CLIMATE

6.9.1 INTRODUCTION

One of the most important factors in climate change is atmospheric aerosols, the product of a complicated totality of physical and chemical processes taking place in the environment as well as the results of direct injection of dust, ash, or smoke particles into the atmosphere. This may happen by natural processes such as volcanic eruptions, sand storms, etc. or it may occur as a result of human activities. Part of these processes is determined by the growing anthropogenic loading on the environment. Human impact on atmospheric aerosols shows itself through changes in the intensity of formation (or disintegration) and in the physical and chemical properties of aerosol particles, which, in its turn, affect the climate (KC98, page 270). The three principal mechanisms through which aerosols affect climate are as follows.

1. The direct effect of aerosols on the radiation budget of the surface–atmosphere system through a re-distribution of shortwave solar radiation and thermal infrared emissions in this system, caused by scattering and absorption by aerosol particles non-uniformly distributed in the atmosphere.
2. The crucial role of aerosol particles in the phase transformations of water in the atmosphere and in particular in cloud formation, which is more important for the energetics of the atmosphere than the direct effect of aerosols. In this connection, of great concern is the role of gas-to-particle reactions of aerosol formation (condensation nuclei, first of all) connected with natural and anthropogenic emissions to the atmosphere of compounds such as dimethyl sulphide and sulphur dioxide.
3. Heterogeneous chemical processes, in particular the processes of disintegration of ozone molecules on the surface of aerosol particles, which affects the change in the atmospheric gas composition.

Complex field programmes of the aerosol–radiation experiments accomplished in various global regions within the CAENEX and GAAREX programmes (Kondratyev 1972a, 1976, 1991, Jennings 1993) showed that the aerosol can substantially absorb the solar radiation and greatly affect the atmospheric radiation budget. Major problems include the impact of dust outbreaks from the arid regions of the surface of the Earth on the radiative regime of the atmosphere over the continents and the oceans, the transport of aerosols to high latitudes leading to the formation of the arctic haze and also the extended stratified cloudiness in the Arctic. A special role belongs to aerosol perturbations due to volcanic eruptions (see Section 6.8).

As we can see, there are various sources of aerosols in the atmosphere, some of natural origin and some of anthropogenic origin. And, of course, in any serious consideration of anthropogenic effects on the climate, one needs to try to separate the natural effects and the consequences of human activities, However, we should not fall into the trap of assuming that the natural effects are unchanging with time – they are not. From the point of view of modelling, the effects of volcanic eruptions are particularly not susceptible to modelling and there are several reasons for this. The locations and times of the eruptions themselves are (largely) unpredictable – there are many volcanic areas in the world and we know, from the historical records, that many of them erupt from time to time, but like earthquakes (and for the same reason) predicting when any given volcano will erupt is (almost) a non-starter. Secondly, the amount and detailed chemical and physical properties of the material ejected can be expected to vary from one site and occasion to another. Thirdly the dispersal pattern of the ejected material will vary according to the location, particularly the latitude, of the volcano and the global wind patterns, which of course may also be disturbed by the eruption itself.

6.9.2 MODELLING AEROSOL PROPERTIES

The accomplishment of various climatological calculations and assessments requires a consideration of the aerosol characteristics, which it is not always possible to measure directly. It is important to note that such characteristics vary strongly, depending on various environmental conditions. For example, the variability of the optical properties and size distribution of aerosol particles even within a certain climatic zone is well known. On the other hand, there are relatively stable aerosol characteristics subject to certain laws, such as particle size distribution and chemical composition of the aerosol substance. The difficult problem of substantiating the representative models is of fundamental importance.

The difficulties in the modelling of the characteristics and properties of aerosol particles are connected with the large number of parameters which influence their variability and with the absence of reliable criteria for the choice of the values of the various parameters. So, for example, the spatial and temporal variability of aerosol properties is determined by a complicated sum of the processes of generation, transformation, and evolution of an ensemble of aerosol particles, their propagation from the sources, and the removal of the particles from the atmosphere. An isolated consideration of these processes, which is often practised, is rather arbitrary, since all of them are interconnected.

The problem of aerosol modelling may be divided into two parts, including:

1. global spatial distribution of aerosols (the field of the aerosol mass concentration), with consideration of the average characteristics of generation, propagation, and removal of particles from the atmosphere;
2. local properties of aerosols resulting also from the effect of the processes of transformation of the spectra of atmospheric aerosols of different origin.

The processes of the generation of particles determine, first of all, their physical and chemical properties and their concentration in the atmosphere and, hence, the subsequent processes of transformation of the spectra of size distribution and the rate of removal of the particles from the atmosphere. The following types of natural aerosol particles can be identified:

1. products of the disintegration and evaporation of sea spray,
2. mineral dust, wind-driven to the atmosphere,
3. products of volcanic eruptions (both directly ejected to the atmosphere and resulting from gas-to-particle conversion),
4. particles of biogenic origin (both directly ejected to the atmosphere and resulting from the condensation of volatile organic compounds, for example terpenes, and chemical reactions taking place between them),
5. smoke from biomass burning on land,
6. products of natural gas-to-particle conversion (for example, sulphates formed from sulphur compounds supplied by the ocean surface).

The principal types of anthropogenic aerosols are particles of industrial emissions (soot, smoke, cement and road dust, etc.), as well as products of gas-to-particle conversion. In addition, some aerosol substance results from heterogeneous chemical reactions, in particular, photocatalytic reactions.

As a result of the sufficiently long life-time of aerosol particles, their transport, mixing, and interaction with each other and with gaseous compounds of various substances, differences in the physical and chemical properties of particles from various sources are smoothed out and the particles form what are called the background aerosols.

In different atmospheric layers, the characteristics of the background aerosols suffer substantial changes. In particular, due to the weak mixing of tropospheric and stratospheric aerosols, the particles in the stratosphere differ strongly in their composition and size distribution from the tropospheric ones. The troposphere can contain regions with particles of different types, in view of a relatively short life-time of the particles in the troposphere, especially in the surface layer, as well as some of the sources markedly dominating the intensity of the remainder.

The processes of transport and removal of aerosols from the atmosphere are determined by both the meteorological factors and the properties of the aerosol particles themselves. The processes which depend on meteorological conditions include organised convective and advective transports of particles as well as mixing due to the turbulent diffusion. The individual properties of particles determine, in particular, such processes of their removal from the atmosphere as sedimentation on obstacles.

In the processes of transport and removal of particles from the atmosphere, their size (mass) and composition are transformed due to coagulation and condensation growth as well as to heterogeneous reactions. These processes are responsible for the spectrum of particle sizes and determine the value of the complex refractive index and the particles' shapes, that is their optical properties (d'Almeida *et al.* KC98, Ivlev and Andreev KC98).

The most substantial gas-to-particle reactions in the formation of aerosol particles are as follows:

1. reactions of sulphur dioxide with hydroxyl radicals which eventually lead to the formation of sulphuric acid molecules, which are easily condensed in the presence of water molecules;
2. reactions of hydrocarbon compounds with ozone and hydroxyl radicals with the subsequent formation of organic nitrates through the reactions of primary products with nitrogen oxides (e.g. the well-known peroxyacetyl nitrate).

All the processes of formation and evolution of aerosol systems depend on periodical variations of solar radiation in the atmosphere and at the surface. The processes of atmospheric heating near the surface, water evaporation, and chemical reactions of aerosol formation have a clear-cut diurnal change resulting in variations of the aerosol characteristics, which in some cases are stronger than the annual change (Galindo 1965). Strong daily variations of anthropogenic aerosols have been documented.

The cycles of aerosol matter are closely connected with the hydrological processes in the atmosphere. On the one hand, clouds and precipitation play an important role in the formation, transformation, and removal of aerosol particles from the atmosphere, while on the other hand, the aerosols are condensation nuclei whose physico-chemical properties determine the microphysical processes in clouds. It is not accidental that both water vapour molecules and aerosol particles have approximately equal life-times. Therefore, a deeper knowledge of the interactions of aerosols and cloud elements is needed to obtain a better understanding of the processes of the formation of aerosols and of the variability of aerosol properties in time and space.

6.9.3 CLIMATIC IMPACTS OF AEROSOLS

The principal problem connected with studies of the climatic impact of aerosols consists in the consideration of their influence on the longwave and shortwave radiation transfer in the atmosphere from the viewpoint of climate change that can take place due to regional and global-scale variations in the spatial distribution of the concentration, size distribution, and chemical composition of aerosols. In this connection, of great concern was the problem of a possible climatic impact of soot particles ejected to the troposphere and the stratosphere by fires caused by nuclear explosions in the atmosphere. During recent years, the assessments of the climatic implications of fires on the oil fields of Kuwait have been of serious concern.

6.9.4 TROPOSPHERIC AEROSOLS

The development of numerical climate modelling necessitates a reliable consideration of the effect of aerosol on climate, based on the development of adequate and realistic aerosol models and assessment of climate sensitivity to various characteristics of atmospheric aerosols.

The following aspects of the problem are of the first priority:

1. impacts of aerosols on regional and global processes,
2. identification of the types and properties of aerosols, most important from the viewpoint of their climatic implications, and
3. assessing the contribution of impacts on the regional and global climate of the variability of aerosols, minor gaseous components (H_2O, O_3, CO_2, CO, etc.), clouds and surface albedo.

The development of models of the dynamics of formation, transformation, and transport of aerosols in interaction with the processes of climate formation is a substantial priority for the consideration of aerosol effects in climate theory. Numerical experiments on climate sensitivity to aerosol effects should be accomplished to reveal impacts of aerosols on the radiation budget of the surface-atmosphere system, heat flux divergence in the atmosphere, heat balance, and the climatic system as a whole. Radiative–convective equilibrium models are useful for approximate assessments of the effects of aerosols on climate. In particular, they have revealed a strong effect of aerosols on the vertical temperature profile, when the optical thickness of aerosols is large, and a weak effect if their optical thickness is below 0.6. Depending on the surface albedo, optical properties of aerosol and aerosol spatial distribution, either cooling or warming of the climate by several degrees can take place.

6.9.5 SULPHATE AEROSOLS

The origins of sulphate aerosols in the atmosphere were described briefly in Section 2.4.3. This included a table (Table 2.6) of the natural and anthropogenic sources of sulphate aerosols in the atmosphere. There has been considerable progress in the understanding of sulphate aerosols since the late 1990s when Table 2.9 was produced (KC98) and this is recorded in Chapter 7 of the IPCC's 5th Assessment Report (Boucher et al. 2013). This is a lengthy specialised document and we simply quote some extracts from the Executive summary.

Clouds and aerosols continue to contribute the largest uncertainty to estimates and interpretations of the Earth's changing energy budget and continue to consume a large amount of time in the formulation of the IPCC Assessments. The latest situation is described at some considerable length in Chapter 7 of the 5th Assessment Report (AR5). This chapter (Boucher et al. 2013) focuses on process understanding and considers observations, theory, and models to assess how clouds and aerosols contribute and respond to climate change. It concludes as follows.

Many of the cloudiness and humidity changes simulated by climate models in warmer climates are now understood as responses to large-scale circulation changes that do not appear to depend strongly on sub-grid scale model processes, increasing confidence in these changes. Climate-relevant aerosol processes are better understood, and climate-relevant aerosol properties better observed, than at the time of the previous IPCC assessment (AR4) in 2007. However, the representation of relevant processes varies greatly in global aerosol and climate models and it remained unclear what level of sophistication was required to model their effect on the climate. It has been determined that cosmic rays enhance new particle formation, but the effect on the concentration of cloud condensation nuclei is thought to be too weak to have had any detectable climatic influence over the last century. No robust association between changes in cosmic rays and cloudiness has been identified.

Recent research has stressed the importance of distinguishing between forcing, i.e. instantaneous change in the radiative budget, and rapid adjustments which modify the radiative budget indirectly through fast atmospheric and surface changes as a result of feedbacks. These feedbacks operate through changes in climate variables that are caused by a change in surface temperature.

The quantification of cloud and convective effects in models, and of aerosol-cloud interactions, continues to be a challenge. Climate models are incorporating more of the relevant processes than at the time of AR4, but confidence in the representation of these processes remains weak. Cloud and aerosol properties vary at scales that are significantly smaller than those resolved in climate models, and cloud-scale processes respond to aerosol in nuanced ways at these scales. Until sub-grid scale parameterisations of clouds and aerosol-cloud interactions are able to address these issues, model estimates of aerosol-cloud interactions and their radiative effects will continue to carry large uncertainties. Satellite-based estimates of aerosol-cloud interactions remain sensitive to the treatment of meteorological influences on clouds and assumptions on what constitutes pre-industrial conditions.

Precipitation and evaporation are expected to increase on average in a warmer climate but also to undergo global and regional adjustments to carbon dioxide (CO_2) and other forcings that differ from their warming responses. Moreover, there is high confidence that, as climate warms, extreme precipitation rates on, for example, daily timescales will increase faster than the time average.

The net feedback from water vapour and lapse rate changes combined, as traditionally defined, is extremely likely positive (amplifying global climate changes). The sign of the net radiative feedback due to all cloud types is less certain but likely to be positive. Uncertainty in the sign and magnitude of the cloud feedback is due primarily to continuing uncertainty in the impact of warming on low clouds.

Other issues, including studies of aerosol-climate feedbacks through changes in the source strength of natural aerosols or changes in the sink efficiency of natural and anthropogenic aerosols, the ERF due to aerosol-radiation interactions that takes rapid adjustments into account; the total ERF due to aerosols excluding the effect of absorbing aerosol on snow and ice, and persistent contrails from aviation are still areas of active research.

Maps of the global distribution of the concentration of sulphates in the lower 500 m layer of the atmosphere, drawn from numerical modelling data, reveal distinctly the effect of anthropogenic emissions in the northern hemisphere with maxima of more than 1 ppb in the regions of Europe, North America, and China. The sufficiently long life-time of SO_4^{2-} (about 5 days) allows the possibility of long-range transport of sulphate aerosol, which shows itself east of North America and China, as well as east and south-east of Europe. In the southern hemisphere, the anthropogenic impact can be observed in Africa, South America, and Australia. In the case of the absence of anthropogenic emissions, a maximum of the concentration of sulphates takes place in the tropical belt, where the natural emissions of DMS are most intensive. The calculations of the ratios of the pre-industrial and present concentrations of sulphates in the layer 0.5–1.5 km, where the lower-level clouds form (mainly stratus clouds), have led to the conclusion that in the most polluted European regions the concentration of sulphates has increased by two orders of magnitude, and in the northern hemisphere the overall increase of the concentrations in July (January) reached 51 per cent (0.8 per cent).

6.9.6 STRATOSPHERIC AEROSOLS

The effect of stratospheric aerosols on climate is quite different from the case of tropospheric aerosols. It is relevant, first of all, to the impact of volcanic eruptions on climate. Volcanic eruptions have contributed to the formation of the land surface, the oceans and the atmosphere throughout the history of the Earth and still do so, but on a long timescale. On a shorter timescale, they affect the Earth's climate. The products of explosive volcanic eruptions reach the stratosphere and remain there for a year or longer, changing the chemical composition of the stratosphere and thereby affecting the Earth's radiation budget. We shall discuss the post-eruption climate changes variations caused by volcanic aerosol.

Volcanic eruptions generate clouds of particles and gases into the atmosphere and these materials disperse to distant parts of the Earth and cause changes in the atmospheric conditions which affect the radiation budget, alter the distribution of greenhouse gases and cause changes in the hydro-logical cycle with consequent effects on the weather and the climate. Benjamin Franklin proposed that an eruption of the Hekla volcano in Iceland could have been a reason for the severe winter of 1783–1784 in Europe. Since then, numerous studies have been undertaken in a search for correla-tions between individual volcanic eruptions and adverse weather conditions or between serial erup-tions and climate anomalies. It has been possible to establish correlations between historical records of adverse weather or climate conditions with volcanic eruptions that were in distant places and which were unknown to the writers at the time. These studies have revealed that volcanic eruptions may be one of the important factors of climate changes.

Of all the possible causes of climate changes, volcanic eruptions are among the most adequately documented and understood. Following the 1815 Tambora volcanic eruption in Indonesia, 'a year without sun' was observed, and a summertime temperature decrease in New England and western Europe (with respect to the climatic mean) reached 1–2.5 °C

Mass and Schneider (KC98) analysed the consequences of numerous eruptions causing an increase of several tenths in the aerosol optical thickness in the stratosphere in the visible wave-length range. They found that a statistically reliable temperature decrease a year after the eruption would constitute 0.3 °C and after 2 years 0.1 °C and more.

A trend towards a stratospheric temperature increase was observed after the 1963 Agung erup-tion in Indonesia. This trend was superimposed on a stratospheric temperature variation that was quasi-biennial and increasing in amplitude and could be explained by the Agung eruption. This increase was clearly pronounced in the southern hemisphere, where the variability of the amplitude of quasi-biennial oscillations is small, whereas in the northern hemisphere (with a small input from the eruptive cloud) there was no stratospheric temperature increase. These observational data agree with theoretical estimates, from which it follows that a stratospheric warming is mainly caused by absorption of the upward thermal emission flux and, to a smaller degree, by the solar radiation absorption. In the troposphere and at the surface, there was a global-scale temperature decrease (larger in the northern hemisphere, 0.6 °C). This decrease could not, however, have been totally caused by the eruption.

To substantiate the most essential properties of stratospheric aerosol affecting the climate, Lacis *et al.* (KC98) calculated the global-mean vertical temperature profile based on the use of a one-dimensional radiative-convective model with the cloud amount assumed to be 50 per cent, and verti-cal temperature lapse rate constituting 6.5°C km^{-1}. In this connection, calculations were made of the extinction coefficient Q_{ext}, the single-scattering albedo, ϖ_c and the coefficient of asymmetry of the phase function $\cos \theta$ in the wavelength region 0.3–30 μm for the droplets of a 75 per cent water solution of H_2SO_4, as well as spectral reflection, transmission, and absorption with an optical depth τ at the wavelength 0.55 μm equal to 0.1 and 1.0.

The size distribution has been prescribed for the May and October stratospheric aerosol observed, respectively, 1.5 and 6.5 months after the El Chichón eruption (the special feature of the May aero-sol is the presence of a large number of large particles with radius $\geqslant 1$ μm, which results in a consid-erable enhancement of the absorption of the thermal emission by the eruptive cloud).

As a measure of climate forcing by stratospheric aerosol, one can consider a change of the radiation budget ΔF_{net} and its components ΔF_{solar} and ΔF_{IR} at the tropopause level (i.e. for the surface–troposphere system). Calculations of ΔF_{solar} and ΔF_{IR} as a function of the effective radius of particles, r_{eff}, have shown that the infra-red forcing depends strongly on the radii of the par-ticles, increasing with increasing radius, within the interval 0.5–3.5 μm, whereas the albedo for the shortwave radiation is almost independent of the radii of particles, provided they are larger than the effective wavelength (0.5 μm). Therefore, the aerosol with a critical radius of particles $r_c \geqslant 2$ μm causes a strong heating of the troposphere (due to the attenuation of the longwave radia-tive cooling).

It turns out that r_c depends weakly on the composition and shape of particles. Though the small aerosol particles ($r_{eff} < 0.05$ μm) also determine the warming, the effect is weak because of the small content of such aerosols. The width of the particle size distribution characterised by the parameter v_{eff} markedly affects ΔF_{net}. Thus, it can be reliably concluded that the impact of net radiative forcing on the troposphere depends mainly only on two parameters, r_{eff} and v_{eff}, and the aerosol size distribution can be approximated by a γ distribution. With low values, $\tau < 0.2$, the simplest approximations are justified:

$$\Delta F_{net} \sim 30\tau\,\mathrm{W\,m^{-2}} \text{ and } \Delta T \sim 9\tau\,°C \qquad (6.4)$$

Of great importance is the presence of admixtures in the droplets of the H_2SO_4 aerosol, causing changes in the single-scattering albedo. Even for the critical value $\varpi_c = 0.99$, the stratospheric cooling due to backscattering is halved, as compared with $\varpi_c = 1$, and the temperature of the stratosphere increases substantially. However, this does not impact on the troposphere.

The principal conclusion is that the radiative forcing of the stratospheric aerosol on the troposphere can be simulated precisely enough with only one parameter taken into account – the optical depth of stratospheric aerosol, with the exception that it is necessary to consider the effect of particles sizes, when they exceed 1 μm. An enhancement of the large-particle-induced greenhouse warming can be comparable to or even exceed the albedo effect of cooling, provided $r_{eff} > 2$ μm.

To assess climate changes, it is necessary to substantiate the minimum combination of optical parameters of atmospheric aerosols needed. The calculation schemes are usually based on the following optical characteristics: the volume coefficient of attenuation (extinction), σ_e, (the availability of the data on the vertical $\sigma_e(z)$ profile makes it possible to calculate the optical depth of the atmosphere τ), the volume coefficient of scattering, σ_{sc} (or the single-scattering albedo $\varpi_c = \sigma_{sc}/\sigma_e$), and the volume phase function, which is approximately simulated by introducing the coefficients g of asymmetry of the scattering function. The representative characteristic of the effect of aerosols on climate is the ratio of the coefficients of the aerosol attenuation of radiation in the longwave $(\lambda = 10 \ \mu m)$ and shortwave $(\lambda = 0.55 \ \mu m)$ spectral regions. Some of these parameters for different types of aerosols are shown in Table 6.10.

Several points should be noted. First, this is only a one-dimensional model. Secondly the accuracy of the values of these parameters leaves much to be desired; more reliable up-to-date values may be available now. Thirdly the concentrations of these various types of aerosol at any given place and time are largely unknown. The result of any computer calculations based on this, or any more sophisticated, model should be regarded with extreme caution.

TABLE 6.10

Typical values of the optical parameters of different types of aerosol

	Wavelength				Horizontal	Vertical
	0.55 μm		10 μm			
Aerosol origin	T	ϖ_c	τ	ϖ_c	variation	structure
Desert	0.054	0.890	0.0052	0.44	Yes	ED in ABL at $H = 3$ km km km
Stratospheric	0.045	1.000	0.0049	0.06	No	Homogeneous
Oceanic	0.033	0.988	0.0083	0.76	Yes	ED in ABL at $H = 1$ km
Continental	0.031	0.890	0.0030	0.44	Yes	ED in ABL at $H = 1$ km
Tropospheric	0.030	0.890	0.0029	0.44	No	Homogenous
Urban	0.008	0.647	0.0003	0.15	Yes	ED in ABL at $H = 1$ km
Volcanic	0.007	0.943	0.0002	0.15	No	Homogenous

Source: KC98 Table 4.42).

Note: τ = optical thickness ϖ_c = single scattering albedo; ABL = atmospheric boundary layer; H = scale height; ED = exponential decrease.

For any model, it is necessary to have either complete information on the distribution and chemical composition of aerosol particles (number density, size distribution, complex refractive index of the aerosol matter, and shape of particles), or a large statistically substantiated data base of the optical measurements. Describing the information available in the 1990s from the results of a number of studies of the optics of atmospheric aerosols, Kondratyev and Cracknell (1998a, b, page 284) expressed the view that "the available information should be considered inadequate." They went on to say: "The results of *in situ* measurements of the size distribution and chemical composition of aerosols, which contain rather reliable information, cannot be considered representative for large spatial and temporal scales. Therefore new sets of results of ground, aircraft, balloon and satellite observations should form the basis for the climatological aerosol models, with emphasis on the accomplishments of integrated programmes with the use of all observational means available. Also, it is necessary to develop both theoretical and empirical models of the processes of formation, evolution and removal from the atmosphere of the aerosol particles, as well as various models of their size distribution and optical characteristics."

In this respect, a new climate research facility, the Atmospheric Radiation Measurement (ARM) programme, was started by the U.S. Department of Energy on the basis of a broad international cooperation (Ellingson et al., 2016). The following text is taken from the ARM Fact Sheet:

"The Atmospheric Radiation Measurement (ARM) Climate Research Facility is a U.S. Department of Energy scientific user facility for researchers to study the effects and interactions of sunlight, infrared radiation, aerosols, and clouds to better understand their impact on temperatures, precipitation, and other aspects of weather and climate.

A central feature of this user facility is a set of heavily instrumented field research sites located at climatically diverse regions around the world. These sites obtain continuous measurements of atmospheric radiation and the properties controlling this radiation, such as the distribution of clouds, aerosols, and water vapour. Measurements from the fixed sites are supplemented through field campaigns using the ARM Mobile Facilities (AMF) and ARM Aerial Facility (AAF). Data collected through these capabilities, collectively referred to as the ARM Facility, are stored in the ARM Data Archive. Selected data sets are additionally analysed and tested to create enhanced data products, and software tools are provided to help open and use these products. All data and enhanced products are freely available to the science community via the ARM Data Archive (www.archive.arm.gov) to aid in further research.

Researchers can use the ARM Facility in several ways:

- Make an in-person or virtual visit to a site
- Access data through the ARM Data Archive
- Propose and conduct a field campaign.

Each year, a call for proposals to use the ARM Facility is issued via advertisements in scientific news publications and on the ARM website (www.arm.gov/campaigns)."

In the early days, there were attempts made to identify a correlation between volcanic eruptions and the ENSO (El Niño/Southern Oscillation) phenomena for five years, 1902, 1907, 1912, 1963, and 1982 (KC98, page 285). However, modelling studies have so far reached no consensus on either the sign or physical mechanism of El Niño response to volcanism (Khodri et al., 2017). In the late 1990s, it was clear that there was a great deal of work needing to be done to improve the understanding of the effects of volcanic eruptions on climate, see Section 4.6.5 of Kondratyev and Cracknell (1998a, b) and it was becoming apparent that remote sensing was able to play an important role in tackling this problem. It is now possible to carry out an extensive complex of observations of volcanic emissions and their propagation in the atmosphere with the use of conventional means (e.g. ground-based, aircraft, balloon) as well as Earth-observing satellites. Substantial progress has been reached in the development of the techniques for numerical modelling of climate and large-scale transport. Of great interest are attempts to reveal a volcanic signal in the observed climate changes.

The difficulty of the solution to this problem is that the interannual variability of surface air temperature is determined by various processes which can be classified as connected with the external forcings, internal regular oscillations, and random processes, which appear to be largely due to the dynamic instability of the atmosphere. The most probable external factors are large-scale volcanic eruptions and increased CO_2 concentrations in the atmosphere (to date, there is no convincing evidence for the effect of solar activity on climate). A considerable contribution to atmospheric variability is made by its internal instability. Because of the varying intensity and location of cyclones and jet streams, a random variability of heat fluxes occurs in the atmosphere, leading to a random variability in air temperature.

6.10   REMOTE SENSING AND VOLCANIC ERUPTIONS

In Section 6.8.1, we considered the large volcanic eruption of circa A.D. 535 and the widespread effect it is believed to have had on the climate at that time. There is an excellent review article on *Volcanic Eruptions and Climate* by Alan Robock (2000) and Table 6.11 is taken from that article, with details of two additional recent large eruptions. This table goes back to when Benjamin Franklin suggested that the eruption of Grimsvotn (Lakagigar) in Iceland in 1783 might have been responsible for the exceptionally cold summer of 1783 and the cold winter of 1783-4. This can be taken as the beginning of the scientific study of volcanoes. The official names at the volcanoes in Table 6.12 are taken from Simkin and Siebert (1994). The relative importance of the volcanoes is indicated by the values of the volcanic explosivity index (VEI) (Newhall and Self 1982) which are taken from Simkin and Siebert [1994], while the dust veil index (DVI/E_{max}) is from Lamb (1970, 1977, 1983a), updated by Robock and Free (1995) and the ice core volcanic index (IVI) from Robock and Free (1995, 1996).

Volcanic eruptions are responsible for releasing molten rock or lava, from deep within the Earth, causing destruction and often death, covering the ground and forming new rock on the Earth's

TABLE 6.11
Major Volcanic eruptions of the past 250 years

Volcano	Year of Eruption	VEI	DVI/E_{max}	IVI
Grimsvotn [Lakagigar], Iceland	1783	4	2300	0.19
Tambora, Sumbawa, Indonesia	1815	7	3000	0.50
Cosiguina, Nicaragua	1835	5	4000	0.11
Askja, Iceland	1875	5	1000	0.01*
Krakatau, Indonesia	1883	6	1000	0.12
Okataina [Tarawera], N. Island, New Zealand	1886	5	800	0.04
Santa Maria, Guatemala	1902	6	600	0.05
Ksudach, Kamchatka, Russia	1907	5	500	0.02
Novarupta [Katmai], Alaska, United States	1912	6	500	0.15
Agung, Bali, Indonesia	1963	4	800	0.06
Mount St. Helens, Washington, United States	1980	5	500	0.00
El Chichón, Chiapas, Mexico	1982	5	800	0.06
Mount Pinatubo, Luzon, Philippines	1991	6	1000	...
Eyjafjallajökull, Iceland	2010	4	...	...**
Arak, Krakatoa, Indonesia	2018	6	...	...***

Source: (Robock 2000, except for the last two volcanoes).

* Southern Hemisphere signal only; probably not Askja.

** Caused the worst flight disruption over Europe since the Second World War.

*** A major eruption triggered a tsunami that killed at least 426 people and injured 14,059 others. As a result of the landslide, the height of the volcano was reduced from 338 m to 110 m. This event caused the deadliest volcanic eruption of the 21st century.

TABLE 6.12

Values of coefficients of polynomial approximation of the vertical aerosol profile

Season	a	b	c	d
March–May	−3.60	$−3.59 \times 10^{-2}$	6.30×10^{-3}	$−3.17 \times 10^{-4}$
June–August	−3.78	$−1.79 \times 10^{-2}$	$−6.66 \times 10^{-4}$	$−1.27 \times 10^{-4}$
September–November	−3.67	$−3.26 \times 10^{-2}$	$−2.99 \times 10^{-4}$	$−1.20 \times 10^{-4}$
December–February	−3.50	$−6.42 \times 10^{-2}$	4.01×10^{-4}	$−1.21 \times 10^{-4}$
Mean global	−3.64	$−4.77 \times 10^{-2}$	$−1.46 \times 10^{-3}$	$−1.71 \times 10^{-4}$

Note: See Equation (6.5) for the polynomial approximation.

surface. These effects are restricted to the immediate vicinity of the volcano. However, apart from the ash and lava that cover the ground around a volcano after it has erupted, the eruptions also affect the atmosphere. The material injected into the atmosphere spreads out rapidly over enormous areas and so affects the weather and climate over enormous areas too and can have long-lasting direct and indirect effects, see Section 6.3. Satellite remote sensing enables observations to be made over enormous areas compared with what was possible previously. Because of atmospheric circulation patterns, eruptions at mid or high latitudes can be expected to affect the climate over the whole of the hemisphere that they are in. Eruptions in the tropics generally affect the climate in both hemispheres.

The gases and dust particles thrown into the atmosphere during volcanic eruptions influence the climate in various ways. Most of the particles spewed from volcanoes cool the planet by shading it from incoming solar radiation. The cooling effect can last for months or years depending on the characteristics of the eruption. Volcanoes have also caused global warming over millions of years during times in Earth's history when there have been large amounts of volcanic activity, releasing greenhouse gases into the atmosphere. The main materials finding their way into the atmosphere from eruptions include particles of dust and ash, sulphur dioxide, and greenhouse gases like water vapour and carbon dioxide. Volcanic ash or dust released into the atmosphere during an eruption shade sunlight and cause temporary cooling. Larger particles of ash have little effect because they quickly fall out of the air. Small ash particles form a dark cloud in the troposphere that shades and cools the area directly below. Most of these particles fall out of the atmosphere or are washed out by rain a few hours or days after an eruption. But the smallest particles of dust get into the stratosphere and are able to travel vast distances, often worldwide. These tiny particles are so light that they can stay in the stratosphere for months, i.e. they are described as aerosols, and they block the sunlight, causing cooling over large areas of the Earth.

The sulphur dioxide moves into the stratosphere and combines with water to form tiny particles of sulphuric acid aerosols. Like the dust, the sulphuric acid aerosols block the sunlight causing cooling of the Earth's surface. The SO_2 aerosols can stay in the stratosphere for up to three years, moved around by winds and causing significant cooling worldwide. Eventually, the droplets grow large enough to fall to Earth. Volcanoes also release large amounts of greenhouse gases and these gases simply add to the greenhouse gases that have reached the atmosphere from other sources.

Dust ejected into the high atmosphere during explosive volcanic eruptions has been considered a possible cause for climatic change. Dust veils created by volcanic eruptions can reduce the amount of light reaching the Earth's surface and can cause reductions in surface temperatures. These climatic effects can be seen for several years following some eruptions and the magnitude and duration of the effects depend largely on the density of tephra (i.e. dust) ejected, the latitude of injection, and atmospheric circulation patterns. Lamb (1970) formulated the Dust Veil Index (DVI) in an attempt to quantify the impact on the Earth's energy balance of changes in atmospheric composition due to explosive volcanic eruptions. The DVI is a numerical index that quantifies the impact of a particular volcanic eruption's release of dust and aerosols over the years following the event.

Lamb's dust-veil index is an index of the amount of finely divided material suspended in the atmosphere after great volcanic eruptions, and of the duration of an effective veil intercepting the Sun's radiation. It can be calculated from estimates of the amount of solid matter thrown up, from the reduction of intensity of the solar beam, or from the reduction of temperatures prevailing at the surface of the Earth. The great eruption of Krakatoa in Indonesia in 1883, which ejected about 17 km^3 of particulate matter into the atmosphere, where it remained for three years, was defined by Prof. Lamb to be an index value of 1000.

The tools of satellite remote sensing began quite early on, following the launch of TIROS-1 in 1960, to be applied to an extensive complex of observations of volcanic emissions and their propagation in the atmosphere with the use of conventional means (e.g. ground-based, aircraft, balloon) as well as Earth-observing satellites.

The first eruption for which satellite remote sensing data were seriously collected was an extensive complex of observations of volcanic emissions and their propagation in the atmosphere has been carried out with the use of conventional means (e.g. ground-based, aircraft, balloon) as well as satellites. Substantial progress has been made in the gathering of input data for large-scale transport studies and computer modelling. However, the computer modelling of volcano-induced climate change is still at the research and development stage and this is now an area of very active research.

Of great interest are attempts to reveal a volcanic signal in the observed climate changes. The difficulty of the solution to this problem is that the interannual variability of surface air temperature is determined by various processes which can be classified as connected with the external forcings, internal regular oscillations, and random processes, which appear to be largely due to the dynamic instability of the atmosphere. The most probable external factors are large-scale volcanic eruptions and increased CO_2 concentrations in the atmosphere (to date, there is no convincing evidence for the effect of solar activity on climate). A considerable contribution to atmospheric variability is made by its internal instability. Because of the varying intensity and location of cyclones and jet streams, a random variability of heat fluxes occurs in the atmosphere, leading to a random variability in air temperature.

Work on the use of satellite data in volcano studies initially concentrated on two major eruptions, one of Mount Agung in Indonesia in 1963 and the second of El Chichón in Mexico in 1982. The eruption of Mount Agung (8.3433° S, 116.5071° E) in 1963 was one of the largest and most devastating volcanic eruptions in Indonesia's history. The strong 1963 Agung eruption on the island of Bali (8° S, 115° E) was, apparently, second in power to the Krakatoa eruption of 1883 and was adequately documented by data from numerous observations. The post-eruption global distribution of aerosol can be identified through observations of the aerosol optical effects. According to Wikipedia on 18 February 1963, local residents heard loud explosions and saw clouds rising from the crater of Mount Agung. On 24 February lava began flowing down the northern slope of the mountain, eventually travelling 7 km in the next 20 days. On 17 March, the volcano erupted, sending debris 8 to 10 km into the air while lava flows devastated numerous villages, killing an estimated 1,100–1,500 people. Cold lahars caused by heavy rainfall after the eruption killed an additional 200 people. Lava flows caused by a second eruption on 16 May killed another 200 people while minor eruptions and flows followed and lasted for almost a year. The eruption of El Chichón (17.36° N, -93.23° W) which occurred on 29 March and 3, 4 April 1982 is the largest volcanic disaster in modern Mexican history. The powerful 1982 explosive eruptions devastated an area extending about 8 km around the volcano and a new 1 km wide 300-m-deep crater was created.

Some discussion of pre-1963 work will be found in Section 6.3 of Kondratyev and Cracknell (1998a, b).

The latitude at which an eruption takes place is an important factor which determines the volcanic signal. If the high-latitude eruptions are strongest in high-and mid-latitudes, then the effect of low-latitude eruptions is confined, largely, to the tropical and middle latitudes. Thus, it seems that middle latitudes are much more subject to the effect of both low-latitude and high-latitude explosive eruptions. The eruptions with a volcanic eruption index of about 4 somehow affect the surface air temperature field only during the first post-eruptive month. Depending on the season, the temperature decrease varies between 0.05 and 0.1 °C. The effect of large explosive eruptions on the decrease of surface air temperature over the land for the last 100 years (except for the strongest five eruptions) has been characterised by the fast formation and short lifetime of the volcanic signal.

McCormick and Wang (KC98) processed the aerosol remote-sounding data from the SAGE-1 satellite for March 1979 to February 1980, a period without strong volcanic eruptions and by which time the eruptive products of the Mount Agung 1963 eruption would have settled or been washed out of the atmosphere. They proposed a base model of the background stratospheric aerosol in the form of zonally averaged (monthly means and zonal means) vertical profiles of the coefficient of aerosol extinction at a wavelength of 1 μm in the tropics, middle and high latitudes of both hemispheres (the corresponding data have been tabulated and graphed). The similarity of the seasonal mean vertical profiles of the coefficient of aerosol extinction β in the latitudinal bands (75° S to 40° S; 40° S to 20° S; 20° S to 20° N; 20° N to 40° N; 40° N to 75° N) has made it possible to suggest the following polynomial approximation:

$$\log \beta = a + bz + cz^2 + dz^3 \tag{6.5}$$

where z is the height with respect to the troposphere in km. The values of the coefficients for different seasons are given in Table 6.12.

Although volcanoes have long been considered a cause of glaciations, this hypothesis remains unsubstantiated. Analysis of the aerosol particle content in the Greenland ice cores has not revealed any increase in the aerosol content during the early stages of the Wisconsin glaciation cycle.

Assessment of post-eruption variations in the Earth's radiation budget is a key aspect of the climatic impact of volcanic eruptions.

When discussing the results of observations of volcanic aerosols, it is important to assess possibilities for quantitative characteristics of the eruption intensity. In this regard, Kelly and Sear (1982) analysed the adequacy of Lamb's dust veil index (DVI), which was introduced to characterise quantitatively the impact of volcanic aerosols on the Earth's radiation budget and climate during the several years following an eruption. An estimate of the DVI, taking into account observational data and empirical and theoretical studies of the possible climatic implications of the dust veil, is based on the nature of such implications as understood in 1960. The DVI is calculated by averaging over the maximum possible number of DVI estimates determined with the use of various techniques.

Five available techniques were mutually calibrated (normalised) to give the DVI value 1000 for the 1883 Krakatoa eruption. Techniques B and C have been favoured. According to technique B, the mean global DVI is

$$\text{DVI} = 0.97 R_{max} E_{max} t_{mo} \tag{6.6}$$

where R_{max} is the post-eruption maximum extinction of direct solar radiation determined as a monthly mean for the mid-latitudes of the given hemisphere (maximum extinction can sometimes be reached only two yr after the eruption), E_{max} is the geographical extent of the dust veil in conditional units as a function of the volcano's latitude ($E_{max} = 1.0$ for the band 20° N to 20° S and 0.3 for latitudes 40°), and t_{mo} is the lifetime of the dust veil (in months).

Technique C gives

$$\text{DVI} = 52.5 T_{D\,max} E_{max} t_{mo} \tag{6.7}$$

where $T_{D\,max}$ is the mean temperature decrease (°C) in the mid-latitudes of the given hemisphere during the year when the effect of an eruption is at a maximum. Approximate estimates of the DVI, which may be considered as showing only an order of magnitude, are explained by different available observational data and hence by different techniques chosen for the estimation.

Recently specified data on the chronology of eruptions and a more adequate understanding of the nature of their climatic impact (e.g. the role of the secondary H_2SO_4 aerosol) necessitate a revision of the understanding of the DVI.

A detailed study of the meteorological data over the period 1980–1990, covering the Mount Agung and El Chichón eruptions is given by Kondratyev and Cracknell (KC98, page 457). From the data of the global network of 63 aerological stations for the period winter 1957–1958 to summer 1983, Angell 1986) and Angell and Korshover (KC98) analysed the variability of tropospheric temperature for seven climatic zones in both hemispheres (the representativity of data for the southern hemisphere needs further studies) and for the world as a whole (after the 1963 Agung and 1982 El Chichón eruptions). For extra-tropical latitudes, information has been obtained on surface air temperature and the temperature of the 850–300 hPa layer (1.5–9 km), and for the tropics calculations have been made of the temperature of the 300–100 hPa layer (9–16 km).

Results obtained by Angell and Korshover (KC98) show that during the first year after the eruption of Agung, a decrease of average surface air temperature for the northern hemisphere by 0.34 °C was observed and the El Chichón eruption was followed by a temperature increase of 0.37 °C (at the 0.5 per cent level of statistical significance according to the Student criterion t). These changes in the hemispherical mean temperature are largely caused by the contribution of the mid-latitude band, where the respective variations reached −0.36 °C (Agung) and +1.27 °C (El Chichón). Apparently, the trend of temperature increase is connected with the effect of an abnormally strong positive sea surface temperature anomaly in the eastern equatorial Pacific (ENSO) observed in 1982–1983. The connection between changes of the sea surface temperature anomaly and surface air temperature in the tropics is direct and in mid-latitudes is indirect.

Analysis of meteorological observations made after the 1982 El Chichón eruption showed a warming of the lower stratosphere or near the surface, as it had been after the 1963 Agung eruption (8° S). The troposphere in both hemispheres was abnormally warm in 1983 (a year after the eruption).

If we use the results obtained to filter out the ENSO signal, it turns out that after the El Chichón eruption, the same (by amplitude) cooling of the atmosphere (about 0.5 °C) took place as after the Agung eruption but it was more short-lived. The use of the same correction procedure as applied to observational data after six strong volcanic eruptions in the tropics for the last century (Krakatoa, Soufrière, Pelée, Santa Maria, El Chichón, and Agung) has led to the conclusion that there was a decrease of the mean surface temperature of the northern hemisphere continents of about 0.3 °C, whereas after the Katmai eruption in Alaska it was only 0.1 °C. As for the cooling of the whole Earth's surface, in this case the data on the volcanic signal in the past are uncertain and controversial, being strongly dependent on the volcanically induced sea surface temperature increase in the eastern equatorial Pacific.

The importance, for model calculations, of a knowledge of the nature of aerosol particles is illustrated by Table 6.13. This Table shows a comparison of calculated and observed (in Australia) temperature changes at the 60 and 100 hPa levels that revealed a satisfactory agreement, bearing in mind that part of the observed temperature increase could have been determined by quasi-biennial variability. The calculated change of the tropospheric mean temperature in the band 30° N to 30° S, from the moment of eruption to the beginning of 1966, agreed with the data of observations: namely, a temperature decrease of about several tenths of a degree Celsius, with a time constant of about 1 year. The significance of these results is that sulphate aerosols weakly absorb solar radiation while silicate aerosols are strong absorbers of solar radiation. The considerable difference between the results illustrates the importance of having an adequate knowledge of the chemical composition, and therefore of its optical properties, when using the aerosol data as input for a computer model.

A thick gas-dust cloud of volcanic origins, under the influence of diffusion and circulation, propagates rapidly, covering a vast band of latitudes. Simultaneously, a gas-to-particle conversion leads to a strong increase of the sulphate aerosol concentration in the stratosphere.

TABLE 6.13

Variations in stratospheric and tropospheric temperature due to sulphate and silicate aerosols

	Time (days)							
	30	60	120	180	360	540	720	1000
Sulphates								
$T_{55\,hPa}$	0.63	2.15	4.77	6.34	3.75	1.90	0.84	0.02
T_{tr}	−0.01	−0.03	−0.12	−0.23	−0.48	−0.54	−0.51	−0.44
Silicates								
$T_{55\,hPa}$	3.8	8.1	12.8	13.1	10.5	7.4	6.2	2.8
T_{tr}	−0.01	0.03	0.08	−0.13	−0.22	−0.23	−0.19	−0.13

Source: (KC98 Table 6.23).
$T_{55\,hPa}$ = stratospheric temperature; T_{tr} = tropospheric mean temperature.

The first volcanic eruption to which these tools began to be applied seriously was that of El Chichón. For the 1982 El Chichón eruption in México, a rather complete satellite monitoring of the eruptive aerosol propagation was accomplished for the first time. A considerable amount of data regarding the nature and dispersal pattern of the eruptive products was obtained. However, in terms of providing input data for computer modelling of the climatic effects of the eruption, this data was of limited value. The situation changed in 1991, when the eruption of Mount Pinatubo occurred in the Philippines. By this time the scientific community had deployed a variety of satellite, aircraft, balloon, and surface instruments which could be used to monitor the consequences of eruptions not only from the viewpoint of global propagation of erupted products but also a diverse monitoring of the evolution of volcanic aerosol properties and their impact on the atmospheric radiation regime and climate.

It should be emphasised that of principal importance was a combination of conventional and satellite observational means, the conventional means having been used not only for the remote sounding of the eruptive cloud but also for direct measurements of the characteristics of volcanic aerosol. No doubt, an experience of the complex monitoring of the Pinatubo eruption can be considered as a convincing and, at that time, most complete illustration of the fruitful combination of both conventional and satellite observational means to study the environmental variability.

An explosive eruption of the andesite island volcano Mount Pinatubo (16.14° N; 120.35° E) on 15 June 1991 in the Philippines (the island of Luson) affected most of the stratosphere, compared with the previous data of satellite observations (the previous eruption of Mount Pinatubo took place about 635 years ago). Gigantic ejections of SO_2 and other products of the eruption led to the formation of a stratospheric aerosol layer at altitudes of 30 km and higher, reliably recorded by satellite observations (McCormick 1992). At altitudes above 20 km, the aerosol cloud moved rapidly westward, and its front made a complete revolution round the Earth in three weeks.

Three weeks after the eruption (7 July 1991), aircraft measurements were begun of the characteristics of the eruptive cloud, covering (during six flights from 7 to 14 July) a region 4.5° S to 37° N; 80° W to 45° E. The complex of the airborne instruments included: zenith-looking lidar and correlation spectrometer, a side-looking Fourier spectrometer, and a multi-channel radiometer to measure upward and downward fluxes of total, scattered, and direct solar radiation.

A database was obtained from synchronous satellite observations using the TOMS spectrometer to map the total ozone content (these data were processed to retrieve the content of SO_2), the AVHRR, and the occultation sounder SAGE. A preliminary analysis of the observational data showed that the mass of ejected products was about 20 Mt, which exceeded three times the power of the 1982 El Chichón eruption. During the first several months, the basic aerosol mass remained within the tropical band 30° N to 20° S.

From the data of optical measurements, the total mass of eruptive H_2SO_4 aerosol constituted 20–30 Mt (these estimates agree with the data on SO_2). The aerosol optical thickness at the wavelength 0.5 μm varied between 0.2 to 0.4. An additional absorption of solar radiation by eruptive aerosol has caused a considerable increase of stratospheric temperature. So, for example, in September, the temperature rise at the 30 hPa level in the band 0–30° N reached 3.5 °C. A decrease of NO_2 content was observed over New Zealand, which agreed with the hypothesis on stimulating the process of ozone depletion under the influence of heterogeneous reactions on the surface of aerosol particles. In view of the great significance of the Mount Pinatubo eruption studies, we shall discuss the relevant results in some detail.

6.10.1 SATELLITE OBSERVATIONS

The available satellite remote-sensing sources that can be used to monitor the consequences of volcanic eruptions are rather diverse, which allows the acquisition of rich information. Not only are digital images in the visible and infra-red spectral intervals transmitted from meteorological satellites available, but also there are data on the content in the stratosphere of minor gaseous and aerosol components (the instrumentation SAGE and TOMS), aerosol optical thickness of the atmosphere (AVHRR), etc.

The processing of the Nimbus-7 TOMS data (this instrumentation had been successfully operating since the launch of Nimbus-7 in 1978) made it possible to retrieve the total SO_2 content in the stratosphere with an error of about +30 per cent. Initial data on the SO_2 content refer to several days (11–14 June 1991) prior to the cataclysm, when several weak eruption-precursors had taken place.

The first eruptive cloud about 100 km^2 was observed on 12 June and contained about 25 kt of SO_2. By 13 June, it had moved westward at a distance of 1100 km (appearing over Vietnam), and its mass had reached 110 kt. On 14 June, this cloud was hardly seen at all. On 13 June, a second similar cloud (100 km^2, 15 kt) was observed over the western edge of the island of Luson, but it was observed only for 24 hours. On 15 June, a third cloud was recorded (7500 km^2, about 450 kt) stretching for 1600 km west of the volcano Pinatubo to south Vietnam.

The 15 June cataclysmic eruption began before noon and continued until the morning of 16 June. It was discovered from the TOMS data only on 16 June (at which time the satellite had been over the zone of eruption). The mass of eruptive SO_2 at that moment was estimated at 16.5 Mt (the value could be underestimated in view of some malfunction of the instruments). By 23 June the eruptive cloud had covered a region of about 15 million km^2 stretching a distance of 10000 km from Indonesia to central Africa, being rather uniformly distributed. Two weeks after the eruption (30 June), the length of the cloud, located in the latitudinal band 10° S to 20° N, grew to 16000 km, and the mass of SO_2 reached 12 Mt. Thus, two weeks after the eruption, the eruptive cloud contained about 60 per cent of the initially ejected SO_2. This gaseous cloud could be observed during its motion round the globe for 22 days.

Table 6.14 contains comparative data characterising various eruptions. Bearing in mind the large power of the Pinatubo eruption, one can suppose a possibility of serious climatic consequences of this eruption.

Continuous functioning of the occultation photometer SAGE-II carried by the ERBS satellite has provided a long observation series on the content of some minor gaseous components and aerosol in the atmosphere, beginning in 1984. From the data on aerosol extinction at the wavelength 1020 nm for June-August 1991, McCormick and Veiga (1992) analysed the evolution of spatial distribution of eruptive aerosol (both its H_2SO_4 and solid components).

Approximately one month after the eruption, these data referred largely to middle and tropical latitudes. In this period and later on, the aerosol was observed at altitudes up to 29 km, being mainly concentrated in the layer 20–25 km. At first, the aerosol optical thickness reached a maximum in the band 10° S to 30° N, but by the end of July, high values of aerosol optical thickness had been observed at least to 70° N, with maxima north of 30° N in the layer below 20 km.

TABLE 6.14

Comparative characteristics of various eruptions from the TOMS data

Characteristics of eruptions	Nevarupta/Katmai 1912	El Chichón 1982	Nevado del Ruiz 1985	Pinatubo 1991
SO_2 amount (kt)	$(6.2–20) \times 10^3$	7×10^3	750	2×10^4
Ash ejections (g)	3.4×10^{16}	3.0×10^{15}	4.8×10^{13}	1.0×10^{16}
Gas/ash ratio	$(2–6) \times 10^{-4}$	2.3×10^{-3}	0.015	1.9×10^{-3}
VEI	6	4–5	3	5–6

Source: (KC98 Table 6.25).
VEI = volcanic eruption intensity.

In both hemispheres, there was observed a correlation of anticyclonic systems of high pressure with an advection of the volcanic material of the sub-tropics, located at the level 21 km (16 km) to the zones of mid-latitude jet streams in the southern (northern) hemisphere. By August, the aerosol optical thickness in the southern hemisphere atmosphere had increased by an order of magnitude, compared with early July, at the expense of increasing aerosol content above 20 km. The mass of eruptive aerosol was estimated at 20–30 Mt, which considerably exceeded the estimates (about 12 Mt) obtained for the El Chichón eruption.

A successful development of the technique to retrieve the aerosol optical thickness of the atmosphere at the wavelength 0.5 μm from the NOAA AVHRR data (the retrieval error being 0.03–0.05) has made it possible, starting from July 1987, to perform a regular global mapping of aerosol optical thickness over the world's oceans. The availability of such information has also opened up possibilities to monitor the dynamics of such phenomena as aerosol outbreaks from the Sahara desert and Arabia.

To analyse the variability of the optical properties of the atmosphere after the eruption of Pinatubo, Stowe *et al.* (1992) undertook a combined daily mapping of aerosol optical thickness for 12 weeks after the eruption. The maps have shown that the outbreaks of desert aerosol mentioned above are overshadowed by the scales of eruption consequences. The aerosol optical thickness retrieval technique is based on comparing measured outgoing shortwave radiation values with results of calculations for various optical models of the atmosphere for pixels 1 km by 4 km in size with the use of a multi-spectral (channels 0.63 and 3.7 μm) algorithm to recognise the clear-sky cases, similar to the sea surface temperature retrieval algorithm (the aerosol optical thickness is mapped over the grid 1° latitude by 1° longitude).

Analysis of a series of global maps of aerosol optical thickness has shown that the eruptive aerosol revolves round the globe in 21 hours. The distribution of aerosol is characterised by horizontal inhomogeneity even two months after the eruption (in August, the aerosol optical thickness maxima were observed near Kamchatka and over the Bering Sea, as a result of forest fires in Siberia). If the criterion for recognition of the eruptive aerosol layer is an aerosol optical thickness value exceeding 0.1, it turns out that two months after the eruption it covered 42 per cent of the Earth's surface, which is twice as much as after the El Chichón eruption.

By the end of the first two months, the aerosol layer had been confined to the band 20° S to 30° N, with individual clusters in higher latitudes. A maximum of aerosol optical thickness was observed on 23 August and was 0.31. The global mass of emitted SO_2 equal to 13.6 Mt corresponds to observed aerosol optical thickness values (the mass of H_2SO_4 aerosol is larger than that of SO_2 by a factor of 1.8).

A decrease of the Earth's radiation budget due to the Earth's albedo increase at the expense of scattering on eruptive aerosols (by about 1.3 per cent) reached 2.5 W m^{-2}, which was equivalent to a global mean surface air temperature decrease of not less than 0.5 °C, provided aerosols were uniformly distributed over the globe during the next 2–4 yr. The value of the Earth's radiation budget decrease by 2.5W m^{-2} is obtained from the outgoing shortwave radiation increase by 4.3 Wm^{-2} and outgoing longwave radiation decrease by 1.8 W m^{-2}. Of course, such global cooling may not manifest itself, in view of multifactor surface air temperature variability (for example, the ENSO). A consequence of the growth of stratospheric aerosol content can also be its destructive impact on the ozone layer due to the ozone-depleting heterogeneous chemical reactions and a weak oxidation of precipitation (with the H_2SO_4 aerosol particles depositing onto the troposphere).

A flavour of what can be achieved in general terms in monitoring the effect of volcanic eruptions by remote sensing can be gained from what was done following the Mount Pinatubo eruption of 1991 in the Philippines. It should, however, be noted that these results and conclusions were rather general; it is quite another story to try and determine input data for computer climatological models. This is a reflection of the fact that modelling the effect of CO_2 increases is rendered easy (modellers may disagree) due to the fact that the CO_2 is a uniformly mixed gas in the atmosphere. Aerosols are not uniformly mixed nor are the concentrations and constitutions independent of time.

6.10.2 AIRCRAFT OBSERVATIONS

In addition to satellite data, there was aircraft data too. Three weeks after the Mount Pinatubo eruption, NASA scientists undertook an extensive programme of aircraft observations of the characteristics of the eruptive products plume (Valero and Pilewskie 1992). Six flights in the period 7–14 July 1991 covered the region 34° N to 4.5° S, 76° W to 46° E. The flying laboratory "Electra" has a flight ceiling of 6.5–7.0 km, which in some cases generated serious difficulties in the remote sounding of the stratosphere through tropospheric layers.

The complex of the airborne instruments included: a zenith-looking lidar and correlation spectrometer, a side-looking Fourier spectrometer and a multi-channel radiometer to measure upward and downward fluxes of total, scattered, and direct solar radiation. The data on direct solar radiation made it possible to find the aerosol optical thickness, with the Rayleigh component being calculated from measured atmospheric pressure and values of extra-atmospheric fluxes of solar radiation found by extrapolation (using the Langley method) of the data of surface observations at Mauna Loa observatory, with an error of below 0.5 per cent. The errors of radiation flux measurements did not exceed 1–2 per cent.

Analysis of observational data revealed an aerosol optical thickness increase near to the volcano (16.15° N), but aerosol optical thickness maxima were recorded near 10° N, which reflected specific motion of the eruptive cloud. A maximum of aerosol optical thickness was recorded on 13 July at the wavelength 412 nm and reached 0.49.

The available data on the spectral dependence of aerosol optical thickness, characterised by a rapid decrease of aerosol optical thickness with increasing wavelength in the near infra-red, allowed the retrieval of the effective size of particles as a parameter of the aerosol size distribution. Here it was necessary to classify the aerosol optical thickness spectra into two groups, corresponding to the data for 7, 8, 12, and 13 July, when the aerosol was mainly monomodal, and for 10 and 14 July, when the aerosol was bimodal.

From the data for 8 and 13 July, the values of the effective radius were 0.18 and 0.35 μm, and the masses of aerosol in the air column were 35 and 80 mgm^{-2}, respectively. The bimodal size distribution observed on 10 July was characterised by particle radii 0.1 and 0.8 μm with the mass of aerosol 40 mgm^{-2}. Apparently, the bimodal size distribution reflects the process of gas-to-particle conversion of the fine-disperse fraction of H_2SO_4 droplets in the presence of larger particles of volcanic origin.

Calculations of the increase of the surface-atmosphere system albedo as a function of Sun elevation, made in various suppositions, gave maximum values up to 0.12, for surface albedo = 0.05, Sun elevation = 10°, and aerosol optical thickness = 0.6.

Lidar sounding from "Electra" in July 1991 (Nd: YAG lidar, wavelength 532 nm) revealed a multi-layer structure and horizontal inhomogeneity of the aerosol plume in the stratosphere three to four weeks after the Pinatubo eruption (Winker and Osborn 1992a). The scattering ratio was determined as the ratio of the sum of the coefficients of aerosol and Rayleigh backscattering to the coefficient of Rayleigh back-scattering, with the use of normalisation from the data for the level near the tropopause at an altitude of 16 km.

The thickness of aerosol layers between the altitudes 17 and 26 km varied from less than 1 km to several kilometres. The layers north of 20° N were relatively weak (scattering ratio 3 and less), but south of 15° N the scattering ratio values exceeded 10, as a rule. A maximum of scattering ratio recorded on 12–13 July at an altitude of about 25 km reached 80 and more, but north of 11°N this thick aerosol layer suddenly disappeared.

Assessments of the mass of aerosol in the vertical atmospheric column, made in supposition that the aerosol is pure H_2SO_4, have given 95mgm^{-2}, which (with a conditional recalculation) is equivalent to a global mass of 8 Mt. Apparently, by the moment of observation, only half the eruptive SO_2 had been converted into H_2SO_4 aerosol; that is, finally, the mass of sulphate aerosol could reach 16 Mt (the El Chichón eruption had given about 12 Mt of aerosol).

The polarisation lidar carried by "Electra" generated a linearly polarised signal at the wavelength 532 nm. The signal of backscattering on stratospheric volcanic aerosol was recorded for two orthogonal polarisations with subsequent calculation of both the scattering ratio and depolarisation ratio determined as a ratio of signals for perpendicular and parallel components of polarisation (aircraft observations of depolarisation ratio for the volcanic plume were made for the first time). The observation data, with vertical resolution 15 m, were averaged to the resolution 150 m.

Analysis of the depolarisation ratio data revealed aerosol layers with different depolarisation signatures, which is determined by special chemical composition, physical state, and shape of particles (Winker and Osborn 1992b). The depolarisation of backscattered light is known to reflect the presence of scattering on anisotropic particles. In this case, possible reasons of depolarisation can be the presence of crystallised H_2SO_4 droplets, aggregates of frozen droplets, and silicate particles of an irregular shape. The data for 12–13 July testify to the existence of a layer with an intensive depolarisation at an altitude of 22 km and a layer with a low depolarisation near the 25 km level. Apparently, in the latter case, there was almost pure H_2SO_4 aerosol with a small amount of ash or other non-spherical particles, and in the first case there was a mixed aerosol. For a reliable explanation of the causes of depolarisation, data are needed of direct measurements of the aerosol size distribution, the shape, and chemical composition of particles.

Since the global mass of SO_2 ejected by Pinatubo was estimated from the TOMS data at 20 Mt, it follows that the transformation of all SO_2 into H_2SO_4 aerosol should result in 41 Mt of aerosol. Hoff (1992) performed aircraft observations with the use of the zenith-looking ultraviolet correlation spectrometer COSPEC-V (measurements of scattered radiation intensity were made at seven wavelengths in the interval 297–315 nm) to estimate the SO_2 content in the air column.

The results of observations for different atmospheric masses gave values of SO_2 content from 8.9 to 64.2 μm atm m. Maximum values, corresponding to minimum masses, were observed near 14° N. Apparently, the most reliable value should be between the value averaged over all data (24.6 μm atm m) and data of near-noon observations (40.9 μm atm m).

The SO_2 content increased between 12 and 14 July, which could be explained by the input of the most powerful plume of eruption products at an altitude of 26 km, revealed from the lidar sounding data. The SO_2 global mass was estimated at 8.8 Mt (with a global mean content in air column of 25 μm atm m) to 14.4 Mt (41 μm atm m). By the moment of the observations discussed, about 25 Mt of eruptive aerosol should have formed.

Using the airborne infra-red Fourier spectrometer, Mankin *et al.* (1992) measured the Sun spectra at Sun elevations from 0° to 15° after the Pinatubo eruption to retrieve (by comparing measured absorption spectra with those calculated) the total content of SO_2, HC1, and O_3 in the stratosphere from the data of wavenumbers 1364–1368, 2926, and 1141 cm^{-1}, respectively; the resolution of the spectrometer constituted 0.06 cm^{-1}. The errors in retrieving the total content of SO_2, HC1, and O_3 were, respectively, 25 per cent, 20 per cent, and 16 per cent.

Analysis of the results obtained has led to the conclusion that measured values of the total ozone content did not differ considerably from the normal ones and, in contrast to the data of observations after the El Chichón eruption of 1982, the HCl content did not increase markedly. Thus, the Pinatubo eruption has not led to any increase of chlorine in the stratosphere. The estimates of the SO_2 content are within those obtained from the data of the correlation spectrometer. A maximum of SO_2 was recorded near 15° N, with considerable amounts of SO_2 in the band 15–30° N.

With SO_2 supposed to be uniformly distributed over the layer 20–30 km, its average mixing ratio turns out to be 32 ppbv. Proceeding from the fact that the SO_2 plume covers the latitudinal band 20° S to 20° N, the mass of gas in the plume 4 weeks after the eruption should be 6.5 Mt. Such estimates should be considered very approximate. It follows from them, however, that three to four weeks after the eruption, a large quantity of eruptive SO_2 remains in the stratosphere. Aircraft observations at mid-tropospheric altitudes did not permit one to obtain data on volcanically induced changes in the stratospheric water vapour content.

6.10.3 BALLOON OBSERVATIONS

Balloon observations were also of great importance for the analysis of the eruption consequences. In the period 16 July to 29 August, seven launches of the aerosol radio-sonde were made at Laramie (41° N, Wyoming, USA) to monitor the dynamics of stratospheric aerosol before and after the Pinatubo aerosol products had got to the region of Laramie. Apart from the sensors for atmospheric pressure, temperature, and ozone, the complex of aerosol radiosonde instruments included a condensation nuclei counter (the radii of particles exceeding 0.01 μm), a fast-operating eight-channel aerosol particle counter (the radii of particles exceeding 0.15, 0.25, 0.50, 1.0, 2.0, 3.0, 10.0 μm) and a special two-channel ($\geqslant$ 0.15 and 0.25 μm) particle counter to assess the volatility of particles. In some launchings, the aerosol radiosonde carried also a three-cascade impactor to sample the aerosol for its subsequent electron-microscopic analysis. The aerosol radiosonde ceiling varied within 36.5–38.8 km.

Analysis of the vertical profiles of condensation nuclei from the data of all launches, performed by Deshler et al. (1992), showed that first the eruptive cloud could be observed on 16 July. (An increase of condensation nuclei concentrations near the tropopause due to, probably, oilfield fires in Kuwait, took place on 17 June.) By 26 July, the aerosol number concentration below the level of the east-west transport in the upper stratosphere had increased to a record value of about 50 cm^{-3} at a height of 17 km. This persistent lower aerosol layer, determined by a rapid northward transport of particles, was also revealed from the data of all other launches of aerosol radiosondes.

The first signs of the layer formed above 20 km (directly in the zone of east–west transport) were noticed in the form of a thin layer at a level of 23 km (this layer was also recorded from the data of lidar sounding in Boulder, Colorado on 29 July to 2 August). By 6 August, it had ceased to exist but was again observed over Laramie on 29 August with a decreased aerosol concentration after the aerosol particles had circled the Earth. The vertically stretched layer of eruptive aerosol remained until the end of August at altitudes from the level of the tropopause to 22 km, and in mid-September the aerosol was observed at altitudes up to 30 km.

An increase of condensation nuclei concentrations at altitudes 15–20 km, due to processes of sulphate aerosol gas-to-particles formation, exceeded the saturation level for the condensation nuclei counter. This means that the condensation nuclei concentration was more than 500 cm^{-3}; that is, it exceeded by two orders of magnitude the background value for the altitudes 18–20 km.

Experiments with heating the aerosol to 150°C showed that 95–98 per cent of it was volatile, which permitted one to consider it as droplets of concentrated H_2SO_4 water solution, and the characteristic time for SO_2 conversion into H_2SO_4 to be equal to 38 days (previous observations of smoke aerosol from Kuwait had given about 50 per cent of volatile aerosol).

The aerosol size distribution was characterised by large particles above the level of reversal of wind direction in the stratosphere (in the layer 22–23 km), compared with the layer 16–17 km. The altitudes below 20 km are characterised by particles with a modal radius of about 0.07 μm, whereas above 20 km the radius increases to 0.35 μm (in the case of a monomodal log-normal size distribution).

In the lower aerosol layer and on the boundaries of the upper layer the aerosol size distribution was first (from the data for 2 August) bimodal. The high concentration of small droplets of H_2SO_4 aerosol identified from the relevant observation data suggested the conclusion that the droplets had formed either due to homogeneous nucleation or through condensation on small ions.

During three launches of the aerosol radio-sonde in Laramie on 26 and 30 July and on 14 August 1991, the volcanic aerosol was sampled at altitudes between 16.5 and 37.5 km using a three-cascade impactor ensuring a separation of particles with a diameter of more than 4, 1, and 0.25 μm (Sheridan *et al.* KC98). These samples fixed on a thin film substrate were analysed with the use of an electron microscope, which showed that almost all (>99 per cent) small size aerosol particles were submicrometre H_2SO_4 droplets which later (under the influence of ammonia) transformed into $(NH_4)_2SO_4$ particles. The remaining particles – eruptive aerosol – were much larger sulphite and sulphate ones with an addition of the Earth crust components up to 10 μm in size.

Data from the aerosol radio-sonde optical counter showed that a maximum of the July size distribution of particles larger than 0.15 μm radius in the layers of stratospheric aerosol exceeded that in August. This fact was verified; a layer of sub-micrometre aerosol deposited onto the film in July was 20–30 per cent thicker (heavier) than in August. A detailed analysis of the small particles of sulphate aerosol suggested the conclusion that almost no investigated particle revealed the presence of either solid or dissolved condensation nuclei. From 98 per cent to more than 99 per cent of particles turned out to contain H_2SO_4. On this basis, one can suppose that the H_2SO_4 aerosol resulted from homogeneous nucleation. This supposition agrees with experimental assessments of the volatility of particles, which testify that 95–98 per cent of volcanic aerosol consists of H_2SO_4 water solution droplets. The 30 July samples revealed a quantity of large (super-micrometre) particles in the form of soil particles covered with a layer of H_2SO_4, and the 26 July data showed that the small sulphate aerosol particles were, as a rule, surrounded by one or numerous rings of smaller sulphur-containing particles (probably, these particles had resulted from small droplets deposited onto the film).

It would probably be fair to say that in the mid to late 1980s, when the IPCC's First Assessment report was in preparation, remote sensing data was able to play a significant role in evaluating the atmospheric/meteorological consequences of a major volcanic eruption, in this case El Chichón. However, it would also be fair to say that in terms of providing aerosol data as input to climatological computer models the data left much to be desired and this has remained an active research interest since then, including studying more recent eruptions.

6.10.4 SURFACE OBSERVATIONS

So far, only part of the results of surface observations of the eruptive cloud has been published. However, these preliminary results are an important addition to the data discussed above. So, for example, the surface lidar soundings performed at Mauna Loa observatory (in Hawaii: 19.53°N, 156.58°E; 3400m above sea level) detected on 1 July 1991 the Pinatubo-ejected volcanic plume. In this connection, Goldman et al. (1992) processed the Sun spectra recorded with the Fourier spectrometer, to retrieve the SO_2 content in the air column (Table 6.15).

The retrieval was made by comparing measured and calculated absorption spectra for the v_3 band of SO_2 centred at wavenumber 1362.06 cm^{-1} (a stronger v_1 band at 1157.71 cm^{-1} could not be used because of considerable overlapping with water vapour and CH_4 absorption bands).

As can be seen from Table 6.15, by September, the SO_2 content had dropped to the threshold of the spectrometer sensitivity (ca. 10^{13} mol cm^{-2}); that is, by this time, practically all eruptive SO_2 had transformed into H_2SO_4 aerosol.

Naturally, lidar sounding was the most intensive means of surface observations. Since there had not been large-scale volcanic eruptions since 1986, at the latest, which could have led to an

TABLE 6.15

Summarised results of SO$_2$ content assessment

	Spectral resolution	Content
Date 1991	(cm^{-1})	(10^{16} molecules cm^{-2})
11 May	0.004	<0.9
9 July	0.010	(6.1 ± 0.5)
10–24 September	0.004	<0.9

Source: (KC98 Table 6.26).

increase of the aerosol content in the stratosphere, one may consider that by June 1991, when Mount Pinatubo erupted, the aerosol content was close to the background level. Lidar sounding (ruby-laser) performed at Mauna Loa (DeFoor *et al.* 1992) showed that the post-eruption stratospheric aerosol plume had reached Hawaii on 1 June and had been concentrated in the layer 21.5–22.8 km, with maximum scattering ratio 22.5 (in June the scattering ratio was below 1.5). At that time the tropopause was at an altitude of 14.2 km.

Analysis of the data of subsequent soundings in July-August showed that the plume was, as a rule, below 26–27 km (the height of the tropopause varied within 14–16 km), whereas after the El Chichón eruption the eruptive aerosol had reached the 30 km level. Both a general increase of the aerosol backscattering coefficient and its cyclic variability, reflecting apparently the movement of the plume round the globe with a zonal mean velocity of 70–80 km h^{-1}, have been observed during the evolution of the volcanic plume.

An increase of the total (for the layer 16.8–33 km) normalised backscattering coefficient NRBS in July-August can be approximated by the following dependence on the day of the year D, counted from 1 July, when $D = 180$:

$$\text{NRBS} = (0.374 \pm 0.87)(D - 180) + (1.16 \pm 0.76) \times 10^{-3} \tag{6.8}$$

The correlation coefficient in this case is 0.67. The volcanically induced decrease of the solar radiation transmission by the atmosphere constituted 13 per cent and approximately corresponded to that observed after the El Chichón eruption.

Analysis of the results of lidar soundings (Nd:YAG laser) performed in Garmisch-Partenkirchen (Germany: 47.6° N, 11.1° E) revealed a first arrival of the stratospheric plume of Pinatubo-erupted products on 1 July 1991, when the aerosol backscattering coefficient exceeded the background level, but on the following day there was no increase, and during the next months rather variable vertical profiles of backscattering coefficient were recorded (Jäger 1992).

The vertical stratification of aerosol can be explained by special features of atmospheric circulation. The aerosol, concentrated in the layer 15–17 km and observed in early July, moved to the observation location by the shortest trajectory. The two-layer structure of the volcanic plume recorded in August agrees with typical features of the transport in the summertime mid-latitudes: the west-to-east transport at altitudes below 20 km and the transport in the opposite direction above 20 km. Upon the transition to the characteristic wintertime circulation regime (beginning in September), with the prevailing west-to-east transport through the thickness of the eruptive aerosol layer, one thick layer formed at altitudes from the tropopause to about 28 km and the northward transport of erupted products intensified. All these regularities are similar to those observed after the El Chichón eruption.

Lidar aerosol soundings of the troposphere and stratosphere performed by Post *et al.* (1992) at three locations near Boulder (Colorado, USA) with the use of three lidars at wavelengths 10.591 μm

(CO_2 laser), 0.694 μm (ruby-laser), and 0.574 μm (Nd:YAG laser) detected an arrival of Pinatubo-ejected aerosols, beginning on 27 July 1991 (probably the persistent stratified cloudiness taking place during five days prior to this date had prevented an earlier detection of eruptive aerosols).

In 1982, the El Chichón eruptive plume was first observed over Boulder only four months after the eruption, whereas the Pinatubo eruptive cloud reached this place after only 1.4 months, though Pinatubo is located 2° lat. south of El Chichón. No doubt, these differences are determined by special circulation features in the stratosphere and (or) peculiar ejections of volcanic matter. The difference is that after the Pinatubo eruption the backscattering in the stratosphere immediately intensified, whereas in the case of El Chichón, it only happened eight months after the eruption.

The eruptive cloud from Pinatubo was characterised by strong variability, which necessitated frequent observations during the initial phase of the cloud evolution. The calculations of the ratio of total backscattering (for the layer from the tropo-pause to 30 km) for CO_2 lidars and in the visible revealed a decrease of this ratio with time reflecting a decrease of the effective size of aerosol particles.

In December 1980, Johnston et al. (KC98) started daily observations in Lauder (New Zealand: 45° S, 170° E) of the twilight sky brightness in zenith at sunrise and sunset, to retrieve the total content of NO_2 in the atmosphere from the data at wavelengths of about 450 nm, where the NO_2 absorption band is located. An analysis of the total content of NO_2 observation series for 11 yr revealed, apart from diurnal and annual changes, a long-term variability determined, apparently, by the El Chichón eruption (1982) and solar activity (during an 11-yr cycle). On the whole, the total content of NO_2 was less in the mid-1980s than at the beginning and the end of that decade.

During the whole period of observations, except for 1991, a minimum of total content of NO_2 in the annual change took place in June. However, in the period July-September 1991, a gradual decrease of monthly mean morning quantities of total content of NO_2 took place, which had not been observed earlier.

Since part of the maximum decrease of total content of NO_2 in October, amounting to 34–45 per cent (sunrises) and 30–40 per cent (sunsets), compared with the data for 1981, 1989, and 1990 (with a similar level of solar activity), could be determined by the contribution of an additional scattering on volcanic aerosol, a conclusion can be made that the real total content of NO_2 decrease constituted more than 20 per cent and was probably determined by heterogeneous chemical reactions with participation of aerosol. It is known that the reaction

$$N_2O_5 + H_2O \rightarrow 2HNO_3 \tag{6.9}$$

on the surface of aerosol particles must lead to a decrease of the NO_2 and O_3 content and to an increase of HNO_3. Since there was no observed decrease of ozone after the Pinatubo eruption, the possibility of HNO_3 increase must be checked.

6.10.5 CLIMATIC CONSEQUENCES OF THE MOUNT PINATUBO ERUPTION

The stratospheric heating is known to be the most clear-cut manifestation of climatic consequences of volcanic eruptions (Kondratyev 1985, Kondratyev and Galindo 1997 in KC98). In this connection, based on the data of aerological sounding, Labitzke and McCormick (KC98) undertook a comparison of stratospheric temperatures at the 30 and 50 hPa levels in the northern hemisphere, beginning from June 1991, with their monthly means over 20 yr (1965–1984) and 26 yr (1964–1989). This comparison revealed a considerable increase of temperature in July, August, September, and October in the latitudinal band 0–30° N.

In September and October, there was an increase of temperature at the 30 hPa level, reaching in some locations 3.5 °C at the level of statistical significance from 2σ to 3σ (in June the temperature was below and in July at the level of the multiannual mean). Monthly mean, zonal mean temperatures at the 30 hPa level near 20° turned out in September and October to be approximately 2.5 °C, above average over 26 yr, with diurnal mean temperatures increasing almost by 3 °C. A still greater increase of stratospheric temperature occurred south of 20° N.

It is natural to suppose that all these changes are determined by enhanced solar radiation absorption in the stratosphere by volcanic aerosol. One should also expect a simultaneous warming of the southern hemisphere stratosphere, especially in the band 0–20° S, taking into account data of satellite and lidar soundings on the spatial distribution of eruptive aerosols, which show that maximum heating of the stratosphere should take place in the layer 23–26 km in the band 35° N to 20° S.

An important peculiar feature of the Mount Pinatubo eruption is the fact that it took place in a year when, in the course of quasi-biennial oscillations as a major manifestation of the interannual variability of stratospheric temperature, a transition took place in the lower stratosphere, from west-to-east to east-to-west transport (cold phase of the quasi-biennial oscillation), whereas during the eruptions of Agung in 1963 and El Chichón in 1982 an opposite situation took place. It is also important that if during those two eruptions the Sun moved from the equator to the sub-tropics, during the Pinatubo eruption it was located farther north and moved approximately southward. Apparently the transport of eruptive aerosol to high latitudes of both hemispheres, which started in the autumn of 1981, ought to have reduced the stratospheric heating in the tropical band and intensified the global warming, see also Robock and Mao (KC98).

The 15 June 1991 Pinatubo eruption was followed, apparently, by ejections of gases and dust to altitudes above 25 km, the most powerful in this century, and it can be expected to produce significant climatic consequences. Hansen *et al.* (KC98) obtained preliminary estimates of such consequences with the use of the GISS climate model.

With this aim in view, three scenarios of volcanically induced changes of the stratospheric aerosol optical thickness have been prescribed: (1) El: aerosol properties correspond to those prescribed for the conditions of the El Chichón eruption (aerosols are droplets of a 75 per cent H_2SO_4 water solution; aerosol optical thickness decreases from 1 immediately after the eruption to 0 six months later; eruptive aerosols, initially uniformly distributed in the band 0–30° N, then propagate over the globe, with aerosol optical thickness in the band 30–90° N being doubled, compared with the band 30°N to 90°S); (2) 2 × El: prescribed doubled aerosol optical thickness, compared with El; (3) P: the same temporal change of aerosol optical thickness as in the models El and 2 × El (exponential decrease 10 months after the eruption with a time constant of 1 yr), but aerosol optical thickness is greater by a factor of 1.7 than for El and there is a more realistic global distribution of aerosol optical thickness than before (hemispherically uniform distribution of aerosol had been reached by January 1992).

Numerical modelling has shown that in the case of scenario P the Pinatubo-induced global mean change of the Earth's radiation budget in early 1992 could reach about 4 W m^{-2}. This means that the effect of the volcanically induced cooling considerably exceeds the greenhouse warming due to an accumulation of all greenhouse gases during the period from the beginning of the industrial revolution. A decrease of global mean surface air temperature, which by the end of 1992 should have constituted about 0.5 °C, exceeds the global mean surface air temperature mean standard deviation by a factor of approximately three.

Calculations of surface air temperature changes for scenarios El and 2 × E1 have demonstrated that the eruption-induced cooling is too weak to change the situation of 1991 as one of the warmest years in the current century, in view of the strong inertia of the climate system and weak initial perturbing effect. However, for scenario 2 × E1 (as in the case of P), the decrease of surface air temperature reaches 0.6 °C, and, if this estimate is realistic, the predicted cooling will considerably exceed the warming by about 0.2 °C, characteristic of the El Niño effect.

Though regional estimates of surface air temperature changes are unreliable, they have shown that if, without account of the effect of eruptions, the probability of unusually cold winters in Moscow ought to have decreased from 33 per cent in the 1950s to 15–20 per cent at present, then, bearing in mind the contribution of the Pinatubo eruption, the probability of severe winters in the period 1992–1994 should increase to 30–50 per cent. The reliability of the numerical modelling results is still substantially limited by imperfection of the climate model applied, which does not take into account, for example, the interaction with deep layers of the ocean and thermal inertia of the climate system. The properties of clouds and aerosol require a more adequate consideration.

The key directions of prospective studies on volcanoes and climate may be formulated as follows:

1. The complex global monitoring of gas and aerosol products of eruptions, with the use of all available conventional and satellite techniques, bearing in mind detailed studies of the chemical composition and physical properties of eruptive products near the eruption and in the whole global atmosphere (sulphur and nitrogen compounds claim special attention).
2. Further studies of gas-to-particle conversions responsible for the formation of stratospheric aerosol.
3. Numerical modelling of regional and global impacts on climate (first of all, air temperature and precipitation).
4. A search for the most sensitive indicators of the effect of eruptions on climate.
5. Palaeoclimatic reconstructions that characterise the effect of eruptions on climate.

A reliable assessment of the effect of volcanic eruptions on climate is only possible based on numerical modelling with the use of complete climate models based on adequate empirical information.

So far, we can state that none of the strong volcanic eruptions has been adequately documented. Therefore, there is a need for a specialised international programme which would combine studies on numerical modelling and observational subprogrammes, aimed at (1) regular investigations of the active volcanoes and their effect on the atmosphere and (2) alert observations with the use of regular networks and specialised observational means after strong volcanic eruptions.

6.11 PALEOCLIMATOLOGY

The big debate in the early days of the IPCC was between the western climatologists on the one hand, who placed great faith in computer modelling of the climate system, and the soviet scientists, who placed great faith in palaeoclimate studies. This section is devoted to a consideration of palaeoclimatology. Our task is understanding global climate change. And I suppose one question is what sort of timescale are we concerned with. Discussions of climate change are very often concerned with trying to influence political decisions made by governments or by supranational organisations presumably with a view to making life better for ourselves, our contemporaries or our not too distant descendants. Given that the Earth is about 4.5 billion years old, that life evolved perhaps at least 3.7 billion years ago it is possibly an interesting intellectual exercise to consider what we can determine about the climate a long time ago. But such knowledge is not to be regarded as useful knowledge. While it may be of academic interest to consider ancient climate, in practical terms there is probably little point in going back more than about 11,000 years before the present (BP) in terms of understanding the present climate. Our hunter-gatherer forefathers are likely to have had some empirical understanding of some important features of the climate but they are not likely to have had a formal knowledge of the science of climatology.

6.11.1 DESERTIFICATION

One important issue is that of desertification. There are a number of areas, of which perhaps the Sahara is the best known example, which are now desert but which at some stage in the past were fertile agricultural land. One question, to which there is no easy answer, is why did this desertification take place? Was it from climatic deterioration due to natural causes or was it from over grazing and if it was from over grazing why did it occur? Or was it from some other human actions? It is worth quoting as an example some work carried out by Henry *et al.* (2017) in southern Jordan. Henry *et al.* conducted a multidisciplinary study of a dry lake bed in southern Jordan, which covered the last 11,000 years before the present (BP). This was divided into two stages. The first was from about 11,000 to about 5,600 BP in which proxies of plant cover and sheep/goat stocking co-varied rather directly with climate cycles. Traditionally, herding had been thought to have been introduced to the arid zone of the southern Levant relatively late in the Neolithic Age after about 8,900 BP and thus considerably later than in the northern Levant and Anatolia. However, according to Henry *et al.* it now appears that Desert Neolithic groups herded sheep and goats as early as 9,500 years ago but that stock rates were maintained during the period up to about 5,600 BP below the carrying capacity of the pasturage in the region so that over grazing did not occur. However, round about 5,600 BP, near the beginning of the early Bronze Age, this pattern changed leading to the onset of over grazing.

About 5,600 this pattern changed and evidence was found to suggest over grazing and the use of fodder supplements. There is archaeological evidence that at about this time there was a marked rise in regional population and emergence of widespread trade. Henry *et al.* argue that desertification was a consequence of socio-economic factors (e.g. high stocking rates) associated with a shift from a subsistence economy to a market economy. In addition, they contend that there is a stone artifact, a chert, or flint tabular scraper, that appeared in huge numbers during this period which was directly connected to the emergent market economy and its secondary product (wool). Moreover, in that these changes took place largely concurrent with local and regionally recognised evidence of a moist interval, they conclude that the desertification of the southern Levant was induced more by anthropogenic than climatic factors in this second period.

While the immediate cause of desertification of the southern Levant 5,600 years ago appears to be over grazing, what was the ultimate cause? Could climatic deterioration have been the ultimate cause with excessively high stocking rates simply accelerating the tipping point for the onset of desertification or were other forces at play? Henry *et al.* conclude that a climatic explanation appears less compelling than one associated with a rise in regional population density tied to intensified economic activity in the region. If climate deterioration is not the ultimate cause of desertification, then it is likely that anthropogenic forces were the ultimate driver of economic development and the accompanying population growth and expansion, see Figure 6.14.

As time went on, groups became more settled and "the growth in population of these settled communities was tied to economic development involving the mining, production and trade of copper and cortical flint flakes for the fabrication of tabular scrapers and the handling of ornaments of shell and stone and various other exotic items in an exchange system that stretched as far away as Egypt". The Timnian pastoral groups who inhabited this region quarried chert (flint) from local outcrops sometimes at an industrial scale for the fabrication of tabular scrapers.

Let us consider a second example, the desertification of the Sahel in more recent times (Brough and Kimenyi, 2004). The Sahel is, of course, very much larger than the site in southern Jordan that was studied by Henry *et al.* and, inevitably it is much more complicated. It can be taken to be that part of north Africa with a mean annual rainfall of between 150 mm and 700 mm. The Merriam-Webster

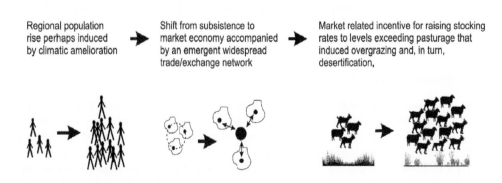

Regional population rise perhaps induced by climatic amelioration → Shift from subsistence to market economy accompanied by an emergent widespread trade/exchange network → Market related incentive for raising stocking rates to levels exceeding pasturage that induced overgrazing and, in turn, desertification.

FIGURE 6.14   Schematic tracing the evolutionary factors leading to the desertification in southern Jordan as discussed in the text (Henry *et al.* 2017).

dictionary definition of a desert is of a warm area with less than 25 cm of rainfall per year, by which token the 150-700 mm rainfall can be regarded as enough to scrape a living from agriculture with some difficulty. The Sahel region along the southern border of the Sahara desert was once, say from the 7th to the 18th centuries (Collins 2018), the home of vast trading empires. Although drought and famine were unavoidable components of life in this harsh region, the people were relatively prosperous and developed agricultural and livestock practices that allowed local populations to endure and recover from the extremes of nature. That era contrasts sharply with the stark poverty and the barren lands so prominent in the Sahel today. During the 20th century, this fragile ecological zone was unable to sustain its growing population. Increased pressure on the land made the inevitable droughts more ruinous, and the dramatic famine of the 1970s prompted urgent calls to reverse the devastating toll of desertification. Desertification is still often viewed as an irreversible process triggered by a deadly combination of declining rainfall and destructive farming methods.

As we have already noted, it is not easy to decide whether desertification is a consequence of climate change or whether it arises from increased population pressure. The failure to cope with drought in the Sahel originated in the French colonial disruption of the indigenous system of property rights and market interactions, followed by independent governments' policies and aid from other nations that intensified pressure on the land.

Traditionally, two distinct populations have inhabited the region, on the one hand the pastoralists (or nomads) and on the other hand sedentary farmers, each with their own cultures but inextricably bound together by trade. Nomads raised cattle and migrated across the Sahara down to the savanna. An intricate structure of markets and division of labour allowed them to use existing resources without destroying the environment.

The Tuareg, for example, were a nomadic tribe that derived income from cattle and trans-Saharan trade. Water wells were owned by the clan that dug them, and the use of water was strictly regulated. A clan's leader determined the length of time spent at the well and contracted with other clans, granting them rights to use its wells in exchange for rights to use theirs. This provided the Tuareg with a network of wells to support their cattle as they moved along their trade routes.

Although the wells and pasture lands were controlled by the clans, cattle were privately owned. Communal ownership of the pasture led to overgrazing, but limits on the length of time spent at each well constrained the number of cattle that individual households could own. While a system of fully defined property rights was lacking, the system was relatively efficient and insulated its people from natural catastrophes.

The arrival of the French in West Africa in the late 19th century altered the agricultural patterns of the Sahel. French policies that emphasised export crops and east-west trade from the interior to Atlantic port cities led more merchants and farmers to abandon trans-Saharan trade. By the 1920s, the region was showing signs of stagnation. The French implemented a three-pronged development scheme to revitalise the area: digging more wells, conducting veterinary and medical campaigns, and opening new markets in the south (Swift 1977).

As the French dug new wells, they established no clear ownership rights, which led to overgrazing (Sterling 1974). The veterinary and medical campaigns increased the populations of both humans and animals, putting further pressure on the land. With no one to regulate the use of new wells, the larger populations intensified the level of overgrazing. The French hoped that nomads would slaughter more cattle for the market. But to hold on to what wealth they had, the nomads tended to maintain the largest herds possible.

Collectively owned land led to deforestation as well as overgrazing. Forests were depleted as individuals collected wood for burning. In the 1930s, in an effort to regulate the use of wood, the French nationalised ground cover. The result was the tragedy of the commons. The ground cover was overused and no individuals had an incentive to plant anything more. On gaining independence, the African states maintained this system of nationalised ground cover, with enforcement by forestry officials at the national level. The nomads were forced to move farther south in search of better lands, and the slash-and-burn methods they used damaged more land as they advanced. Increased pressure on the land in turn affected the farmers' ability to grow food. Droughts began to take a greater economic and human toll.

Western aid to the region became more prominent after the drought of 1968 to 1973. Again, medical aid programmes intensified the pressure on the land without providing any monitoring system. Wells continued to be a popular aid project. Thousands of wells were dug increasing the size of herds and the problem of overgrazing. Hundreds of square miles of land were lost from overgrazing and trampling by cattle in search of water and food. Elsewhere, poor policy making and indiscriminate Western aid have weakened the property rights that promote healthy land-use practices.

A conclusion that one rapidly reaches in connection with desertification is that it is not simple. There is a space dependence involved; while there are some areas where the desert is advancing or has taken over, there are other areas where the desert would appear to be retreating. Likewise, there is also a time dependence involved; there are some times when, in a given place, the desert may be advancing and other tines when it may be retreating. It is possible that advances in the desert may be temporary, with vegetation returning as the rains return. It would be naïve to suppose that the same thing is happening at a given time all over the whole of this vast area. The question of desertification in the Sahel since the droughts of the 1970s has been studied extensively over the last few decades. It is pointed out by Rasmussen *et al.* (2015) that "One intriguing feature is that an agreement on the overall trends of environmental change does not appear to emerge: questions such as whether the Sahel is greening, cropland is encroaching on rangelands, drought persists remain contested in the scientific literature and arguments are supported by contrasting empirical evidence." It would not be appropriate to reproduce all the details of the arguments of Rasmussen *et al.* but we shall summarise their more important points; the full details can be seen by referring to their paper. Rasmussen *et al.* explore the generic reasons behind this situation in a systematic manner. We distinguish between divergences in interpretations emerging from (1) conceptualisations, definitions, and choice of indicators; (2) biases, for example, related to selection of study sites, methodological choices, measurement accuracy, perceptions among interlocutors, and selection of temporal and spatial scales of analysis. The analysis of the root causes for different interpretations suggests that differences in findings could often be considered as complementary insights rather than mutually exclusive. This will have implications for the ways in which scientific results can be expected to support regional environmental policies and contribute to knowledge production.

Following the drought in the early 1970s in the Sahel, the UN spurred an intensive interest in the issue of dryland degradation/desertification, most prominently marked by the UN Conference on Desertification in 1977, followed by the UN Convention to Combat Desertification (emerging from the Rio-conference in 1992). This fuelled a significant increase in the scientific efforts to provide an empirically supported understanding of both climatic and anthropogenic factors involved, as well as a surge of, often alarmist, popular accounts of desertification.

On the one hand, the environmental constraints, such as highly erratic rainfall and poor soil quality, limit the potential for agricultural and pastoral production, which has been and still is an important component in people's livelihood. On the other hand, land use practices have been tuned to cope with and adapt to the environmental constraints. The linkages between the biophysical and human subsystems sometimes take the form of positive feedbacks: soil degradation due to "overcultivation" may be expected to trigger expansion of cropland in order to compensate for low yields, which in turn will cause accelerated soil degradation in the newly included marginal land. Such unsustainable trajectories have been proposed to represent the archetype of a "Sahel syndrome."

Despite decades of intensive research on human-environmental systems in the Sahel, there are a range of conflicting observations and interpretations of the environmental conditions in the region and the direction of changes. The disputes have evolved around especially the three themes

- "land degradation/desertification,"
- "land use and land cover change" and
- "climate change and variation."

With respect to all three broad themes, conflicting evidence and interpretations are presented in the literature, which obviously raises the question: Why is it so difficult for the scientific community to agree on the environmental changes taking place in the Sahel? Rasmussen *et al.* (2015) explore such conflicting evidence and interpretations of Sahelian environmental change by

1. Providing a general overview of possible generic sources of contrasting evidence
2. Discussing whether the apparent contradictions are "real" or if the contrasting evidence may be considered complementary and gives a better understanding of environmental change in the Sahel.

Key concepts in the literature on environmental change in the Sahel are not always defined and used in a consistent manner. Disagreements may in some cases be traced back to inconsistencies of this sort.

6.11.2 Land degradation

Contrasting evidence has indeed been visible within the theme of land degradation. The massive number of studies showing or assuming that land degradation is ongoing in the Sahel, summarised in meta-studies and summaries, such as (Geist 2005; Geist and Lambin 2004), stand out in sharp contrast to a substantial, yet still much smaller, number of studies, demonstrating that since 1981, the Sahel has been characterised by a "greening" trend (Anyamba and Tucker 2005; Eklundh and Olsson 2003; Fensholt et al. 2012; Fensholt *et al.* 2013; Herrmann *et al.* 2005; Heumann *et al.* 2007; Olsson et al. 2005). Both categories of studies go further from observing land degradation to explaining them, by reference to a long range of explanatory variables and models. The question to be discussed here is whether the contrasting evidence finds some of its explanation at the conceptual level: Are there conceptual incongruences between these two categories of studies that result in the different findings? We argue that this is the case, and that the confusion lies in the definitions of land degradation used, and in the widely different interpretations of these definitions found in the individual studies. This can be illustrated by the choice of "indicators" of land degradation selected in each study.

Many different definitions of desertification/land degradation have been proposed and used. The closest we get to a common standard is the current definition suggested by the United Nations Convention to Combat Desertification (UNCCD, (www.unccd.int)) in which desertification is equated with land degradation in drylands:

Desertification means land degradation in arid, semiarid, and dry sub-humid areas resulting from various factors, including climatic variations and human activities.

Land degradation means reduction or loss, in arid, semi-arid, and dry sub-humid areas, of the biological or economic productivity and complexity of rainfed cropland, irrigated cropland, or range, pasture, forest, and woodlands...

For operational use, the definition must be translated into a measurement protocol, assuring that research produces results that allow comparison. The definition cited points to several indicators, e.g. biological productivity, economic productivity, and ecosystem complexity. This is not necessarily problematic, but some of these are difficult to measure in ways assuring comparability. The conceptually simplest one, biological productivity, is actually very difficult to estimate in situ for any sizeable area, while EO does offer such possibilities. Economic productivity, on the other hand, may be easier to estimate in situ, whereas EO techniques are of limited use. Finally, ecosystem complexity may be difficult to define and measure, and indicators may vary greatly between studies. Indicators representing these three components of land degradation are not necessarily correlated: there are numerous cases where an increase in biological productivity is accompanied with decreases in economic productivity and/or ecosystem complexity (Herrmann and Tappan 2013). One example is conversion of semi-natural savannah grasslands into cultivated land. This will probably increase economic productivity yet decrease ecosystem complexity, while effects on biological productivity may be positive or negative. For these reasons, it is not surprising those studies of land degradation, whether based on in situ observations or satellite image analysis, show incomparable and/or inconsistent results.

It is, however, not fair to claim that most inconsistencies are caused by the ambiguities of the UNCCD definition on land degradation. Many empirical studies of land degradation go beyond this to develop radically different operational definitions. In the meta-study of 132 case studies by Geist (2005), the choices of land degradation indicators (most of which are in situ studies) are summarised. This useful exercise demonstrates that the land degradation indicators used include many that do not conform to the UNCCD definition. As concerns the EO-based studies, those aiming at identifying long-term trends on the basis of time series of coarse resolution data can be divided into two main groups: one uses indicators of biological productivity, such as the "normalized difference vegetation index" (NDVI), while the other associates land degradation with changes in "rain use efficiency" (RUE) (see e.g. Fensholt and Rasmussen 2011; Prince *et al.* 2007). The latter group attempts to eliminate the effect of rainfall change (by a "normalisation" procedure) on biological productivity in order to isolate the impact of non-rainfall-related changes, e.g. human impacts. The results of these two groups are not easily comparable. Confusion obviously arises, when studies of NDVI trends explain these trends by rainfall change (Fensholt *et al.* 2012), while studies of RUE trends for the same area and period and based on the same EO data by definition point to non-rainfall explanations. This contradiction is only a product of the indicators used, not a real case of contrasting evidence.

6.11.3 THE CONCEPTS OF LAND COVER AND LAND USE

We can cut short quite a long discussion by going straight to the definitions.

The term land cover has most widely been used and defined as the surface cover on the ground, such as millet fields or forest. The remote sensing research community has played a key role in land cover change detection. On the other hand, land use has traditionally been a concern primarily of social scientists, such as economists, human geographers, anthropologists, and planners as land use refers to the activities undertaken on the land and the purpose the land serves. Although the two terms denote areas of study that have historically been separate, there is nowadays a tendency to use the two terms interchangeably as observed (Turner and Meyer 1994). Land cover change detection carried out by remote sensing specialists have, for example, been confused with land use despite the fact that land use and land use change mapping assessments through remote sensing techniques still remain a major challenge (Martinez and Mollicone 2012).

In his meta-study of land degradation, Geist (2005) relates the causal complexes, identified in each case study, to the disciplinary background of the researchers involved. Natural scientists tend to find biophysical causes, while scientists with a social science background tend to find human causes of land degradation. Apparent discrepancies can therefore sometimes be found between the conclusions that different specialist scientists draw from a common set of data. This may be attributed to either that scientists find what they are trained to look for or that scientists select study sites in a biased manner. Similarly, Rasmussen (1999) discusses disciplinary biases in land degradation research, often associated with the specific definitions and sets of indicators they chose to use: soil scientists tend to interpret land degradation as soil degradation, and geomorphologists tend to focus on erosion as the key indicator of land degradation, while botanists tend to interpret it as loss of ecosystem complexity, vegetation impoverishment or reduction in diversity. Anthropologists, on the other hand, focus on human perceptions of degradation, rather than on biophysical indicators, while economists are interested in the economic productivity of land. Interestingly, all will have some basis for claiming that the UNCCD definition of land degradation, cited above, justifies their particular focus.

Another reason behind conflicting evidence may have to do with strategic bias to the answers by interviewees. Discourses related to development agendas and priorities make it to most villages in the region (Nielsen et al. 2012). High dependency on foreign aid, stories of need, poverty, land degradation, desertification, and in recent years, climate change vulnerability is often communicated by villagers to researchers (Olwig 2013; Nielsen et al. 2012). Such stories are tightly entangled with a hope that some benefits such as development projects can be obtained by communicating vulnerability. In Burkina Faso, projects have become one of the most prominent sources of income (Nielsen et al. 2012).

Methodological bias by researchers is also a problem. Rasmussen *et al.* (2015) point out that diverging or conflicting results may also originate from differences in methodological choices and strategies adopted by the researchers. There are several studies based on coarse resolution Earth observation data that have reported an increase in vegetation productivity, referred to as the greening of the Sahel. But this narrative is not necessarily confirmed by the other types of analysis, including high-resolution satellite-based studies. Three major methodological choices may lead to such differences: (1) choice of temporal and spatial scale and resolution, (2) choice of the data set (or data provider), and (3) choice of the biophysical variable(s).

In this regard, the conclusions that were drawn by Rasmussen *et al.* (2015) are briefly discussed. Certain cases of conflicting evidence on environmental change in the Sahel reflect differences in the use of concepts and in the methodological choices. The apparent conflicts may be partly resolved, or at least understood, if the differences are brought to the front. Once this is achieved, the conflicting evidence may sometimes be transformed into complementary perspectives, enriching and nuancing our understanding. A few examples illustrate this.

Seemingly conflicting evidence on desertification/land degradation from EO-based studies, based on different interpretations of the land degradation concept, utilising different indicators and leading to contrasting conclusions as concerns whether climatic or human factors are the most important, maybe transformed into complementary information by taking the conceptual differences and the various methodological choices into account.

Claims of land degradation, based on in situ observations of reduced tree cover and disappearance of species, are not necessarily in conflict with claims of the absence of land degradation based on observation of increased vegetation productivity using time series of EO data. Rather, they rely on two different interpretations of the UNCCD definition of land degradation, illustrating the ambiguities of the definition.

When generalisations about land use and land cover changes in the Sahel, based on small-scale and short-term field studies, appear to be in contrast with the evidence acquired from wall-to-wall analyses using satellite data, it illustrates on the one hand that extrapolation of findings in time and space should be done with caution. On the other hand, it shows that large-scale and long-term trends do not say much about environmental change processes at micro-scale and over shorter periods.

In order to use the conflicting evidence constructively, the causes of the discrepancies must be identified. Hence, the definitions of central concepts, the methodological choices made, and the spatial and temporal scales considered must be transparent to the reader, and possible alternatives to the choices made should be outlined. While a number of apparent discrepancies may be resolved by doing so, we do not claim that all scientific controversies about environmental change in the Sahel will disappear. Many scientific questions remain open, such as the question of the effects on greening trends of changes in species composition (Mbow et al. 2013) and the importance of changes in grazing intensity and atmospheric CO_2 concentration in explaining changes in vegetation productivity and composition (Bond and Midgley 2012; Higgins and Scheiter 2012; Scheiter and Higgins 2009).

To conclude, it was shown that seemingly conflicting findings as regards whether or not deserti-fication/land degradation is a general feature of the Sahel, whether agricultural land use is generally expanding, and whether drought is continuing are commonly found in the literature on Sahel. We trace these conflicting findings to differences and inconsistencies in the definition of concepts and to disciplinary, strategic, methodological, and sampling biases. If these differences are taken into account, the results are shown often to be complementary rather than contradictory. It is suggested that in an interdisciplinary field such as this, special attention must be paid to making conceptual and methodological choices explicit, also to scientists from other disciplines with different languages, traditions, and norms.

Environmental change in the Sahel is of great policy concern at local, national, regional, and global levels, yet many policy documents tend not to be firmly based on research results. Resolving or reducing scientific controversies would place science in a stronger position to inform policies.

6.11.4 PALAEOCLIMATIC INFORMATION: CATASTROPHIC CHANGES

There are quite a few laboratories around the world which are involved in running their own climate models on large and powerful computers and inevitably they do not all produce exactly the same results. Working Group 1 of the IPCC, as we have indicated in Section 2.3, attempted to take into account the results of all these models and to make the best estimates possible of the anticipated rise in temperature, and the changes in precipitation and in soil moisture for various different sce-narios of future greenhouse gas emissions. Kondratyev was critical of the way these models took into account atmospheric aerosols and cloud screening. Palaeoclimatic information is an important source of data for comparative analysis of the present and palaeo-climate. Analysis of the data of palaeoclimatic observations reveals large-scale abrupt climate changes taking place in the past in conditions when the climate system had exceeded certain threshold levels.

What the IPCC does is based on the idea of slow gradual responses in the climate. However, one of the problems with the assumption of gradual change and the use of computer models to predict the future climate is the inability to predict sudden changes. Take, for example, the melting of ice (Gore 2006, Pearce 2006). It is commonly assumed that a glacier or an ice shelf just melts from absorbing radiation at its surface, i.e. rather slowly. However, in reality, cracks develop in the ice, meltwater pours into the cracks, and the whole melting process accelerates and the water pressure in the cracks acts like wedges and forces the ice to break up. Spectacular situations occur like the break-up of the Larsen B ice shelf in Antarctica in early 2002. The Larsen B ice shelf was about 150 miles (270 km) long and 30 miles (54 km) wide and it suddenly broke up and floated away in fragments over a period of about one month and released around 500 billion tonnes of ice into the ocean. Following the melting of the ice, the water surface uncovered has a much lower albedo (reflectivity) than the ice, it therefore absorbs more heat and provides a positive feedback mecha-nism that enhances the global warming. A second example of positive feedback is associated with drought. The withering or death of plants causes a decrease of evapotranspiration and, hence attenu-ates precipitation which further increases drought.

In the conclusion to his book "The Last Generation: How Nature Will Take Her Revenge for Climate Change" Pearce (2006) says that he "called this book 'The Last Generation', not because I believe we humans are about to become extinct, but because we are in all probability the last generation that can rely on anything close to a stable climate in which to conduct our affairs." What Pearce is saying is that people have been overlooking positive feedback mechanisms, of which we have just mentioned two examples, and this positive feedback can lead to sudden precipitous swings in the climate. While the models – and the IPCC's general approach – can handle a cata-strophic change *after* the event by making adjustments to the parameters in the models, they cannot *predict* such sudden events. With so much emphasis having been placed on climatic implications of the growth of greenhouse gases concentrations in the atmosphere, less effort has been made to study possible sudden climate change of natural origin and intensified by anthropogenic forcings.

Pearce claims that there are many instances of sudden swings in the past – and therefore presumably this can happen in the future too. He argues that whereas the IPCC is talking about a rise in global mean surface temperature of 3–4 °C in 100 years, it could be far worse than that; it could be 10 °C which would take us way beyond the changes that we have outlined in Section 18.4.1. The most important aspect of these problems are the potential effects of abrupt climatic change on ecology and economy since past estimates were generally based, as a rule, on the assumption of slow and gradual change. Pearce's general thesis is that nature often flips suddenly from one state to another and that therefore the consequences of climate change may be quite different from – and much more serious than –the rather simple kind of gradual changes that are generally predicted by computer models.

Apart from possible human-induced rises that may very much exceed the 3-4 °C rise in the next 100 years predicted by the climate models, there is the possibility of a substantial decrease in the temperature that would correspond, as a result of natural causes, to a return to a new ice age. Until recently, and in the absence of any evidence to the contrary, geologists and palaeoclimatologists had assumed that climate changes in the past had always been slow and, therefore, in relation to human-induced global warming the opposite effect, of a natural cooling leading to a new ice age has been assumed to be on a much longer timescale than a century or so. However, it has recently emerged that rather than slow and gradual transitions between ice ages and inter-glacial periods, there have been many abrupt changes in the climate. General discussions will be found, for instance in the books by Cox (2005) and Pearce (2006). What Cox is concerned with is the recent evidence that has been found of sudden, rather than gradual, changes in the climate in the past. The terms sudden and gradual need to be clarified a little; in the framework of geological time, a sudden transition from a warm inter-glacial period such as the present to a full-blown ice age in a period of 100 years would be regarded as sudden, whereas gradual would imply a period of 1000 years or several thousand years. In "Climate Crash – Abrupt Climate Change and What it Means for Our Future," Cox (2005) examines the records of past climate changes. Evidence of past climate variations has been obtained from ice cores drilled from the Greenland ice sheet, supported by some other ice cores from elsewhere, and from ocean sediment cores from various parts of the world, see Section 6.8.

Considerable abrupt changes of regional climate in the last ice age have been detected from palaeoclimatic reconstructions that manifested themselves as changes in the frequency of occurrence of hurricanes, floods, and especially droughts. Evidence of more than 20 oscillations, known as Dansgaard-Oeschger oscillations, has been observed in the Greenland ice-cores record of the last ice age (between 110,000 and 23,000 years before the present). In each of these Dansgaard-Oeschger oscillations, there was a sudden sharp rise in temperature of between 2 and 10 °C over a period of a decade or so and this was followed by a slow cooling over several centuries, on average about 1,500 years. These changes in climate appear to have occurred suddenly in the past, over a few years or perhaps a decade or two and certainly on a different timescale from the 3 or 4 degrees per century predicted by climate models. While the evidence for abrupt changes is quite clear, the mechanisms driving these changes are less clear and are still the subject of very active research. Even if the causes of these changes were known, it seems unlikely that computer models would ever predict sudden changes.

6.12 MODELS

We start with a Concise Oxford Dictionary definition of the word "model" where there are five meanings of the noun and four meanings of the verb:

> *Model* noun 1 a three-dimensional representation of a person or thing, typically on a smaller scale. 5 a simplified description, especially a mathematical one, of a system or process, to assist calculations and predictions.
> *Model* verb 1 fashion or shape (a figure) in clay, wax, etc. .....4 devise a mathematical model of.

We might almost define the work of Arrhenius (see Section 6.13.1) as a simple climate model. He sought to estimate the rise in the Earth's temperature due to an increase of CO_2 in the atmosphere. It could be regarded as a zero-dimensional model, because he was not considering the spatial variation in either of the two horizontal directions or in the vertical direction.

6.12.1 WEATHER FORECAST MODELS

What we have seen over the last few decades has been a transition from the traditional methods of weather forecasting to forecasts based on models run on computers. According to Weart (2008), the idea of applying numerical modelling to weather forecasting can be traced back to 1922 when Lewis Fry Richardson proposed a complete numerical system for weather prediction. His idea was to divide up a territory into a grid of cells, each with a set of numbers for air pressure, temperature, etc. as measured at a given time. He would then apply the basic physics equations that described how air would respond. He could calculate, for example, a wind speed and direction according to the difference in pressure between two adjacent cells. This would give the pressures and temperatures in the grid cells an hour later, which would serve for next round of computations, and so forth. The number of computations was so great that Richardson scarcely hoped his idea could lead to practical weather forecasting. Even if someone employed tens of thousands of clerks with mechanical calculators, he doubted that they would be able to compute weather faster than it actually happened. "Perhaps someday in the dim future" he wrote "it will be possible to advance the calculations faster than the weather advances, …. But that is a dream." (Richardson 1922) Still, if he could make a numerical model of a typical weather pattern, it would help show how the weather worked.

"So Richardson attempted to compute how the weather over Western Europe had developed during a single eight-hour period, starting with the data for a day when there had been coordinated balloon-launchings that measured the atmosphere at various levels. The effort cost him six weeks of pencil work and ended in complete failure. At the centre of Richardson's pseudo-Europe, the computed barometric pressure climbed far above anything ever observed in the real world. Taking the warning to heart, for the next quarter century meteorologists gave up any hope of numerical modelling." (Weart 2008 page 54).

It was not until 1946 that von Neumann began to advocate using computers for numerical weather prediction. Weart (2008) describes how first weather forecast models and then climate models developed, leading up to the climate models that were used in the preparation of the IPCC's First Assessment Report in 1990.

The objective of a weather forecast model is to be able to predict the weather at a particular place and time. The idea is that one writes down the mathematical equations that describe the various physical processes that occur and use the present situation as the starting conditions and tries to solve these equations for some future time. In an atmospheric model, the behaviour of the atmosphere is represented by the values of appropriate parameters specified on a three-dimensional grid of points. In an early model, the UK Meteorological Office used both a global model operating with a grid with spacing of 90 km (i.e. of the order of 10° of latitude or longitude) in the horizontal and also a limited area model with a grid spacing of about 40 km; in each model, there were about 20 levels in the vertical. The atmosphere knows no national boundaries, which is why meteorological data is shared even between states which are rather hostile to one another. Thus, in theory, one would prefer to use a global model rather than local or regional weather forecast model. The other side of the coin is that there is a trade-off between, on the one hand, the grid spacing which modellers would like to be as small as possible and, on the other hand, processing speed which modellers would like to be as fast as possible. As computing power has increased, following Moore's Law, so models have been developed to exploit the improved computing facilities. The physics involved include the following:

- horizontal momentum equations
- the hydrostatic equation

- the equation of continuity (i.e. matter is neither created nor destroyed)
- the equation of state
- the thermodynamic equations

along with parametric descriptions of the following processes:

- evaporation
- condensation
- cloud formation and dispersal
- radiation and convection in the atmosphere
- the exchange of momentum, heat, and water vapour at the Earth's surface.

It is also necessary to take into account effects or processes that occur on a scale which is smaller than the grid spacing that is used in one's model. We suppose that the results of meteorological observations allow the values of the atmospheric parameters to be specified at some starting time, t = 0. The forecasting problem then is to solve the various equations describing the physical principles and processes that we have just mentioned for a subsequent time t. That is, one seeks to determine new values of these atmospheric parameters at a subsequent time, in practice up to a few days ahead. Because there are several of these equations, and especially because they are non-linear equations, the problem has to be tackled by numerical methods using a computer. Hence the terms "computer model" or "numerical model".

The operational use of numerical weather forecast models began about 50 years ago. Although their usefulness at that stage was limited, there have been rapid developments since then and within ten years the models were able to provide better forecasts of the basic motion field than could be achieved by an unaided human forecaster. These improvements have come about for two reasons. First, the power of the computing facilities available has increased enormously. Secondly, the amount of data available to describe the present atmospheric conditions has been greatly increased in recent years. In addition to conventional surface measurements and radiosonde measurements, there are now also satellite observations including satellite soundings and satellite-derived wind speeds obtained from cloud tracking from geostationary satellite images, see Figure 6.15.

FIGURE 6.15   Satellite data for models. Hurricane Forecast Models Get a Boost From New Satellite Data (Source: NOAA).

If at present the weather forecast models are successful in providing forecasts a few days ahead, then one can ask what improvements we can expect to see in the future. The European Centre for Medium-Range Weather Forecasting (ECMWF) produces forecasts for four weeks ahead.

The use of models along the lines we have outlined is basically a deterministic approach. We have the present conditions, we have a set of mathematical equations describing the various atmospheric processes and we rely on the mathematics to predict the future conditions. At present, this can be done successfully a few days ahead; the specification of what one means by a "few" days is not very easy but, at present, a figure of five days might be a reasonable one. It is not a simple cut-off situation, where forecasts up to a certain time ahead are accurate and those beyond that time are not. Rather one should say that accuracy, as specified by rms values in the errors of forecasts of particular parameters, is improving all the time. This is illustrated by Figure 6.15 alpha; this shows that three-day forecasts of surface pressure now are, on average, as good as the two-day forecasts were in 1980. The question arises as to whether this improvement can go on indefinitely. It is held to be likely that the ultimate limit of forecasts from deterministic models is about 2-3 weeks ahead. In making this statement is one simply being pessimistic, or realistic, about possible future developments of computers or is there some more fundamental reason for this limitation? This brings us to the idea of chaos which we have already discussed in Section 1.1.3.

6.12.2 CLIMATE-FORECAST MODELS

One is entitled to ask why there is apparently so much interest in computer climate models. We have argued in Chapter 1 that it was a result of the green agenda that led up to the establishment of the IPCC with an agenda to demonstrate that there was a major human-induced global warming arising from the combustion of fossil fuels with the emission of increasing amounts of CO_2 into the atmosphere. To do this, computer models were introduced which attempted to demonstrate the existence of such a warming arising from the increasing emissions of CO_2.

Climate models are basically developments of weather forecast models, but the timescales are quite different. Climate models are concerned with periods of the order of decades or centuries, whereas weather forecast models are used to make predictions a few hours or days ahead and possibly a few weeks ahead.

The use of models along the lines we have outlined is basically a deterministic approach. We have the present conditions, we have a set of mathematical equations describing the various atmospheric processes, and we rely on the mathematics to predict the future conditions. These forecasts are quite reliable for a few days ahead; the specification of what one means by a "few" days is not very easy but, at present, a figure of five days might be a reasonable one. It is not a simple cut-off situation, where forecasts up to it certain time ahead are accurate and those beyond that time are not.

In the light of the above, we are entitled to ask the question as to whether there is any hope of ever being able to make meaningful climate-forecast predictions over the scale of a century or two if one has no hope of being able to make reliable weather forecasts for more than a few weeks ahead. This question is particularly pertinent since we are saying that if we take the numerical-modelling approach to climate prediction then we are using GCMs of the atmosphere which have been derived from weather-forecast models and which are based on the same physical mechanisms and equations as are used in the weather-forecast models. This brings us to the idea of chaos theory. If it is surprising that any success can be achieved with climate models, then it may help to appreciate this if we recall that in looking at climate we are looking at an average of the weather over a reasonably long time (e.g. one year or 10 years or more). The chaotic elements, or the statistical fluctuations, in the climate are the weather systems. In looking at climate change, we are looking at long-term changes in this average. Viewed over one or two centuries or a millennium, we see that climatic parameters (temperature, rainfall, etc.) are basically stable and vary only slowly. It is the nature of these slow long-term variations that are of concern to us in climate studies after the local short-term fluctuations have been smoothed out. It is the stability of the long-term components that makes climate prediction possible.

In a climate model, we consider the effect of various forcing conditions on a given mean state. In spite of randomness or of chaos at the small-scale level, we can nevertheless make useful and meaningful calculations at the macroscopic level based on a transition from one mean (or quasi-equilibrium) state to another one. Thus, we commonly assume in climate prediction that the climate system is in equilibrium with its forcing. This means that as long as its forcing is constant and the slowly varying components alter only slightly on the timescale considered, the mean state of the climate system will be stable. If there is a change in the forcing, the mean state will change until it is again in balance with the forcing. The timescale of the transition period until an equilibrium state is re-established is determined by the adjustment time of the slowest climate component (i.e. the ocean). The stable ("quasi-stationary") behaviour of the climate system gives us the opportunity to detect changes by taking time averages. Because the internal variability of the system is so high, the averaging interval has to be long compared with the chaotic fluctuations if one expects to be able to detect a statistically significant signal which can be attributed to the external forcing.

Studies of the completed change from one mean state to another are called equilibrium-response studies. Studies of the time evolution of the climate change as a result of an altered forcing, which might also be time dependent, are called transient-response experiments. The main interest in climate modelling probably lies in predicting the effects of human activities, principally the release of large amounts of CO_2 into the atmosphere, over timescales that, in climatic terms, are very short (i.e. decades or one or two centuries). This does not mean that climatologists are not interested in other causes of climate change. It just means that CO_2 and the reasonably short-term effects are the aspects which are of great interest to society at large, and therefore to politicians.

The climate system consists of five components:

- the atmosphere;
- the oceans:
- the cryosphere;
- the biosphere; and
- the lithosphere.

There are different timescales associated with these components. While the atmosphere reacts very rapidly to changes in its forcing (on a timescale of hours or days), the ocean reacts more slowly (on timescales ranging from days [at the surface layer] to millennia at greatest depths). The ice cover reacts on a timescale of days for sea-ice regions to millennia for ice sheets. The land processes react on a timescale of days up to months, while the biosphere reacts on timescales from hours (plankton growth) to centuries (tree growth). The ideas of numerical modelling of the climate may be regarded as a development from numerical weather-forecast models (see Chapter 3 of Houghton et al. 1990 or for more detailed descriptions, the books by Washington and Parkinson 2005 and Trenberth 1992). Figure 6.16 is a much-reproduced diagram illustrating the atmosphere-ocean-ice-land climatic system. The arrows in this figure indicate internal interactions.

The fundamental processes driving the global climate system are the heating by incoming radiation of rather short wavelengths and the cooling by the loss of longer wavelength radiation into space. The heating is strongest in tropical latitudes and the cooling is predominantly in winter in the polar regions. The temperature gradients between equatorial and polar regions drive both atmospheric circulation and oceanic circulation; these circulation mechanisms then transport large quantities of heat from equatorial to polar regions. The five components of the climatic system listed above have already been considered briefly in Chapter 1. However, it is useful to include some additional discussion of the handling of the geosphere in a model because it illustrates just how

FIGURE 6.16 The five components of the climate system all interact (Images by NASA. Collage by Producercunningham at Wikimedia Commons. Images from unsplash.com, created using Visme).

complicated things are and it indicates how much we have to simplify things by making approximations and taking averages (see, in particular Chapter 5 of Trenberth 1992).

The Earth's surface has a dual role in the climate system. First, it acts as a lower boundary for about 30 per cent of the atmosphere and it is involved in the exchange of moisture, momentum, and heat at that boundary. Secondly, from the practical point of view for human requirements, the land is crucial. However, in some ways, the land may be considered to be less important, in terms of general circulation, than the oceans – or the cryosphere. Its area is smaller than that of the oceans. It provides less heat storage than the oceans and it provides negligible horizontal transport. But, on the other hand, the land is much more difficult to model than, say, the oceans. It has a complicated topography and relief, while the surface cover is complicated, both geometrically and in terms of its physical properties. Its nature changes very rapidly spatially. For a climate model, we are concerned with grid spacings of 300 km to 1,000 km in the two horizontal directions. Thus, we have to have one value of a parameter, such as the mass of water exchanged, or the albedo, or the energy flux, or the momentum transfer, which is (a) a spatial average over the whole area of one grid cell and (b) a temporal average, probably over the whole year. With regard to (b), the best we are likely to do is to have a few (e.g. four) seasonal values to use. This coarse spatial resolution of a climate model also means that the land-sea boundary is only very roughly defined too.

6.12.3 THE SURFACE BOUNDARY CONDITIONS

Apart from the need to populate the cells x, y, z, t with values of various meteorological parameters, it is also necessary to specify the nature of the floor of the lowest cell of each column because the transfer of energy and of mass between the atmosphere and the ground is an important boundary condition of the model. Thus, if the cell is over land, it is important to know whether this boundary is bare rock, bare soil, vegetation (and if so what kind of vegetation), urban, water, snow or ice, etc. The transfer of energy or mass across this boundary will be governed by the nature of the surface. In other words, we need a global land cover map.

We can illustrate this by considering the question of global land cover/land use maps. This is usefully summarised in the introduction to a rather important paper on finer resolution global cover mapping by Gong et al. (2013) (with about 50 co-authors). The Gong et al. (2013) paper cites six stages in the development of global land cover maps at increasingly fine scale spatial resolution before their own which involves work done on the global Landsat database.

The first two started at 1 km spatial resolution and they were based on NDVI (Normalised Difference Vegetation Index) data derived from the AVHRR. Later ones involved MODIS data at 500 m resolution and MERIS data at 300 m resolution:

- 1 km IGBP (International Geosphere-Biosphere Programme) AVHRR NDVI data
- 1 km University of Maryland AVHRR NDVI data
- 1 km SPOT NDVI data
- 500 m MODIS (Moderate Resolution Imaging spectrometer) data
- 300 m GlobCover MERIS (Medium Resolution Imaging Spectrometer) data
- 1 km MODIS land-cover map MODIS data.

What Gong et al. have done is to make use of the recent release of the Landsat archive to derive a 30 m resolution Finer Resolution Observation and Monitoring of Global Land Cover (FROM-GLC) database. This involved using 8929 TM or ETM+ scenes to produce a global land cover map, following the workflow shown in Figure 6.17. Although the map is only shown as an image here, the full resolution digital data at 30 m resolution, which should be more than adequate for any existing climate model, is available from the authors of Gong et al. (2013). Indeed, the same group, at Tsinghua University in Beijing, is now working on developments including 10 m resolution land cover maps and annual 30 m resolution global land cover maps. While in many areas the land cover does not change with time, there are some areas which do change either just once over a long period or seasonally.

If the boundary surface is the sea or a large freshwater lake the situation is not quite so complicated; the main issue is whether the water is ice-covered or not. Some areas are never covered by ice, some are permanently ice-covered, and some areas are ice-covered for part of the year and ice-free at other times of the year. Maps of ice cover are routinely generated from satellite data, e.g. since about 1980 by NOAA's National climatic Data Center (ncdc.noaa.gov. accessed 5 August 2019) which has now been absorbed into NOAA's National Centers for Environmental Information.

Brightness contrasts in the visible spectrum recorded with AVHRR and multichannel scanning radiometers on the Landsat and SPOT satellites make it possible to identify the zones of melting snow and ice cover. Microwave measurement data can also be used for this purpose due to a strong contrast between the emissivities of water and ice (or snow).

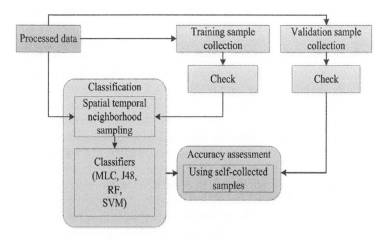

FIGURE 6.17   The overall classification workflow (Gong et al., 2013).

However, microwave thermal images have a very low spatial resolution for monitoring ice drift (icebergs, in particular). For this purpose, radar data are more useful. Another possibility is to monitor the drift of sea buoys.

The requirements for monitoring of the snow cover extent can be met with available data. The combined use of visible and near-infra-red imagery provides information on variations in the structure of snow cover and, in particular, snow melting. Previous results have revealed a good correlation between the dynamics of snow cover extent and river run-off. So far, no remote sensing technique has been developed to assess the snow cover thickness. Techniques for remote sensing of soil moisture are at an initial stage of development. To solve this problem, data from multichannel microwave measurements are the most useful. Techniques to estimate surface albedo from satellite data must be improved. Of principal importance is ensuring that all data obtained can be processed efficiently. Monitoring the sea ice cover is of particular importance in connection with the climate problem, not only in providing reliable input to climate models but also because both ice and snow cover are the most sensitive indicators of climate change.

The fraction of solar radiation reflected (i.e. the albedo) varies with the land-surface cover. Thus, any value used to describe the albedo or any other surface parameter is clearly a crude average. Moreover, these values also vary with time as well. Temporally the land surface is far more variable than the oceans – and on a number of different timescales too. Because of the relatively low thermal capacity of the land the local thermal conditions are much more responsive to net radiation from or through the atmosphere than are the oceans. The presence or absence of clouds has a substantial effect. The temperature of the land surface varies by a much greater amount during the diurnal cycle ($\sim 10°$ or $20°C$ or more) than is the case for the ocean.

Consider the exchange of moisture. When the land is wet, it can give up water to the atmosphere more rapidly than the oceans because of its greater roughness. But when it is dry, it gives up nothing. The mechanism for the transfer of water from the land surface is complicated. Some of it is by evaporation, from a wet concrete surface, from wet soil or from a river or lake. Some of the water is lost by run-off into the sea. But a great deal of it is by evapotranspiration; that is, the water is absorbed from the soil via the roots of a plant, finds its way through the plant, and finally finds its way into the atmosphere via the leaves. The rate of evapotranspiration will clearly depend on the actual processes within the plant.

Water is supplied from the atmosphere by precipitation as water, ice (hail), or snow. Soils act as reservoirs for this water but in the long run, the precipitation is balanced by the evaporation, the evapotranspiration, and the run-off. Climatologists have often viewed run-off as simply the residual after evapotranspiration requirements have been satisfied. Hydrologists have often perceived run-off as a direct response to precipitation, with evaporation and evapotranspiration as a residual. In reality, they both affect each other. Hydrologists with their studies of run-off on local and basin scales have demonstrated the importance of the rates of water infiltration into soils and the distribution of hill slopes in determining the rate of water movement from soils into streambeds and subsoil storage. Micrometeorologists and plant scientists have established the important controls of surface boundary layers and stomatal resistance. Climatologists have shown that, under well-watered conditions, water fluxes are controlled primarily by net radiation or a temperature surrogate.

Moreover, in winter in high latitudes, additional processes come into play. Seasonally or permanently, frozen soil impedes the flow of water into the soil. Snow accumulates during the winter season and acts as a good insulator, allowing little heat to be exchanged between the soil and the atmosphere. It enhances the surface albedo and, by this reflection of solar radiation, delays spring warming. This effect can be weakened by the presence of overlying vegetation. The parameterisation of the transfer of heat and water within the soil (e.g. the balance between evaporation and precipitation, snow melt, storage of water in the ground, and river run-off) is of extreme relevance for climate-change predictions, since it shows how local climates may change from humid to arid and vice versa depending on global circulation changes. It furthermore reflects, in some of the more sophisticated schemes, the changes that could occur through alterations in surface vegetation and land use.

Most soil-moisture schemes used to date are based either on the so-called bucket method or the force-restore method. In the former case, soil moisture is regarded as being available from a single reservoir, or thick soil layer. When all the moisture is used up, evaporation ceases. In the latter method, two layers of soil provide moisture for evaporation, a thin near-surface layer which responds rapidly to precipitation and evaporation and a thick deep soil layer acting as a reservoir. If the surface layer dries out, deep soil moisture is mostly unavailable for evaporation and evaporation rates fall to small values. However, in the presence of vegetation, realistic models use the deep soil layer as a source of moisture for evapotranspiration. At any given grid point over land, a balance between precipitation, evaporation, run-off, and local accumulation of soil moisture is evaluated. If precipitation exceeds evaporation, then local accumulation will occur until saturation is achieved. After this run-off is assumed and the excess water is removed. The availability of this run-off water input to the ocean has been allowed for in ocean models only recently. Most models differ in the amount of freshwater required for saturation, while few treat more than one soil type. The force – restore method has recently been extended to include a range of soil types (Noilhan and Planton 1989).

6.12.4 FEEDBACK MECHANISMS

An important feature of the climate system is the presence of feedback. Feedback may, of course, be either positive or negative (i.e. it may amplify or diminish the climate response to a given forcing). One example of positive feedback arises if the surface area of the Earth covered by ice expands; the expansion of the area covered by ice leads to conditions which cause the area of ice to expand still further until some other process becomes more significant and finally causes the expansion to come to a halt. We can look at it from the opposite viewpoint – if there is a warming of the Earth, this leads to less snow and ice cover. Then the surface of the Earth becomes (on average) less reflecting and more absorbing. This causes further warming and a further reduction in the snow and ice cover (i.e. this is also an example of positive feedback). The real situation is, of course, more complicated and eventually the amplification of the warming by positive feedback (hopefully) comes to a halt. There are two other important feedback mechanisms, one involving water vapour and the other involving clouds. We shall consider them in turn.

Let us suppose that an increase in concentration of CO_2, a greenhouse gas, occurs. This will lead to some warming of the atmosphere. As a result of this warming, the amount of water vapour in the atmosphere can be expected to increase. The water vapour is also a greenhouse gas and so we see that an increase in the concentration of one greenhouse gas (CO_2) leads to an increase in another greenhouse gas (water vapour) and thus provides a positive feedback mechanism. However, the question of cloud feedback is very complicated. Let us consider first the effect of clouds in the present climate before we consider changes in the cloud situation. The presence of clouds heats the climate system by about 31 W m^{-2} by reducing the infrared emission at the top of the atmosphere. Although clouds contribute to the greenhouse warming of the climate system, they also produce a cooling because they reflect incoming solar radiation and reduce the intensity of the solar radiation reaching the surface of the Earth and this produces a cooling of about 44 W m^{-2}. The cooling dominates and so the net effect of clouds is a radiative cooling of about 13 W m^{-2}.

Although clouds at present produce net cooling of the climate system, this must not be construed as a possible means of offsetting global warming because of increased concentrations of greenhouse gases. It is far from obvious what would be the effect on the cloud systems as a result of a human-induced warming arising from increased atmospheric CO_2 concentration. Let us suppose that we start from a certain dynamic equilibrium situation in which the cooling mechanisms and the heating mechanisms balance exactly. Cloud feedback arises from changes in the net cloud radiative forcing. However, this feedback arises as a result of several separate effects and it is far from obvious whether to expect the feedback to be positive or negative. Cloud feedback is not a simple matter: it depends on a variety of factors which we shall now enumerate.

- Cloud amount: If the amount of cloud decreases because of global warming, as occurs in typical general-circulation model simulations, then this decrease reduces the infrared greenhouse effect attributed to clouds. Thus, as the Earth warms, it is able to emit infrared radiation more efficiently, moderating the warming and so providing a negative climate feedback mechanism. But there is a related positive feedback; the solar radiation absorbed by the climate system increases because the diminished cloud amount causes a reduction of reflected solar radiation by the atmosphere. Thus, there is no simple way of determining the sign of this feedback component.
- Cloud altitude: A vertical redistribution of clouds will also induce feedbacks. For example, if global warming displaces a given cloud layer to a higher and colder region of the atmosphere, this will produce a positive feedback because the colder cloud will emit less radiation and thus have an enhanced greenhouse effect.
- Cloud water content: There has been considerable recent speculation that global warming could increase cloud water content, thereby resulting in brighter clouds and hence a negative component of cloud feedback. However, it has also been demonstrated that this negative solar feedback induces a positive infrared feedback. The net result of these two feedback mechanisms of opposite sign may be positive or negative. The question of cloud feedback is clearly very complex and what actually happens in any given situation will be the result of the various factors we have just outlined.

6.12.5 The use of general: circulation models

So far we have outlined many natural phenomena and human activities which affect the weather and the climate. We have tried to indicate what is involved in modelling the climate. The general principles are fairly clear but when it comes to implementing or representing these principles in a model there are inevitably many approximations that have to be made. Generally, the more one improves the models to reduce the approximations, then the heavier the computing requirements become. A great deal of effort has gone into climate modelling in recent years and some very useful results have been obtained. Given what we said above about weather forecasting, the difficulties of producing weather forecasts for more than a few days ahead, and the whole idea of chaos versus determinism, it is perhaps both surprising and encouraging that some success in climate modelling can be achieved.

In using an atmospheric GCM for weather-forecasting purposes, the surface temperature over the oceans provides a very important set of boundary conditions on the model and it is very important to have accurate values of sea-surface temperatures as input data for the model. However, because oceanic conditions change very much more slowly than atmospheric conditions, one does not make elaborate provision to take into account the circulation of heat within the oceans in a weather-forecast model. However, in making calculations of predicted climate change we are dealing with much longer periods of time and the circulation of water and of heat in the oceans is important. For climate modelling, therefore, one needs to use GCMs for both the atmosphere and the ocean and to allow these models to be coupled together. Because this is so much more complicated than using only an atmospheric GCM for weather forecasting. the spatial resolution used in practice in climate modelling tends to be coarser than that used in weather forecasting. Grid spacings of 90 km or even 40 km in the horizontal surface may be used for numerical models for weather forecasting. For climate modelling, typical models have a horizontal resolution of 300–1,000 km and between two and nineteen vertical levels.

There are a number of groups of scientists working on climate modelling in various different laboratories around the world. Each group has its own model and there are considerable differences among the results obtained by these different groups. These differences have sometimes been emphasised by those (usually politicians) who – for whatever reason – would like to claim that there is no evidence for human-induced global warming and that, therefore, there is no need to restrain our behaviour in terms of curtailing emissions of greenhouse gases into the atmosphere.

Differences among the various models arise at various stages:

a) modelling of the physical processes (i.e. turning what we have said qualitatively into a set of mathematical equations) – there are different ways of doing this;
b) choice of grid spacing;
c) boundary conditions and their parameterisation;
d) the amount of computing power available – this will affect (b) and (c) (the computer power required increases very rapidly with increasing spatial resolution);
e) assumptions made about the future.

The parameterisation of radiation is possibly the most important issue for climate-change experiments, since it is through radiation that the effects of the greenhouse gases are transferred into the general circulation. A radiation parameterisation scheme calculates the radiative balance of the incoming solar radiation and the outgoing terrestrial long-wave radiation and, as appropriate, the reflection, emission, and absorption of these fluxes in the atmosphere. Absorption and emission are calculated in several broad spectral bands (for reasons of economy) taking into account the concentration of different absorbers and emitters like CO_2, water vapour, O_3, and aerosols.

One sensitive part in any radiation scheme is the calculation of the radiative effect of clouds. We have already mentioned that determining the effect of clouds is very complicated. In early work with numerical GCMs clouds were prescribed using observed cloud climatologies and were not allowed to alter during the experiments with, for example, changed CO_2 concentration. Later schemes contained interactive cloud parameterisations of various sophistication, but mostly based on an estimate of the cloud amount from relative humidity experiments. Only the most advanced schemes involve calculating the variation of cloud optical properties by cloud water-content experiments.

The seasonal variation of solar insolation is included in almost all experiments, but a diurnal cycle is omitted in many simulations. Climate experiments run without a seasonal cycle are limited in scope and their reliability for climate-change experiments is therefore doubtful. The inclusion of the diurnal cycle improves the realism of some feedback mechanisms and therefore the quality of the climate simulations.

There are two important consequences of climate modelling work:

1. faced with all the approximations and averaging (over space and time) it is, perhaps, surprising that any kind of realistic predictions can come out of a model;
2. many processes occur on a scale that is quite small compared with the grid-point spacing (we can therefore never expect to obtain predictions out of model calculations that relate to these processes at these scales).

It should be apparent from the above discussion that there are many uncertainties and problems associated with any given model. It would be rather risky to rely on absolute predictions made with one particular model. We should therefore consider the manner in which these models are used and this is basically done by attempting to eliminate the effects of the uncertainties or errors and trying to establish a consensus. We shall return to this issue in the next chapter.

In the early days leading up to 1990, a popular model was to double the concentration of CO_2 and run the model until equilibrium was re-established. In the First IPCC Assessment Report (IPCC 1990), there is a summary (Table 3.2) giving the results of model calculations to determine the effect of doubling the present concentration of CO_2 in the atmosphere and assuming that equilibrium is re-established. The whole of the next chapter is devoted to the Intergovernmental Panel on Climate Change (IPCC) which is so important because it offers advice to the UNFCC (United Nations Framework on Climate Change) which came into force on 22 March 1994 and now has near-universal membership of very nearly all the countries in the world. The countries which have ratified the Convention are called the Parties to the Convention and they hold annual conferences (Conferences of the Parties – COPs). As we shall see in Chapter 7, the First IPCC Assessment Report was produced in 1990 and one of the criticisms was that it relied very heavily on the modelling results that were available at that time. The results presented in IPCC 1990 were for 22 global mixed-layer ocean atmosphere models. The mixed-layer approach falls short of considering full ocean circulation but it is an approximation which is useful for reasons of economy, or necessity, in terms of limited computing power. In this approach, one considers the uppermost layer of the ocean where the oceanic temperature is relatively uniform in depth. It is frequently modelled as a simple slab for which a fixed depth of the mixed layer is prescribed and the oceanic heat storage is calculated; the oceanic heat transport is either neglected, or is treated as being carried only within the mixed layer, or is prescribed from climatology. Variations of mixed-layer depth, oceanic heat flux convergence, and exchanges with the deep ocean, which would entail an additional storage and redistribution of heat, are all neglected as well. Sea ice extents may be determined interactively, but some models already include dynamical effects such as sea ice drift and deformation caused by winds and ocean currents. The simple mixed-layer model has strong limitations for studies of climate change, particularly as it does not allow for the observed lags in heat storage of the upper ocean to be represented, and clearly a coupled ocean-atmosphere model is, in principle, to be preferred. One can ask, quite reasonably, what is the value of carrying out equilibrium calculations such as those we have just mentioned. Climate is in equilibrium when it is in balance with the radiative forcing. Thus, as long as greenhouse gas concentrations continue to increase, the climate will not reach equilibrium; moreover, even if concentrations of greenhouse gases are eventually stabilised, it would be a long time before equilibrium would be established. Thus, equilibrium calculations cannot be used directly to provide forecasts of climate changes. The first reason given for carrying out equilibrium simulations (usually with double the concentration of CO_2 in the atmosphere) is that they are done with atmospheric-oceanic mixed-layer models which ignore both the deep ocean and changes in the ocean circulation; these require less computer time than time-dependent simulations which include the influence of the deep ocean. Secondly, because these simulations are relatively inexpensive (in terms of computer time and therefore of money) to carry out, they can be used fairly easily for intercomparison of different models and for studying the effects of different parameterisations, e.g. of cloud. Thirdly, it has been claimed (Kondratyev and Cracknell, 1998a, b, page 392) that apart from areas where the oceanic thermal inertia is high, equilibrium solutions can be scaled and used as approximations to the time-dependent response.

Apart from the IPCC Reports (1990, 1992), there was another comparison of atmospheric general circulation models by Randall *et al.* (1992) in which they described the responses of the surface energy budgets and hydrological cycles of 19 models to a different forcing, namely a uniform increase of 4 deg K in the sea surface temperature, rather than doubling the concentration of CO_2. The responses of the simulated surface energy budgets were extremely diverse and were closely linked to the responses of the simulated hydrological cycles. The response of the net surface energy flux is not controlled by cloud effects; instead, it is determined primarily by the response of the latent heat flux. The prescribed warming of the oceans led to major increases in the atmospheric water vapour content and the rates of evaporation and precipitation. The increased water vapour amount drastically increased the downwelling infra-red radiation at the Earth's surface, but the amount of the change varied dramatically from one model to another.

Perhaps 1990 is a point at which to pause briefly in the consideration of climate models. At that stage the models were relatively simple and were all somewhat similar to one another, and the computing power and data handling facilities were relatively limited. We note, in particular, the comment that predictions for smaller than continental scales should be treated with great caution and there is still a considerable amount of truth in this. Developments of models after 1990 included (1) the development of models to simulate more adequately past and present climates, the calibration, and verification of models based on observational data; (2) a comparison of models of different complexity; (3) experiments on the predictability of models useful for climate change forecasts on the scales of months, years, and decades as well as assessments of the predictability of natural variability; (4) studies on climate sensitivity aimed at analysis of the models' response to prescribed changes in boundary conditions and various types of forcing; and (5) the determination of the models' sensitivity to numerical algorithms and parameterisation of physical processes.

In the period from circa 1990 to the present – a period of about 30 years – climate modelling has developed into a major scientific activity. This was very much as a result of the reception and endorsement of the IPCC's First Assessment Report by the UNFCC and the signing of the Kyoto Protocol (see Section 7.6). However, it was also a result of the rapid development of computing power, software systems, and data storage and handling facilities. These things have expanded out of all recognition over the last few decades, but it is very difficult to quantify this expansion in each of these areas A convenient way to get some appreciation of his expansion is to consider a graph which describes Moore's Law, see Figure 6.18. This Figure shows the number of transistors which it has been possible to fit on to an integrated circuit chip, as a function of time. One should not be misled by the fact that this looks like a simple straight line because it is a log-linear plot. On a linear scale, the y axis of this graph would be 20 km long. This just relates to the number of transistors on an IC chip so that it does not translate directly into any of the three things, data storage, computer power, or software but it does give some sort of feel for what is involved. Data storage has been revolutionised. In the early days of satellite remote sensing in the 1970s, the data for one Landsat scene, 185 km × 185 km, could be stored on one 2400 ft (731.5 m) magnetic tape which contains about 46 Mbyte of data, see Figure 18. In terms of modern-day storage available to the general public, this means that one 64 Gbyte memory stick (flash disk) is equivalent to about 1390 of those tapes. This discussion was intended to convey something of the vast increases in data storage that have occurred in the last 50 years. One might attempt to do the same sort of thing for computing speed/power and for software. Of course, Moore's Law is relevant not just to climate modelling but to everyday life. And, of course, then there is the Internet too.

To understand the development of the principles of the science of climate modelling, since 1990, there is a valuable compendium edited by Kevin E. Trenberth (1992) and the IPCC Assessment Reports 3, 4, and 5 (not the summaries for policymakers which are politically tainted).

6.13   CARBON DIOXIDE AND CLIMATE

6.13.1   Numerical modelling for CO_2 increase

Over the last nearly 50 years, an enormous amount of effort has gone into computer modelling of the effect of increasing atmospheric CO_2 on the climate, in particular the temperature and the various consequences of the increasing temperature. We shall leave it to the next chapter to consider what this has achieved in terms of benefits for humanity.

Suppose one seeks to construct a computer model to investigate the extent to which human activities are responsible for global warming, then we have seen from the previous sections that there are various mechanisms which are responsible for affecting the radiation budget in the atmosphere that need to be considered. Principally these involve greenhouse gases and aerosols. The major greenhouse gases are water vapour, carbon dioxide, and various minor or trace gases. The major greenhouse gas, in terms of its quantity in the atmosphere, is water vapour; this is followed by CO_2 and then a whole variety of trace gases (see above) which although they are present in much lower quantity than the water vapour and the CO_2 are, in some cases – molecule for molecule – much more powerful greenhouse gases. The effect of the aerosols varies enormously with place and time and also with the chemical nature of the aerosol and the size distribution of the aerosol particles. In terms of the work done on model computations and the popular publicity that is given to the results, it is the CO_2 that receives by far the most publicity. Why is this? We should bear in mind that CO_2 is fundamentally a uniformly distributed gas, even though the spatial distribution will vary slightly, there will be a slight seasonal variation depending on whether it is northern hemisphere or southern hemisphere summer, and there is also a small steady annual increase. This is not the case for water vapour, the spatial and temporal distribution of which is changing all the time and for which data to use as boundary conditions for the model are not available. The combustion of fossil fuels of course produces CO_2 and water vapour. The water vapour adds to the amount of water vapour that is already present in the atmosphere. But its amount is small compared with the amount of water vapour already there and it is indistinguishable from that. So we simply do not know the extent of the greenhouse effect of the *additional* water vapour in the atmosphere generated by the combustion of fossil fuels. The effect of the trace gases is, by their nature, less important and their effect is often considered by simply assuming that they can be replaced by a CO_2 equivalent. When it comes to the aerosol contribution, the situation is even more difficult. Aerosols may be composed of different chemical substances, the aerosol particles are of various sizes, and aerosols come from different sources at different times and survive for various durations. Therefore, simply obtaining data on the aerosols is a very difficult task before one even tries to put the details of the nature and distribution of the aerosols into a model. A consequence of all this is that people expend much more effort on trying to model the effect of the human generated CO_2 than the effect of the other greenhouse gases and of the aerosols and the conclusions of this work on CO_2 have been given disproportionate attention by the IPCC and the organisations that rely on the IPCC for advice.

As far back as 1863, J. Tyndall clearly formulated the notion of the greenhouse effect. Comparatively transparent to solar radiation, atmospheric water vapour strongly absorbs the thermal emission of the Earth's surface. Some years later, Arrhenius and Chamberlain drew attention to a significant contribution of the CO_2 thermal emission to the formation of the heat balance and, consequently, of the temperature field in the atmosphere. Arrhenius' estimates have shown that a doubled CO_2 concentration should lead to a global warming by about 6 °C. Callender was the first who suggested the possible impact of anthropogenic releases of CO_2 on climate.

At present, the anthropogenic growth of CO_2 concentrations in the atmosphere is clearly established, and a successful development of climate theory has created the basis for comparatively realistic estimates of the impact of this growth on climate. A group of experts headed by the late J. Charney, formed by the US National Academy of Sciences, prepared and published in 1979 a report in which the results from studies on the problem of CO_2 impact on climate were summarised. Some years later, the Academy published a second report (National Research Council 1982) supervised by J. Smagorinsky. In this report, together with a discussion of the problem as a whole, the following aspects were considered in detail: (1) the role of the ocean, (2) the reliability of the conclusions about the impact of CO_2 on climate, drawn from analysis of local observational data on the land surface heat balance, (3) the contribution of minor optically active components and aerosols, and (4) the perspectives of the identification of the CO_2 signal in climate change from observational data. Subsequently documents were prepared by the US Department of Energy (MacCracken and Luther KC98) and a detailed annotated bibliography of the early work on climate modelling was compiled by Handel and Risby (KC98).

It would not be appropriate in this book to attempt to summarise 50 years of scientific effort by hundreds of climate modellers. The IPCC's reports, leaving aside the politically tainted Executive Summaries, have done an excellent job of this and we shall continue to consider these reports on the next chapter. In the meantime, we seek to consider the present state of climate modelling by considering the "Fifth Workshop on Systematic Errors in Weather and Climate Models – Nature, Origins, and Ways Forward" which was held in Montreal, Canada, in 2018 with over 200 participating scientists. The workshop's primary goal was to increase understanding of the nature and cause of systematic errors in numerical models across timescales. The workshop offered a forum to identify systematic errors and physical processes that are not well represented in current weather and climate models. The programme was organised around six themes:

- the coupled atmosphere-land-ocean-cryosphere system
- errors in the representation of clouds and precipitation

- resolution issues, including the representation of processes in the so-called grey zones
- model errors in ensembles
- errors in the simulation of teleconnections between the high/midlatitudes and tropics and
- novel metrics and diagnostics.

There is a meeting report by Zadra et al. (2018) which gives some details of the material presented. The following conclusions and ways forward were noted:

"All model evaluation efforts reveal differences when compared to observations. These differences may reflect observational uncertainty, internal variability, or errors/biases in the representation of physical processes. The following list represents errors that were noted specifically during the meeting:

- convective precipitation – diurnal cycle (timing and intensity); the organisation of convective systems; precipitation intensity and distribution; and the relationship with column-integrated water vapour, sea surface temperature, and vertical velocity
- cloud microphysics
- precipitation over orography – spatial distribution and intensity errors
- Madden-Julian oscillation (MJO) modelling – propagation, response to mean errors, and teleconnections
- subtropical boundary layer clouds – still underrepresented and tending to be too bright in models; their variation with large-scale parameters remains uncertain; and their representation may have a coupled component/feedback
- double intertropical convergence zone/biased ENSO – a complex combination of westward ENSO overextension, cloud-ocean interaction, and representation of tropical instability waves (TIW)
- tropical, cyclones – high-resolution forecasts tend to produce cyclones that are too intense although moderate improvements are seen from ocean coupling; wind-pressure relationship errors are systematic
- surface drag – biases variability, and predictability of large-scale dynamics are shown to be sensitive to surface drag CMIP5 (Coupled Model Intercomparison Project Phase 5) mean circulation errors are consistent with insufficient drag in models
- systematic errors in the representation of heterogeneity of soil
- stochastic physics – current schemes while beneficial, do not necessarily/sufficiently capture all aspects of model uncertainty
- outstanding errors in the modelling of surface fluxes; errors in the representation of the diurnal cycle of surface temperature
- errors in variability and trends in historical external forcings
- challenges in the prediction of midlatitude synoptic regimes and blocking
- model errors in the representation of teleconnections through inadequate stratosphere-troposphere coupling

and

- model biases in mean state, diabatic heating, sea surface temperature; errors in meridional wind response and tropospheric jet stream impact simulations of teleconnections."

Our purpose in quoting these details is to convey the message that climate modelling is an advanced scientific discipline that is very active with lots of specialised questions to be addressed. The other side of the coin is that in a book of this nature we cannot go into all the fine details of the latest results of climate modelling. We see no sign that the IPCC plans to discontinue its studies of the latest results.

7 The IPCC and its recommendations

7.1 INTRODUCTION

This chapter is devoted to the events that led up to the establishment of the Intergovernmental Panel on Climate Change (IPCC) and its activities once it had been set up. In looking for a picture to start this chapter, we chose Humpty Dumpty, see Figure 7.1.

In talking to Alice "When I use a word" Humpty Dumpty said, in rather a scornful tone, "it means just what I choose it to mean – neither more nor less." (from "Through the Looking-Glass, and What Alice Found There" by Lewis Carroll first published in 1872, page 130).

The discussion of human-induced global warming has almost assumed the dimension of a religious controversy, with neither side actually listening to what the other side is saying. We can divide the protagonists into two sides, the activists and the sceptics. Our line, as we have stated already, is that "yes, human–induced global warming is a problem that threatens our way of life but there are at least a dozen other things that equally threaten our way of life and it is foolish to concentrate on one of them to the exclusion of all the others."

The story of the IPCC can be argued to begin in the Cold War, the confrontation between the West, the USA, and the former Soviet Union, that is generally considered to span the 1947 Truman Doctrine to the 1991 dissolution of the Soviet Union. We leave aside for the moment the threat of the Nuclear Winter; we shall return to this in Section 8.6."

There are a number of issues:

1. Is humanly generated CO_2 causing global warming?
2. How serious is human-induced global warming in relation to other threats to our well-being or even our existence?
3. To what extent have the scientific results (of modelling, etc.) been used to further a particular political agenda.

With regard to 1, if one accepts the very simple arguments of Arrhenius (1896) the answer is "yes," but then the next question is "by how much"? And the answer to that is something that climate modellers have been struggling to find for around 30 years. With regard to 2, we have discussed this to a considerable extent in Chapter 1 already. In this chapter, we seek to summarise what the IPCC has been doing for the last 30 years and relate it to the UN climate programme and the politics of that programme and what we might loosely call the "Green Agenda" or "Sustainable Development."

7.2 SOVIET CLIMATOLOGY IN THE SECOND HALF OF THE 20TH CENTURY

7.2.1 THE COLD WAR PERIOD

We have included this section because the contribution of Russian/Soviet scientists to the global warming debate has been largely ignored in much of the recent literature on the subject of climate change. There is an interesting paper published by Jon Oldfield (2016) in which he looked at the

HUMPTY DUMPTY ON THE WALL.

FIGURE 7.1   Alice and Humpty from Lewis Carroll," Through the Looking-Glass, and What Alice Found There" illustrations by John Tenniel, first published by Macmillan & Co., London, 1872.

Soviet contribution to climatology from about 1945 up to the publication of the First Assessment Report of the Intergovernmental Panel on Climate Change (IPCC) in 1990. His thesis is that "Soviet climate science is given short shrift…emerging as a somewhat marginalised 'other' in the context of the Cold War." There is a common perception that humankind is responsible for important changes to the climate and that an important component of this responsibility is associated with the combustion of fossil fuels and the increasing concentration of atmospheric CO_2. One aspect of the involvement of Soviet/Russian scientists is to say that they took the view that while these are important aspects of the situation things are far more complicated than that and things are in danger of being oversimplified. It is important to appreciate the limitations of oversimplified constructions of the climate change debate, placing an emphasis on the need to engage with the multiple knowledge involved in comprehending the climatic processes, as well as the significance of different scales of analysis for interpreting the past, present, and future climates. Oldfield aims to offer a detailed synopsis of the character of Soviet climate science and to reflect on the hitherto largely overlooked contribution of Soviet scientists to the international debate on anthropic climate change as it developed from the late 1950s to the first report of the IPCC in 1990. He argues that a purposeful evaluation of the Soviet Union's efforts in this area as they developed post-WWII promises a more nuanced understanding of the history of climate change science, one that recognises the existence of marginalised layers of understanding within the broader international discourse of anthropogenic climate change. Soviet scientists made significant contributions to the natural science view of anthropogenic climate change that emerged so strongly after 1945 and were responsible for some of the earliest forays into predicting future climates.

Oldfield identifies a number of broad trends in Soviet scientific commitment to the climate change issue.

- Innovative advances were made with respect to physical and quantitative climatology, and particularly the functioning of the heat-water balance at the Earth's surface, which provided a basis from which to deepen understandings of the global climate system more generally.

- Certain Soviet climatologists and cognate scientists engaged progressively with the notion of society's growing influence on the climate system from the early 1960s onwards, integrating such understanding with work on future climate predictions.
- Soviet scientists were influential participants in the evolving international agenda, taking an active role in initiatives such as the formative International Geophysical Year event as well as the activities of the WMO and IPCC.
- Soviet work at the international level concerning climate change forecasts tended to be dominated by a relatively small group of scientists who became increasingly marginalised by the Western consensus that emerged during the foundational work of the IPCC in the late 1980s.

Oldfield identifies four key protagonists. These are the climatologist M.I. Budyko (1920–2001), the geophysicist E.K. Fedorov (1919–1981), the atmospheric physicist K.Ya. Kondratyev (1920–2006) and the geophysicist Yu. A. Izrael' (1930–2014). Among them, they produced a large amount of work devoted to climate change as well as broader global environmental concerns. They were all highly visible on the international scene and played significant roles in the WMO and related initiatives. It should be borne in mind that the dominance of the protagonists highlighted, particularly on the international stage, drew away from the more involved domestic debate in this area.

One feature of the climate discussions of the Cold War period was the nuclear winter which speculated on the climatic consequences of a full-scale nuclear war between the USA and its allies on the one hand and the USSR and its allies on the other. We shall discuss this later on (see Section 8.3) but seen from this distance it seems that the climate change would be one of the more minor consequences of a full-scale nuclear war. Another feature of this period is that relatively little was known about global geophysical processes during the early post-war years. Exploration activities were key features of the twentieth century, with attention devoted to the polar regions in addition to the atmosphere and the oceans. The International Geophysical Year (IGY) (1957–1958) was a key event which led to large-scale scientific collaboration across the ideological divide. The IGY also gave a significant boost to the use of rocketry and related technologies in exploring the more inaccessible regions of the Earth. A landmark was the Soviet Union's success in launching Sputnik 1 in October 1957. Sputnik marked the beginning of the space age and has led to the development of the numerous Earth-orbiting remote sensing satellite systems which have revolutionised all sorts of studies of the surface of the Earth at all sorts of scales. Indeed, we can say that two things in particular have revolutionised climate studies in the last 50 years; these are (1) the vast amounts of remote sensing data collected and (2) the enormous expansion of computing power and climate modelling.

The first beneficiary in the environmental sciences from the twin developments of remote sensing and of computing power was weather forecasting. Then weather forecast computer models slowly developed into climate models. Effective weather forecasting challenged national boundaries as well as ideological ones. It was required to assist economic planning, shipping activities, air flights, and a host of other socio-economic activities. In time, systematic weather data would also provide effective input for climate modelling. The twin concerns of short-term weather forecasting and climate system understanding encouraged the formation of a more effective global system of weather data generation grounded on advances in computing technology. The linked history of climate modelling was dominated by US and European scientists, although Soviet academics such as Budyko produced influential semi-empirical climate models during this period. Furthermore, the Soviet Union was not short of skilled climate modellers, building on the country's strong traditions in mathematics, physics, and related fields, and there was evidence of a robust exchange of expertise between East and West in this area. Nevertheless, explicit mention is made in many Western accounts of the relative backwardness of Soviet computing technology, thus helping to retard their development of large-scale computer modelling systems such as General Circulation Models (GCMs).

The establishment of large-scale data initiatives such as the WMO's World Weather Watch (WWW), and the persistent desire to grasp the complexities of the global climate system, moved the climate debate in some notable directions. In particular, it encouraged the development of an appreciation of the Earth system, which transcended regional, international, and ideological divisions; creating an "infrastructural globalism." These developments functioned in tandem with a plethora of other globalising trends linked to advances in remote sensing, a burgeoning environmental movement, and the emergence of concepts such as the biosphere and sustainable development. These globalising trends laid the foundations, at least in part, for an understanding of climate change which relied increasingly on the use of statistical modelling, while tending to marginalise other ways of interpreting the climate evident in the social sciences and humanities. Tensions also emerged within the natural sciences community, with individuals such as the British meteorologist Hubert Lamb (1913–1997) arguing for an expansive, interdisciplinary approach to climate; one that included the examination of historical sources, in order to ensure a comprehensive observational base for related modelling activity. The new breed of climate modellers was typically from a physics and mathematics background and their work embraced notions of scientific precision, which contrasted with the more interpretative science characteristic of Lamb's historical reconstruction of past climates. Nevertheless, this purported precision was of a particular kind, associated primarily with the techniques of statistical analysis and extrapolation. As such, the underlying assumptions framing any given model could vary markedly, Simon Shackley utilises the notion of "epistemic lifestyles" in order to draw attention to the way in which different groups of modellers go about the task of constructing their climate models. For Mike Hulme, the natural science-driven approach to climate change established itself firmly as the dominant interpretive framework during the period 1985–1992, effectively encompassing the early activities of the IPCC and the lead up to the 1992 Rio Conference on Environment and Development. He terms the approach "climate reductionism," conceptualising it as a form of neoenvironmental determinism with climate abstracted from the complexities of the broader socio-cultural context and thus removing human agency, among other things, from the debate. Soviet scientists formed a significant element of this general initiative as will be discussed in more detail below.

Oldfield observes that English-language critiques of Soviet climate science and associated understanding are relatively limited, although there was interest in the ideas and concepts of Soviet climate scientists among communities of Western physical scientists post-1945. He points to a long history of meteorological and climatological science within the Soviet and Russian context including tracing it back to the pre-revolutionary, i.e. tsarist, context. He also notes the emphasis "due to state coercion" on channelling activity into ways in which meteorology and climate science might benefit agriculture and other parts of the economy. We have also noted that it had been suggested that for a cold country like Russia there could be some advantages from a warming of the climate (Neumann et al., 2020). Oldfield claims that:

> M.I. Budyko is arguably the most well-known Soviet climatologist from a Western perspective. He began working at the Main Geophysical Observatory (GGO) in Leningrad during the 1940s, rising quickly through the ranks, and acting as its director from 1954 to 1972. Budyko participated in the international debate around anthropogenic climate change as it developed from the late 1960s onwards. His intellectual output can be usefully divided into three chronological stages: first, his early conceptual and applied work on the heat balance at the Earth's surface; second, a shift towards the global level via an interest in climate change and human influence on climate systems during the course of the 1960s; and third the development of a complex global understanding of climate situated within the all-encompassing concepts of the biosphere and global ecology from the late 1970s onwards.

Oldfield (2016) has written a review of Budyko's contributions to global climate science.

Budyko's shift into work concerning climate change and human influence on climate systems was part of a broader, albeit restricted, trend evident within the Soviet scientific literature during the late 1950s and early 1960s. A paper by E.K. Fedorov in 1958 can be considered an early effort to reflect on humankind's influence on meteorological processes, ranging from land-use changes, nuclear explosions, and shifts in the chemical state of the atmosphere.

"Fedorov and Budyko were both key instigators of a specially convened meeting on the transformation of climate which took place in Leningrad during April 1961. This meeting, together with a related workshop the following June, represented the first focussed Soviet discussions concerning anthropogenic climate change. A 1962 paper by Budyko noted the growing influence of humankind on the climate system linked in particular to increased energy use. This early intervention in the general debate reflected on the potential consequences of such trends for the Arctic region, a theme that Budyko would return to repeatedly in subsequent years. Budyko's general approach to the issue of climate change was one of cautious optimism, and this would remain a feature of his work during the course of the next three decades. For example, the complex role of the Arctic region's ice cover in regional and global climate processes was acknowledged. At the same time, he determined that if science suggested ice removal was feasible," while having limited adverse consequences for broader natural systems, then potential socio-economic benefits could be envisaged.

Generally speaking, Soviet scientists in the early days (say up to 1990) published their papers in Russian although books and papers sometimes were translated into English and published in the West. An exception was a paper which was published in English in *Tellus* by Budyko (1969). This paper is entitled "The effect of solar radiation variations on the climate of the Earth." For anyone who has the time, this is an interesting paper to read. It highlights some important points and illustrates what could be achieved without the massive computing power that is now available and without access to the vast amount of satellite remote sensing data that is now available.

It is open to the criticism that it starts with a general discussion of the great Quaternary glaciations over millions of years but when it comes to observations there is no data on long-term factors such as variations in intrinsic energy generated in the Sun, variations in the Earth's orbit (the Milankovich cycles, see Section 1.1.4.1), sunspots, etc. But when it comes to detailed calculations, Budyko is restricted by the data available which is from the (Russian) Main Geophysical Observatory (located in St. Petersburg) and to an 80-year period from 1881 to 1960. His first figure is reproduced in Figure 7.2.

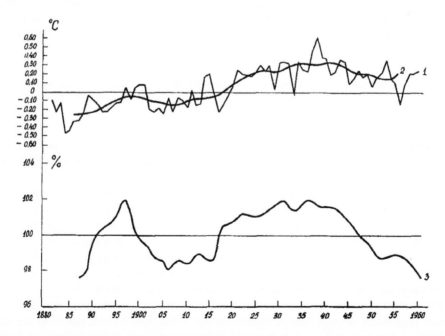

FIGURE 7.2   Secular variation of temperature and direct radiation. Line 1 illustrates the values of anomalies that are not smoothed, line 2 the anomalies averaged by ten-year periods, and line 3 the values of solar radiation smoothed for ten-year periods. (Source: Budyko, 1969).

The upper half of the Figure shows the variation of the northern hemisphere temperature about the mean, while the lower half shows the variation of the direct solar radiation. Budyko attempts to explain why there is a variation of the incoming solar radiation. On the timescale involved, 80 years not millions of years, it was suggested that it was due to changes in the transparency of the atmosphere due to volcanic dust and, in the later part of the study period – after 1940 – to an increase of dust in the atmosphere due to human activity. An atlas of the heat balance of the Earth edited by Budyko was published in Moscow (in Russian) in 1963. At the time of the *Tellus* paper data from meteorological remote sensing satellites was beginning to become available from which it was apparent that the albedo (i.e. the reflectivity) of areas covered by ice is greater than that of ice-free areas with a consequent important effect on the thermal regime. Budyko also considered the effect of latitude on the thermal regime. Several other Budyko papers are referenced in the 1969 *Tellus* paper.

In 1972, Budyko published a monograph, *Influence of Humankind on Climate* [Vliyanie Cheloveka na Klimat], which was subsequently regarded by Soviet scientists as an early attempt to provide a realistic prognosis of future global temperature increases resulting from human activity. This publication is also important for the emphasis it placed on understanding past climates in order to anticipate future climatic conditions. As the 1970s progressed, Budyko published a number of papers with K.Ya. Vinnikov over-viewing work related to climate change and the role of humankind in such change. These papers supported the idea of humankind's growing influence on the climate system and demonstrated a detailed engagement with Western scholarship concerning the potential warming consequences of human CO_2 emissions. For Budyko, anthropogenic influence on the wider environment was generally reducible to three main causal factors, namely increased concentrations of CO_2, changing levels of aerosol pollution from industrial sources, and increased levels of heat/energy output from anthropogenic sources. He considered the first two factors to be the dominant ones for the late twentieth century, with human energy output having the potential to gain in significance as the twenty-first century progressed. He also suggested that in certain regions characterised by large urban development anthropogenic heat generation was already comparable with certain natural flows of heat. Budyko's ideas in this area have been revisited in recent years. For example, E.J. Chaisson (2008) opened his paper on the subject by noting that:

> Even if civilization on Earth stops polluting the biosphere with greenhouse gases, humanity could eventually be awash in too much heat, namely, the dissipated heat by-product generated by any nonrenewable energy source

While Budyko's general work appears to be back in vogue, his projections have nevertheless been questioned, with Chaisson advancing, for example, a much slower growth in human heat emissions in comparison with incoming solar radiation.

During the course of the 1970s and early 1980s, Budyko developed his work on climate in order to place it more firmly within a broader conceptual framework of global ecology. Indeed, his 1977 book on this theme identified global ecology as an emerging area of great potential. This, and later work, advanced the global environment as a relatively fragile entity that required careful oversight of human activity in order to ensure its continued functioning. At the same time, Budyko's approach was also characterised by a persistent technocratic belief in the ability of humankind to understand the Earth's complex natural systems thus resulting in its effective management.

Budyko published a number of articles on the general theme of anthropogenic climate change and future predictions of such change during the late 1980s which coincided with the activities of the IPCC. A key text, and one which would feature prominently in the findings of the IPCC's Working Group II report (see below), was co-edited with Yuri Izrael' and entitled *Anthropogenic Climate Change*. This volume was referred to by Budyko in a later article as providing "the fullest description [to date] of forthcoming climate change," with its projections of temperature and rainfall maps for much of the Northern Hemisphere during the first half of the twenty-first century. Importantly, the map predictions were based primarily on the use of past climate reconstructions, which were then used as analogues for understanding possible future climates, although certain findings from climate theory were also incorporated. As will be discussed below, the general methodology of utilising past climate analogues would come under significant pressure from the Western modelling fraternity during the IPCC process due to concerns over the robustness of resultant future climate predictions. In response to such criticism, Budyko intimated that Soviet scientists were aware of the limitations of relying too heavily on past climate analogues and, furthermore, had progressed a relatively expansive approach to the issue since the 1970s, which combined meteorological observations and certain elements of climate theory together with palaeoclimatic data from past warm epochs. He went on to suggest that the promotion of an empirical approach "wholly independent from the conclusions of climate theory" had the potential to provide an important check with respect to the findings of modelling activities. Furthermore, the use of both empirical and modelling approaches promised to "increase the reliability of information concerning climatic conditions of the future." Budyko's defence of an empirical approach, and in particular the use of palaeoclimatic reconstructions, was restated again in a 1991 article in the Russian-language journal *Meteorologiya i Gidrologiya*, noting that such work allowed for an appreciation of the inherent sensitivity of the Earth's climate to shifts in the chemical composition of the atmosphere.

Budyko was joined in his global level theorising by the geophysicist Kirill Ya. Kondratyev. Kondratyev gained domestic and international recognition for his work in areas linked to solar radiation, atmospheric physics, satellite meteorology, and remote sensing. Kondratyev also worked extensively with the international scientific community and this included strong links with the WMO. During the 1980s, Kondratyev began to give global environmental change a significant amount of attention, and this included the specific issue of global climate change. His work in this area was characterised by a number of general themes. First, while he considered the debate concerning anthropogenic climate change of significance, he was at the same time wary of oversimplifying the issue as well as the inadequacies of available data sets. In particular, he understood the climate system as just one facet or expression of the Earth's global physical system. Second, this expansive understanding of the climate system ensured that he placed emphasis on the functioning of the biosphere as a whole and resisted reducing the climate issue down to single factors such as an increase in CO_2 emissions. This general approach was evidenced in his support for the work of V.G. Gorshkov concerning biotic regulation. This concept is predicated on the belief that the fundamental tension between society and the wider environment can only be addressed through the restoration and long-term conservation of significant parts of the biosphere. The notion of biotic regulation refers to the self-regulating properties of the biosphere, which it is postulated have developed over the long term. It is suggested that there is a threshold level of "anthropogenic perturbation" beyond which the integrity of this regulating function is compromised. Third, his work during the late 1980s emphasised the importance of utilising data from various analogues in order to provide deeper insight into global climate change. In addition to the value of examining past climates of the Earth, Kondratyev also advocated an analysis of the climate systems of other planets. His interest in utilising analogues was also evident in his work on the effects of nuclear war on the atmosphere.

In the light of the above, it can be suggested that Soviet engagement with the international discussion around climate change was characterised by a growing awareness of the scope and nature of society's influence on climate systems, particularly with respect to CO_2 and aerosol pollution, a concerted effort to establish future scenarios of climate change, and a positive engagement with similar Western science in this area. At the same time, Soviet work related to future climate scenarios became increasingly defined, at least on the international level, by a dependence on palaeoclimatic analogues, and this emphasis would precipitate a marked stand-off between the Soviet contingent and the Western-dominated community of climate modellers during the formulation of the first IPCC report.

7.2.2 Soviet climate change dialogue with the West

The relation between Soviet scientists and their Western counterparts in the early days of the IPCC regarding the deepening awareness of anthropogenic climate change is described by Boehmer-Christensen and Cracknell (2009) and we rely heavily on that article. From this period, there are three areas of particular note for the advancement of climate change understanding: the Nuclear Winter debate, US-USSR environmental collaboration and associated initiatives, and finally the activities leading up to and including the publication of the first IPCC report in 1990. We discuss the nuclear winter debate elsewhere (see Section 8.6). The nuclear winter discussions were an interesting intellectual exercise, but in terms of practical reality, the consequences for the climate of a nuclear war would be as nothing compared to the destruction of lives, homes, infrastructure, and life-support systems/infrastructure by such a war.

As background, it is worth acknowledging the relative difficulties characterising the movement of scientific knowledge during the Cold War period linked to ideological as well as linguistic barriers. Post-1945, climate science became entangled with broader issues such as climate modification and the effects of nuclear fallout, and the urgency of these concerns encouraged some to pursue scientific conduits between East and West in spite of the various administrative and political barriers in place. International bodies such as the WMO facilitated the exchange of ideas by supporting short fact-finding visits to the Soviet Union as well as encouraging collaborative work amongst its members. Nevertheless, the level of exchange was relatively low and the situation was aggravated by the language barrier, which reduced Western engagement with Soviet science and thus only served to heighten the uncertainty and suspicion underpinning the relationship. In the area of climatology, efforts were made by bodies such as the American Geophysical Union (AGU) to facilitate the exchange of ideas, in this case via the activities of its Russian Translation Board. The American Meteorological Society was also active in publishing a series of reviews of Soviet meteorological science in its *Bulletin* during the course of the 1950s and 1960s. Such initiatives were buttressed by the focused activity of "think-tank" organisations including the RAND Corporation, which emerged as an influential mediator of Russian language materials.

7.2.3 US-USSR climate science collaboration

US-USSR interaction with respect to the science of climate change was also evident as part of more purposeful collaboration between the two countries. For Stephen Brain, the specific context of the Cold War, with its emphasis on rivalry and competition, encouraged favourable engagement with the global environment (Oldfield, 2018). This assertion finds expression in the emergence of initiatives such as the International Institute for Applied Systems Analysis (IIASA) in Austria, which was driven primarily by the two superpowers. A further long-term environmental cooperation programme between the Soviet Union and the US was initiated in the early 1970s. Robinson and Waxmonsky (1988) suggest a combination of politically expedient as well as "mutually beneficial reasons were behind the relative success of this bilateral initiative." The two presidents (Nixon and Brezhnev) supported the signing of the US-USSR *Agreement on Cooperation in the Field of Environmental Protection* on 23 May 1972. It consisted of eleven thematic areas and this included a focus on air pollution as well as the influence of the environment on climate (Working Group VIII, for the formulation of response strategies: IPCC-1, TD-No. 267, 4). E.K. Fedorov and Yuri Izrael' were installed as chair and coordinator of the Soviet contribution in view of their leading roles in the State Hydrometeorological Service (Izrael' would become co-chair from 1974). At this time, the State Hydrometeorological Service was a relatively weak institution within the context of the broader Soviet political machinery, but it had been suggested that as its power grew, Izrael' became less interested in the collaborative endeavour.

There was a general recognition that the Soviet Union and the USA shared a range of common environmental problems, in spite of their ideological differences. In addition, the prevailing rhetoric suggested that both sides anticipated benefits from the agreement beyond the general political aspects of the initiative. Soviet prowess was said to reside in their system of protected areas centred on the *zapovedniki* (areas given high levels of protection within the Soviet Union), as well as in fields such as ecological assessment (Izrael and Kuvshinnikov, 1975). At the same time, the Soviet Union gained from US applied and technical expertise linked to national park management and environmental monitoring equipment. The various initiatives were complemented by a series of workshops and symposia which brought together scientists from both sides around themes such as the comprehensive analysis of the environment and nature reserves as well as climate change.

More specifically, Working Group VIII of IIASA (see also, Bruckmann, (1980) was established to examine "the influence of environmental changes on climate." According to Kelley, the US side had initially hoped to gain from Soviet understanding in the area of climate science. However, while collaborative work around climate change was considered a general success, the US participants concluded that the USSR had been "at least a decade behind the United States in the area of climate research." The specifics of this assessment are not clear, but they are likely to refer primarily to the Soviet Union's computer modelling capabilities. At the same time, other reports are suggestive of US interest in at least some areas of Soviet modelling activity.

Working Group VIII was co-chaired by Budyko and the American Alan D. Hecht (US Environmental Protection Agency) (Hecht and Döös, 1988). It pursued a relatively extensive joint programme over the course of more than 15 years enroling a significant number of scientists and institutions on both sides (Meleshko and Wetherald, 1981). It also facilitated the exchange of scientists between the US and the Soviet Union and resulted in a number of joint publications. The work of the group was predicated on a general acceptance on both sides that "global warming would be inevitable as a consequence of the on-going perturbations to atmospheric composition." The main emphasis of the activities carried out by the working group was to establish a firm basis on which to predict future changes in climate associated with a warming trend.

Two main directions of research were pursued with respect to predicting future climates, one examined past climates and reflected on the value of such knowledge for current understanding, and the other pathway focused on theory and modelling of contemporary climates. As noted, this latter area was one in which Soviet science trailed its American counterpart. While the report itself does not make explicit reference to the nature of the input from the two sides, the two pathways reflected the relative strengths of the Soviet and US contributors. In reviewing the book that emerged out of the long-term collaboration between the two countries (entitled *Prospects for Future Climate*), the US climatologist William W. Kellogg noted that:

In general, the United States scientists rely heavily (but not exclusively) on the theoretical results of experiments with a hierarchy of climate models, whereas the Soviet scientists emphasise a more empirical approach that relies on reconstructions of past climates and studies of current trends (Kellogg, 1992).

While the differences in approach evidenced by the two parties were an accepted part of the US-USSR collaboration, they would take on a much more divisive hue within the high politics of the early IPCC discussions.

7.2.4 Soviet involvement in the activities of the WMO and IPCC

The Soviet Union had a visible presence within the World Meteorological Organisation (WMO) reflecting in part its historical role in the preceding International Meteorological Organisation, as well as the importance of such a vast country for the success of a world meteorological initiative. Key positions were held by Soviet academics including E.K. Fedorov (vice-president, 1963–1971) and Yu.A. Izrael' (first/second vice-president, 1975–1987), and individuals such as Budyko, Kondratyev and G.S. Golitsyn were also heavily involved in different aspects of the organisation's work over the years.

The 1979 World Climate Conference was an event of significance with respect to the developing momentum around climate change. The WMO executive committee's decision to hold such a meeting was taken in 1977 based on the need to understand and plan for climatic variability as well as the "strong evidence that the climate itself may be influenced by the activities of mankind." The first week of the conference was devoted to the delivery of invited papers on current knowledge, and week two provided a forum for discussion and reflection in order to determine recommendations for future areas of focus. More than 20 Soviet representatives attended the meeting. Furthermore, four of the main discussion papers were delivered by Soviet scientists in the first week and included a paper by Fedorov (Climate change and human strategy) in the opening session, as well as papers by the physical scientists Cl. Marchuk (Modelling of climatic changes and the problem of long-range weather forecasting) and Yu. A. Izrael' (Climate monitoring and climate data collection services for determining climatic changes and variations). Fedorov's paper is particularly noteworthy in view of its delivery in the opening session. With reference to human influence on climate, he noted three main mechanisms: changes to the land surface with consequences for heat exchange, changes to the water balance again impacting linked physical systems, and changes to the transparency of the atmosphere leading to shifts in the Earth's energy balance.

As noted, 1985–1992 is considered of key importance by Mike Hulme (2009), with the suggestion that it was during this period that the current dominant framing of climate change in both science and policy circles was fashioned. In particular, it was a period during which modelling emerged as the key "epistemological authority" and this approach was utilised in order to abstract climate from its complex entanglements with a range of physical as well as social phenomena, to be used as a main predictor of the future. Hulme goes further in his critique, referring to it as "epistemological slippage – a transfer of predictive authority from one domain of knowledge to another without appropriate theoretical or analytical justification."

The period was opened by the activities of the Second joint UNEP/ICSU/WMO International Assessment of the Role of Carbon Dioxide and of Other Greenhouse Gases in Climate Variations and Associated Impacts held in Villach, Austria in 1985, and concluded with the 1992 United Nations Framework Convention on Climate Change (UNFCCC)) Sonja Boehmer-Christiansen draws attention to the pivotal nature of the Villach conference in helping to propel the climate change debate into the political sphere (Boehmer-Christiensen, 1994a, 1994b). The conference gave rise to the Advisory Group on Greenhouse Gases (AGGG), which paved the way for the IPCC in 1988, with the latter emerging in part due to apparent US concerns that the climate change issue should remain more firmly under the control of a state-driven body.

The first session of the UNEP/WMO sponsored IPCC took place in Geneva in November 1988. The Soviet Union sent a principal delegate (A.P. Metalnikov) and four advisors (S.S. Hodkin, E. Koni-gin, V. Blatov, and B. Smirnov) and its supporting statement to the opening session highlighted Soviet enthusiasm for the work of the panel and the need for greater understanding of the climate change issue. The opening meeting established the basic outline of the panel's activities which were to be structured around three main areas: an assessment of the science concerning climate change (Working Group I), an assessment of the potential socio-economic impacts of climate change (Working Group II), and the formulation of response strategies (Working Group III). Yuri Izrael' was nominated as chairman of Working Group II. Furthermore, Soviet representation was formally allocated to the other two working groups. By the next meeting of the general IPCC body in February 1989, Working Group II had delineated five areas for further work (agriculture, forestry and land use; natural ecosystems; hydrology and water resources; energy, industry, transport, settlements and human health; world oceans, cryosphere and human health). In addition to Izrael' as chairman, each of the five sub-groups was allocated a Soviet specialist. A steering group was also established during the first meeting of Working Group II in Moscow (February 1989) and this incorporated a Soviet representative. The focus of Working Group II on an assessment of the impacts of climate change ensured that it required robust scientific insight into the anticipated extent and regional character of climate warming. However, this aspect of its work was undermined by the apparent slow emergence of relevant scenario recommendations from Working Group I, a point that was underlined at the third plenary meeting in Washington D.C. in early 1990.

The IPCC process was clearly a difficult one, stemming from the complexity of the climate change issue, the complicated working practices of the panel's working groups, the political connotations of its work, and the inevitable clash of views and personalities that became evident as the process evolved. During the preparatory phase of the IPCC's first report, an international ministerial conference was held at Noordwijk (The Netherlands, November 1989) on *Air Pollution and Climate Change*. For Boehmer-Christiensen (1994a, 1994b), this event was part of a marked politicisation of the climate change debate and provided evidence of "the deep split between Europe and the USA, as well as within the IPCC itself." She goes on to note that:

> Battle lines were now clearly drawn inside the IPCC, then in the process of drafting its first report. It could not afford to offend major governments or its sponsors. Born into the controversy over response strategies, it had already become a target for conflicting pressures. One of its first actions would be to discredit the Soviet view, stated by Professor Izreal [sic] at home, that global warming was a good thing, and reducing Soviet influence in WGI [Working Group I].

For Alan Hecht, writing in the foreword to the English-language edition of Izrael' and Budyko's book *Anthropogenic Climate Change* (1987), the notion of a possible favourable future climate for parts of the northern hemisphere was grounded on the results of the application of palaeoclimatic analogues outlined in the book. Budyko's insistence on the potential beneficial impacts of climate change, primarily through anticipated increased levels of precipitation and the so-called "fertilizer effect" of heightened CO_2 levels (enhancing crop growth), clashed with Western climate modellers as well as the emerging international consensus that anthropogenic climate change was an issue to be addressed with growing urgency. The somewhat crude and dogmatic character of Budyko's pronouncements during the late 1980s differed from his more cautious and measured statements in early years, and his views were understandably treated with scepticism by many.

The evident marginalisation of the Soviet contingent persisted along overtly scientific lines, framed by the growing dominance of a predictive, law-based modelling approach within Working Group I of the IPCC. In particular, doubt was cast over the future climate predictions of the palaeo-analogue approach (pushed strongly by the Soviet representatives), due to uncertainties over the underlying mechanisms and the robustness of data sets, as well as suggestions that past climates would be unable to accurately predict the specificities of a rapidly warming global climate in the near future. In view of the importance of Working Group I for the whole IPCC enterprise, its emphasis on a modelling approach had negative consequences for the relative standing of Soviet scientific input. In recounting this period, the then chair of Working Group I, John Houghton (former chief executive of the UK Meteorological Office), highlights a more personal consequence of this shift in emphasis. He suggests that Budyko was greatly affected by the critique of his work on the use of past climate analogues at a special meeting of Working Group I that took place in Bristol. At the same time, it is important to note that the move towards the use of General Circulation Models (GCMs) was not total as evidenced by the findings of the aforementioned US-USSR collaborative endeavour around climate change (published in 1990), which emphasised the initiatives "innovative approach to projecting future climate change based on the hypothesis that we can combine the strengths of what we have learned about past warm periods and what we can simulate with our models." While the politics of international collaboration were clearly at play here, it is also reasonable to conclude that the scientists involved were generally supportive of this general conclusion. Crucially, however, in the final report of Working Group I, it was explained that GCMs were "the most highly developed tool which we have to predict future climate." The value of palaeoclimates for predictive purposes was also highlighted, although with qualification:

We cannot therefore advocate the use of palaeo-climates as predictions of regional climate change due to future increases in greenhouse gases. However, palaeo-climatological information can provide useful insights into climate processes and can assist in the validation of climate models.

The slow emergence of conclusions from Working Group I prompted the incorporation of palaeoclimate analogues into the predictive scenarios utilised by Working Group II. The final report for Working Group II made a clear distinction between the Soviet palaeoclimate approach and the Western use of GCMs. The related discussion suggested that the relative merits of both approaches were open to ongoing debate. The summative work of Izrael' and Budyko with respect to palaeoclimatic analogues, published in *Anthropogenic Climate Change,* was referenced extensively within the report. In Izrael's account of events, he suggested that the recommendations of Working Group I were both late and also characterised by uncertainties linked primarily to predictions of rainfall and soil moisture, as well as extreme events at the regional scale. He also promoted the potential value of palaeoclimatic analogues, highlighting the fact that they provided an additional "validation" of climate models and possessed the capability to deliver regionally sensitive predictions concerning rainfall.

7.2.5 THE DISTINCTIVE SOVIET CONTRIBUTION

The Soviet case study advanced here provides insight into a hitherto little studied aspect of the evolving climate change debate during the Cold War period, one which was nevertheless influential in helping to advance a natural science understanding of the phenomenon at the international level. Indeed, shaped by a handful of highly visible climatologists and geophysicists, it is evident that Soviet science was a significant presence with respect to the strengthening international consensus around anthropogenic climate change. Allied to this, the scientific record is suggestive of a distinctive Soviet engagement with the issue, grounded on a long tradition of interest in the role of climate, in the functioning of complex physical and biological systems. This tradition was further characterised by a general emphasis on what might be termed a holistic approach towards understanding climate change, one that emphasised the complex character of the climate system, as well as the potential of humankind to use its understanding of the climate system for the wider benefit of society. Elements of this particular feature of Soviet climate science remain relatively underdeveloped in Western accounts of the climate change debate during the 1970s and 1980s.

Soviet climate science appears to have been held in high regard by Western scientists throughout much of the Cold War period. For example, the Soviet Union's work concerning physical climatology, led by Budyko and his colleagues at the GGO, was well received, and Soviet work on climatic analogues proved significant within the work of bodies such as the WMO. At the same time, there was a growing sense that Soviet developments in climate science, and particularly climate modelling, lagged behind achievements in the West as the 1970s and 1980s unfolded, linked to the slowdown in the Soviet economy and associated technological inadequacies rather than any fundamental scientific shortcomings. While Soviet scientists certainly made progress in the formulation of climate models, the purported limitations in computing power are likely to have been a key factor behind the increased emphasis placed on palaeoclimatic analogues by the Soviets on the international stage; an emphasis which contrasted with the growing Western interest in computer modelling. Budyko's evident support of this approach, as a key player internationally, was undoubtedly a further reason for Soviet prominence in this area. Importantly, both sides recognised the value of the two approaches for advancing the climate change issue. However, there was growing debate at the international level as to their relative value and this was framed by a complex mix of scientific coalitions and high politics. It also seems likely that the dominance of the Soviet Union's international agenda by a relatively small group of ageing scientists resulted in the gradual ossification of their collective input and an increasingly dogmatic approach to their own work, thereby helping to undermine relationships with Western colleagues and disguise more varied domestic debates. Ultimately, the marked shift towards the use of GCMs as the key science of climate change forecasting during the late 1980s, allied to scientific uncertainties over aspects of the palaeoanalogue approach as well as personality clashes, appears to have resulted in the relative marginalisation of Soviet scientific input during the formative stages of the first IPCC report. Thus, Soviet contributions to the scientific debate around climate change at the international level moved from a position of significance during the 1960s–1980s to a position of relative isolation by the start of the 1990s. This state of affairs was amplified by a tendency for the climate debate to revolve around the science of future predictions and the growing significance of advanced climate modelling activity.

7.3 THE IPCC REPORTS

7.3.1 BACKGROUND

We mentioned in Chapter 1 that the IPCC reviews and assesses the most recent scientific, technical, and socio-economic information produced worldwide relevant to the understanding of climate change. It does not conduct any research nor does it monitor climate-related data or parameters. The IPCC produced its first report in 1990 and has produced several reports since then, making five in all so far:

First Assessment Report	1990
Second Assessment Report	1995 Full Report
Third Assessment Report (TAR)	2001 TAR Climate Change: Synthesis Report
Fourth Assessment Report (AR4)	2007 AR4 Climate Change: Synthesis Report
Fifth Assessment Report (AR5)	2014 AR5 Climate Change: Synthesis Report.

The Sixth Report is expected to be finalised in 2022.

For each of these Assessments, there are individual reports of the three working groups:

Working Group 1 (WG1) The Physical Sciences
Working Group 2 (WG2) Impacts, Adaptation and Vulnerability
Working Group 3 (WG3) Mitigation of Climate Change

In addition to these Reports the IPCC has published a number of Special Reports, including (starting with the latest at the time of writing):

"Global warming of 1.5 °C" 2018
"Managing the Risks of Extreme Events and Disasters to Advance Climate Change Adaptation" 2012
"Renewable Energy Sources and Climate Change" 2011
"Safeguarding the Ozone Layer and the Global Climate System" 2005
"Carbon Dioxide Capture and Storage" 2005
and so on.

It is now 30 years since the first IPCC Assessment Report was published; altogether five have now been published and the sixth is due soon. There is no doubt that a very great deal of intellectual activity has been devoted to this task by a large number of highly intelligent people with a great deal of experience in the subject matter of climatology. The result is not surprisingly a set of long and very detailed reports that policy makers, at national and UN levels, do not have time, or generally the expertise, to assimilate and understand. Consequently starting with the First Assessment Report and continued ever since then it has been the practice to produce Summaries for Policy Makers (SPMs).

An enormous amount of time and effort has gone into the production of the individual Working Group Reports, the various assessments, and the individual special reports. It is one thing to produce copious advice, but what happens next?

The Assessment Reports are available for anyone to read who cares to do so. We clearly do not have the space to summarise these reports. There have been 25 Conferences of the Parties to take note of all this advice. Some of it will have been taken, and some of it will have been ignored. Some of it may yet still be taken into account by a future Conference of the Parties.

7.3.2 Why are the IPCC assessments so important?

The reason that the IPCC Assessments are so important is two-fold. First, it is because the IPCC is the organisation to which the United Nations looks for scientific advice on climate change. This advice is provided within the United Nations Framework Convention on Climate Change (UNFCCC); this was set up and came into force after the First IPCC Assessment Report. The second reason is that inevitably, as a result of the first reason, media attention automatically attends the pronouncements of the IPCC.

The United Nations Framework Convention on Climate Change (UNFCCC) is an international environmental treaty. Initially, an Intergovernmental Negotiating Committee produced the text of the Framework Convention during its meeting in New York from 30 April to 9 May 1992. The Framework was adopted on 9 May 1992 and opened for signature at the Earth Summit in Rio de Janeiro from 3 to 14 June 1992. It then entered into force on 21 March 1994, after a sufficient number of countries had ratified it. The UNFCCC objective is to "stabilize greenhouse gas concentrations in the atmosphere at a level that would prevent dangerous anthropogenic interference with the climate system." It is important to understand that the framework only sets non-binding limits on greenhouse gas emissions for individual countries and contains no enforcement mechanisms. Instead, the framework outlines how specific international treaties (called "protocols" or "Agreements") may be negotiated to specify further action towards the objective of the UNFCCC.

The UNFCCC had 197 parties as of December 2015. The Convention enjoys broad legitimacy, largely due to its nearly universal membership. One of the first tasks set by the UNFCCC was for signatory nations to establish national greenhouse gas inventories of greenhouse gas (GHG) emissions and removals, which were used to create the 1990 benchmark levels for accession of Annex I countries to the Kyoto Protocol and for the commitment of those countries to GHG reductions. Updated inventories must be submitted annually by Annex I countries. Loosely speaking, the Annex I countries are the countries of Europe (including the Russian Federation), North America, Japan, Australia, and New Zealand.

The parties to the Convention have met annually from 1995 in Conferences of the Parties (COP) to assess progress in dealing with climate change. It has to be hoped that the advice offered by the IPCC to the UNFCCC gets implemented. There is no supra-national police force. The United Nations can pass Resolutions but there is no way to enforce them except by applying political pressure to individual states. Sometimes that works and sometimes it does not work. There have, at the time of writing, been 25 of these Conferences, with the latest being the 25th COP, which was held in 2019 in Madrid, Spain.

In 1997, at the 3rd Conference of the Parties which was held in Kyoto, Japan, the Kyoto Protocol was concluded, see Section 7.6; this established legally binding obligations for developed countries, the Annex I countries, to reduce their greenhouse gas emissions in the period 2008–2012. By this time the IPCC's Second Assessment Report had been published. The 2010 United Nations Climate Change Conference produced an agreement stating that future global warming should be limited to below 2.0°C (3.6°F) relative to the pre-industrial level. The Kyoto Protocol was amended in 2012 to encompass the period 2013–2020 in the Doha Amendment, which as of December 2015 had not entered into force. In 2015, the Paris Agreement was adopted, governing emission reductions from 2020 on through commitments of countries in Nationally Determined Contributions (NDCs), with a view of lowering the target to 1.5 °C. The Paris Agreement entered into force on 4 November 2016.

7.3.3 THE IPCC FIRST ASSESSMENT

We consider very briefly the first IPCC Assessment. The report comes with a "Policy-maker's Summary" of about 20 pages which was prepared "to meet the needs of those without a strong background in science who need a clear statement of the present status of scientific knowledge and the associated uncertainties." It is quite remarkable that from the contributions of so many scientists, who surely must have many disagreements among themselves on matters of details, there emerged some very clear conclusions. The conclusions of the first Report of WG1 (IPCC 1 1990) are widely but not universally accepted. We shall turn to the international response to this and the later IPCC reports later in this chapter.

It is worth quoting verbatim from the policy makers summary of this first assessment, because it was a landmark document, notwithstanding the fact that some of its evidence and conclusions have subsequently been refined in the later assessments and also notwithstanding the fact that the assessments have not been universally accepted (see Sections 7.8 and 7.9):

"We are certain of the following:

- There is a natural greenhouse effect which already keeps the Earth warmer than it would otherwise be.
- emissions resulting from human activities are substantially increasing the atmospheric concentrations of the greenhouse gases: CO_2, CH_4, CFCs, and N_2O nitrous oxide. These increases will enhance the greenhouse effect, resulting on average in an additional warming of the Earth's surface. The main greenhouse gas, water vapour, will increase in response to global warming and further enhance it.

We calculate with confidence that:

- Some gases are potentially more effective than others at changing climate, and their relative effectiveness can be estimated. CO_2 has been responsible for over half the enhanced greenhouse effect in the past and is likely to remain so in the future.
- atmospheric concentrations of the long-lived gases (CO_2, N_2O, and the CFCs) adjust only slowly to changes in emissions. Continued emissions of these gases at present rates would commit us to increased concentrations for centuries ahead. The longer emissions continue to increase at present day rates, the greater reductions would have to be for concentrations to stabilise at a given level.
- The long-lived gases would require immediate reductions in emissions from human activities of over 60% to stabilise their concentrations at today's levels; CH_4 would require 15-20% reductions.

Based on current model results, we predict:

- Under the IPCC Business-as Usual (Scenario A) emissions of greenhouse gases, a rate of increase of global mean temperature during the next century of about 0.3°C per decade (with an uncertainty range of 0.2°C to 0.5°C per decade); this is greater than that seen over the past 10,000 years. This will result in a likely increase in global mean temperature of about 1°C above the present value by 2025 and 3°C before the end of the next [now present] century. The rise will not be steady because of the influence of other factors.
- Under the other IPCC emission scenarios, which assume progressively increasing levels of controls, rates of increase in global mean temperature of about 0.2°C per decade (Scenario B), just above 0.1°C per decade (Scenario C) and about 0.1°C per decade (Scenario D).
- that land surfaces warm more rapidly than the ocean, and high northern latitudes warm more than the global mean in winter.

- regional climate change is different from the global mean, although our confidence in the prediction of the detail of regional changes is low. For example, temperature increases in Southern Europe and central North America are predicted to be higher than the global mean, accompanied on average by reduced summer precipitation and soil moisture. There are less consistent predictions for the tropics and the southern hemisphere.
- Under the IPCC Business as usual emissions scenario, an average rate of global mean sea level rise of about 6 cm per decade over the next century (with an uncertainty range of 3-10 cm per decade), mainly due to thermal expansion of the oceans and the melting of some land ice. The predicted rise is about 20 cm in global mean sea level by 2030, and 65 cm by the end of the next [now the present] century. There will be significant regional variations."

It is worthwhile to make one or two points about this.

The executive summary only talks about the mean global temperature, see Section 7.8.2 – models give far more information than just that. We have already got to the point of details which are unclear. Different models give different answers.

However, the executive summary continues by admitting the uncertainty level, to attempt to give a balanced judgement and to make recommendations; the quote continues:

There are many uncertainties in our prediction: particularly with regard to the timing, magnitude and regional patterns of climate change, due to our incomplete understanding of sources and sinks of greenhouse gases, …clouds, …oceans, …[and]…polar ice sheets.

A great deal of work was done over the next 30 years, the construction and operation of climate computer models became big science and many details were refined. We hope to see what was actually achieved.

After the first report of WGl of the IPCC was produced, there were a number of subsequent attempts to update it to include the results of later work in a supplementary report, followed by the second, third, fourth, and fifth assessments. The sixth assessment is due to be published in 2022. However, in many ways, it was the first report which made the big impact and it is more in matters of detail rather than in general terms that it has been updated. We shall only note a few salient points from these later assessments.

The next report (IPCC 2) was a supplementary assessment to IPCC 1. The production of this report involved 118 scientists from 22 countries, with reviews carried out by a further 380 scientists from 63 countries and 18 UN or non-governmental organisations (NGOs). Its production was timed to coincide with the final negotiations of the UN Framework Convention on Climate Change (UNFCCC) which was one of the five basic documents of the Rio Convention (the Second United Nations Conference on Environment and Development (UNCED-2)) (see Section 7.4). This report reviewed and reaffirmed the basic conclusions of the 1990 report; it did, however, provide more detail on two sources of negative radiative forcing, namely the depletion of stratospheric ozone and the effect of aerosols derived from anthropogenic emissions. The Preface, however, included a statement "The text of the Supplement was agreed in January 1992 at a plenary meeting of WG1 held in Guangzhou, China, attended by 130 delegates from 47 countries. It can therefore be considered as an authoritative statement of the contemporary views of the international scientific community." This shows a dangerous trend in the thinking of the IPCC, namely that scientific truth can be established by consensus – we shall return to this issue later in this chapter (see Section 7.12).

Rather than include all the detailed points made in this supplementary assessment, we just reproduce section 13 of the 1992 Supplementary Report on "Key uncertainties and further work required":

"The prediction of future climate change is critically dependent on scenarios of future anthropogenic emissions of greenhouse gases and other climate forcing agents such as aerosols. These depend not only on factors which can be addressed by the natural sciences but also on factors such as population and economic growth and energy policy where there is much uncertainty and which are the concern of the social sciences. Natural and social scientists need to cooperate closely in the development of scenarios of future emissions.

Since the 1990 report, there has been a greater appreciation of many of the uncertainties which affect our predictions of the timing, magnitude, and regional patterns of climate change. These continue to be rooted in our inadequate understanding of:

- sources and sinks of greenhouse gases and aerosols and their atmospheric concentrations (including their indirect effects on global warming)
- clouds (particularly their feedback effect on greenhouse-gas-induced global warming, also the effect of aerosols on clouds and their radiative properties) and other elements of the atmospheric water budget, including the processes controlling upper level water vapour
- oceans, which through their thermal inertia and possible changes in circulation, influence the timing and pattern of climate change
- polar ice sheets (whose response to climate change also affects predictions of sea level rise)
- land surface processes and feedbacks, including hydrological and ecological processes which couple regional and global climates

Reduction of these uncertainties requires:

- the development of improved models which include adequate descriptions of all components of the climate system
- improvements in the systematic observation and understanding of climate-forcing variables on a global basis, including solar irradiance and aerosols
- development of comprehensive observations of the relevant variables describing all components of the climate system, involving as required new technologies and the establishment of data sets
- better understanding of climate-related processes, particularly those associated with clouds, oceans, and the carbon-cycle
- an improved understanding of social, technological, and economic processes, especially in developing countries, that are necessary to develop more realistic scenarios of future emissions
- the development of national inventories of current emissions
- more detailed knowledge of climate changes which have taken place in the past
- sustained and increased support for climate research activities which cross national and disciplinary boundaries; particular action is still needed to facilitate the full involvement of developing countries
- improved international exchange of climate data

In parallel with the work of the IPCC climate questions are also being studied by other major international programmes, in particular by the World Climate Research Programme (WCRP), the International Geosphere Biosphere Programme (IGBP) and the Global Climate Observing System (GCOS). Adequate resources need to be provided both to the international organisation of these programmes and to the national efforts supporting them if the new information necessary to reduce the uncertainties is to be forthcoming. Resources also need to be provided to support on a national or regional basis, and especially in developing countries, the analysis of data relating to a wide range of climate variables and the continued observation of important variables with adequate coverage and accuracy."

7.3.4   THE SECOND AND SUBSEQUENT ASSESSMENT REPORTS

Following the success of the first report and its 1992 supplementary report, in terms of its input into the UNFCCC, the IPCC set itself the task of producing a second assessment. The Second Assessment Report (SAR) (IPCC 2 1992) was particularly concerned with radiative forcing and the relative importance of human and natural factors that give rise to radiative forcing as drivers of climate change. The Second Assessment Report runs to 63 pages and can be downloaded from the IPCC website. The IPCC completed its Second Assessment Report (SAR) in December 1995.

What is particularly interesting is that the IPCC was obviously feeling sensitive to criticism and in the Foreword to the report took the opportunity to inform the reader on how the IPCC conducts its assessments:

1. The Panel at the outset decides the content, broken down into chapters, of the report of each of its Working Groups. A writing team of three to six experts (on some rare occasions, more) is constituted for the initial drafting and subsequent revisions of a chapter. Governments and intergovernmental and non-governmental organisations are requested to nominate individuals with appropriate expertise for consideration for inclusion in the writing teams. The publication record of the nominees and other relevant information are also requested. Lists of such individuals are compiled from which the writing team is selected by the Bureau of the Working Group concerned (i.e. the Co-Chairmen and the Vice-Chairmen of the Working Group). The IPCC requires that at least one member of each writing team be from the developing world.

2. The reports are required to have a Summary for Policymakers (SPM). The SPM should reflect the state-of-the-art understanding of the subject matter and be written in a manner that is readily comprehensible to the non-specialist. Differing but scientifically or technically well-founded views should be so exposed in the reports and the SPMs, if they cannot be reconciled in the course of the assessment.

3. The writing teams draft the chapters and the material for inclusion in the SPMs. The drafts are based on literature published in peer-reviewed journals and reports of professional organisations such as the International Council of Scientific Unions, the World Meteorological Organization, the United Nations Environment Programme, the World Health Organization, and the United Nations Food and Agriculture Organization. Sometimes, the IPCC holds workshops to collect information that is otherwise not readily available; this is particularly done to encourage information-gathering on and in the developing countries.

    • Each draft chapter is sent to tens of experts worldwide for expert review. The reviewers are also chosen from nominations made by governments and organisations. The mandated time for this review is six weeks. The draft, revised in the light of the comments received, is sent to governments and organisations for their technical review. The mandated time for this (second) review is also six weeks. In some cases, the expert and government reviews are conducted simultaneously when the time factor would not permit sequential reviews.

    • The draft is revised a second time in the light of the reviews received from governments and organisations. It is then sent to governments (and organisations) one month in advance of the session of the Working Group which would consider it. The Working Group approves the SPM line by line and accepts the underlying chapters: the two together constitute the Report of the Working Group. It is not practical for the Working Group to approve its Report which usually runs to two hundred pages or more. The meaning of the term acceptance in this context is that the underlying chapters and the SPM are consistent with each other.

    • When the Working Group approves the SPM, selected members of the writing teams – from the developing as well as the developed worlds – are present and the text of the SPM is revised at the session with their concurrence. Thus, in reality, the Reports of the Working Groups are written and revised by experts and reviewed by other experts.

    • The Report of the Working Group (with the approved SPM) is sent to governments and organisations one month before the session of the IPCC which would consider it for acceptance.

    • The reader may note that the IPCC is a fully intergovernmental, scientific-technical body. All States that are Members of the United Nations and of the World Meteorological Organization are Members of the IPCC and its Working Groups. As such, governments approve the SPMs and accept the underlying chapters, which are, as stated earlier, written and revised by experts.

To illustrate this, we note that in the First Assessment 175 scientists from 25 countries contributed while a further 200 were engaged in the peer review process. For the Second Assessment 118 scientists from 22 countries contributed and a further 380 were engaged in the peer review process. For the Fourth Assessment Review, 559 contributors were selected from about 2000 nominations; for the Fifth Assessment Review, 831 contributors were selected from around 3,000 nominations; and in preparation for the Sixth Asset Review, a list of 232 contributors to the WG1 Report have now been selected.

The Second Assessment updated the original report in many matters of detail; it also spoke rather more strongly than in the earlier report in terms of the evidence of the effects of human activities on the climate. Thus, the 1990 report said "the observed increase (in global-mean air surface temperature) could be largely due to (the) natural variability (of the climate); alternatively this variability and other human factors could have offset a still larger human-induced greenhouse warming. The unequivocal detection of the enhanced greenhouse effect from observations is not likely for a decade or more." The second assessment (the 1995 report) said "The balance of evidence suggests a discernible human influence on global climate." There is a huge wealth of detail in the second assessment but it is not appropriate to consider all that detail here.

Since the 1990 IPCC Report, considerable progress had been made in attempts to distinguish between natural and anthropogenic influences on climate. This progress has been achieved by including effects of sulphate aerosols in addition to greenhouse gases, thus leading to more realistic estimates of human-induced radiative forcing. These have then been used in climate models to provide more complete simulations of the human-induced climate-change "signal." In addition, new simulations with coupled atmosphere-ocean models have provided important information about decade to century timescale natural internal climate variability. A further major area of progress is the shift of focus from studies of global-mean changes to comparisons of modelled and observed spatial and temporal patterns of climate change.

The most important results related to the issues of detection and attribution are:

- The limited available evidence from proxy climate indicators suggests that the 20th century global mean temperature is at least as warm as any other century since at least A.D. 1400. Data prior to 1400 are too sparse to allow the reliable estimation of global mean temperature.
- Assessments of the statistical significance of the observed global mean surface air temperature trend over the last century have used a variety of new estimates of natural internal and externally forced variability. These are derived from instrumental data, palaeodata, simple and complex climate models, and statistical models fitted to observations. Most of these studies have detected a significant change and show that the observed warming trend is unlikely to be entirely natural in origin.
- More convincing recent evidence for the attribution of a human effect on climate is emerging from pattern-based studies, in which the modelled climate response to combined forcing by greenhouse gases and anthropogenic sulphate aerosols is compared with observed geographical, seasonal, and vertical patterns of atmospheric temperature change. These studies show that such pattern correspondences increase with time, as one would expect, as an anthropogenic signal increases in strength. Furthermore, the probability is very low that these correspondences could occur by chance as a result of natural internal variability only. The vertical patterns of change are also inconsistent with those expected for solar and volcanic forcing.
- Our ability to quantify the human influence on global climate is currently limited because the expected signal is still emerging from the noise of natural variability, and because there are uncertainties in key factors. These include the magnitude and patterns of long-term natural variability and the time-evolving pattern of forcing by, and response to, changes in concentrations of greenhouse gases and aerosols, and land surface changes. Nevertheless, the balance of evidence suggests that there is a discernible human influence on global climate.

As far as the lengthy detailed discussions in the 3rd, 4th and 5th Reports are concerned, it would be fair to regard them as providing fine tuning of the conclusions of IPCC 1. These reports can all be downloaded from the IPCC website. The level of confidence attached to the various conclusions generally increased as time went on. Suffice it to say that there have always been those who have opposed the Reports of the IPCC and its claims to operate a consensus; we shall examine this in the next sections and study the concept of consensus in Section 7.12.

We have argued in Chapter 6 that, since the launch of Sputnik in 1957, Earth orbiting satellites have played an important role in gathering climate data and we edited a special issue of the International *Journal of Remote Sensing* (*IJRS*) on "Remote Sensing and Climate Change" that was published in February 2011 (volume 32, no. 3). This was after the IPCC's Fourth Assessment Report (AR4) and in the Preface to that Special Issue we reflected on the AR4 (Cracknell and Varotsos 2011). Remote sensing is playing an increasingly important role in climate research, and as the archives of remote sensing observations extend over longer periods of time, we are obtaining increasing evidence of trends in the climate system. Moreover, a growing array of more accurate sensors and measurements is providing more and more information about the Earth's climate parameters and providing additional data for input to climate models. A great deal of effort in climate studies goes into modelling in an effort to predict future climatic conditions. These models which have evolved from weather-forecast models involve massive computing facilities and there are a number of laboratories around the world running such models. The IPCC makes extensive use of the outputs from such models in producing its reports, which are intended to guide policy makers at national and international level, although as we shall describe below there are people who are critical of the IPCC and its very heavy reliance on modelling projections.

One can raise the question as to whether climate models are, or should be, deterministic, chaotic, or stochastic. The difference between these three situations is very simply:

- In deterministic mechanisms, one cause corresponds to one effect.
- In stochastic processes, one cause leads to more than one effect; probabilities are needed to predict one particular effect.
- In chaotic systems, the ability to predict a certain plausible result is extremely low, so low that the probability of any particular result is close to zero; i.e. there is no prediction (the sensitivity of the system is very high). In other words, chaos is when the present determines the future, but the approximate present does not define the approximate future, see also Section 1.1.3.

Models generate time series according to pre-specified rules while real data always possess a stochastic component due to omnipresent dynamical noise. In this context, Wold (1938) proved that any (stationary) time series can be decomposed into two different parts. The first (deterministic) part can be exactly described by a linear combination of its own past; the second part is a moving average component of a finite order. Hence, it may seem superfluous to ask whether a time series generated by "natural processes" is deterministic, chaotic, or stochastic.

The chaotic character of the atmospheric – oceanic – lithospheric – cryospheric – biospheric dynamics limits the reliability of climate projections. Although being of a quite different physical origin, time-series arising from chaotic systems share with those generated by stochastic processes several properties that make them almost indistinguishable: (1) a wide-band power spectrum, (2) a delta-like autocorrelation function, (3) an irregular behaviour of the measured signals, etc. (Rosso et al. 2007). Thus, the reliable prediction of global climate change or of one of its components (e.g. the ozone sphere) is impossible without consideration of the complexity of all the interactive processes.

Nowadays, the most satisfactory analysis of the climate system is that in which its underlying mechanisms can be modelled from first principles. For example, the approach by the general circulation models (GCMs) allows a deeper understanding of causes and effects, but the conclusions drawn from their simulations will be revised as these models are continuously improved; this constitutes a weakness of the approach of the IPCC as pointed out particularly strongly in chapter 2 of Lovelock's book "The Vanishing Face of Gaia A Final Warning" (2009). An alternative modelling path is given by data-driven (statistical) techniques. According to this philosophy, the evolution of the climate system is studied by recording time series, which are sets of observations collected at regular intervals of time. In this framework, Verdes (2007), using two independent driving force reconstruction techniques, showed that the combined effect of greenhouse gases and aerosol emissions has been the main external driver of global climate during the past few decades.

In 2007 Working Group I of the IPCC, based on the ground-based and remotely sensed climate variables, released the Fourth Assessment Report, which describes estimations of projected future climate change (IPCC 2007a, 2007b). As is well known, since it was set up by WMO and UNEP, with the help of other bodies, in 1988, the IPCC has been very successful in terms of raising the profile of carbon dioxide emissions and the enhanced greenhouse effect in scientific circles and making the public and politicians aware of the hazards of global warming. This was acknowledged by the award of the Nobel Peace Prize in December 2007 to the IPCC, for its relentless pursuit of the scientific evidence, digesting it and presenting it, and to Al Gore, the former Vice-President of the USA for his writings (Gore 1992, 2006) and other presentations on the subject. However, as we have pointed out previously (Cracknell and Varotsos 2007a,b, Cracknell et al. 2009 , chapter 2), there is a downside to this success due to four components. (1) The first is that it leads people to think that the emission of greenhouse gases and the consequent global warming is the major threat, or even the only threat, to the continuation of human life, or at least of our highly sophisticated society, on the planet. This is not the case, as we have already stressed in Section 1.3.12. (2) The second component of the downside to the IPCC's success is the fact that there are genuine scientific concerns about the climate models that are currently in use and on which the IPCC relies so heavily, see Section 6.13.1 which lists many of the problems still facing climate modellers. (3) The third point concerns the question of consensus and the manipulation by politicians, which is criticised very heavily by, for example, James Lovelock (Lovelock 2009, chapter 2). (4) Fourthly there have been one or two lapses from the IPCC's usual high standards in checking their sources, particularly in relation to the melting of Himalayan glaciers. We shall consider the last three of these points in a little more detail.

7.3.4.1   Concerns about the IPCC's climate models

What the IPCC does is based on the idea of slow gradual responses in the climate. However, one of the problems with the assumption of gradual change and the use of computer models to predict the future climate is the inability to predict sudden changes. Take, for example, the melting of ice (Gore 2006, Pearce 2006). It is commonly assumed that a glacier or an ice shelf just melts from absorbing radiation at its surface, i.e. rather slowly. However, in reality, cracks develop in the ice, meltwater pours into the cracks and the whole melting process accelerates and the water pressure in the cracks acts like wedges and forces the ice to break up. Spectacular situations occur in Antarctica like the break up of the Larsen B ice shelf in early 2002; this shelf was about 270 km long and 54 km wide and it suddenly broke up and floated away in fragments over a period of about one month, releasing around 500 billion tonnes of ice into the ocean. Following the melting of the ice, the water surface uncovered has a much lower albedo (reflectivity) than the ice; it therefore absorbs more heat and provides a positive feedback mechanism that enhances the global warming. A similar situation has occurred in the case of the large decrease in the area of floating ice in the Arctic in recent summers. A second example of positive feedback is associated with drought. The withering or death of plants causes a decrease of evapotranspiration and hence reduces precipitation which further increases drought.

Pearce (2006) claims that people have been overlooking positive feedback mechanisms, of which we have just mentioned two examples, and that such positive feedback can lead to sudden precipitous swings in the climate. While the models – and the IPCC's general approach – can handle a catastrophic change after the event by making adjustments to the parameters in the models, they cannot predict such sudden events. Whereas the IPCC is talking about a rise in global mean surface temperature of 3-4°C in 100 years, it could be far larger than that; it could be 10°C in as little as 10 years. The most important aspect of these problems is the potential effects of abrupt climatic change on ecology and economy since past estimates were generally based, as a rule, on the assumption of slow and gradual change. Pearce's general thesis is that nature often flips suddenly from one state to another and that therefore the consequences of climate change may be quite different from – and much more serious than – the rather simple kind of gradual changes predicted by the IPCC. However, it has recently emerged that rather than slow and gradual transitions between ice ages and inter-glacial periods, there have been many abrupt changes in the climate in the past. We noted in the discussion of the evidence from ice cores, in Section 6.11.4, for the existence of the Dansgaard-Oeschger oscillations in which there is a sudden rise in temperature over a period of a decade or so and this is followed by a slow cooling over several centuries, on average about 1,500 years. General discussions are given by Alley et al. (2002) and in the books by Cox (2005), Pearce (2006), and Lovelock (2009). While the evidence for abrupt changes is now quite clear, the mechanisms driving these changes are less clear and they are still the subject of very active research. Moreover, even if the causes of these changes were known, it seems unlikely that computer models would be able to predict sudden changes.

One problem of GCMs, both weather forecast and climate models, lies in the spacing of the grid points or the size of the grid cell that is used, e.g. 1° by 1° or 0.5° by 0.5° in the horizontal plane and 100 or 50 mb in the vertical direction. Inevitably average values of the various parameters over these grid cells have to be used, e.g. for cloud cover, land surface albedo, sea surface temperature, etc. The variations within a unit cell (due to small clouds, small parcel sizes of the land, etc.) are not able to be taken into account. Aerosols and, even more so, clouds vary spatially and temporally and are therefore notoriously difficult to take into account, see the later sections of Chapter 2.

7.3.4.2 Political manipulation

We can distinguish between two different types of consensus (Cracknell and Varotsos 2007a, b). The first is a consensus obtained from the "good and the great," i.e. relying on their status and authority, and the second is a consensus based on the assessment of the evidence by scientific workers in the field. The IPCC has represented itself as following the second of these and in consequence has been criticised by some people as being over-cautious in its conclusions. However, James Lovelock, who has developed the hypothesis or theory of Gaia (i.e. the theory that the Earth behaves like a living system), dissents from that view of the IPCC. "Do not suppose that conventional wisdom among scientists is similar to consensus among politicians or lawyers. Science is about the truth and should be wholly indifferent to fairness or political expediency" (Lovelock 2009, p. 7). Lovelock's criticism of the IPCC consensus is principally directed at the lack of wisdom among managers and politicians who forced scientists to present the conclusions of different national and regional climate centres this way. He quotes from Schneider and Lane (2005); in chapter 2, Schneider recalls in some detail "his part in a session at the UN in Geneva during the development of the IPCC Working Group II report of 2001, describing how the good science presented at the session was manipulated until it satisfied all of the national representatives present". Schneider "makes clear that the words used to express the consequences of global heating were blurred until they were acceptable to representatives from the oil-producing nations, who saw their national interests threatened by the scientific truth."

7.3.4.3 Himalayan glaciers

There was considerable interest in the media in late 2009 and early 2010 on the question of the retreat of Himalayan glaciers. This centred around the Asia chapter of the IPCC Working Group II report of 2007 which included the statement that the glaciers in the Himalayas "are receding faster than (those) in any other part of the world and, if the present rate continues, the likelihood of them disappearing by the year 2035 and perhaps sooner is very high if the Earth warms at its present rate."

This statement led to considerable controversy, the details of which need not concern us here; it was discussed fairly objectively in Science in November 2009 (Bagla 2009). The upshot was "an unprecedented apology" from the IPCC "over its flawed prediction that Himalayan glaciers were likely to disappear by 2035" (Page 2010a). Further errors related to the IPCC's report on these glaciers are cited by Page (2010b) and by Cogley et al. (2010). To set the record straight, we invited Dr Anil Kulkarni from the Space Applications Centre in India to prepare a contribution that appears in the special issue of the IJRS which we have mentioned (Kulkarni et al. 2011).

Shortly after the controversy surrounding the IPCC's false claim about the melting of Himalayan glaciers broke out, another false claim made by the IPCC was discovered and reported in the Netherlands newspaper *NRC Handelsblad* on 4 February 2010. The 2007 IPCC Report incorrectly states that 55% of the Netherlands lies below sea level, whereas, in fact, only 26% of the Netherlands is below sea level. According to *NRC Handelsblad*, the IPCC based its claim about the vulnerability of the Netherlands to rising sea level on data it received from the Netherlands environmental assessment agency PBL. But this agency admitted that it delivered incomplete wording to the IPCC. "It should have said 55 percent of the Netherlands is vulnerable to floods; 26 percent of the Netherlands is below sea level and another 29 percent can suffer when rivers flood," the PBL said in a statement after the mistake was uncovered.

It is, perhaps, understandable that an organisation that produces such voluminous reports involving so many people should occasionally get something wrong. But while errors of this magnitude do not invalidate the general message of the IPCC, they are unfortunate because they damage the credibility and the public image of the IPCC and can easily be latched onto by climate-change sceptics to discredit the whole general message of the IPCC. Some notice of these problems was taken and at a press conference held on 10 March 2010, at UN Headquarters in New York, the UN Secretary-General Ban Ki-moon, and the chairman of the IPCC, Rajendra Pachauri announced that the Inter-Academy Council (IAC), which is the umbrella organisation for various national academies of science from countries around the world, had been asked to conduct an independent review of the IPCC's processes and procedures to further strengthen the quality of the Panel's reports on climate change. The IAC reported on 30 August 2010 and recommended various measures and actions to strengthen the IPCC's processes and procedures so as to be better able to respond to future challenges and ensure the ongoing quality of its reports.

7.3.5 Predicted consequences of climate change

Some of the more direct predicted consequences of climate change are described in the later chapters of the First IPCC WG1 report of 1990 (1990 1) or in the later reports or in the more popular book by Houghton (1985).

We see from the results of the model calculations that the likely climate change will vary a great deal from place to place (e.g. in some regions precipitation will increase, in others it will decrease). Not only is there a large amount of variability in the character of the likely change, there is also variability in the sensitivity of different systems to climate change. Houghton et al. (1997) defines the terms sensitivity, adaptability, and vulnerability:

- *sensitivity* to climate change refers to the degree to which a system responds to climate change;
- *adaptability* to climate change refers to the degree to which adjustments are possible in practices, processes, or structures of systems to projected or actual changes of climate;
- *vulnerability* to climate change refers to the extent to which climate change may damage a system. It depends both on a system's sensitivity and on its ability to adapt to a given change.

Both the magnitude and the rate of climate change are important in determining the sensitivity, adaptability, and vulnerability of a system. Different ecosystems will respond very differently to changes in temperature or in precipitation.

There will be a few impacts of likely climate change (i.e. warming) which will be positive as far as humans are concerned. For instance, in parts of Siberia or northern Canada, increased temperature will tend to lengthen the growing season with the possibility of growing a wider variety of crops.

However, because over the centuries human communities have adapted their lives (and themselves) to the present climate, most changes in climate will have an adverse effect. If changes now occur rapidly a quick and costly adaptation to the new climate will need to be made by the affected community. The alternative of migrating to another region where less adaptation would be necessary becomes less and less of an option as the world becomes more and more crowded.

The questions that arise include:

- How much will sea level rise and what effect will that have?
- How much will water resources be affected?
- What will be the impact on agriculture and food supply?
- Will natural ecosystems suffer damage?
- Will human health be affected?

The answers are far from simple. One could consider the question of sea level or of water resources fairly easily in isolation if nothing else were to change. But other factors will change. Some adaptation for both ecosystems and human communities may be relatively easy to achieve; in other cases, it may be difficult, expensive, or impossible.

Global warming, enhanced or reduced precipitation, and changes in soil moisture are also not the only human-induced effects in the environment. There is the loss or impoverishment of soil, the over-extraction of groundwater, the damage because of acid rain, etc.

7.3.5.1   Sea-level rise

During the warm period before the last ice age, about 120,000 years ago, the global mean temperature was a little higher (perhaps 2°C) than at present. Average sea level was about 5 or 6 m higher than now. When the ice cover was at its maximum, about 18,000 years ago, sea level was over l00 m lower than now.

According to the IPCC First Assessment Report, the average rate of rise of sea level over the last 100 years was between 1.0 and 2.0 mm per year. This may not sound very large, but for areas that are close to sea level, they can become significant over time. Half the human population lives near the coast. Particularly vulnerable areas are places such as the Netherlands, Bangladesh, and many small low-lying islands if sea level rises by about l m by 2050 and nearly 2 m by 2100 (Davis et al., 2018; Best, 2019). The figure of 1 m is made up of 30 cm as a result of global warming and 70 cm as a result of subsidence; the 2 m is made up of 70 cm as a result of global warming and 1.2 m as a result of subsidence. It is impractical to consider full protection of the long and complicated coastline of islands. Much of the Netherlands is already below sea level and they have centuries of experience and of investment and presumably could enhance their defences to cope with a rise in sea level of 12-50 cm.

But the situation is rather different for Bangladesh, where Davis et al. (2018) studied the sea level rise (SLR) projections based on four Representative Concentration Pathway (RCP) scenarios from the IPCC's Fifth Assessment Report (IPCC, 2013). Davis et al. (2018) also examined more extreme conditions with a mean SLR of 1.5 and 2.0 metres to investigate more rapid SLR by 2100. They did not consider higher levels of mean SLR, normal high tides, or storm surges as either the likelihood of their occurrence or the long-term response of inhabitants to their effects remains unclear. They found that mean SLR will induce displacements in 33% of Bangladesh's districts under the considered RCP scenarios and 53% under the more intensive conditions (Figure 7.3).

By mid-century, it is estimated that nearly 900,000 people are likely to migrate as a result of direct inundation from mean SLR alone. Under the most conservative and extreme scenarios that

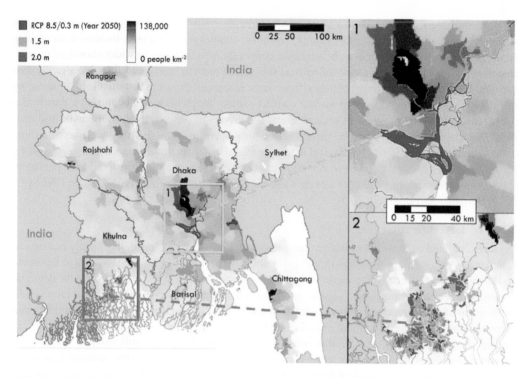

FIGURE 7.3 Inundation of mean sea level rise scenarios for Bangladesh. Scenarios presented in the map represent the smallest (RCP 8.5; year 2050) and largest (2.0m) areas inundated among the scenarios of this study. The areas inundated under the other scenarios correspond by and large to that of RCP 8.5 year 2050. Population density map is for the year 2010. (Modified from Davis et al., 2018).

we consider (0.44 m and 2 m mean SLR, respectively), the number of estimated migrants driven by direct inundation could range from 731,000 people to as many as 2.1 million people by the year 2100. In large part because of the generally high population density across Bangladesh, SLR migrants will likely not search far for an attractive destination. Indeed, the chosen destination tends to reflect a trade-off between the employment opportunities that the destination district can give (for which population serves as a proxy), its distance from the migrant's origin, and its own exposure to SLR impacts. To assess the relocation of these migrants, Davis et al. (2018) first demonstrated that their model predictions showed good agreement with bilateral census data on internal migration. This indicated that their statistical mechanics approach based on a parameter-free diffusion model was able to capture the fundamental aspects of the displacement process. After accounting for the effect of SLR, Davis et al. (2018) were able to predict the likely destinations of migrants and evaluate their additional food, housing, and employment needs. This approach therefore offers an effective and flexible alternative to empirical and agent-based models typically used to predict climate-driven migrations.

7.3.5.2 Freshwater resources

Water availability varies from 1,000 m^3 to 50,000 m^3 per head per annum in different countries. A rise in temperature means more evaporation. This may be cancelled out by increased precipitation, but in some areas, there is expected to be decreased precipitation. Thus, there could be less run-off. Changes in land use, e.g. deforestation and replacing the forest by agriculture, can have an effect on rainfall.

Unsustainable use of freshwater resources worldwide creates enormous challenges for human societies populating these natural systems, and these challenges are likely to grow with climate change. Will societies respond with increased cooperation to manage freshwater resources more sustainably or will there be more conflict over this scarce but vital resource? The research on conflict and cooperation over transboundary freshwater resources shows that, thus far, the prevailing response is cooperation, albeit non-violent conflict is quite frequent, too. The current research also documents substantial progress in understanding the drivers of water-related cooperation and conflict. Key knowledge gaps remain, particularly with respect to transboundary water conflict and cooperation in the past 10 to 15 years and in terms of local water-related events. The key prerequisite for filling these gaps is that the research community engages in a joint effort to address persistent shortcomings in existing event data sets on water cooperation and conflict (WMO, 2018; UNESCO, 2019; Bernauer and Böhmelt, 2020).

7.3.5.3 Desertification

Drylands, defined as those areas where precipitation is low and where rainfall typically consists of small, erratic, short, high-intensity storms, cover about 40 per cent of the total land area of the world and support over one-fifth of the world's population. Desertification in these drylands is the degradation of land because of decreased vegetation, reduction of available water, reduction of crop yields, and erosion of soil. It results from excessive land use generally because of increased population and increased human needs, or political or economic pressures, e.g. the need to grow cash crops to raise foreign currency. It is often triggered or intensified by a naturally occurring drought.

The rate of desertification is currently about 60,000 km^2 per year or 0.1 per cent per year of the total area of drylands. It is a potential threat to 70 per cent of these drylands (i.e. to about 25 per cent of the world's land area).

Drylands may undergo relatively rapid changes in land cover, plant communities, erosion rates, and hydroclimatic conditions that result in losses of ecosystem services and livelihoods. These socio-environmental changes are often termed "desertification," which literally means "transition to desert-like conditions." Because deserts occur in different forms and shapes, desertification is a rather ambiguous term that does not uniquely point to a specific environmental process, final outcome, or underlying driver of environmental change in drylands. Indeed, desertification can lead to, among other outcomes, a loss of vegetation cover, accelerated soil erosion, bush encroachment in xeric grasslands, displacement of perennial native grasses by exotic annuals, and an increase in soil salinity or toxicity. These changes are relatively irreversible at timescales spanning human generations. In other words, desertification is the environmental "deterioration" of drylands that leads to a persistent and, in the most extreme case, an irreversible transition to desert-like conditions (D'Odorico et al., 2019).

7.3.5.4 Agriculture and food supply

Temperature and rainfall are key factors in making decisions about what crops to grow. Thus, agriculture will need to adapt to changes in climate. This is not a serious problem for crops that mature over one or two years. It may be a problem for trees with periods of maturing of decades or even a century or so.

The case of Peruvian farmers adjusting their choice of crops according to forecasts based on the presence or absence of El Niño is cited by Houghton et al. (1997). There is also the carbon-dioxide fertilisation effect (i.e. an increased concentration of CO_2, in the atmosphere could be expected to increase the rate of photosynthesis – with other factors assumed to be unaltered – and therefore to increase the yield of a crop). But there does not seem to be much evidence on the magnitude of this effect.

There is no simple answer to "what is the effect of climate change on agriculture?" The answer is that agricultural success depends on human decisions, human management operating within the natural environment. Changing climate is one factor that will affect these decisions. However, if the changes are so dramatic that agricultural regions shift (because of sea-level rise, desertification, etc.) then human populations will need to move and we may see large numbers of environmental refugees.

Since 1990, the IPCC has produced five Assessment Reports (ARs), in which agriculture as the production of food for humans via crops and livestock have featured in one form or another. A constructed database of the ca. 2,100 cited experiments and simulations in the five ARs was analysed with respect to impacts on yields via crop type, region, and whether adaptation was included. Quantitative data on impacts and adaptation in livestock farming have been extremely scarce in the ARs. The main conclusions from impact and adaptation are that crop yields will decline, but that responses have large statistical variation. Mitigation assessments in the ARs have used both bottom-up and top-down methods but need better to link emissions and their mitigation with food production and security. Relevant policy options have become broader in later ARs and included more of the social and nonproduction aspects of food security. The recent overall conclusion is that agriculture and food security, which are two of the most central, critical, and imminent issues in climate change, have been dealt with in an unfocused and inconsistent manner between the IPCC five ARs. This is partly a result of not only agriculture spanning two IPCC working groups but also the very strong focus on projections from computer crop simulation modelling. For the future, we suggest a need to examine interactions between themes such as crop resource use efficiencies and to include all production and nonproduction aspects of food security in future roles for integrated assessment models (Porter et al., 2019).

Very recently, Bentham et al. (2020) proposed the index of overall change in food supply, which combines changes in the four scores and shows clear regional patterns (Figure 7.4). By definition, the Index of change equals to (0.46 × absolute change in animal source and sugar score) + (0.21 × absolute change in vegetable score) + (0.18 × absolute change in starchy root and fruit score) + (0.15 × absolute change in seafood and oilcrops score).

The greatest changes in food supply from 1961–1965 to 2009–2013 occurred in east and southeast Asia, especially in South Korea, China, and Taiwan, and in parts of the former Soviet Union and the Middle East. In high-income Western countries, the largest changes took place in six southern European countries (Cyprus, Portugal, Greece, Spain, Malta, and Italy), and in some high-income English-speaking countries (for example, Australia and Canada). The countries with the smallest

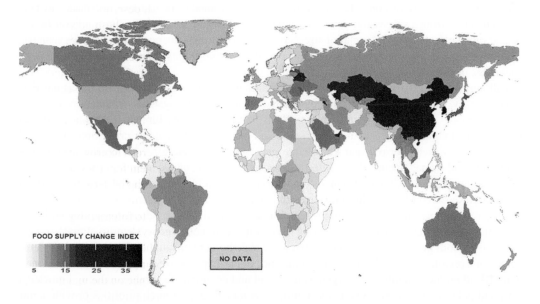

FIGURE 7.4 This index is a weighted sum of the absolute values of change in the four food supply scores. The weights are the proportion of the total variance explained by each score, normalised to add to one. No data were available for the countries shown in grey (Modified from Bentham et al., 2020).

changes in their food supply were in sub-Saharan Africa (for example, Mali, Chad, and Senegal), Latin America (for example, Argentina), and south Asia (for example, Bangladesh).

Food systems are increasingly globalised and interdependent, and diets around the world are changing. Characterisation of national food supplies and how they have changed can inform food policies that ensure national food security, support access to healthy diets, and enhance environmental sustainability. Here we analysed data for 171 countries on the availability of 18 food groups from the United Nations Food and Agriculture Organization to identify and track multidimensional food supply patterns from 1961 to 2013. Four predominant food-group combinations were identified that explained almost 90% of the cross-country variance in food supply: animal source and sugar, vegetable, starchy root and fruit, and seafood and oilcrops. South Korea, China and Taiwan experienced the largest changes in food supply over the past five decades, with animal source foods and sugar, vegetables, and seafood and oilcrops all becoming more abundant components of the food supply. In contrast, in many Western countries, the supply of animal source foods and sugar declined. Meanwhile, there was remarkably little change in the food supply in countries in the sub-Saharan Africa region. These changes led to a partial global convergence in the national supply of animal source foods and sugar, and a divergence in those of vegetables and of seafood and oilcrops. Our analysis generated a novel characterisation of food supply that highlights the interdependence of multiple food types in national food systems. A better understanding of how these patterns have evolved and will continue to change is needed to support the delivery of healthy and sustainable food system policies.

7.3.5.5   Natural ecosystems

Over history, natural systems have obviously adapted to climate change. The problem is that we are now looking at quite rapid changes of climate. Changes in temperature that previously occurred over thousands of years are now occurring over decades and many ecosystems cannot change or migrate that fast. Trees, especially, are long-lived and take a long time to reproduce. They are also surprisingly sensitive to the average climate in which they develop. For the likely changes in climate in the next few decades, a substantial proportion of existing trees will find themselves subject to unsuitable climatic conditions.

There has been a decline in the health of many forests in recent years and this has often been attributed to acid rain. While that is certainly a contributing factor, it now seems that climatic changes are also partly responsible. In several regions of Canada, for instance, die-back has been related to stress from a succession of warmer winters and drier summers. Sometimes, of course, it is the combined effect of pollution and of stress as a result of climate change that causes the damage.

As to marine ecosystems, while it is clear that the climate is important, there is little detailed knowledge of the effects involved.

Williams et al., (2020) stressed the point that climate change poses significant emerging risks to biodiversity, ecosystem function, and associated socioecological systems. Adaptation responses must be initiated in parallel with mitigation efforts, but resources are limited. As climate risks are not distributed equally across taxa, ecosystems, and processes, strategic prioritisation of research that addresses stakeholder-relevant knowledge gaps will accelerate effective uptake into adaptation policy and management action. After a decade of climate change adaptation research within the Australian National Climate Change Adaptation Research Facility, we synthesise the National Adaptation Research Plans for marine, terrestrial, and freshwater ecosystems. We identify the key, globally relevant priorities for ongoing research relevant to informing adaptation policy and environmental management aimed at maximising the resilience of natural ecosystems to climate change. Informed by both global literature and an extensive stakeholder consultation across all ecosystems, sectors, and regions in Australia, involving thousands of participants, we suggest 18 priority research topics based on their significance, urgency, technical and economic feasibility, existing knowledge gaps, and potential for co-benefits across multiple sectors. These research priorities provide a unified guide for policymakers, funding organisations, and researchers to strategically direct resources, maximise stakeholder uptake of resulting knowledge, and minimise the impacts of climate change on natural ecosystems.

7.3.5.6 Impact on human health

The direct effect of global warming can be handled relatively easily. On the other hand, deaths as a consequence of heat stress in times of extremely high temperatures or of hypothermia in times of extremely low temperature (since the amplitude of fluctuations about the mean seem set to increase) are likely to increase.

There are less direct effects as well – pollution of the atmosphere, pollution of the water supplies, poor soil (leading to poor crops and inadequate nutrition), droughts and floods, enhanced activity of carriers of diseases, etc.

Recently, Scovronick et al., (2019) investigated the distribution of health co-benefits by region using the full Regionalized Integrated Climate Economy + Aerosol Impacts and Responses (RICE+AIR) model and the results obtained are displayed in Figure 7.5(a). In line with recent scenarios, many of the co-benefits accrue in India and China in early periods, attributable to their large populations and high capacity for mitigation-related reductions in $PM_{2.5}$. China's benefits decline by mid-century due to relatively rapid economic development and a stabilising population – which both act to constrain emissions – whereas those in India persist and are the major driver of increased decarbonisation relative to the reference case (see below for a sensitivity test which corroborates India's importance). Towards the end of the century, sub-Saharan Africa replaces China as the second-largest beneficiary, as air pollution remains problematic due to lagging economic

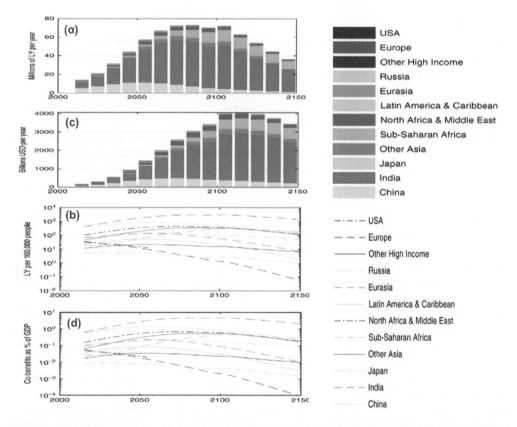

FIGURE 7.5 Health benefits of carbon mitigation. Life-years gained (a) overall and (b) per 100,000 population by region from the air quality improvements associated with the optimal decarbonisation in RICE+AIR. (c), (d) show the resulting monetised benefits in total, and as a percent of GDP, respectively. Note that if a region's $PM_{2.5}$ exposure (concentration) drops below 5.8 $\mu g/m^3$, health benefits no longer accrue – this threshold assumption is common in other global air quality assessments and is tested in the sensitivity analyses (Modified from (Scovronick et al., (2019).

development accompanied by the world's largest population. Other regions also stand to benefit, including less populous regions that show important benefits per capita and/or per Gross Domestic Product (GDP) (Figure 7.5(b)).

In the results presented above, the averted premature mortality from aerosol co-reductions produces annual monetised benefits in the hundreds of billions of dollars over the next few decades, rising to several trillion annually at the end of the century (Figure 7.5(c)). To derive these numbers, we multiply, for each region-time pair, the total life-years gained by two years of per capita consumption. This approach produces different life-year monetisations for each region, which leads to the slight change in composition of monetised benefits (compare Figure 7.5(a), (c)). However, this does not imply that we assign life-years in poorer regions less value in the objective function, because the optimisation accounts for the diminishing marginal utility of consumption through a concave relationship between wealth and well-being.

In conclusion, Scovronick et al., (2019) found that when both co-benefits and co-harms are taken fully into account, optimal climate policy results in immediate net benefits globally, which overturns previous findings from cost-benefit models that omit these effects. The global health benefits from climate policy could reach trillions of dollars annually, but their magnitude will importantly depend on the air quality policies that nations adopt independently of climate change. Depending on how society values better health, we show that economically optimal levels of mitigation may be consistent with a target of 2 °C or lower.

7.3.5.7   Costs

Attempts have been made to estimate the cost of global warming and the first impression is that we might be able to buy our way out of the problem.

However, there are two factors that are important:

1. we have only tended to look at the next 50 years or so and with the hope that greenhouse-gas emissions will not go too far out of control:
2. the effects are likely to be very serious for many people currently living at subsistence level in developing countries. If as a result of desertification or sea-level rise their land becomes uninhabitable, they will wish to migrate and become environmental refugees. Houghton (1997) quotes possible numbers of 3 million per year or 150 million before 2050. Where will they go?

It is a truism that worldwide, landslides incur catastrophic and significant economic and human losses. Previous studies have characterised the patterns in landslides' fatalities, from all kinds of triggering causes, at a continental or global scale, but they were based on data from periods of <10 years. The recent research presented hypotheses that climate change associated with extreme rainfall and population distribution is contributing to a higher number of deadly landslides worldwide. The recent study by Haque et al., (2019) maps and identified deadly landslides in 128 countries and it encompasses their role, for a 20 years' period from January/1995 to December/2014, considered representative for establishing a relationship between landslides and their meteorological triggers. A database of geo-referenced landslides, their date, and casualties' information, duly validated, was implemented. A hot spot analysis for the daily record of landslide locations was performed, as well as a percentile-based approach to evaluate the trend of extreme rainfall events for each occurrence. The relationship between casualty, population distribution, and rainfall was also evaluated. For 20 years, 3876 landslides caused a total of 163,658 deaths and 11,689 injuries globally. They occurred most frequently between June and December in the Northern Hemisphere, and between December and February in the Southern Hemisphere. A significant global rise in the number of deadly landslides and hotspots across the studied period was observed. Analysis of daily rainfall confirmed that more than half of the events were in areas exposed to the risk of extreme rainfall. The relationships established between extreme rainfall, population distribution, seasonality, and landslides provide a useful basis for efforts to model the adverse impacts of extreme rainfall due to climate change and human activities and thus contribute towards a more resilient society.

7.3.5.8 Consensus and Validation

What we have been talking about is largely based on the IPCC WG1 report and we have mentioned that this report was based on a consensus of the majority of the world's climatologists. But:

a) this does not mean that it is right – there is an alternative view expressed in Section 7.9 and 7.12 (see also Gribbin 1990);
b) it is still necessary to do what we can to validate the model predictions.

Let us consider briefly the question of validation. If climate models are to be used to make predictions of climatic change, whether natural change or change induced as the result of human activities, then it is obviously important to attempt to assess the accuracy of the results produced by models. To carry out the validation of climate models, we need appropriate observed historical climatic data. Climate models can be tested by running them to predict historical climatic conditions from an earlier historical starting point and some considerable success has been obtained; undoubtedly there has been some fine tuning of the models in this work, but that is fair enough.

A long-term commitment is now needed by the world's national weather services to monitor climate variations and change. Changes and improvements in observational networks should be introduced in a way which will lead to continuous, consistent long-term data sets sufficiently accurate to document changes and variations in climate. Most sets of satellite observations are not yet of long enough duration to document climate variations. In order for these data to be most useful, it is very important that they can be analysed and interpreted with existing in situ data. High priority should be given to the blending or integration of space-based data and in situ data sets in such a way as to build upon the strengths of each type of data. We are beginning to reach the stage at which calibrated long-term data sets from remote sensing (Earth observation) satellites may be considered useful in terms of seeking evidence of climate change, whether regionally or globally. There are some examples of ongoing projects that have used mixed satellite data sets, careful cross-calibration, and co-ordinated data processing to produce global data sets; these include the World Climate Research Programme's (WCRP) International Satellite Cloud Climatology Project (ISCCP), the Global Precipitation Climatology Project (GPCP), and the NASA Pathfinder Program.

Quantifying the consensus on anthropogenic climate change and its communication have become a controversial research subject in recent years. Jankó et al. (2020) utilised a reference list from a climate sceptic report and a previously published quantitative method of consensus research to revisit the theoretical and methodological questions. Beyond rating the abstracts according to their position on anthropogenic global warming (AGW), Jankó et al. (2020) classified the strategic in-text functions of the references. Results not only showed the biased character of the literature set but also revealed a remarkable AGW endorsement level among journal articles that took a position concerning AGW. However, Jankó et al. (2020) do not argue for modified consensus numbers but instead emphasise the role of "no position" abstracts and the role of rhetoric. Our quantitative results provided evidence that abstract rating is a suboptimal way to measure consensus. Rhetoric is far more important than it appears at first glance. It is important at the level of scientists, who prefer neutral language, and at the level of readers such as report editors, who encounter and re-interpret the texts. Hence, disagreement appears to stem from the disparate understanding and rhetorically supported interpretation of the research results. Neutral abstracts and papers seem to provide more room for interpretation.

7.4 THE CONCEPT OF GLOBAL ECOLOGY

In this section, we address the question of global ecology which was pioneered particularly by the two Russian Academicians K.Ya. Kondratyev and M.I. Budyko and largely ignored otherwise. Kirill Ya Kondratyev is described by Prof. V.K. Donchenko (2008) as the "the Founder of Global Ecology," but the title ought to be shared with M.I. Budyko. Kondratayev in particular had very definite views on the IPCC and its Reports. We include a few details of their backgrounds.

Kondratyev was born on 14 June 1920, in the town of Rybinsk, about 300 km north east of Moscow, went to school in Leningrad (now once again St Petersburg) and in 1938 he was admitted to the Faculty of Physics at the Leningrad State University. His study at the University was interrupted by World War II in 1941. He was drafted to the Red Army and after a third wound, which was found to be a serious one, he was discharged from the army. In 1944, when the Leningrad blockade was raised, he resumed his studies at the University. In 1946, Kondratyev graduated from the Department of Atmospheric Physics and, as a gifted graduate, was offered a job in the same Department.

During the first 30 years of his academic career, Kondratyev worked at Leningrad University (later renamed as the St Petersburg State University), where he developed into a scientist, educator, and science organiser, while progressing along the path from an assistant, professor, and head of the Department of Atmospheric Physics to the rector, i.e. the head of the University. From 1958 to 1981, he was Head of the Department of Radiation Studies at the Main Geophysical Observatory (GGO) (in Leningrad), and from 1982 to 1992, he was the Head of the Remote Sensing Laboratory at the Institute for Lake Research.

Academician Kirill Ya. Kondratyev (1920–2006) was a full member of the Russian Academy of Sciences, prominent scientist, and outstanding geophysicist; a world renowned expert in research of solar radiation, satellite meteorology, remote sensing of atmosphere and earth surface, global environmental change, and climate and ecology dynamics; an author (and coauthor) of more than 1260 papers published in the leading domestic and foreign magazines, and more than 120 monographs published both in Russia and abroad. He was a Counsellor of the Russian Academy of Sciences and he joined the Research Centre for Ecological Safety of the Russian Academy of Sciences in Saint-Petersburg from 1992. The Centre of Ecological Safety was established within the Russian Academy of Sciences in order to carry out interdisciplinary research aimed at trying to understand the large body of information concerning the environment when it is exposed to technological and human activities. Ecological safety is an interdisciplinary area of knowledge. The Centre's activities included theoretical work and field experiments on numerous issues pertaining to ecological safety. He helped to create the Nansen International Environmental and Remote Sensing Centre (NIERSC) in Saint-Petersburg. Kirill Kondratyev died on 1 May 2006.

A very valuable source of information about Kirill Kondratyev and his work will be found in an interview published in an issue of the WMO Bulletin (WMO 1998). Kondratyev's 60-year academic, scientific, and organisational activity was multifarious and extremely fruitful and it is described in English in some detail by Buznikov (2008). It has also been described in English in chapter 1 of our book on Global Climatology and Ecodynamics (Cracknell, 2009; Cracknell et al., 2009).

Budyko was born on 28 July 1920 in Guomel, Byelorussian SSSR. Budyko spent much of his academic career in Leningrad (St Petersburg). He studied at Leningrad Polytechnic Institute graduating from the Engineering and Physics Faculty in 1942. He then moved to the Voeikov Main Geophysical Observatory (GGO) where he worked on his postgraduate Candidate of Science (approximately equivalent to a PhD) on military meteorology, which he subsequently defended in 1944 at the relatively early age of 24. He pioneered studies on global climate and calculated the temperature of the Earth considering a simple physical model of equilibrium in which the incoming solar radiation absorbed by the Earth's system is balanced by the energy re-radiated to space as thermal energy. Budyko's groundbreaking book, *Heat Balance of the Earth's Surface*, published in 1956, transformed climatology from a qualitative into a quantitative physical science. These new physical methods based on heat balance were quickly adopted by climatologists around the world. In 1963, Budyko directed the compilation of an atlas illustrating the components of the Earth's heat balance. From this point on, his work and intellectual interests became more focused around inter-disciplinary concerns of the relationship between the lower atmosphere and natural processes at the Earth's surface. This interest provided the basis for his later initiatives regarding the global climate system and the Earth's biosphere. In terms of his overall career, Budyko remained at the GGO for the next 30 years, becoming assistant director in 1951 and director in 1954, a position he would hold until 1972. From the mid-1970s, he took up a new position leading work on climate change at the State Hydrological Institute. And, from 1989 he became involved with the Research Centre for Interdisciplinary Cooperation, Academy of Sciences (INENKO). Mikhail Budyko's contributions to global climate science are described in a review by Oldfield (2016). Budyko died on 10 December 2001. Both men, Budyko and Kondratyev, were widely honoured and became full members of the Russian Academy of Sciences in 1992.

Kondratyev's early work was concerned with the transfer of radiation in the atmosphere. Many of Kondratyev's studies were devoted to the development of the theory of radiative infrared (thermal) transfer in the atmosphere which has important applications not only in atmospheric physics, but also in practical problems. He also performed numerical modelling of atmospheric absorption spectra as well as some numerical modelling of the greenhouse effect for various planetary atmospheres (Earth, Mars, Venus, Jupiter, Saturn, and Titan).

Following the launch of Sputnik on 4 October 1957, there was a rapid development of polar-orbiting and geostationary meteorological satellites; see Section 5.2, and Kondratyev and his co-workers rapidly moved into the field of satellite meteorology, or the remote sensing of the atmosphere. A curious fact is that in 1958 Kondratyev published in the Finnish journal "Arkhimedes" the very first scientific paper on satellite observations of the upper atmosphere

A significant part of his work was connected with the development of satellite meteorology and environmental observations. In particular, the principles of the interpretation of meteorological satellite data were developed. One significant difference between Earth observation (remote sensing) from space in the former USSR and in the West was the greater involvement of manned Earth-orbiting space missions in the former USSR. A publication by Kondratyev (1965) contained for the first time a discussion of the meteorological importance of the Earth's pictures obtained by Soviet cosmonauts. The original results of later studies in the field of remote sensing from manned spacecraft have been discussed in the monograph by Kondratyev (1972a). The available observational data were used to study the basic factors of the Earth's radiation balance and to determine the net radiation of the Earth as a planet. Kondratyev was the first scientist to propose and substantiate a statistical approach to the analysis of satellite measurement results on the Earth radiation budget. The results of these studies have been summarised in various monographs (Kondratyev, 1961, 1969, 1972b, 1983; Kondratyev and Timofeyev, 1970). All these books have been published in English as NASA Technical Translations.

A number of years were spent in designing and manufacturing balloon instrumentation (solar spectrometers, pyrheliometers, pyranometers, aerosol impactors and filters, etc.) and this made it possible to conduct during the 1960s a series of 22 high-altitude large balloon (about 800 kg weight) flights launched from a test site in the middle Volga River region. The principal purpose was to obtain data on vertical profiles (up to 30-33 km) of the spectral transparency of the atmosphere, total direct solar radiation, and downward and upward shortwave radiation fluxes with simultaneous information on aerosol properties (number concentration, size distribution, chemical composition, i.e. complex refractive index). This work was backed up by an IL-18 aircraft flying laboratory which was packed with instruments which performed three main functions:

(a) to test prototypes of satellite instrumentation;
(b) to test and apply remote sensing instrumentation (mainly for the microwave wavelength region); and

(c) to investigate radiation processes in the free atmosphere which are responsible for climate formation (first of all, from the viewpoint of aerosol and cloud impact on climate).

Kondratyev's work on clouds and atmospheric greenhouse gases necessarily led him into the study of climate change (Kondratyev and Binenko, 1981, 1984; Kondratyev and Zhvalev, 1981; Kondratyev, 2001). However, he was an active advocate of the "multidimensional global change" principle which implied an analysis of the interaction between dynamical processes going on in society and the environment.

A very productive stage was the participation in various activities of the World Meteorological Organization (WMO, 1998), especially its Advisory Committee (in the 1960s) which was also responsible for the development of the Global Atmospheric Research Programme (GARP) as a precursor of WCRP (World Climate Research Programme) (GARP-climate). The memorable events were receiving the WMO gold medal and the delivery of a lecture for the WMO Congress with the subsequent publication of the WMO monograph (Kondratyev, 1972a). A similar honour was obtained later from the International Astronautical Federation; in the 1960s, Kondratyev initiated the organisation of the IAF Committee on Application Satellites which functioned successfully for more than a decade. A significant part of his international efforts was connected with various activities of the IAMAP International Radiation Commission, where Kondratyev served as a member for a long time and during the period 1964–68 he was a President. A notable event in the International Radiation Commission's history was the International Radiation Symposium in 1964 in Leningrad; similar events were the COSPAR Symposium in 1970 in Leningrad and the IAF Congress in 1974 in Baku, Azerbaidjan.

We have already mentioned in Section 7.2.3 the bilateral Soviet-American agreements on environmental cooperation and space research and the environmental agreement signed in 1972 survived successfully for more than 20 years of the cold war era. Kondratyev's participation was connected with the Working group on climate studies and included a number of joint Soviet-American expeditions in the USSR and USA (see above). Later on, Kondratyev served as co-chairman (with Dr. S. Tilford from NASA as another co-chairman) of the Soviet-American Working Group on Remote Sensing during the five-year period 1988–1993. A rather broad cooperative programme included studies of the Kamchatka volcanoes, remote sensing of Siberian forests, preparations to install American TOMS ozone instrumentation on board the Russian meteorological satellite Meteor-3M and the accomplishment of this task in 1992, preparations of an international Earth's resources module for the space station Mir (the module was launched in 1996).

In the later decades, Kondratyev concentrated his efforts largely on global change issues, including the issue of global climate change. He emphasised the clarification of complex interactivity and the multiple-scale nature of processes in the "nature-society" system, as well as on nonlinearity of dynamics in this system. A new area of knowledge was generated – global ecology. We start with the Oxford Dictionary definition of ecology "the branch of biology concerned with the relation of organisms to one another and to their physical surroundings." Fundamentals of this knowledge came from the pen of Kondratev at high speed. Global ecology was created by the labour of an international team, where Kondratyev was the leader. The late 1990s and the early 2000s were enormously fruitful years for Kondratyev. Owing to his creative energy, global ecology has grown into an independent area of knowledge that has been recognised throughout the world. In the monographs published by him independently and as a co-author, Kondratyev provided rationales for major trends in modern thermodynamics and made an attempt to determine a plan for the development of terrestrial civilisation in the third millennium. He stated key aspects for this development. The major aspect determining the meaning of life of every human being is preservation and continuation of life on Earth. This aspect is exactly the goal of sustainable ecologically safe, social, and economic development. That was the goal for the sake of which Kondratyev lived and worked.

An important strand to his work in later years was his concern that people had become obsessed with global warming and climate change and that not enough attention was being paid to various other changes that are being brought about by human activities and which threaten various eco-logical systems and the viability of the future standards of living – and indeed the very contin-ued existence – of human life (Grigoryev and Kondratyev, 2001a, 2001b, 2001c; Kondratyev and Grigoryev, 2002; Kondratyev et al., 2006).

At the beginning of the 1970s, when the Club of Rome was developing its programme of study-ing global change, Kondratyev organised regular seminars to discuss relevant problems. Steadily independent research efforts were being pursued with the purpose of determining key issues of global change and the requirements for observations. A cornerstone aspect was the development by Gorshkov (1990) of a concept of biotic regulation of the environment. These efforts resulted in two books (Kondratyev, 1989, 1990a). An important aim of these writings was an analysis of concep-tual issues of such international programmes as the World Climate Research Programme (WCRP), the International Geosphere-Biosphere Programme (IGBP), as well as the outcome of the Second U.N. Conference on Environment and Development (UNCED) in the context of the concept of the biotic regulation of the environment (Kondratyev, 1982). An important step was the completion of the monographs (Kondratyev, 1998; Kondratyev et al., 1997, 2003b, 2003c, 2005, Krapivin and Kondratyev, 2002) of which the principal aim was an analysis of the interaction between societal and environmental dynamics. Special emphasis was placed on the analysis of the role and place of global climate change studies in the context of global change; this was necessary in the light of certain overemphasis in UNCED documents and Intergovernmental Panel on Climate Change (IPCC) reports on climate change and greenhouse gases reduction. Another conceptual aspect is connected with the problem of optimising global environmental observing systems of combined conventional and satellite observations (Kondratyev et al., 1996, 2002a; Marchuk and Kondratyev, 1992; Kondratyev and Krapivin, 2004; Kondratyev and Galindo, 2001; Kondratyev and Moskalenko, 1984; Kondratyev and Cracknell, 1998a,b).

As he noted in his interview (WMO Bulletin Interview, (WMO, 1988), the critical objective of global change research is to examine the interaction of dynamic processes proceeding in the society and environment. As early as 1970, he had organised a seminar within the Club of Rome operation, which formulated the key problems involved in global change, and the demands placed in this con-nection on observations and exploration. He provided active support to the concept of biological regulation of the environment, which had been developed by V.G. Gorshkov. Kondratyev wrote that, whenever it comes to global carbon circulation, everybody becomes concerned about an increase in carbon dioxide concentration in the atmosphere implying the most disastrous scenarios of fur-ther development. However, the biosphere assimilates a huge amount of carbon dioxide emitted to atmosphere, which is to ensure ecological safety in the future. If we destroy the biosphere, which is acting as a reservoir for carbon, we shall actually have an ecological catastrophe. Kondratyev has always emphasised that the global change problem cannot be solved without applying a systems approach to allow for all existing processes. Such research shall be carried out in the context of a general problem.

The main point that we would like to make in this Section is to emphasise that the study of anthro-pogenic causes of climate change owes many of its origins to the lifelong work of the great Soviet and Russian scientist Academician Prof. Dr Kirill Yakovlevich Kondratyev. His work provided the initial stimulus for much of the work that is described later in this chapter. He was responsible for the development of various important national and international research programmes in meteorol-ogy and atmospheric physics. In the field of satellite meteorology, he made remarkable efforts in connection with environmental observations and the interpretation of data. He was the first scientist to propose and substantiate a statistical approach to the analysis of satellite measurements of the Earth's radiation budget. In the field of climate change, he was a fervent advocate of the principle of "multidimensional global change" (Kondratyev et al., 2003a, 2004a, 2004b), which aims at an analysis of the interaction between societal and environmental dynamics.

There are three main stages in the development of Budyko's work related to climate systems and global ecology (late 1940s-mid 1980s). The first period encompasses his early efforts devoted to understanding and quantifying the interrelationship between the lower atmosphere and the Earth's surface. This stage of his career was also characterised by a growing interest in regional- and global-scale processes and was underpinned by collaborative work involving climatologists, physical geographers, and other cognate scientists. The second stage highlights the broadening of his global interest in order to engage more deeply with both natural and anthropogenic climatic and environmental change.

A further intellectual strand evident in the work of Budyko, and which emerged particularly strongly during the course of the 1970s and 1980s, is associated with the activities of the pedologist V.V. Dokuchaev and his student the biocheohemist V.I. Vernadsky. Both Dokuchaev and Vernadsky were interested, among other things, in the relationship between the physical environment and living matter. For example, Dokuchaev famously highlighted the intimate connections between soil formation and a range of "soil-forming" factors of which climate was central, noting the consequent latitudinal regularity of soil type and natural regions more generally. Vernadsky developed this general theme through his work on the biosphere in order to advance an understanding of the various ways in which living matter, understood as the totality of all life on Earth acting in unison, shaped the physical world. The later work of Vernadsky acknowledged the growing ability of collective human activity to influence the state of the biosphere. Vernadsky utilised the concept of the noosphere in order to try and capture the nature of this trans-formative process, identifying human intellectual and scientific abilities as key driving forces behind the noted changes. This combination of environmental concern allied to a belief in the ability of society to address the emergent issues via a sound understanding of the physical processes at work would find a strong echo in Budyko's later publications.

Budyko's early years at the Leningrad Polytechnic Institute and the GGO were influenced by the twin pressures of war and late Stalinism, both of which encouraged an applied focus to his work. During the immediate postwar period, the situation for Soviet scientists changed rapidly. The relative openness of the war years was replaced by a period of marked uncertainty framed by a renewed emphasis on ideological conformity led by Stalin's lieutenant Andrei Zhdanov. As a result of this, during the late 1940s and early 1950s, an emphasis was placed on indigenous science as well as the need for applied science that could contribute to the advancement of Soviet society. This phase of Soviet history was a traumatic one for many scientists across a range of disciplines. However, in general, it would appear that Budyko was able to navigate the difficulties, at least with respect to his scientific endeavours, relatively successfully. This can be attributed to the aforementioned applied nature of his work and perhaps also to his intellectual links to pre-revolutionary Russian scientists such as Voeikov and Dokuchaev.

At the heart of Budyko's applied work was his interest in understanding the interrelationship between the lower levels of the atmosphere and the Earth's surface. His first two major monographs reflected this emphasis. The first, entitled "Evaporation under natural conditions," was published by Budyko (1948) when he was just 28 years old. The second was his highly regarded 1956 publication "Heat Balance at the Earth's Surface." The 1948 monograph aimed "to work out physical methods for calculating evaporation rates from the Earth's surface" and simultaneously address the gap that existed between the hydrological and meteorological sciences. In retrospect, the monograph was deemed to be an innovative contribution to work in this area. Among other things, it helped to develop general laws for such phenomena as the rate of evaporation from soil, insight which would lay the basis for his later more developed work on the heat and water balance. His 1956 publication was the culmination of several years of collaborative work with colleagues from the GGO as well as individuals such as A.A. Grigor'ev at the institute of Geography. Interest in the heat balance at the earth's surface has a long history in both Russia and the West. For example, writing in 1965, Miller (1965) noted the earlier work of Voeikov as well as that of the 19th century American environmentalist George Perkins Marsh (1801–1882) in this regard. Both scientists had drawn attention to the potential importance of determining the energy budget at the Earth's surface in recognition of the fundamental importance of heat energy and associated transformation processes for the understanding, of the climate system. While both scientists were able to articulate the problem, they lacked the necessary data and methods with which to advance insight into the issue. These lacunae were addressed during the course of the early to mid 20th century. Budyko was certainly aware of this intellectual heritage and in reviewing work in the field he made explicit reference to the pioneering efforts of Voeikov. Furthermore, Budyko, as well as his colleagues at the GGO involved in heat balance work, was also abreast of relevant work carried out in the West during the first half of the 20th century.

At the beginning of his 1956 publication, Budyko outlined the significance of the work from his perspective. Investigations of the heat balance at the Earth's surface are now occupying an important place in all hydrometeorological disciplines.... The main purpose of these investigations is the study of the causal principles which determine the meteorological and hydrological regimes in various geographical regions which could be used for the forecasting and calculation of important hydrometeorological processes and phenomena. (Budyko, p. 3).

This work garnered favourable attention for Budyko both within the Soviet Union and beyond. Aided by a relatively rapid translation into English, much of this positive reception was linked to its emphasis on a quantitative approach and accompanying efforts to understand global-level characteristics. For example, writing in the preface to Budyko's English-language edition of "Climate and Life," the US climatologist, David Miller (1974) noted:

> In many areas of geophysics in North America (including climatology, hydrology, and meteorology) energy mass budget work then under way was powerfully strengthened by the methods and the global-scale data published in the English translation of 'The Heat Balance of the Earth's Surface,' Washington, 1958. It has since repeatedly served as the base level on which further investigations have built – some at the micro- and mesoscales, others at the world scale.

From the early to mid-1960s, Budyko began to develop, further lines of work drawing from his energy budget research and linked to understandings of global climate change and the role of humankind in such changes. This included, among other things, a number of focused papers on historical climate change, the relationship between polar ice and the global climate, and the relative instability of the current global climate regime. In terms of the causal mechanisms behind historical shifts in global climate regimes, and particularly those in the recent past, he drew attention to natural factors including fluctuations in solar radiation as well as changes in levels of atmospheric transparency linked to natural events such as volcanic eruptions. With respect to human influence, Budyko reflected on the extent to which humankind's growing technological capabilities and collective ability to produce large quantities of heat promised to play a role in modifying local and regional climates in the near future. In order to gain a feel for his general work as it developed during the 1960s, it is instructive to reflect on the content of his concise 1969 semi popular publication entitled Climate Change (Izmeneniya ku-nara). It was divided into four main sections covering: contemporary climate change, quaternary glaciations, pre-quaternary climates, and climates of the future. He opened with a consideration of the lessons to be drawn from empirical work on climates of the past and in doing so noted the relative instability of the earth's climate over both long- and short- (last 100 years) time periods. In considering climate change evident during the late 19th and 20th centuries, and particularly within the mid and high latitudes, Budyko (1969) made reference to theories of varying atmospheric transparency in addition to changing levels of carbon dioxide, solar fluctuations, and so on, while simultaneously noting the difficulties of verifying such hypotheses. Budyko then moved on to consider the links between climate and glaciation processes. He rehearsed his earlier findings in which he concluded that full glaciation of the earth did not necessarily require a significant change in the current level of incoming solar radiation. With respect to future climates, Budyko opened by acknowledging the need to take into account both natural and anthropogenic influences. He then highlighted humankind's varied and growing influence on climate regimes through construction activities as well as ameliorative work such as afforestation, before moving on to focus on the possible consequences of a marked increase in artificial energy and heat production for meteorological processes. Budyko ended by highlighting once again the importance of humankind's emergence as a major climate player:

> "...the contemporary stage of biological evolution, linked with the appearance of humankind, is a significant factor for the future development of climate in so far as the activity of humankind opens up the prospect of substantial climate change in the near future. Thus, in our time, natural changes of climate are to be gradually replaced by changes created and regulated by humankind"
>
> (Budyko 1969)

The third stage reflects on the development of his expansive and evolutionary approach to the biosphere, and his insight into the formative role of climate with respect to the functioning of physical and biological processes. Furthermore, this later work also exhibited a strong belief in the ability of humankind to reflect wisely on its growing influence on the physical environment and respond appropriately.

There has been limited attention devoted to the work and activities of Soviet climatologists in the English-language literature. Budyko and Kondratyev are something of an exception in this regard due to the translation of a number of their texts into English. Nevertheless, Budyko and his compatriots remain subdued voices within the broader published work concerning the science of global climate change.

As noted, Budyko's name is relatively well-known within the Anglo-US context due to the significant number of his monographs, edited books, and papers translated into the English language. Furthermore, he featured on the international stage through conference attendance and his participation in the activities of organisations such as the WMO. Budyko was involved with a joint US-USSR Agreement on Cooperation in the field of Environmental Protection which emerged in 1972 and he also provided input to the scientific work of the Intergovernmental Panel on Climate Change (IPCC).

The Soviet period witnessed a significant physical expansion and reorganisation of meteorological and climatological institutions and activities following the disruption of war and revolution in the early part of the 20th century. Understandably, Soviet assessments of such developments post-1917 emphasised the energising influence of the revolutionary endeavour. At the same time, there was certainly some truth in the hyperbole. The fact that weather and climate concerns were intimately linked to broader economic considerations, and most obviously in the case of agricultural output, ensured state resources were channelled towards improving the existing system. For example, the Main Observatory was able to focus more of its attention on research activities during the 1920s and 1930s as administrative functions were absorbed by newly created entities such as the Hydrometeorological Committee and Central Weather Bureau. In general, the early- to mid-Soviet period witnessed the expansion of the physical monitoring system and the focused development of meteorological and hydrometeorological activities. Indeed, a relatively significant number of new research centres was established from the 1930s onwards.

During the course of the 1970s and 1990s, Budyko published a series of more developed pieces of work, many of which were subsequently translated into English. Common themes underpinning these publications included an interest in contemporary climate change (both natural and anthropogenic), and an effort to place climate change within a broader framework of global ecological change. In his 1973 publication "Atmospheric carbon dioxide and climate," there is an attempt to consider the extent to which relatively small changes in atmospheric carbon dioxide levels might trigger significant shifts in global climate regimes (Budyko 1973). In order to provide some sense of the main features of his work during this period, we focus on three influential publications. First, in 1971, he published "Climate and Life," which was translated into English in 1974. This book explored the relationship between climate and a range of natural processes. For David Miller, writing in the preface to the English-language edition, the book's value resided in two specific areas, as an authoritative statement of the new concepts of climatic analysis that are based on "the laws of conservation of mass and of energy…" and in its application of "energy-budget concepts to important questions in the biology of the planet" Miller (1978, pp. vii). The book represented an attempt to imbue the relationship between climate and a range of global physical and biological processes with greater precision. Budyko reflected on humankind's growing influence on climate formation through energy production and pollution emissions (and this included carbon dioxide production) picking up on a range of themes advanced in earlier publications.

Budyko's 1977 (English edition 1980) publication *Global Ecology* (Global'naya ekologiya) engaged purposefully with what he termed the "distinct scientific discipline" of global ecology. Writing in the preface, he noted the relative immaturity of the discipline and yet underlined its importance for addressing the growing influence of humankind on natural processes. For Budyko, the focus on global ecology was a natural progression of his earlier work on energy flows at the Earth's surface and developed many of the themes advanced in his monograph Climate and Life. The approaches and insights of both Dokuchaev (soil and natural historical zones) and Vernadsky (biosphere concept) loomed large in this particular work. A key conclusion of the work concerned the relative instability of the global ecological system and, linked to this, the rapid growth of humankind's influence on the system and particularly its atmospheric processes. As with his earlier work, Budyko emphasised the need to ensure effective management of humankind's activities and on this occasion highlighted the Soviet Union's proactive stance both domestically and internationally in this regard.

In 1984, he published a further monograph, Evolution of the Biosphere (Evolyutsiya biosfery). This extended his discussion of the future of the biosphere in view of humankind's growing influence; it also drew heavily on Vernadsky's intellectual legacy and his earlier work on the biosphere. This included a more developed discussion of his noosphere concept which, as noted above, envisaged a qualitative shift in the nature of the biosphere underpinned by humankind's intellectual and scientific capabilities. In keeping with his earlier emphasis on humankind's collective potential to manage effectively the state of the biosphere in the future, Budyko outlined his hope for the establishment of a regulated global ecological system, thus helping to prolong the existence of the biosphere.

Budyko and his work would also feature significantly in the pioneering activities of the IPCC. He contributed to section five of the first Scientific Assessment concerned with "equilibrium climate change and its implications for the future" and was a named peer reviewer for this report together with three other Soviet scientists, namely G.S. Golitsyn, I.L. Karol, and V. Meleshko. His jointly authored work with Yuri Izrael, entitled Anthropogenic Climate Change (Budyko and Izrael 1987), was also heavily cited in the accompanying volume dealing with Impacts Assessment. Writing in the foreword to the English-language translation of this book, Alan D. Hecht of the US Environmental Protection Agency noted that their analyses suggested that anthropogenic climate change may have some benefit for significant parts of the northern hemisphere, thus encouraging Budyko "to argue that international measures to reduce the emissions of greenhouse gases are not justified."

7.5   ACADEMICIAN KIRILL KONDRATYEV AND THE INTERGOVERNMENTAL PANEL ON CLIMATE CHANGE (IPCC)

We have outlined the scientific work of Academician Kirill Yakovlevich Kondratyev in the fields of atmospheric physics, meteorology, and the pioneering of remote sensing methods in these sciences, work which occupied a period of nearly 50 years from the mid-1940s. It involved considerable international cooperation and led to widespread international recognition. However, in the last 15-20 years, when he was no longer involved in front-line fundamental scientific research, he turned his attention to ecology, climate change, and global change. His research work at the St Petersburg Scientific and Research Centre of Ecological Safety and at the Nansen International Environment and Remote Sensing Centre was then concerned with environmental problems in general and especially those that might be arising from human activities. As far as Budyko was concerned, global warming as a result of increasing emissions of greenhouse gases was not necessarily unwelcome to people in far northern climates. Kondratyev's view was slightly different; he believed that too much attention was being given to greenhouse gas emissions to the neglect of any other serious threats to human well-being. We now examine briefly Kondratyev's relationship with the Intergovernmental Panel on Climate Change (the IPCC) and the Kyoto Protocol. By the late 1980s, Kondratyev had achieved widespread international recognition, including by the WMO, for his scientific work. Therefore, one might have supposed that he would play a leading role in the IPCC, but that was not so. In this section, we shall confine ourselves to examining Kondratyev's relationship with – and his views of – the IPCC, of which he was an intelligent and informed critic. The United Nations Framework Convention on Climate Change (UNFCCC) is an international environmental treaty which was produced at the United Nations Conference on Environment and Development (UNCED) in Rio de Janeiro in 1992. The stated objective is "to achieve stabilization of greenhouse gas concentrations in the atmosphere at a low enough level to prevent dangerous anthropogenic interference with the climate system." However, the treaty itself sets no mandatory limits on greenhouse gas emissions for individual nations; limits, enforcement conditions, and penalties are provided for in updates, of which the principal update is the Kyoto Protocol. Kondratyev, as we shall see, was highly critical of much of the work of the IPCC, of what is generally pronounced to be its scientific consensus, and therefore of the Kyoto Protocol. It may be of interest, however, to stress at the outset that he was not a critic of the global emission reduction effort from an "anti-environmentalist" perspective but from the deeper "green" or Gaia side. For him the postulated enhanced global warming due to increasing greenhouse gas emissions as a result of human activities remained an unproven hypothesis and was in any case not the most serious to human life on Earth (Kondratyev et al., 2004a, 2004b).

He had become a lone scholar who had turned his very active mind away from fundamental research on atmospheric physics towards encouraging younger researchers and synthesising the available knowledge of ecology, of humanity, and of understanding the Earth, in order to manage it sustainably. His publications on this subject are numerous, both in English and Russian.

Dr Sonja Boehmer-Christiansen, the editor of *Energy and Environment*, met Kirill Kondratayev four times before his death in May 2006. In the summer of 2001, she visited his Institute in St. Petersburg to make a presentation on the politics of climate change. One year later, she accompanied him to a meeting he had organised in Rostov-on-Don ("Round Table: Global Environmental Dynamics Now and in 21st Century," Chairman Thor Heyerdahl, May 2001, Rostov-on-Don, Russia) and they both attended the Third World Climate Change Conference in Moscow (29 September – 3 October 2003). She last met Kirill Kondratyev in the spring of 2005 on a research trip to Moscow to explore, rather unsuccessfully because of bad timing, Russia's climate policy. In Rostov-on-Don, there was ample evidence of the decline of infrastructure and of industrial activity and also the opportunity to learn from younger Russian environmental scientists and economists who all, at that time at least, bemoaned the decline of Russian research and their growing dependence on funding from abroad, or even going abroad to find work. Obtaining grants for any research at all had become the overriding issue, and at that time environmental research money came mainly from collaboration with the EU or North America. In Moscow in September 2003 at the World Climate Change conference in Moscow, Dr Boehmer-Christiansen and Prof. Kondratyev addressed a press conference together after Prof. Kondratyev had addressed a large crowd of scientists in front of President Putin. Kondratyev warned against taking precipitate action against fossil fuels because of the lack of evidence for man-made climate change, pointed to serious uncertainties, and encouraged the assembled scientists to read his books. Mankind would have to work much harder to understand ecological damage and then regulate itself according to ecological principles and targets defined by research. The biosphere needed protection rather than mere emissions reduction! The evolution of the Russian attitude to the Kyoto Protocol will be discussed in the next section. He was obviously a grand old man among Russian scientists, highly respected including by a considerable number of people from the West, some of whom had made their peace with Working Group I of the Intergovernmental Panel on Climate Change (IPCC) either because they believed its scientific consensus, or because public opposition would have endangered their funding and cordial relations with national governments. The policies advocated by the UN and hence major governments in the West were not, he argued, scientific enough or directed to the main issues, which included – for him – ecological damage, pollution, depletion of resources, overpopulation, etc. He was a man with a deep belief in ecological principles and the power of science to shape human behaviour in a "top down" fashion. He hoped that the UN, advised by scientists from many countries, would and could decide in the interest of all humanity. Advocating this with much passion and learning, as well as intellectual consistency, meant that it seemed to some that Kondratyev paid too little attention to the realities of politics and economics, and especially to the deep divisions of humanity.

In a succession of reports over the period since 1990 (1992, 1995, 2000, 2001, 2007, 2012, 2014, 2018, 2019), the IPCC has come more and more firmly to the view that human activities are contributing significantly to global warming and the report, which was published in February 2007, says that it is 90% likely that global warming is due to human activities. In December 2007, the IPCC shared the Nobel Peace Prize with the former American Vice-President Al Gore. A whole double issue of the journal *Energy and Environment* (volume 18, nos 7-8, December 2007) was devoted to "The IPCC: Structure, Process and Politics" and the first article in that issue gives a particularly good account of the IPCC, its method of working in producing its assessments, and also summarises both sides of the arguments of those who support the IPCC and those who are against it (Zillmann 2007).

It is important that IPCC Report (2018) states that its key finding is that meeting a 1.5 °C (2.7 °F) target is possible but would require "deep emissions reductions" and "rapid, far-reaching and unprecedented changes in all aspects of society." Furthermore, this report finds that "limiting global warming to 1.5 °C compared with 2 °C would reduce challenging impacts on ecosystems, human health and well-being" and that a 2 °C temperature increase would exacerbate extreme weather, rising sea levels, and diminishing Arctic sea ice, coral bleaching, and loss of ecosystems, among other impacts.

The main statements of the last IPCC Report (2019) are the following:

- since 1970, the "global ocean has warmed unabated" and "has taken up more than 90% of the excess heat in the climate system." The rate of ocean warming has "more than doubled" since 1993. Marine heatwaves are increasing in intensity and since 1982, they have "very likely doubled in frequency." Surface acidification has increased as the oceans absorb more CO_2. Ocean deoxygenation "has occurred from the surface to 1,000 m (3,300 ft)."
- Global mean sea levels rose by 3.66 mm per year which is "2.5 times faster than the rate from 1900 to 1990." At the rate of acceleration, it "could reach around 30 cm to 60 cm by 2100 even if greenhouse gas emissions are sharply reduced and global warming is limited to well below 2°C, but around 60 cm to 110 cm if emissions continue to increase strongly.
- There has been an acceleration of glaciers melting in Greenland and Antarctica as well as in mountain glaciers around the world, from 2006 to 2015. This now represents a loss of 720 billion tons (653 billion metric tons) of ice a year.
- Carbon Brief said that the melting of Greenland's ice sheets is "unprecedented in at least 350 years." The combined melting of Antarctic and Greenland ice sheets has contributed "700% more to sea levels" than in the 1990s.
- The Arctic Ocean could be ice free in September "one year in three" if global warming continues to rise to 2 °C. Prior to industrialisation, it was only "once in every hundred years."
- Future climate-induced changes to permafrost "will drive habitat and biome shifts, with associated changes in the ranges and abundance of ecologically-important species." As permafrost soil melts, there is a possibility that carbon will be unleashed. The permafrost soil carbon pool is much "larger than carbon stored in plant biomass." "Expert assessment and laboratory soil incubation studies suggest that substantial quantities of C (tens to hundreds Pg C) could potentially be transferred from the permafrost carbon pool into the atmosphere under the Representative Concentration Pathway 8.5" projection.

Kondratyev was out of sympathy with much of the work of the Intergovernmental Panel on Climate Change (IPCC), especially its heavy reliance on computer climate models for predicting future climates, while neglecting several factors in the modelling. It is worth noting that Russian scientists at that time did not have access to large powerful computers (only the Russian military had that) and so they could not participate in the computer modelling experiments.

Fairly soon Kondratyev was largely marginalised by many of the leading figures in the IPCC. This was because he was no longer involved directly and personally in front-line experimental or theoretical research. But this cannot have been the main reason – there were several people in very senior positions in the IPCC who were also not themselves conducting front-line research in person. It seems (Boehmer-Christensen and Cracknell 2009) that there are three reasons why there seemed to be such a wide gulf between Kondratyev and the IPCC.

First, he claimed that there were various processes etc. that were not included in the computer climate models on which the IPCC relies so much. While models are being improved all the time, as computing power is continually being increased, it can be argued that while the models can deal quite successfully with gradual change they are not able to predict abrupt changes that suddenly occur. For instance, changes in feedback due to the changes in albedo arising from the collapse of an antarctic ice sheet can be accommodated after the event, but such a collapse would not be predicted by the models.

Secondly, Kondratayev strongly believed that global warming induced by excessive carbon dioxide production by the burning of fossil fuels is only one, and possibly even only a minor one, of a large number of serious threats facing humanity. These threats have already been mentioned in Section 1.3.12. They may come from over population, pollution of the atmosphere, pollution of the water sources, contamination and degradation of the land, damage to the biosphere and the extinction of many species, the depletion of fossil fuel sources, the depletion of non-fuel mineral resources, the destruction of stratospheric ozone, etc. (Kondratyev et al. 2004a, 2004b). It can seriously be argued that the success of the IPCC in making people generally aware of the threat of global warming induced by the burning of fossil fuels has led to these many other threats being largely ignored at various levels of policy making and human behaviour. Kondratyev laboured to bring to people's attention the whole question of global change in general and its threats to human life. Kondratyev's warnings in this area still go largely unheeded. It is the purpose of this book to address this issue.

The third point of dispute with the IPCC was that Kondratyev was sceptical about the interpretation of the experimental evidence that was adduced for global warming. He speculated that it was selected to confirm a hypothesis already assumed as true for political reasons. He was rightly cautious, as many other people have been.

Although the main arguments outlined in the five IPCC Assessment Reports are very widely accepted, there has been a succession of people who have not been convinced. Kondratyev was one of the very first of these, but there have been others over the last 30 years since the IPCC was set up and we shall discuss some of these in Section 7.9.

Kondratyev developed and expounded his views on global change and the threats to our way of life in various monographs. To understand his views on the IPCC, it is best to consider those of his writings that were more specifically concerned with the IPCC and its dependence on computer models and its use of climate-related observations. These writings are mostly to be found in some articles of his that were published in the journal *Energy and Environment* and elsewhere.

We must consider Kondratyev's position with respect to global warming, the IPCC and the Kyoto Protocol from his own writing. Following the Second World Climate Conference in Geneva in 1990, the United Nations Conference on Environment and Development (UNCED), informally known as the Earth Summit, was held in Rio de Janeiro in 1992. The stated objective was "to achieve stabilisation of greenhouse gas concentrations in the atmosphere at a low enough level to prevent dangerous anthropogenic interference with the climate system." This is vague, it does not say how the stabilisation is to be achieved, nor does it define what is meant by the "low enough level" to which it refers. This conference produced the United Nations Framework Convention on Climate Change

(UNFCCC), which came into force in 1994 but made virtually no demands on any country. As far as emission reduction policies were concerned, only three countries did reduce their emissions, the UK by a massive switching from coal to gas, Germany through re-unification and the collapse of the East Germany energy demand, and Russia as a result of its de-industrialisation following the collapse of Communism.

The UNFCCC was criticised, quite forcefully, by Kondratyev (1997) for concentrating so much on greenhouse gas emissions. In his paper on "Key issues in global change" (Kondratyev 1997) he wrote "The most discussed problem is global warming – it is more appropriate to call it global climate change – and ... specifically the growth of greenhouse gas emissions into the atmosphere. World carbon emissions from fossil fuel burning are still growing although some countries have undertaken certain measures to reduce emissions." He then went on to study in detail the carbon dioxide emissions of various countries. He thought that people were devoting far too much of their attention to the increase of CO_2 in the atmosphere and a predicted catastrophic scenario of global warming. But we know that the biosphere assimilates a great deal of carb CO_2 emitted in the atmosphere and helps to guarantee future ecological safety. "If we destroy the biosphere which functions as a sink for carbon, we create an ecological catastrophe..." he said in the interview with the WMO (1998). However, he stressed that carbon dioxide emissions and global warming are not the only problem, or even the most serious problem facing the future of mankind. "Undoubtedly, one of the most worrying features of the present time is the continuing growth of the global population. Two specific features of this growth have been the concentration in developing countries and the growth of urban populations...." "An important question in this context is the adequacy of the UNFCCC recommendation to reduce greenhouse gas emissions. On the one hand, it is obvious that, generally speaking, the reduction of greenhouse gas emissions is a very useful measure. But on the other hand, it is equally clear that such a measure is not a panacea against global change dangers." (Kondratyev 1997). "The problem of global change cannot be solved without using a system's approach comprising all processes involved. Studying carbon dioxide or ozone in isolation," he said, "will serve little purpose. Such studies should be made in the context of the overall problem." (WMO 1998).

"As far as global change science is concerned," (Kondratyev 1997) "it is important to recognize that present-day numerical climate modelling (even in the case of 3-D coupled global models) remains far from being able to reliably simulate real climate change and, consequently, to identify the contributions of various climate-forming factors, including the enhanced greenhouse effect. Though it is well known that climate change results from interaction between all components of the climate system, the relative influence of various factors cannot be defined precisely and 'new' influences are still being added to the climate equation." To be specific, aerosols are highly variable, both spatially and temporally, and it is very difficult to build their effect reliably into the models; Kondratyev had himself done a lot of work earlier on atmospheric aerosols and this work was persistently ignored. He continued "As far as climate change is concerned, the key task must be to study climate in all its complexity without an overemphasis on certain individual factors such as the greenhouse effect. But it is also necessary to identify the place and the role of climate change within the more general framework of global change." He argued that it had been shown by Gorshkov (1995) that the basic processes which regulate environmental dynamics are founded on the principle of the biotic regulation of the environment. If we accept such a concept then the priority order given in Table 2.1 was suggested as a basis for further discussion. This preliminary scheme of priorities demonstrates a subordinate role for climate change within a much more general framework of concern about global change.

7.6   THE UNFCCC AND THE KYOTO PROTOCOL

From 1750 till now, the CO_2 concentration in the atmosphere has increased by a little over one third, reaching the highest level for the last 420 thousand years (and, probably, for the last 20 million years), which is illustrated by the data of ice cores (IPCC Third Assessment Report 2001). About two thirds of the growth of CO_2 concentration in recent years is explained by emissions to the atmosphere from fossil fuel burning and the remaining one third is due to deforestation and cement manufacture. It is of interest that by the end of 1999, CO_2 emissions in the USA exceeded the 1990 level by 12%, and by 2008, their further increase should raise this value by 10% more (Victor 1998). Meanwhile, according to the Kyoto Protocol, emissions should be reduced by 7% by the year 2008 with respect to the 1990 level which requires their total reduction by about 25% which is of course utterly unfeasible. According to the IPCC Third Assessment Report (2001), the probable levels of CO_2 concentration by the end of the century will range 540–970 ppm (pre-industrial and present values are, respectively, 280 ppm and 385 ppm).

As mentioned at the beginning of this chapter, the Kyoto Protocol was the first attempt to implement the stabilisation of greenhouse gas emissions referred to in the United Nations Framework Convention on Climate Change (UNFCCC). Once a sufficient number of countries had ratified the UNFCCC and it therefore came into force in 1994, there have been annual Conferences of the Parties (COPs). In December 1997 in Kyoto (Japan) the third Conference of the Representatives, COP-3, of the countries that had signed the UNFCCC (over 160) met and engaged in lengthy and hot debates on the need to recommend a 5% CO_2 emissions reduction by 2008–2012 for industrially developed countries (relative to the 1990 level). It was at this conference that the Kyoto Protocol was adopted. However, before it could become legally binding, it had to be ratified by a required number of countries and there was a considerable time lapse before that occurred.

The text of the Kyoto Protocol can conveniently be found in the book by Grubb *et al.* (1999). Updating information is always available, for instance, from the Wikipedia website (http:// en.wikipedia.org/wiki/Kyoto_Protocol). The following summary is adapted from the article in Wikipedia:

- The Kyoto Protocol is underwritten by governments and is governed by international law enacted under the aegis of the United Nations;
- Governments are separated into two general categories: developed countries, referred to as Annex I countries (which have accepted greenhouse gas emission reduction obligations and must submit an annual greenhouse gas inventory); and developing countries, referred to as Non-Annex I countries (who have no greenhouse gas emission reduction obligations but may participate in the Clean Development Mechanism);
- Any Annex I country that fails to meet its Kyoto obligation will be penalised by having to submit 1.3 emission allowances in a second commitment period for every ton of greenhouse gas emissions they exceed their cap in the first commitment period (i.e. 2008–2012);
- By 2008–2012, Annex I countries have to reduce their greenhouse gas emissions by a collective average of 5% below their 1990 levels (for many countries, such as the European Union member states, this corresponds to some 15% below their expected greenhouse gas emissions in 2008). While the average emissions reduction is 5%, national limitations range from an 8% average reduction across the European Union to a 10% emissions increase for Iceland; but since the European Union's member states each have individual obligations, much larger increases (up to 27%) are allowed for some of the less developed European Union countries. Reduction limitations expire in 2013;
- Kyoto includes "flexible mechanisms" which allow Annex I economies to meet their greenhouse gas emission limitation by purchasing greenhouse gas emission reductions from elsewhere. These can be bought either from financial exchanges, from projects which reduce emissions in non-Annex I economies under the Clean Development Mechanism, from other Annex I countries under the Joint Implementation (see below), or from Annex I countries with excess allowances. Only Clean Development Mechanism Executive Board-accredited Certified Emission Reductions can be bought and sold in this manner. Under the aegis of the United Nations, the Bonn-based Clean Development Mechanism Executive Board was established to assess and approve projects (CDM Projects) in Non-Annex I economies prior to

awarding Certified Emission Reductions. (A similar scheme called the Joint Implementation scheme applies in transitional economies mainly covering the former Soviet Union and Eastern Europe).

Given that the objective of the UNFCCC is "to achieve stabilisation of greenhouse gas concentrations in the atmosphere ... " (see above) the Kyoto Protocol is a step in that direction. But the controversy did not end with the conference in Kyoto. Opposition to the Kyoto Protocol has come from various directions. There is the position of the developing countries. Naturally, the position of the developing countries gives primary consideration to socio-economic development, including the overcoming of poverty and its consequences. They argued, not unreasonably, that it is the industrialised countries which have caused most of the human-induced global warming so far, and that their own development or progress towards industrialisation should not be held back because of a problem that they have not themselves created. Developing countries are not prepared to accept greenhouse gases emissions reduction; their point of view was respected and they were not required by the Kyoto Protocol to accept reductions in their emissions. Opposition to the Kyoto Protocol has come from some people who see it as an attempt to reduce the growth of the world's industrial economies. The former Australian Prime Minister, John Howard, refused to ratify the Kyoto Protocol on the grounds that it would curtail development and cost Australian jobs; his successor, Kevin Rudd, ratified the Kyoto Protocol in December 2007. U.S. President G.W. Bush rejected the Kyoto Protocol because: 1) ostensibly this document lacks scientific substantiation; 2) its adoption would cause serious economic damage to the USA (whose energy supply is based mainly on the use of hydrocarbon fuels) without providing any marked positive impact on the environment. Of these two reasons, it is fairly clear that the second one, which is naked self-interest on the part of the USA, was the dominant reason.

To come into force, the Kyoto Protocol needed to be ratified by countries responsible for at least 55% of global carbon dioxide emissions. Since the USA had refused to ratify the Protocol, this minimum could only be achieved if Russia decided to ratify it. A problem arises from the choice of 1990 as the baseline for calculating reductions of carbon dioxide emissions. In 1990, the former Soviet Union had done little to raise its energy efficiency; shortly after that came the collapse of communism and the downturn in the economy and a consequent reduction in energy consumption and greenhouse gas emissions. On the other hand, Japan, as a net importer of oil and other raw materials, had become very energy efficient by 1990. Such factors were ignored and the subsequent inactivity of the former Soviet Union, following the collapse of communism, meant that it could then look forward to generating an income by trading its surplus emissions allowance. This did not prevent many Russians from seeing the Kyoto Protocol as an attempt to hold back the regeneration of their economy. At the Moscow World Conference on Climate Change (29 September – 3 October 2003) the Kyoto Protocol was attacked on two fronts that were rather similar to President Bush's points. President Putin's economic adviser, Andrei Illarionov, said that ratification would stall Russia's economic growth, it would "doom Russia to poverty, weakness and backwardness." The Kyoto Protocol calls for countries to reduce their level of greenhouse gas emissions by certain amounts which are specified individually for the various countries. If a country exceeds the emissions level, it could be forced to cut back industrial production. This would be likely to conflict with President Putin's goal of doubling Russia's gross domestic product by 2010.

According to recent studies (e.g. Li and Jiang, 2019) Russia's energy-related carbon emission decreased by roughly 30% between 1992 and 2017. Previous studies reported that economic recession led to carbon emission reduction in Russia during 1990s. The results show that not economic recession, but improving energy efficiency is the most significant contributor to decreasing Russia's carbon emission from 1992 to 2017. Economic recession is the major contributor to the decrease in Russian carbon emission only before the new century and then reversed to the leading contributor to the increase in carbon emission. The research by Li and Jiang (2019) also found that a shift to less carbon-intensive fuel and decrease in population also contributed to offsetting carbon emission in Russia. Thus, this research argues that the cause for the decline in Russia's carbon emission for 1992–2017 is not economic recession. Indeed, Russia's economic activity and change in carbon emission have been delinked since the new century. It can be concluded that the reduction in Russia's carbon emission during 1992–2017 arises from a combination of improving energy efficiency, a shift to less carbon-intensive fuel, and decrease in population

The economic concerns were supported at the Moscow Conference by several top Russian climate scientists, including Kondratayev. His long paper, on "Key Aspects of Global Climate Change," was submitted just prior to the World Climate Change Conference in Moscow in 2003 and it was published in the following year (Kondratyev 2004). This paper defines his almost entirely scientific objections to climate models and the Kyoto Protocol and also demonstrates the aim of his work during the last years of his life. President Putin told the conference that his Cabinet had not yet decided whether or not Russia would ratify the Protocol. Eventually, about a year later, Russia did ratify the Kyoto Protocol. Finally, it came into force in February 2005, following its ratification by Russia. Although it was adopted nearly seven years before that, the Kyoto Protocol had until then remained a statement of intent, rather than a legally binding document. Once Russia had signed the Protocol, it then became a legally binding document on the signatories. Countries which failed to meet the target cuts in carbon dioxide emissions would face penalties and have to cut back on their production. Thus, eventually the US failure to ratify the Kyoto Protocol has not prevented its adoption, with the requirements to reduce the greenhouse gases emissions.

Kondratyev's second article published in *Energy and Environment* (Kondratyev, 2004) was – as has already been mentioned – prepared in anticipation of the Moscow World Climate Change Conference (29 September – 3 October 2003). This is a lengthy article and it is not possible to recount here all the detail it contains. He was concerned with the question of whether the Kyoto Protocol should be considered as a scientifically justified document: "Confusion reigns and is caused, in particular, by the lack of sufficiently clear and agreed terminology." Ignoring the very complicated notion of climate itself (which needs a separate discussion), one should remember, for instance, that in the UNFCCC climate change was defined as being anthropogenically induced. One of the main unsolved problems is the absence of convincing quantitative estimates of the contribution of anthropogenic factors to the formation of global climate, though there can be no doubt that anthropogenic forcings of climate do exist.

Some international documents containing analyses of the present ideas of climate refer to the prevalent idea of a consensus with respect to scientific conclusions as enshrined in these documents. This wrongly assumes that the development of science is determined not over time by different views and relevant debates and discussions, but by a general agreement and even voting. Apart from the question of definitions, the issue of uncertain conceptual estimates concerning various aspects of climate problems remains of importance. In particular, this refers to the main conclusion in the summary of the IPCC Third Assessment Report (2001) which claims that "... An increasing body of observations gives a collective picture of a warming world and most of the observed warming over the last fifty years is likely to have been due to human activities."

The Earth's climate system has indeed changed markedly since the Industrial Revolution, with some changes being of anthropogenic origin. The consequences of climate change do present a serious challenge to the policy-makers responsible for the environmental ("ecological" in Russian) policy and this alone makes the acquisition of objective information on climate change, of its impact and possible response, most urgent.

The IPCC had, by the time of the Moscow World Climate Change Conference in 2003, prepared three detailed reports, in 1990, 1996, and 2001, as well as several special reports and technical papers. Griggs and Noguer (2001) made a brief review of the first volume of the IPCC Third Assessment Report (TAR) prepared by WG-I for the period June 1998 – January 2001 with the participation of 122 leading authors and 515 experts, each with their materials. Four hundred and twenty experts reviewed the first volume and 23 experts edited it. Besides, several hundred reviewers and representatives of many governments made additional remarks. With the participation of delegates from 99 countries and 50 scientists recommended by the leading authors, the final discussion of the Third Assessment Report was held in Shanghai on 17-20 January 2001. The "Summary for decision-makers" was approved after a detailed discussion by 59 specialists."

Kondratyev (2004) continued with a discussion of the political challenge and ten questions raised by Prof. A.N. Illarionov, Economic Adviser to President Putin, at the Moscow Conference. He then argued that "the main cause of contradictions in studies of the present climate and its changes is the inadequacy of the available observational databases." He cites in particular surface air temperature, ground surface temperature, the extent of snow and ice cover, sea level and the heat content of the upper layer of the oceans and precipitation. He also alluded to abrupt changes in the climate and the fact that the models do not predict such events. The final section of the paper (Kondratyev 2004) deals with the results of numerical climate modelling and their reliability; hopefully their reliability has improved since that paper was written.

CO_2 is, of course, not the only greenhouse gas. The other major greenhouse "gas" is water vapour and the whole question of anthropogenic effects on the hydrological cycle, atmospheric water vapour, and cloud patterns is very difficult to study. There are also many other greenhouse gases, CH_4, various oxides of nitrogen (collectively referred to as NO_x), H_2S, SO_2, SF_6, DMS (dimethyl sulphide, $(CH_3)_2S$), CFCs (chlorofuorocarbons) etc., some of which occur naturally and some of which are of anthropogenic origin. Climate models are usually run on the basis of taking these gases into account by considering their CO_2 equivalent, in terms of global warming, and adding it to the actual predicted concentration of CO_2 itself. The Kyoto Protocol, however, appears only to concern itself with carbon dioxide emissions and makes no reference to any attempt to restrict the emissions of these other gases.

Comparisons are sometimes made between the Kyoto Protocol and the Montreal Protocol. The Montreal Protocol came about as a result of the scientific evidence for human-induced depletion of the ozone layer, and especially the famous "ozone hole" which appears in the Antarctic each spring. This was rapidly accepted to be a result of the escape of CFCs (chlorofluorocarbons) into the atmosphere. The world's leaders came together and, in the Montreal Protocol, agreed to phase out the production of CFCs and to replace them by other "ozone friendly" substances. The reasons for the relative success of the Montreal Protocol are neatly summarised in box 21.2 of the Stern (2007) Review. Twenty-four countries signed the original Protocol in 1987, and by October 2006, 74 countries had ratified the Protocol and this included the major developing countries. Emissions of CFCs have largely been brought under control, but of course the ozone layer will not recover immediately; it is expected to take up to 100 years to do so.

There were several factors which contributed to the success of the Montreal Protocol. First, there was a high degree of scientific consensus and evidence that there was a problem that required urgent political action and public opinion galvanised politicians. The Protocol used expert advice to establish targets and timetables to phase out the use of ozone depleting chemicals, based on recommendations of expert panels including government and industry representatives. Secondly, developing countries participated partly because of the convincing nature of the science, but also because of the financial support provided to help them to make the transition to phase out harmful substances – albeit at a slower pace than that for developed countries. Thirdly, the Montreal Protocol recognised the importance of stimulating and developing new technologies so that industry could manufacture alternatives to the harmful ozone-depleting chemicals, and providing access to these technologies to developing countries. Finally, groups of like-minded countries came together to provide fora to examine the complex issues involved in and to consider the consequences of taking action.

The Kyoto Protocol has been different for several reasons. First, there was much more hesitation by Governments to accept the need for action to curb CO_2 emissions. This was partly because of doubts about the science and these doubts were stimulated by vested interests. It was also because of fears about the restrictions that the Kyoto Protocol would cause on economic activity and industrial development, both in industrialised countries and in developing countries. Secondly, it has become more and more apparent that the restrictions on CO_2 emissions proposed in the Kyoto Protocol were far too small to deal with the problem of human-induced global warming. Thirdly, there are some countries where some warming would actually be welcome for economic or social reasons.

The Kyoto Protocol commits its signatories to a 5.2% reduction in carbon dioxide emissions, relative to 1990, by 2012. However, it is being argued more and more clearly that such a small reduction is far too small to reduce global warming to what might be regarded as an "acceptable" level. From the data of approximate numerical modelling, even the complete achievement of the Kyoto Protocol recommendations would provide a decrease of the mean global mean annual surface air temperature not exceeding several hundreds of degrees. Perhaps the most extreme evaluation is that of George Mobiot (2006) who proposed that a *reduction* of 90% (note *of* 90%, not *to* 90% (which would be a reduction of 10%)) in carbon dioxide emissions by 2030 is necessary and he examines how in one country, the UK as an example, this target might be able to be achieved. It should be pointed out that Monbiot is not suggesting that for the UK alone to reduce its emissions by this amount will achieve very much in global terms; what he is doing is illustrating – for the example of one country – the likely problems that very many countries would face in meeting such a target. The Kyoto Protocol can only be regarded as a first and very tentative step towards making the necessary reductions in CO_2 emissions to enable us to avoid dramatic climate change. Kondratyev's view was that it was such a tiny first step as to be dangerously misleading in the sense that people might think that the problem had been solved once these targets were met. More realistic targets need to be established. Moreover, governments and peoples have got to learn to work together to tackle this serious problem.

7.7   CLIMATE PREDICTIONS

It is extremely difficult to understand the scientific laws governing the present climate system and even more so to assess potential climate changes in future. This is confirmed by the lack of reliable estimates of the contribution of anthropogenic factors to the formation of the present climate and even more so, to any understanding of why the anthropogenically induced enhancement of the atmospheric greenhouse effect (due to the growth of greenhouse gas concentrations in the atmosphere) should cause certain changes of global climate. In this connection, a primitive understanding of global warming as a general increase of temperature increasing with latitude is rather dangerous. An analysis of the observed data obtained in high latitudes of the northern hemisphere (Adamenko and Ya Kondratyev 1990) has shown that such claims do not correspond to reality.

In order to assess the reality of climate predictions, it is critically important to test the adequacy of models from the perspective of their ability to reproduce the present observed changes and palaeo-dynamics of climate (from proxy data). As for the use of the present day observed data, the situation is rather paradoxical: the experience of testing the adequacy is confined to the use of average temperatures while it would be necessary to use different information and moments of a higher order. Goody (2001) drew attention to the prospects of using the space-based observations of the spectral distribution of outgoing longwave radiation. Unfortunately, the issue of an adequately planned climate observation system has not yet been recognized. (Kondratyev 1998, Kondratyev and Cracknell 1998a, 1998b, Kondratyev and Galindo 1997). The present confused paradoxical situation is characterised by a huge amount of poorly systematised satellite observations combined with the degradation of conventional (in situ) observations as mentioned above.

It is very difficult to test the adequacy of global climate models by comparing the results of numerical modelling with the observational data. Most often, this problem is solved by comparing a long data series on the annual-average global-average surface air temperature. The main conclusion, despite the substantial (sometimes radical) differences in the consideration of the climate-forming processes, is practically always the same: on the whole, results of calculations agree with the observation data. Another characteristic feature of such testing is the invariable conclusion in support of the considerable (or even dominating) climate-forming contribution of anthropogenic factors, above all of the greenhouse effect. Yet the necessary quantitative substantiation remains lacking. Such an approach to the verification of the models cannot be taken seriously because

1) the present climate models are still very imperfect from the viewpoint of an interactive account of biospheric processes, aerosol – clouds – radiation interaction, and many other factors;
2) the only long-term (100-150 years) series of surface air temperature observations is far from being adequate, from the viewpoint of calculations of the annual average global average surface air temperature values.

Beven (2002) discusses the conceptual aspects of the numerical modelling of the environment connected with analysis of possibilities of simulation modelling from the viewpoint of realistic simulation of natural processes. At present, computer modelling is widely developed and is actively used as an instrument of theoretical studies of the environment as well as to solve various practical problems and to substantiate recommendations for decision makers. Of special interest are predictions of potential impacts of global climate changes and of the functioning of ground water use systems, as well as long-term geomorphological predictions and assessments of the impacts of underground repositories of radioactive emissions. In all these cases, it is assumed that the problems being studied can be solved despite non-linearity and the open nature of the natural systems considered as well as various assumptions that serve as a basis for numerical modelling.

Of course, such an assumption is rather naïve, since from the methodical ("philosophic") and scientific points of view, it proceeds from the presumption that the considered systems have been sufficiently studied. Clearly, many natural systems are so complicated that the existing ideas of them are far from being adequate. It always happens that real natural systems are much more complicated compared to their analogues which are described by numerical models. One of the most vivid examples in numerical climate modelling is connected with the use of a sub-grid parameterisation of many climate-forming processes (on land surface, in the atmosphere, etc.). This entails not only sometimes far from real representations of the processes being considered, but also the necessity to introduce a great number of insufficiently reliably determined empirical parameters.

Recent developments associated with the global research programmes GCOS (Global Climate Observing System), GOOS (Global Ocean Observing System), GTOS (Global Terrestrial Observing System), and IGOS (Integrated Global Observing Strategy) are useful, but they still do not contain adequate grounds for an optimal global observing system, as discussed in detail in the monographs of Kondratyev (1998) and Kondratyev and Cracknell (1998a,b) and quite recently by Goody (2001, 2002) and Goody et al. (1998, 2002). The main cause of such a situation is the imperfection of climate models which should serve as the conceptual basis in planning the observations to be specified as the models are being improved. In this connection, it should be emphasised that it is not illusory statements about the sufficient adequacy of the global climate models that are needed, but an analysis of their differences when compared with observations. This would reveal the "weak points" of the models. It is clear that a totality of climate parameters should be considered (and not only surface air temperature), with emphasis on the models' capability to simulate climate changes including, at least, moments of the second order.

Preparations of a strategic plan of the Climate Change Science Programme (CCSP) planned for 10 years were started in the USA in July 2002 and completed in 2003. The programme has five main goals (Climate Change Science Program 2003):

1) To get a deeper knowledge of the past and present climates and the environment, including natural variability as well as to improve an understanding of the causes of the observed climatic variability.
2) To obtain more reliable quantitative estimates of the factors determining the Earth's climate changes and changes of related systems.

3) To reduce the levels of uncertainties of the prognostic assessments of future changes of climate and related systems.
4) To better understand the sensitivity and adjustability of natural and regulated ecosystems as well as anthropogenic systems to climate and to global changes in general.
5) To analyse possibilities to use and recognise the limits of understanding how to control the risk in the context of climate changes.

The CCSP indicates concrete ways of how to reach these goals. In this connection, it was pointed out that the priorities of perspective developments should include a decrease of the levels of uncertainties in the problems such as: properties of aerosol and its climatic implications; climatic feedbacks and sensitivity (first of all, for polar regions); carbon cycle. Of key priorities in the CCSP also are developments concerning the climate observing systems (it was very important to organise an ad hoc Group on Earth observations – GEO) and further development of the numerical climate modelling (first of all, for a more adequate consideration of the physics and chemistry of climate).

7.8   COOLING OFF ON GLOBAL WARMING

7.8.1   THE SOVIET CLIMATOLOGISTS

It would be fair to summarise the position of the prominent Soviet climatologists at the time of the publication of the first IPCC Assessment Report (IPCC 1 1990) as follows. Budyko and Izrael came to an accommodation with the IPCC and involved themselves with WGI. Kondratyev on the other hand took a very critical view of the dominance of the First Assessment Report by the computer modelling and its results. Kondratyev accepted the fact that human activities were leading to a contribution to the greenhouse effect associated with atmospheric gases arising from human activities but his criticisms centred around

(1) the extreme difficulty in determining the relative contributions of natural processes and human activities
(2) the various deficiencies of the models in terms of ignoring some effects, or only taking some effects into account approximately, ...

The first of these is still an issue. As to the second one, some of these issues are less serious now than they were in those early days (30 years ago) of the models. Kondratyev wrote copiously on these topics and was marginalised by the powerful elite of the IPCC. He was less concerned about the politicisation of the findings of the IPCC which became so important later on and remain a problem (see Section 7.9).

The current political interest in global warming probably stems from the Rio Summit which took place in 1992 and has been fuelled by the reports of the Intergovernmental Panel for Climate Change (IPCC). The first reports (IPCC 1 1990, IPCC 2 1992), which were based on the predictions of rather crude models, predicted a warming of between 1.5 and 4.5°C by 2050. By 1995, using slightly more sophisticated models (i.e. including such things as the presence of sulphate aerosols), this had fallen to a predicted warming of between 1.0 and 3.5°C by 2100 (IPCC X 1996). Such temperature rises would be expected to produce rises in sea level of between 15 cm and 1 m.

7.8.2   HUMAN-INDUCED GLOBAL WARMING SCEPTICS

The Rio Convention did not close the debate on the alleged importance of human –induced global warming (see for instance Vaughan 2001). Although the Rio Convention is a legal document in international law, it did not specify particular limitations on the emissions of greenhouse gases (CO_2, etc.) and so the controversy raged on through the Kyoto Conference and beyond.

Although there was considerable scepticism among many scientists at the time of the publication of the IPCC's First Assessment in 1990 and in the decade or so following its publication, this was not organised or financially supported. The theory of enhanced global warming due to increased concentrations of greenhouse gases, notably CO_2, that was established by Arrhenius in 1896 was not seriously disputed. Nor was it disputed that there had been an ongoing increase in the production of greenhouse gases, notably CO_2, over the past century. What reasonably was (and still is) in dispute is the extent to which anthropogenic factors play a significant part in these phenomena. In the 1980s, global warming was high on the political agenda and scary scenarios were driving politicians to make unwise political decisions on, for example, limiting the use of fossil fuels (Vaughan 2001).

> Global warming is a political issue. Information provided by scientists is used to inform policy. Decisions about whether to take action and what should be done are taken in the public arena. Such decisions are political; they are subject to international diplomacy and the democratic process. Public-choice theory recognizes that participants in this process cannot help but bring with them their own private aims and incentives. This is an unavoidable, but not unassailable problem except when it is forgotten: when it is assumed that political action is altruistic. In this way, policies have been adopted in the name of averting damage to the planet from global warming which will not have the desired effect, even if society complies fully, because the aims of those influencing and deciding policy were not those stated.

Thus, begins the Foreword to *The Global Warming Debate* (Emsley 1996). This is one of a series of books published to "put the other side of the case and to refute current misconceptions concerning the existence of global warming." The first was entitled *Global Warming – Apocalypse or Hot Air?* (Bate and Morris 1994). These authors were not scientists but members of the Institute of Economic Affairs (IEA) Environment Unit, which was set up in 1993 to apply market analysis to environmental problems and to bring the results of that work to the attention of the general public. It should be pointed out that this book reflects its authors' own views as the IEA claims to have no corporate view. The authors set out to expose what they call the shoddiness of the apocalyptic predictions, and to try to "breach the barriers to the dissemination of good news and the media's natural preference for reporting only the disasters."

Anthropogenic emissions of greenhouse gases were suggested to be the main cause of global warming, and politically controversial and economically unpopular or unsound measures to reduce or at least to stem them have been proposed (the so-called protocols) at a number of summit meetings. This cause must be seen in context with other natural causes of warming and it is interesting to note the change in the tone of the IPCC over the decade or so since the publication of the First Assessment, as their models were improved – in the 1990–1992 reports, the IPCC was talking about a doubling of the CO_2 concentrations by 2100 being expected to result in a warming of $2°C$. In 1996, the conclusion stated "The balance of evidence suggests a discernible human influence on global climate through emissions of carbon dioxide and other greenhouse gases." But now the 1999 Draft Report of the IPCC contains scenarios in which even a doubling of economic activity does not increase emissions and goes on to say that "the future is so inherently unpredictable that views will differ on which of the scenarios could be more likely." Paul Stalpman of the US Environmental Protection Agency appears even to dismiss the need to enquire whether the Earth is warming. He told "the 1999 Earth Technologies Forum the real debate about climate change should not be about whether it exists or not, but what to do about it."

In the early years of global warming, there was far more scepticism and balance than would be permitted later on. A report in the Financial Times in 1990 on the IPCC's First Assessment Report ran beneath the subheadline "The evidence is much weaker than many pundits say." One IPCC contributor criticised the assessment report summary for not accurately reflecting the scientific discussions. "in the scientific papers, a great deal of care was devoted to pointing out the uncertainties," Andrew Solow of the Woods Hole Oceanographic Institution told *the Financial Times*'s David Thomas. It was a caricature to portray the climate debate as between an overwhelming majority committed to the position outlined in the policymakers' summary and a handful of mainly American-based scientists who regarded it as nonsense. According to Thomas, "A third, and large, group of scientists in the middle fears that the hype and the political pressures have pushed the scientific community beyond the bounds of the evidence."

A 1992 Gallup survey of 400 US experts conducted on behalf of the Washington, D.C.-based Center for Science, Technology & Media confirms this picture. While 60 per cent of those surveyed believed that global average temperatures had increased over the past century, only 19 per cent attributed this to human activities. Sixty-six per cent believed that human-induced global warming was under way but only 41 per cent believed that current scientific evidence supported this, and 70 per cent rated the media coverage as "fair" to "poor." "By following national media coverage, one would not gain an accurate view of the scientific debate over global warming," the center's Mark Mills commented.

The steady flow of IPCC Assessments after 1988 and the ongoing reports in the media kept the issue of enhanced global warming due to anthropogenic emissions of greenhouse gases in the public's eyes and on the international political agenda beyond the end of the 20th century and into the early years of the new century. Irrespective of the scientific justification, or otherwise, of the apocalyptic predictions, the alarms achieved one thing at least – a great deal of various governments' research funding has been diverted towards the environmental sciences, particularly to buy supercomputers and thus to run ever more elaborate models. Perhaps it should be asked whether the large amount of money spent in this way might not have been better spent solving real environmental problems by supporting research in biology, ecology, and agriculture (Emsley 1996).

While Vaughan (2001) did not set out to make a closely argued case against the doom scenarios, he did seek to arouse the awareness of readers to the sizeable sceptical science community whose point of view was rarely reported in the media and whose ideas were being dismissed out of hand by the Establishment. Important points were made in the publications by Bate and Morris (1994), Emsley (1996), and Bate (1998) and in the discussions now taking place on the Internet.

From reading the newspapers, watching television, or listening to politicians in the 1990s, there would appear to have been no doubt that there are going to be rising temperatures and rising sea levels and that these are being caused by increased pollution of the atmosphere by humankind. Anyone who tried to argue that these events will not occur, at least not on the scales predicted, or that maybe the changes would be welcome, or that humans are not mainly to blame, tended to be dismissed as either ignorant or complacent. It would appear that there was a consensus even among scientists. However, if we look more closely there is evidence of considerable disagreement – indeed violent disagreement from some eminent scientists – particularly in some heated web-based discussions. As early as 1992, in the run-up to the Rio Summit, Greenpeace carried out a survey of over 400 of the world's leading climatologists and found that only 15 admitted to believing in global warming although relying on it for their employment. It gave very little publicity to this finding! The Leipzig Declaration of 1995, signed by 1,500 scientists from around the world, disputed the IPCC's assertions about man-made global warming. However, a simple declaration disputing the IPCC's assertions, however many eminent scientists may have signed it, carried little weight and the declaration was not widely publicised. The organised opposition to the IPCC and its assessments only came later. But it was too late and the damage in terms of the politicisation of the IPCC and the domination of the international agenda by the Green Movement had been achieved and it has proved very difficult to attempt its reversal. Following the Leipzig Declaration, there was a period of scientific activity involving a detailed analysis of many of the IPPC's Assessments and of the science behind those Assessments. This activated the formation of the NIPCC (Non-governmental International Panel on Climate Change) and the publication of its first report (Singer 2008) and several subsequent reports; this will be discussed further in Section 6.7. The publication of the NIPCC's first report in 2008 provided an alternative view to that of the IPCC which could be regarded as scientifically based and not based on an appeal to authority and not pre-determined by any political agenda. The NIPCC Report which was published in 2008, 20 years after the first IPCC Assessment was too late and the damage, in terms of the politicisation of the work of the IPCC and the domination of the international agenda by the Green movement had been achieved.

There have been suggestions that opponents of the enhanced global warming hypothesis were deliberately excluded from the inner workings of the IPCC and that the political demand for consensus stifled public debate. For detailed criticism of the politics of the IPCC see, for example, Singer (1996) and Boehmer-Christiansen (1996). Vaughan (2001) quotes two thought-provoking quotations with which we finish this section. One is from Bate and Morris (1994) on the workings of the IPCC: "When confronted with this type of data ... the response by ardent global-warming advocates is typified by Dr Stephen Schneider – looking at every bump and wiggle of the record is a waste of time ... so I don't set very much store in looking at the direct evidence." The other from Dr Chris Folland of the UK Meteorological Office and a contributor to the IPCC policymaker's summary, "who went even further by saying 'the data don't matter . . .besides, we [the UN] are not basing our recommendations (for immediate reductions in CO_2 emissions) upon the data: we're basing them upon the climate models'"!

The IPCC's First Assessment Report had been rushed to provide a scientific imprimatur to the planned Ministerial Declaration directly following the Second World Climate Conference in Geneva. This had required a preparatory meeting several months beforehand to agree a draft text. William Kininmonth, head of Australia's National Climate Center, attended the preparatory meeting and the conference. The Australian delegation to the preparatory meeting was led by foreign affairs officials supported by representatives from environment, energy, and industry departments but was to have no science representation. After strong representations from Zillman, Kininmonth was offered a place as the token scientist as long as the meteorological budget would pay for him to go. "The whole activity was orchestrated to produce a particular outcome: a Ministerial Declaration recommending the UN take action to produce a treaty to reduce CO_2 emissions," Kininmonth recalls. Without actually knowing the outcome of the IPCC report (only the Summary for Policymakers of Working Group I had been circulated to governments), it was generally agreed (with square brackets to indicate provisional text) that the Ministerial Declaration should be for the UN to agree to negotiate a treaty to prevent dangerous climate change.

Copies of the IPCC report were distributed to delegates early in the last week of the conference itself. As it was closing, there was a motion from the podium that the report be accepted. "I doubt whether more than one in twenty had read the voluminous report in the time but nevertheless the report was accepted on a show of hands," Kininmonth says. Detailed differences were settled in back-room discussions. The Ministerial Declaration was then agreed and sent to the UN. The science would continue to be taken care of in the capable hands of Professor Bolin, chosen by consensus to be the IPCC's first chair in reflection, records Zillmann (2007), of his standing as a world leader in climate change.

In its latest Synthesis Report, the Intergovernmental Panel on Climate Change emphasised that human influence on the climate system is clear and growing. Many of the observed changes since the 1950s are unprecedented over decades to millennia and it is "extremely likely" that more than half of the observed increase in global average surface temperature from 1951 to 2010 was caused by the anthropogenic increase in greenhouse gas concentrations and other anthropogenic forcings together. In addition, they stressed that the more the human activities disrupt the climate, the greater the risks of severe, pervasive, and irreversible impacts for people and ecosystems. Recent econometric analysis confirms the above findings of climate science (e.g. Varentsov et al. 2019). By employing a time-varying conditional correlation specification for the first time in climate-related research, it was found that the effect of a change in anthropogenic forcings on global temperature changes started becoming significant only after WWII, with the correlation between the two jumping to 0.5, suggesting that half of the change is passed on to temperature, a finding similar to the IPCC. This identifies a nonlinear relationship between temperature and anthropogenic forcings, which indicates that the effect of humans on climate is very likely to become more important in the future. Thus, the severity of climate change impacts may worsen in the coming decades. Moreover, the analysis sheds light into the short-run effects of human activities on temperature change, a valuable insight for climate scientists which most previous studies have overlooked due to the methodology employed. Our results are robust to alternative specifications and to estimations with a different data set.

In general, the issue of global warming and climate change has been popular since the 1980s in media and politics when environmentalists claimed that the earth was in danger due to rising temperatures because of human activities particularly the emission of carbon dioxide through burning of fossil fuels. This argument was taken up by the UN Environmental program formed IPCC. The IPCC published several reports of global warming and climate change and forecasted very alarming picture of climate change. Since 1990, there have been two groups of scientists, one supporting the IPCC assessments and human-induced causes of global warming and the other disagreeing with the IPCC estimated and projected figures of global warming considering them to be overstated, fabricated, and manipulated based as they were upon computer models. They considered warming to be due to natural processes as had happened in the past (Akhtar et al., 2019), see the Milankovich cycles and Figure 1.7 and also the Dansgaard-Oeschger oscillations.

The debates around global warming between pro and anti IPCC scientists continue.

There is almost a consensus that global warming is taking place.

The differences are:

1. Whether it is natural or man-made (e.g. Figure 7.5).
2. Whether the rate is alarming or slow.
3. Whether the policy of the UN Climate Summit to replace fossil fuels energy resources with solar and wind is practical particularly for poor countries.

7.8.3 HOW DO WE DEFINE MEAN GLOBAL (OR GLOBAL MEAN) TEMPERATURE?

Predictions of global warming have been based mainly on computer simulation and these predictions have been criticised on various grounds. But, just as we may criticise the deficiencies in the models, we must also consider the validity of the scientific measurements. It is not so much the accuracy of individual measurements, though, as the way in which they are obtained, extrapolated, and interpreted that may cause confusion. Let us take as an example the measurement of mean global temperature. The question arises as to how the mean is calculated (Vaughan 2001). Temperatures vary widely both in space and time. There are vertical variations in temperature as well as differences across the surface of the Earth. There are short-term differences as a result of diurnal variations as well as longer-term differences as a result of seasonal effects. But do all measurements have the same significance, and, if not, on whose judgement are they weighted? Do all these measurements measure the same thing? Does "global" temperature have any scientific significance anyway?

The traditional method of measuring temperature is by using thermometers situated at different stations dotted around the world. Problems associated with this method relate to station inhomogeneity (Machin and Ruser, 2019). Historically these stations tended to be densest around the higher populated regions. There were no stations in the whole of Africa prior to 1860, only one from then till 1900 and by 1920 there were still only nineteen. Large portions of the land surface were not covered (and still are not covered). While satellite data does provide better spatial (and temporal coverage), it does not directly record temperature but the temperature is deduced from the observations using an algorithm that needs to be carefully calibrated. Moreover, the length of the records from the satellite data does not stretch far back in time. Most of the historical ground data has tended to be obtained from weather stations in or near urban areas where the heat-island effect is not entirely insignificant but has increased with time. Soon et al. (1999) presented a graph of temperature trends per decade between 1940 and 1996 against the population of the county (in California) over which they were measured. There is a linear increase from about 0.1°C for counties of about 20,000 inhabitants to over 0.5°C for counties having ten million inhabitants. The results from the best-situated rural station in the state gave a corresponding trend of -0.005°C per decade. Rural station coverage has fallen from

20 per cent of the Earth's area in the 1970s to 7 per cent in 1998. These varying factors must present a significant uncertainty in global average calculations. Attempts have been made to factor out some of these variables, but the final result will inevitably depend on human frailty (Letcher 2019).

Another problem arises with the way in which temperatures are measured and their accuracy. Although there are standard heights and conditions which should apply, we cannot be sure how well these criteria are satisfied nor how well equipment is maintained at any given weather station. Then again how are these temperatures obtained? Is a straight average of maximum and minimum taken, or is the average of values at certain times of the day used? Little metadata is available for many of these measurements to enable meaningful comparisons to be made. Different parts of the world have different time series of recorded temperatures, so any inferred historical trends will have inherent regional biases which again must be accounted for (Duchesne et al. 2019; Lin and Huybers, 2019).

In this context, Duchesne et al. (2019) used the world's largest single tree-ring data set (283,536 trees from 136,621 sites) from Quebec, Canada, to assess to what extent growth reconstructions based on these – and thus any similar – data might be affected by this problem. Indeed, straightforward growth rate reconstructions based on these data suggest a six-fold increase in radial growth of black spruce (*Picea mariana*) from ~0.5 mm yr^{-1} in 1800 to ~2.5 mm yr^{-1} in 1990. While the strong correlation ($R^2 = 0.98$) between this increase and that of atmospheric CO_2 could suggest a causal relationship, Duchesne et al. (2019) unambiguously demonstrated that this growth trend is an artefact of sampling biases caused by the absence of old, fast-growing trees (cf. "*slow-grower survivorship bias*") and of young, slow-growing trees (cf. "*big-tree selection bias*") in the data set. Thus, innovation will be needed before such data sets can be used for growth rate reconstructions.

Also, previously reported trends in daily monsoon rainfall since 1950 have been estimated using interpolated weather station observations released by the India Meteorological Department. The number of reporting weather stations changes over time, and poor coverage by weather stations can overlook extreme rainfall events. Lin and Huybers (2019) showed that by applying the interpolation of this changing network to satellite-based rainfall data, the changing coverage of weather stations in the Indian rainfall data leads to spurious increases in extreme rainfall. This suggests that previously reported trends of extreme rainfall are biased positive.

Sea-surface temperatures have traditionally been measured from opportunistic measurements made from ships. The distribution of such measurements will therefore follow closely the populated shipping lanes and virtually ignore inhospitable regions and seasons. The traditional method of measuring the temperature is to throw a bucket over the side and dip a thermometer in it, but nowadays it is more usual to use the temperature of the intake of cooling water. During the Second World War, because of the hazards of stopping ships to take measurements, it was noticed that there was a sudden jump in the temperature records coinciding with the change in measuring technique. It is tempting to use a blanket adjustment parameter in order to relate the two types of measurement – but how universally valid is it? Sea-surface temperatures can also be measured from space, but because of the skin effect the value measured can differ considerably from the bulk value obtained from ships and a great deal of effort has been expended by oceanographers on the determination and validation of sea-surface temperatures from infrared and microwave satellite data.

For longer-term trend determinations, indirect methods must be used. On a geological scale, ice-core records have been used. Gases trapped in air bubbles have been analysed for CO_2 content, the changes in concentration of this in the atmosphere being then related to changes in temperature. Not only is this relationship suspect but so is the chemistry; dubious assumptions have to be made about the solubility of CO_2 and its rate of diffusion in ice. On a shorter timescale, indirect methods – such as tree-ring diameters and records of, for example, the date on which certain flowers bloomed in China 2000 years ago, changes in migratory patterns of birds and diary entries of weather observations – can also contribute to the picture of climate trends (Lamb 1995).

More recently, satellite measurements have been able to provide a real measure of global temperatures. The problem here, though, is that again a different parameter is being measured, namely the temperature of the lower troposphere. However, the fact that satellites can give frequent, uniform global sweeps, oceans included, with no problems because of urban heat-island effects, and accurate to one-hundredth of a degree, suggests that this is the most meaningful and accurate method of all. It is therefore significant that temperatures measured in this way over the last 30 years show small fluctuations about a constant mean value (i.e. the warming is zero!). These measurements are confirmed by those made from radiosonde balloons, in contrast to the small but significant rise inferred from "terrestrial" measurements.

Over the past eight hundred thousand years, glacial–interglacial cycles oscillated with a period of one hundred thousand years ("100k world"1). Ice core and ocean sediment data have shown that atmospheric carbon dioxide, Antarctic temperature, deep ocean temperature, and global ice volume correlated strongly with each other in the 100k world. Between about 2.8 and 1.2 million years ago, glacial cycles were smaller in magnitude and shorter in duration ("40k world"7). Proxy data from deep-sea sediments suggest that the variability of atmospheric carbon dioxide in the 40k world was also lower than in the 100k world, but there are not direct observations of atmospheric greenhouse gases from this period. Yan et al. (2019) recently reported the recovery of stratigraphically discontinuous ice more than two million years old from the Allan Hills Blue Ice Area, East Antarctica. Concentrations of carbon dioxide and methane in ice core samples older than two million years have been altered by respiration, but some younger samples are pristine (Figure 7.6).

The recovered ice cores extend direct observations of atmospheric carbon dioxide, methane, and Antarctic temperature (based on the deuterium/hydrogen isotope ratio δD_{ice}, a proxy for regional temperature) into the 40k world. All climate properties before eight hundred thousand years ago fall within the envelope of observations from continuous deep Antarctic ice cores that characterise the 100k world. However, the lowest measured carbon dioxide and methane concentrations and Antarctic temperature in the 40k world are well above glacial values from the past eight hundred thousand years. The results of Yan et al., (2019) confirmed that the amplitudes of glacial–interglacial variations in atmospheric greenhouse gases and Antarctic climate were reduced in the 40k world, and that the transition from the 40k to the 100k world was accompanied by a decline in minimum carbon dioxide concentrations during glacial maxima.

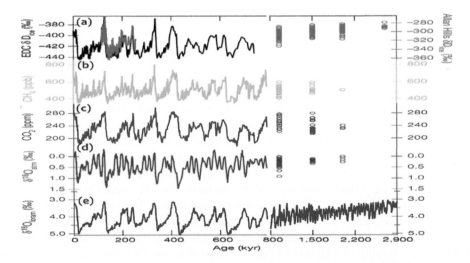

FIGURE 7.6 Health benefits of carbon mitigation. Life-years gained (a) overall and (b) per 100,000 population by region from the air quality improvements associated with the optimal decarbonisation in RICE + AIR. (c), (d) show the resulting monetised benefits in total, and as a percent of GDP, respectively. Note that if a region's $PM_{2.5}$ exposure (concentration) drops below $5.8 \mu g/m^3$, health benefits no longer accrue – this threshold assumption is common in other global air quality assessments and is tested in the sensitivity analyses (Modified from: Yan et al., 2019).

7.9 THE NONGOVERNMENTAL INTERNATIONAL PANEL ON CLIMATE CHANGE (NIPCC)

7.9.1 THE ORIGINS OF THE NIPCC

The Nongovernmental International Panel on Climate Change (NIPCC) is what its name suggests: an international panel of nongovernment scientists and scholars who came together to understand the causes and consequences of climate change. Because they are not predisposed to believe climate change is caused by human greenhouse gas emissions, they have been able to look at evidence that the IPCC ignores. Because they do not work for any governments, they are not biased towards the assumption that greater government activity is necessary. The NIPCC traces its roots to a meeting in Milan in 2003 organised by the Science and Environmental Policy Project (SEPP), a non-profit research and education organisation based in Arlington, Virginia, USA. SEPP was founded in 1990 by Dr. S. Fred Singer, an atmospheric physicist, and incorporated in 1992 following Dr. Singer's retirement from the University of Virginia. NIPCC is currently a joint project of SEPP, The Heartland Institute, and the Center for the Study of Carbon Dioxide and Global Change.

The Heartland Institute is a non-profit research and education organisation based in Arlington Heights, Illinois (USA). Heartland is approximately 5,500 men and women funding a non-profit research and education organisation devoted to discovering, developing, and promoting free-market solutions to social and economic problems. Heartland believes that ideas matter, and the most important idea in human history is freedom. Heartland has a full-time staff of 39. Five centres at the Heartland Institute conduct original research to find new ways to solve problems, turn good ideas into practical proposals for policy change, and then effectively promote those proposals to policymakers and the public. Heartland has a long and distinguished history of defending freedom. It is widely regarded as a leading voice in national and international debates over budgets and taxes, environmental protection, health care, school reform, and constitutional reform. For more information, visit the website at www.heartland.org. The Center for the Study of Carbon Dioxide and Global Change is a non-profit organisation based in Tempe, Arizona (USA). The Center produces a weekly online science newsletter called *CO₂ Science*.

The NIPCC has produced eight reports to date:

S. Fred Singer, ed. "Nature, Not Human Activity, Rules the Climate: Summary for Policymakers of the Report of the Nongovernmental International Panel on Climate Change", Chicago, IL: the Heartland Institute, 2008.

Climate Change Reconsidered: The 2009 Report of the Nongovernmental. International Panel on Climate Change (NIPCC)

Climate Change Reconsidered: The 2011 Report of the Nongovernmental. International Panel on Climate Change (NIPCC)

Climate Change Reconsidered II: Physical Science

Climate Change Reconsidered II: Biological Impacts

Scientific Critique of IPCC's 2013 "Summary for Policymakers"

Commentary and Analysis on the Whitehead & Associates 2014 NSW Sea-Level Report

Idso, R.M. Carter and S.F. Singer "Why Scientists Disagree About Global Warming" Chicago, IL: the Heartland Institute, 2015. This short book was a preliminary version of chapter 2 of the "Climate Change Reconsidered II" series which was subtitled "Benefits and Costs of Fossil Fuels" (2017).

These publications and more information about NIPCC are available at www.climatechangere considered.org.

In 2013, the Information Center for Global Change Studies, a division of the Chinese Academy of Sciences, translated and published an abridged edition of the 2009 and 2011 NIPCC reports in a single volume. On 15 June, the Chinese Academy of Sciences organised a NIPCC Workshop in Beijing to allow the NIPCC principal authors to present summaries of their conclusions.

In the same way that we did not go into all the details of the later IPCC reports, it does not seem appropriate to go into all the details of these reports either. We concentrate on just two of them, the first report (Singer 2008) and the little book on "Why Scientists disagree about Global Warming"

7.9.2 THE NIPCC REPORT OF 2008

We quote from the Foreword by Fred Seitz to the first NIPCC Report (Singer 2008:

"In his speech at the United Nations' climate conference on September 24, 2007, Dr. Vaclav, president of the Czech Republic, said it would most help the debate on climate change if the current monopoly and one-sidedness of the scientific debate over climate change by the Intergovernmental Panel on Climate Change (IPCC) were eliminated. He reiterated his proposal that the UN organise a parallel panel and publish two competing reports."

The present report of the Nongovernmental International Panel on Climate Change (NIPCC) does exactly that. It is an independent examination of the evidence available in the published, peer-reviewed literature – examined without bias and selectivity. It includes many research papers ignored by the IPCC, plus additional scientific results that became available after the IPCC deadline of May 2006.

The IPCC is pre-programmed to produce reports to support the hypotheses of anthropogenic warming and the control of greenhouse gases, as envisioned in the Global Climate Treaty. The 1990 IPCC Summary completely ignored satellite data, since they showed no warming. The 1995 IPCC report was notorious for the significant alterations made to the text *after* it was approved by the scientists – in order to convey the impression of a human influence. The 2001 IPCC report claimed the 20th century showed 'unusual warming' based on the now-discredited hockey-stick graph. The latest IPCC report, published in 2007 (i.e. IPCC 2007a, 2007b) completely evaluates the climate contributions from changes in solar activity, which are likely to dominate any human influence.

The foundation for NIPCC was laid five years ago when a small group of scientists from the United States and Europe met in 2003, during one of the frequent UN climate conferences. But it got going only after a workshop held in Vienna in April 2007, with many more scientists, including some from the southern hemisphere.

The NIPCC project was conceived and directed by Dr. S. Fred Singer, professor emeritus of environmental sciences at the University of Virginia. He should be credited with assembling a superb group of scientists who helped put this volume together.

Singer is one of the most distinguished scientists in the US. In the 1960s, he established and served as the first director of the U.S. Weather Satellite Service, now part of the National Oceanographic and Atmospheric Administration (NOAA), and earned a U.S. Department of Commerce Gold Medal Award for his technical leadership. In the 1980s, Singer served for five years as vice chairman of the National Advisory Committee for Oceans and Atmosphere (NACOA) and became more directly involved in global environmental issues.

Since retiring from the University of Virginia and from his last federal position as chief scientist of the Department of Transportation, Singer founded and directed the nonprofit Science and Environmental Policy Project, an organisation I am pleased to serve as chair. SEPP's major concern has been the use of sound science rather than exaggerated fears in formulating environmental policies.

Our concern about the environment, going back some 40 years, has taught us important lessons. It is one thing to impose drastic measures and harsh economic penalties when an environmental problem is clear-cut and severe. It is foolish to do so when the problem is largely hypothetical and not substantiated by observations. As the NIPCC shows by offering an independent, non-governmental 'second opinion' on the 'global warming' issue, we do not currently have any convincing evidence or observations of significant climate change from other than natural causes."

Since we have summarised some of the findings of the IPCC's reports it is only fair to summarise the NIPCC report too. The Preface to the Report started off:

> Before facing major surgery, wouldn't you want a second opinion?
>
> When a nation faces an important decision that risks its economic future, or perhaps the fate of the ecology, it should do the same. It is a time-honored tradition in this case to set up a 'Team B,' which examines the same original evidence but may reach a different conclusion. The Nongovernmental International Panel on Climate Change (NIPCC) was set up to examine the same climate data used by the United Nations-sponsored Intergovernmental Panel on Climate Change (IPCC).
>
> On the most important issue, the IPCC's claim that "most of the observed increase in global average temperatures since the mid-20th century is *very likely* (defined by the IPCC as between 90 to 99 percent certain) due to the observed increase in anthropogenic greenhouse gas concentrations," (emphasis in the original), NIPCC reaches the opposite conclusion – namely, that natural causes are very likely to be the dominant cause. Note: We do not say anthropogenic greenhouse (GH) gases cannot produce some warming. Our conclusion is that the evidence shows they are not playing a significant role. ...

Just as we have not found it appropriate to recall in enormous detail the findings of the IPCC's various assessments, so also we do not consider it appropriate to recall in enormous detail the findings of the NIPCC. We shall only indicate the outlines. In both cases it is up to the interested reader to consult the original documents for themselves.

The 2008 report by the Nongovernmental International Panel on Climate Change (NIPCC) focuses on two major issues:

- The very weak evidence that the causes of the current warming are anthropogenic
- The far more robust evidence that the causes of the current warming are natural.

It then addresses a series of less crucial topics:

- Computer models are unreliable guides to future climate conditions
- Sea-level rise is not significantly affected by rise in GH gases
- The data on ocean heat content have been misused to suggest anthropogenic warming. The role of GH gases in the reported rise in ocean temperature is largely unknown
- Understanding of the atmospheric carbon dioxide budget is incomplete
- Higher concentrations of CO_2 are more likely to be beneficial to plant and animal life and to human health than lower concentrations
- The economic effects of modest warming are likely to be positive and beneficial to human health

It concludes that our imperfect understanding of the causes and consequences of climate change means the science is far from settled. This, in turn, means proposed efforts to mitigate climate change by reducing greenhouse gas emissions are premature and misguided. Any attempt to influence global temperatures by reducing such emissions would be both futile and expensive (Section 10).

Before commenting on the specific failings of the IPCC's Fourth Assessment Report, it is important to clarify popular misunderstandings and myths:

- For about two million years, ice ages have been the dominant climate feature, interspersed with relatively brief warm periods of 10,000 years or so. Ice-core data clearly show that temperatures change centuries before concentrations of atmospheric carbon dioxide change (Petit, et al. 1999). Thus, there is no empirical basis for asserting that changes in concentrations of atmospheric carbon dioxide are the principal cause of past temperature and climate change.

- The proposition that changing temperatures cause changes in atmospheric carbon dioxide concentrations is consistent with experiments that show carbon dioxide is the atmospheric gas most readily absorbed by water (including rain) and that cold water can contain more gas than warm water. The conclusion that falling temperatures cause falling carbon dioxide concentrations is verified by experiment. Carbon dioxide advocates advance no experimentally verified mechanisms explaining how carbon dioxide concentrations can fall in a few centuries without falling temperatures.
- Carbon dioxide is a minor greenhouse gas and is tertiary in greenhouse effect behind water vapour (WV) and high-level clouds. All other things being equal, doubling carbon dioxide in the atmosphere will increase temperatures by about 1 degree Celsius. Yet, as discussed below, the computer models used by the IPCC consistently exaggerate this warming by including a positive feedback from WV, without any empirical justification.
- In his classic *Climate, History, and the Modern World*, H.H. Lamb (2002) traced the changes in climate since the last ice age ended about 10,000 years ago. He found extensive periods warmer than today and cooler than today. The last warm period ended less than 800 years ago. When comparing these climate changes with changes in civilization and human welfare, Lamb concluded that, generally, warm periods are beneficial to mankind and cold periods harmful. Yet the anthropogenic global warming (AGW) advocates have ignored Lamb's conclusions and assert that warm periods are harmful – without historical reference or knowledge. The second item which we would like to quote is the Key Findings from Idso et al. (2015a) "Why Scientists Disagree About Global Warming".

Key findings of this book include the following:

7.9.2.1 No consensus
- The most important fact about climate science, often overlooked, is that scientists disagree about the environmental impacts of the combustion of fossil fuels on the global climate,
- The articles and surveys most commonly cited as showing support for a "scientific consensus" in favour of the catastrophic man-made global warming hypothesis are without exception methodologically flawed and often deliberately misleading.
- There is no survey or study showing "consensus" on the most important scientific issues in the climate change debate.
- Extensive survey data show deep disagreement among scientists on scientific issues that must be resolved before the man-made global warming hypothesis can be validated. Many prominent experts and probably most working scientists disagree with the claims made by the United Nations' Intergovernmental Panel on Climate Change (IPCC).

7.9.2.2 Why scientists disagree
- Climate is an interdisciplinary subject requiring insights from many fields of study. Very few scholars have mastery of more than one or two of these disciplines.
- Fundamental uncertainties arise from insufficient observational evidence, disagreements over how to interpret data, and how to set the parameters of models.
- IPCC, created to find and disseminate research finding a human impact on global climate, is not a credible source. It is agenda-driven, a political rather than scientific body, and some allege it is corrupt.
- Climate scientists, like all humans, can be biased. Origins of bias include careerism, grant-seeking, political views, and confirmation bias.

7.9.2.3 Scientific method vs. political science
- The hypothesis implicit in all IPCC writings, though rarely explicitly stated, is that dangerous global warming is resulting, or will result, from human-related greenhouse gas emissions.

- The null hypothesis is that currently observed changes in global climate indices and the physical environment, as well as current changes in animal and plant characteristics, are the result of natural variability.
- In contradiction of the scientific method, the IPCC assumes its implicit hypothesis is correct and that its only duty is to collect evidence and make plausible arguments in the hypothesis's favour.

7.9.2.4 Flawed projections

- IPCC and virtually all the governments of the world depend on global climate models (GCMs) to forecast the effects of human-related greenhouse gas emissions on the climate.
- GCMs systematically over-estimate the sensitivity of climate to carbon dioxide (CO_2), many known forcings and feedbacks are poorly modelled, and modellers exclude forcings and feedbacks that run counter to their mission to find a human influence on climate.
- NIPCC estimates a doubling of CO_2 from pre-industrial levels (from 280 to 560 ppm) would likely produce a temperature forcing of 3.7 Wm^{-2} in the lower atmosphere, for about ~1°C of *prima facie* warming.
- Four specific forecasts made by GCMs have been falsified by real-world data from a wide variety of sources.

7.9.2.5 False postulates

- Neither the rate nor the magnitude of the reported late 20th century surface warming (1979–2000) lay outside normal natural variability.
- The late 20th century warm peak was of no greater magnitude than previous peaks caused entirely by natural forcings and feedbacks.
- Historically, increases in atmospheric CO_2 followed increases in temperature, they did not precede them. Therefore, CO_2 levels could not have forced temperatures to rise.
- Solar forcings are not too small to explain 20th century warming. In fact, their effect could be equal to or greater than the effect of CO_2 in the atmosphere.
- A warming of 2 °C or more during the 21st century would probably not be harmful, on balance, because many areas of the world would benefit from or adjust to climate change.

7.9.2.6 Unreliable circumstantial evidence

- Melting of Arctic sea ice and polar icecaps is not occurring at "unnatural" rates and does not constitute evidence of a human impact on the climate.
- Best available data show sea-level rise is not accelerating. Local and regional sea levels continue to exhibit typical natural variability – in some places rising and in others falling.
- The link between warming and drought is weak, and by some measures, drought decreased over the 20th century. Changes in the hydrosphere of this type are regionally highly variable and show a closer correlation with multidecadal climate rhythmicity than they do with global temperature.
- No convincing relationship has been established between warming over the past 100 years and increases in extreme weather events. Meteorological science suggests just the opposite: A warmer world will see milder weather patterns.
- No evidence exists that current changes in Arctic permafrost are other than natural or are likely to cause a climate catastrophe by releasing methane into the atmosphere.

7.9.2.7 Policy implications

- Rather than rely exclusively on IPCC for scientific advice, policymakers should seek out advice from independent, nongovernment organisations and scientists who are free of financial and political conflicts of interest.

- Individual nations should take charge of setting their own climate policies based upon the hazards that apply to their particular geography, geology, weather, and culture.
- Rather than invest scarce world resources in a quixotic campaign based on politicised and unreliable science, world leaders would do well to turn their attention to the real problems their people and their planet face.

The Heartland Institute has published several other reports (see their website), notably one on the *Benefits and Costs of fossil fuels*. However, we take the view that with the IPCC's Assessment and the NIPCC;s report of 2008 most of the important issues between the two sides have been addressed and the subsequent IPCC reports and NIPCC reports are fine-tuning and that neither side is ever going to convince the other side on the fundamental issue of the importance, or otherwise, of global warming due to anthropogenic CO_2 emissions. The whole question has become extremely political and so we now turn to the politics of the situation because our task is to address the understanding of global climate change and this understanding must inevitably include some discussion of the politics but without taking sides in the politics.

The fifth volume in the Climate Change Reconsidered series, *Climate Change Reconsidered II: Fossil Fuels,* produced by the Nongovernmental International Panel on Climate Change (NIPCC), was publicly released on 4 December 2018 in Katowice, Poland – the host city of the 24th session of the Conference of the Parties (COP 24) of the United Nations Framework Convention on Climate Change (UNFCCC).

7.10   POLITICS, MARGARET THATCHER AND JAMES HANSEN

We have already in Chapter 1 paid some attention to the evolution of the green movement with its initial attack on acid rain as an issue and the subsequent replacement of that as an issue by what we might call the "global warming scare" and the attack on fossil fuels, see Section 1.2. We noted in particular the role of Sweden based on the work described in the second book by Rupert Darwall (2017). An earlier book by Darwall (2013) gives a much more general discussion of the green movement and its involvement in the UNFCCC, its conferences of the parties, and the attempts to introduce binding international restrictions on the emissions of CO_2. We shall follow some of his work. Before doing that, we should make it clear that we accept that *homo sapiens* is the dominant species on the planet Earth and subordinates everything to its own interests with consequent disastrous effects on the environment, other species, and ecosystems. But this is a far wider issue than human-induced global warming arising from CO_2 emissions; see also Section 1.3 on sustainability.

In April 2008, Singer's Science and Environmental Policy Project (SEPP) and The Heartland Institute partnered to produce "Nature, Not Human Activity, Rules the Climate", subtitled "Summary for Policymakers of the Report of the Nongovernmental International Panel on Climate Change." The 48-page report listed 24 contributors from 14 countries and included a foreword by Frederick Seitz, one of the world's most renowned scientists. (Seitz passed away on 2 March 2008.) It was released at Heartland's First International Conference on Climate Change (ICCC-1) on 2-4 March 2008. The first full report, produced with a new partner, the Center for the Study of Global Warming and Global Change, was released in 2009. It was titled "Climate Change Reconsidered: The Report of the Nongovernmental International Panel on Climate Change (NIPCC)."

In 2011, the NIPCC produced its third report, "Climate Change Reconsidered: The 2011 Interim Report." The volume summarised new research produced after the deadline for inclusion in the 2009 report as well as some research that had been overlooked when the first volume was produced. In September 2013, NIPCC released *Climate Change Reconsidered II: Physical Science*, the first of three volumes expanding and bringing up-to-date the original 2009 report as well as offering a counter-point to the Intergovernmental Panel on Climate Change's Fifth Assessment Report. This was followed in 2014 by the second volume of *Climate Change Reconsidered II*, subtitled "Biological Impacts." In November, 2015, NIPCC released *Why Scientists Disagree About Global Warming: The NIPCC Report on Scientific Consensus*. We shall return to this issue in Section 7.12.

7.11   THE GREEN MOVEMENT AND HUMAN-INDUCED GLOBAL WARMING

In Section 1.2, we considered the origins of the modern Green movement or sustainable development. The Darwall book (2017) from which we have quoted was particularly concerned with Sweden and Germany. By choosing Sweden as an example, Darwall picked the ideal showcase of a western country where the government significantly managed to shape public opinion about environmentalism over several decades. But even more interesting is the role of Sweden's politicians especially during the 1960s and 1970s, who were critical in setting up various UN organisations that led, among others, to the United Nations Framework Convention on Climate – Change (UNFCCC), the Inter-Governmental Panel on Climate Change (IPCC) and the Kyoto Protocol. In his earlier book, Darwall (2013) was concerned with the politicisation of the IPCC reports much more generally. This book opens with the words:

> Global warming's entrance into politics can be dated with precision – 1988; the year of the Toronto conference on climate change, Margaret Thatcher's address to the Royal Society [in London], NASA scientist James Hansen's appearance at a congressional committee [in the USA] and the establishment of the Intergovernmental Panel on Climate Change (IPCC).

(Darwall 2013, page 1)

The World Conference on the Changing Atmosphere: Implications for Global Security was held in Toronto, Canada, on 27-30 June 1988 (WMO-UNEP 1989). The Brundtland Commission had submitted its report "Our Common Future" (World Commission on Environment and Development, 1987). The Brundtland Commission had been set up as a consequence of the UN General Assembly resolution 38/161 at the 38th Session of the UN in the autumn of 1983. and according to the Chairman's Foreword had been asked to formulate "A global agenda for change." Specifically, it was to propose long-term strategies to achieve sustainable development, to propose ways to lead to cooperation between countries over environmental issues, including the protection of the environment. The Canadian Government's special interest in air pollution led it to organise the Toronto conference in collaboration with the UN Environment Programme (UNEP) specifically to address the management of the atmosphere, which knows no national boundaries, as a global "common" WMO-UNEP 1989). (A common is something to which everyone has access but no-one feels responsible for maintaining or safeguarding it). Air pollution had long been on the environmental campaign agenda, particularly in connection with the acid rain scare and ozone depletion, and the Toronto conference was called to address the question of atmospheric pollution on a global scale. It was only during the conference that the question of the enhanced anthropogenic aspect of the greenhouse effect as a consequence of human activities came to the fore.

Thirty years after the Toronto Conference the *Climate Depot* opined (http://www.thegwpf.com/45646-2/ accessed 8 March 2019) "With the Greenhouse scare turning thirty this month, we remember the conference that launched it onto the global stage as the flagship cause of the Sustainable Development movement. Humanity is conducting an unintended, uncontrolled, globally pervasive experiment whose ultimate consequences could be second only to a global nuclear war. Most climate activists today would be too young to recall where it all began [with the Toronto Conference at which] "'greenhouse' warming exploded onto the global stage, with demands for an immediate policy response. So successful was this event that the 'Toronto Target' [a proposed reduction of 20% in greenhouse gas emissions] remained the benchmark for any government response to the climate emergency until the 'protocol' finally agreed in Kyoto, 1997" nearly ten years later.

Margaret Thatcher's address to the Royal Society in London was delivered on 27 September 1988 (https://www.margaretthatcher.org/document/107346 accessed 14 March 2019). As Darwall (2017, page 92) remarks "Told that the prime minister's speech was going to be on climate change, the BBC decided it wouldn't make the TV news." Darwall (same page) continues by discussing the preparation of this speech. In May 1984, Margaret Thatcher, the prime minister of the UK, asked her officials if any of them had any new policy ideas for the forthcoming G7 summit in London (actually strictly speaking it was still G8 not G7 because at that stage Russia was still a member of the group). Sir Crispin Tickell, whose 1977 book we have already mentioned and who was then a deputy undersecretary at the Foreign Office, suggested climate change and how it might figure in the G7 agenda. The eventual result was to make environmental problems a specific item, and the following statement was included in the London G7 communiqué in June 1984:

> 14. We recognize the international dimension of environmental problems and the role of environmental factors in economic development. We have invited Ministers responsible for environmental policies to identify areas for continuing cooperation in this field. In addition we have decided to invite the Working Group on Technology, Growth and Employment to consider what has been done so far and to identify specific areas for research on the causes, effects and means of limiting environmental pollution of air, water and ground where existing knowledge is inadequate, and to identify possible projects for industrial cooperation to develop cost – effective techniques to reduce environmental damage. The Group is invited to report on these matters by 31 December 1984. In the meantime we welcome the invitation from the Government of the Federal Republic of Germany to certain Summit countries to an international conference on the environment in Munich on 24–27 June 1984.

Darwall (2013), apparently incorrectly, claims that climate change was mentioned in the communiqué. The communiqué from the following Summit in Bonn in 1985 included the following statement in its section on Environmental Policies includes a statement on climate change (and the ozone layer):

> 12. New approaches and strengthened international cooperation are essential to anticipate and prevent damage to the environment, which knows no national frontiers. We shall cooperate in order to solve pressing environmental problems such as acid deposition and air pollution from motor vehicles and all other significant sources. We shall also address other concerns such as climatic change, the protection of the ozone layer and the management of toxic chemicals and hazardous wastes. The protection of soils, fresh water and the sea, in particular of regional seas, must be strengthened.

In 1987, Tickell was appointed the UK ambassador to the UN and informally was acting as Thatcher's envoy on global warming, his position at the UN making him privy to gossip from other nations. During the summer of 1988, Tickell suggested to Thatcher that she should make a major speech on global warming and she chose the Royal Society for it. In her speech, Thatcher addressed the Society as a scientist and a Fellow who also happened to be prime minister. Her main subject was environment policy. She stressed the importance of basic science and the fact that our present way of life is only possible because of the exploitation of the discoveries in basic science.

> Science and the pursuit of knowledge are given high priority by successful countries, not because they are a luxury which the prosperous can afford; but because experience has taught us that knowledge and its effective use are vital to national prosperity and international standing. But we need to guard against two dangerous fallacies: first that research should be driven wholly by utilitarian considerations; and second, the opposite, that excellence in science cannot be attained if work is undertaken for economic or other useful purposes." … "It is mainly by unlocking nature's most basic secrets, whether it be about the structure of matter and the fundamental forces or about the nature of life itself, that we have been able to build the modern world. This is a world which is able to sustain far more people with a decent standard of life than Malthus and even thinkers of a few decades ago would have believed possible. It is not only material welfare. It is about access to the arts, no longer the preserve of the very few, which the gramophone, radio, colour photography, satellites and television have already brought, and which holography will transform further." … "It is only when industry and academia recognise and mobilise each other's strengths that the full intellectual energy of [a country] will be released.

She then devoted a major part of her speech to the environment. She noted that research in medicine and agriculture had led to vast benefits to humanity and that engineering and scientific advances had led to enhanced opportunities for travel for leisure, business, and government. But these benefits have not been obtained without consequential disadvantages in terms of pollution by nitrates and emissions of CH_4, CO_2, and CFCs.

> For generations, we have assumed that the efforts of mankind would leave the fundamental equilibrium of the world's systems and atmosphere stable. But it is possible that with all these enormous changes (population, agricultural, use of fossil fuels) concentrated into such a short period of time, we have unwittingly begun a massive experiment with the system of this planet itself.
>
> Recently three changes in atmospheric chemistry have become familiar subjects of concern" [(a) the increase in greenhouse gases (CO_2, CH_4 and CFCs) leading to global warming, (b) the CFCs leading to damage to the ozone layer and (c) acid rain. In her conclusion she said] When [Sir] Arthur Eddington presented his results to this Society in 1919, showing the bending of starlight, it made headlines. It is reported that many people could not get into the meeting so anxious were the crowds to find out whether the intellectual paradox of curved space had really been demonstrated.

What Eddington's results, which were obtained in an eclipse earlier in 1919, demonstrated was that rays of light from a distant star were bent by the gravitational effect of the Sun as predicted by Einstein's theory of gravitation derived from his theory of general relativity but not predicted by Newton's theory of gravitation which had been tested experimentally by Henry Cavendish in 1797–8 and which successfully described the motions of the planets and the behaviour of the ocean tides on the Earth.

The reference to Eddington's test of the prediction of Einstein's theory of gravitation led Darwall (2013) to recall Karl Popper's discussion of the methodology of science, quoting Popper (2005) who argued that if we look for them it is easy to find confirmations for nearly every theory. "Only a theory which asserts or implies that certain conceivable events will not, in fact, happen is testable. The test consists in trying to bring about, with all the means we can muster, precisely these events which the theory tells us cannot occur." Thus, in this context, Newton's theory of gravitation was shown to be inadequate to describe Eddington's results, whereas Einstein's theory of gravitation based on general relativity did predict the results of Eddington's experiment. Eddington's results did not prove that Einstein's theory was "correct," only that it was superior to Newton's in that it predicted an observed effect which Newton's theory did not. There was little doubt about Arrhenius' work showing that atmospheric CO_2 led to a rise in temperature, the issue was to what extent any observed rise in global temperature was due to human-induced rising CO_2 concentrations and not to natural causes. Thatcher was a wise enough scientist to argue for detailed investigations "We need to consider in more detail the likely effects of change within precise timescales. And to consider the wider implications for policy – for energy production for fuel efficiency, for reforestation," and not to jump to pre-determined conclusions.

Popper (1959) argued that the criterion for assessing the scientific status of a theory should be its capacity to generate predictions that could, in principle, be refuted by empirical evidence, what Popper called its falsifiability, or refutability, or testability. Every good scientific theory is a prohibition. The more a theory forbids, the better it is. Scientists should therefore devise tests designed to yield evidence that the theory prohibits, rather than search for what the theory confirms. If we look for them, Popper argued, it is easy to find confirmation for nearly every theory. "Only a theory which asserts or implies that certain conceivable events will not, in fact, happen is testable. ... The test consists in trying to bring about, with all the means we can muster, precisely these events which the theory tells us cannot occur" In 1988, proponents of the theory that anthropogenic emissions of CO_2 were leading to a relentless rise in global temperature did not provide a similar black and white predictive test of their theory. It is therefore incapable of being falsified. The issue is not the capacity of CO_2 to lead to a rise in temperature, which had been predicted by Arrhenius in 1896, but the effect of anthropogenically increased levels of atmospheric CO_2 and other greenhouse gases on the temperature of the atmosphere. An answer can only be derived from empirical observation.

> Revelle and Suess's characterization of mankind, carrying out a large-scale geophysical experiment, further illustrates global warming's weakness as a scientific statement and its strength as a political idea. While prejudging the results of an experiment constitutes bad science, the proposition simultaneously generates powerful calls to halt the experiment before it is concluded. Yet questioning the science would inevitably be seen as weakening the political will to act. It created a symbiotic dependence between science and politics that marks 1988 as a turning point in the history of science and the start of a new chapter in the affairs of mankind.

(Darwall 2013)

James Hansen's appearance at a US congressional committee was on 23 June 1988, the record-breaking hot day of that year, when he told the Senate Energy and Natural Resources Committee that "The greenhouse effect has been detected and it is changing our climate now." It is interesting to compare his presentation with Margaret Thatcher's speech to the Royal Society in London slightly later in 1988 which we have already discussed. Her speech was a mature scientific and political approach to identifying a problem that needed investigating. The transcript of Hansen's presentation is available (https://www.sealevel.info/1988_Hansen_Senate_Testimony.html accessed 8 March 2019). Hansen presented only the results of some research carried out before then by his colleagues at the NASA Goddard Institute for Space Studies. By comparison with the massive effort that has subsequently gone into climate research in numerous different laboratories throughout the world and which the IPCC has attempted to summarise, this research was preliminary and necessarily inconclusive. Some people accuse the IPCC's First Assessment of not being sufficiently sceptical about the results presented at that stage. There is not a shred of scepticism to be detected anywhere in Hansen's presentation. The language is not even precise enough in talking about "the greenhouse effect" when he meant "the enhanced greenhouse effect" or "human induced global warming" Hansen drew three conclusions from the scanty results that he quoted. (1) "The Earth is warmer in 1988 than at any time in the history of measurements." Whether or not this is true, it proves nothing in relation to the warming being due to increased emission of greenhouse gases as a result of human activities. (2) "The global warming is now large enough that we can ascribe with a high degree of confidence a cause and effect relationship to the [enhanced] greenhouse effect." He went on later to claim 99% confidence in this relationship. (3) He claimed that their "computer climate simulations indicate that the [enhanced] greenhouse effect is already large enough to begin to effect [did he mean affect?] the probability of extreme events such as summer heat waves." Rather than express our own opinions about this presentation, we quote Darwall (2013) once more.

> Global warming's arrival in the world was announced with a blaze of fanfares heralding potential catastrophe. Alarmism went hand-in-hand with predictions of temperature increases that turned out to be excessive. Although warnings that civilization was doomed because economic activity was destroying the biosphere had become something of a routine, two things were different this time. First, mainstream political leaders from across the political spectrum quickly joined and amplified the chorus. Second, an institutional apparatus was constructed to keep attention on the issue. Unlike the 1972 Stockholm conference and the creation of the UNEP far away from the centres of power, the Intergovernmental Panel on Climate Change was an inter-governmental body with close and pervasive relations with its sponsoring governments. The rhythm of the publication of IPCC assessment reports would help feed media interest and keep governments engaged.

(Darwall 2013)

Finally, we have already mentioned the establishment of the IPCC in Section 1.2.6.

The main theme of Darwall's two books (2013, 2017) is that the IPCC's work was hijacked by the Green Movement. It was recognised that the IPCC itself is an advisory body not a regulatory body and so this was done by working through the United Nations and in particular through the UNFCCC (Section 1.2.2) and the establishment of the Kyoto Protocol (Section 6.3). This book is not the place to reproduce Darwall's detailed arguments and the reader who may be interested is recommended to refer to these books themselves. We shall, however, refer briefly to the consequences of this success of the Green movement in terms of the energy policies adopted by various governments in the following section.

The Green movement or sustainable development, had its origins in several different countries, notably the USA with "Silent Spring" (Carson 1962) (see Section 1.2.1) and parts of Europe, notably Sweden and Germany (see Section 1.2.2). We have already discussed the role of Sweden in the context of the acid rain scare in Section 1.2.2 and, in particular, the role played in the acid rain scare by Professor Bert Bolin. Sweden continued to play a leading role in the run up to the establishment of the IPCC. As interest in the acid rain scare declined (see Section 1.2.2) so attention began to switch to CO_2 emissions and once more Sweden played a prominent role. As on acid rain, Olof Palme, the prime minister of Sweden, deployed Bert Bolin to lead the science. In February 1975, Bolin produced *Energy and Climate*, described as a survey of current knowledge about how energy use could affect the Earth's climate. In 1975, a Swedish government bill on energy policy stated, "It is likely that climatic concerns will limit the burning of fossil fuels rather than the size of natural resources." In the early days of the Green movement in Sweden, it had been directed against atmospheric pollution and the burning of coal and thus its attention had been directed at acid rain. At that stage, there was a pro-nuclear power aspect to the Green movement. However, the accident at Chernobyl in 1986 changed all that. On 1 May 1986, Bolin was formally appointed the Swedish prime minister's scientific adviser.

The IGBP was one of the first beneficiaries. "Being at the time scientific advisor to the Swedish prime minister," Bolin (2007) commented "I was able to secure financial support from the Swedish government to develop" the IGBP "and to suggest that the secretariat be located in Sweden." The strands were coming together. From Stockholm, Bolin "informally channelled" the IGBP's preliminary results to the secretariat of the Brundtland Commission. The flow was mirrored reciprocated when Jim MacNeil, secretary of the Brundtland Commission, provided advance briefing to the tenth World Meteorological Congress in May 1987 in Geneva. Important as this was, such dialogue wasn't sufficient for Bolin: "An organ that provided an international meeting place for *scientists and politicians* to take responsibility for assessing the available knowledge concerning global climate change and its possible socio-economic implications was missing." (Darwall 2017).

The article by Zillmann (2007) "provides both an anecdotal and analytical view of the origin, establishment and operation of the IPCC and an evaluation of some of the main lines of criticism that have been directed at its role, its modus operandi and its reports" At this WMO conference, the idea of an intergovernmental panel was suggested. This (according to Darwall) "was exactly what Bolin wanted [and following] informal consultations with UNEP… the WMO and UNEP governing bodies passed resolutions to establish an ad hoc intergovernmental mechanism to carry out scientific assessments on the magnitude, timing, and potential impacts of climate change. Representatives of WMO and UNEP member states were invited to a meeting in Geneva in November 1988 to establish an intergovernmental Panel on Climate Change. By then, preparations were already under way for a climate conference to be held in Toronto in June 1988 immediately following the G7 summit that Canada was hosting. The AGGG was invited to assist. "In this way I was given the opportunity to work with the organizing committee," Bolin (2007).

The soon-to-be-formed IPCC couldn't be left to its own devices. With the institutional structure coming together, it became important to ensure that it would be steered in the right direction and produce the right answers. This was the purpose of two week-long workshops held in the autumn of 1987, the first in Villach at the end of September and the second in Bellagio on Lake Como in November. These meetings constitute important evidence as to the political nature and purpose of the embryonic IPCC. If the function of the IPCC were limited purely to collecting and evaluating scientific papers on climate change, why did its creators and cosponsors convene workshops on developing policy responses to climate change? Run-of-the-mill climate scientists couldn't be entrusted with the future of the project because they lacked an understanding of its deeper purpose. As the Bolsheviks knew, the revolution had to be guided by a revolutionary elite. In truth, the IPCC had a political agenda encoded in its DNA at its conception.

(Darwall 2017)

There are a large number of interest groups involved in the climate-change debate the result of which is a political process driven by diverse and perverse incentives. The general public can base their opinions only on the information which is fed to them by the media, and good news does not sell newspapers. Businessmen, transport managers, insurance companies, etc. have a vested interest in the outcome of the debate, and green pressure groups would have no need to exist if there were no crisis. The people who have most to gain are the politicians and the scientists. Any politician facing credibility problems at home may seek to become a statesman in the international arena where they are less accountable for their actions (Bate 1998). Those politicians fortunate enough to represent countries which will meet their emissions targets can score additional green points and can use the situation as a rod with which to beat their opponents.

Scientists are normally considered to be above suspicion, but it has been suggested that the action of some in this debate has been anything but altruistic. Reference was made to criticism of the IPCC but objectivity comes from open debate, and credibility from peer review. Those many scientists working in disciplines related to global warming have prospered in the last 10 years; indeed, in the UK more is spent on global climate research than on cancer research (Bate 1998). Journals rarely carry articles where the findings are largely negative – the results are just not exciting. An analysis of about 2,000 papers published between 1989 and 1995 picked at random from forty biological journals has shown that fewer than 9 per cent contained "non-significant' results" – the more prestigious journals fared the worst. Since credibility is linked to publishing in these prestigious journals, the temptation is to publish selectively, suggesting that the conclusion to be drawn is that results are biased in favour of global warming. In a zero-sum economy, it is very tempting not to kill the goose that lays the golden egg! Science will remain a servant of politics and should take great care in what it offers and how it responds to opportunities. "Short-termism may not only be the fate of politicians" (Boehmer-Christiansen 1996).

7.12   CONSENSUS

7.12.1   SCIENTIFIC CONSENSUS

One of the points that has come up from time to time so far in this book is the question of consensus and the nature of scientific proof, starting with the work of Kirill Kondratyev who argued that science does not advance on the basis of authority or consensus but on the following of scientific method. We do a piece of scientific work, we publish it, maybe in a refereed scientific journal, but in doing so neither the publisher, nor the editor (if any) of the journal nor indeed the author guarantees that the work is "correct." It is open to anyone else to attempt to repeat the work to demonstrate the validity or otherwise of the conclusions that the author drew. Appeals to authority, or appeals to consensus of "public opinion," have no place in scientific work. We would like to think that science should not be manipulated for political ends, though historically it often has been. Science should be used to provide evidence on which to base political opinions and governmental or intergovernmental actions. However, the question of the extent to which human activities are responsible for causing global warming is so political and so capable of stirring up such deep emotions, that inevitably the question of consensus does raise its head. Despite the fact that we believe it is misguided to seek to justify scientific conclusions by consensus among some sets of scientists (establishing scientific truth by opinion polls) instead of by reporting the results of scientific experiments (which of course include the running of computer models), we cannot ignore the fact that a considerable amount of effort has been devoted to running surveys to try to justify the IPCC's view of human induced global warming. As Idso et al. (2015b) ask in "Why Scientists Disagree About Global Warming", particularly in Chapter 1 from which we shall quote extensively, "Why debate consensus?". The answer is that environmental activists and their allies in the media often characterise climate science as an "overwhelming consensus" in favour of a single view that is sometimes challenged by a tiny minority of scientists funded by the fossil fuel industry to "sow doubt" or otherwise emphasise the absence of certainty on key aspects of the debate. This popular narrative grossly over-simplifies the issue while libeling scientists who question the alleged consensus.

> The global warming debate is one of the most consequential public policy debates taking place in the world today. Billions of dollars have been spent in the name of preventing global warming or mitigating the human impact on Earth's climate. Governments are negotiating treaties that would require trillions of dollars more to be spent in the years ahead.
>
> A frequent claim in the debate is that a "consensus" or even "overwhelming consensus" of scientists embrace the more alarming end of the spectrum of scientific projections of future climate change. Politicians including former President Barack Obama and government agencies including the National Aeronautics and Space Administration (NASA) claim "97 percent of scientists agree" that climate change is both man-made and dangerous.

As Idso et al. (2015b) continue, "the claim of 'scientific consensus' on the causes and consequences of climate change is without merit. There is no survey or study showing 'consensus' on any of the most important scientific issues in the climate change debate. On the contrary, there is extensive evidence of scientific disagreement about many of the most important issues that must be resolved before the hypothesis of dangerous man-made global warming can be validated." We paraphrase this by saying that seeking to establish scientific truth by opinion poll is contrary to scientific method.

The most influential statement of this alleged consensus appears in the *Summary for Policymakers* of the *Fifth Assessment Report* (AR5) from the Intergovernmental Panel on Climate Change (IPCC): "It is extremely likely (95%+ certainty) that more than half of the observed increase in global average surface temperature from 1951 to 2010 was caused by the anthropogenic increase in greenhouse gas concentrations and other anthropogenic forcings together. The best estimate of the human-induced contribution to warming is similar to the observed warming over this period" (IPCC, 2013, p. 17).

In a "synthesis report" produced the following year, the IPCC went further, claiming "Continued emission of greenhouse gases will cause further warming and long-lasting changes in all components of the climate system, increasing the likelihood of severe, pervasive and irreversible impacts for people and ecosystems. Limiting climate change would require substantial and sustained reductions in greenhouse gas emissions which, together with adaptation, can limit climate change risks" (IPCC, 2014, p. 8). In that same report, the IPCC expresses scepticism that even reducing emissions will make a difference: "Many aspects of climate change and associated impacts will continue for centuries, even if anthropogenic emissions of greenhouse gases are stopped. The risks of abrupt or irreversible changes increase as the magnitude of the warming increases" (p. 16)." (Idso et al. (2015b)) The media frequently report the IPCC's claims with sensational headlines.

"Climate science is a complex and highly technical subject so that simplistic claims about what most scientists believe are necessarily misleading. However, this has not prevented some politicians and activists from claiming that there is a "scientific consensus" or even "an overwhelming scientific consensus" that human activities are responsible for global warming and could have catastrophic effects in the future. The claim that "97 percent of scientists agree" appears on the websites of US government agencies such as the U.S. National Aeronautics and Space Administration (NASA, 2015) and even respected scientific organisations such as the American Association for the Advancement of Science (AAAS, n.d.), yet such claims are either false or meaningless"

The key findings of chapter 1 of Idso et al. (2015b) include the following:

- The most important fact about climate science, often overlooked, is that scientists disagree about the environmental impacts of the combustion of fossil fuels on the global climate.
- The articles and surveys most commonly cited as showing support for a "scientific consensus" in favo[u]r of the catastrophic man-made global warming hypothesis are without exception methodologically flawed and often deliberately misleading.
- There is no survey or study showing "consensus" on the most important scientific issues in the climate change debate.
- Extensive survey data show deep disagreement among scientists on scientific issues that must be resolved before the man-made global warming hypothesis can be validated. Many prominent experts and probably most working scientists disagree with the claims made by the United Nations' Intergovernmental Panel on Climate Change (IPCC).

This section reveals scientists do, in fact, disagree on the causes and consequences of climate change.

In May 2014, the US Secretary of State John Kerry warned graduating students at Boston College of the "crippling consequences" of climate change. "Ninety-seven percent of the world's scientists tell us this is urgent," he added . Three days earlier, President Obama tweeted that "Ninety-seven percent of scientists agree: #climate change is real, man-made and dangerous". What is the basis of these claims?

Gillis (2014) asks "What evidence is there for a "scientific consensus" on the causes and consequences of climate change? What do scientists really say? Any inquiry along these lines must begin by questioning the legitimacy of the question. Science does not advance by consensus or a show of hands. Disagreement is the rule and consensus is the exception in most academic disciplines. This is because science is a process leading to ever-greater certainty, necessarily implying that what is accepted as true today will likely not be accepted as true tomorrow. As Albert Einstein famously once said, "No amount of experimentation can ever prove me right; a single experiment can prove me wrong" (Einstein, 1996).

Still, claims of a "scientific consensus" cloud the current debate on climate change. Many people, scientists included, refuse to believe scientists and other experts, even scholars eminent in the field, simply because they are said to represent minority views in the science community. So what do the surveys and studies reveal?

7.12.2 SURVEYS ALLEGEDLY SUPPORTING CONSENSUS

Despite the fact that we place no confidence in attempts to establish scientific truth by opinion polls, we cannot ignore completely the fact that surveys do provide feedback that influences public opinion and we do take note of the surveys that have been carried out. We shall not go into the finer details of the surveys that have been conducted because in doing so it would be impossible to avoid creating the impression of taking sides in this highly contentious subject. Idso, Carter, and Singer note four surveys by Oreskes (2004), Doran and Zimmerman (2009), Anderegg et al. (2010), and Cook et al. (2013). If one has time, one can read the original papers or one can read the discussion of these papers in Idso et al. (2015b). However, these four papers are not all the material that is available and there is a longer list of relevant papers in a "Reply" by Cook et al. (2016) which has a useful list in its Table 1 of the various definitions of consensus which were used in the surveys. In terms of the topic of the present books "*understanding* global climate change," our view is that it is individuals who have to understand the issues and form their own opinions. We are working scientists not manipulators of public opinion or of national or international politics or governments.

The most well-known article claimed to be providing support for the idea of a consensus of scientists supporting human induced global warming is the article by Naomi Oreskes (2004). Oreskes examined abstracts from 928 papers published in scientific journals from 1993 and 2003, obtained by searching on the words "global climate change." She claimed that 75 per cent of the abstracts either implicitly or explicitly supported the IPCC's view that human activities were responsible for most of the observed warming over the previous 50 years while none directly dissented. The other three surveys just mentioned produced similar results, indeed claiming even higher levels of consensus (97%) see Table 1 of Cook et al. (2016). Being basically "bad news," these surveys attracted considerable media attention. One feature that emerged (Anderegg et al. 2010) was a claim that the average sceptic has been published about half as frequently as the average alarmist. The difference in productivity between alarmists and skeptics can be explained by several factors other than merit (Idso et al. 2015):

- Publication bias – articles that "find something," such as a statistically significant correlation that might suggest causation, are much more likely to get published than those that do not;
- Heavy government funding of the search for one result but little or no funding for other results – the US government alone paid $64 billion to climate researchers during the four years from 2010 to 2013, virtually all of it explicitly assuming or intended to find a human impact on climate and virtually nothing on the possibility of natural causes of climate change (Butos and McQuade, 2015, Table 2, p. 178);
- Resumé padding – it is increasingly common for academic articles on climate change to have multiple and even a dozen or more authors, inflating the number of times a researcher can claim to have been published (). Adding a previously published researcher's name to the work of more junior researchers helps ensure approval by peer reviewers (as was the case, ironically, with Anderegg *et al.* 2010);
- Differences in the age and academic status of global warming alarmists versus sceptics – climate scientists who are skeptics tend to be older and more are emeritus than their counterparts on the alarmist side; sceptics are under less pressure and often are simply less eager to publish.

One interpretation of what Anderegg *et al.* discovered is that a small clique of climate alarmists had their names added to hundreds of articles published in academic journals, something which many people would regard as un-ethical.

7.12.3 Evidence of lack of consensus

In addition to surveys that purport to establish a consensus of opinion in support of the IPCC's claim about human-induced global warming, it is important to refer to surveys which failed to find a consensus in favour of the IPCC's claims. Chapter 1 of Idso et al. (2015b) moves on to consider several of these. These surveys and studies generally suffer the same methodological errors as afflict the other surveys that we have described. The first was by Klaus-Martin Schulte (2008), a practising physician i.e. a medical doctor, who observed that "Recently, patients alarmed by the tone of media reports and political speeches on climate change have been voicing distress, for fear of the imagined consequences of anthropogenic 'global warming.' Consequently he reviewed the literature available on 'climate change and health' via PubMed (http://www.ncbi.nlm.nih.gov/sites/entrez)" and then attempted to replicate the conclusions of the Oreskes (2004) report.

He used the same search term on the same database and identified abstracts of 539 scientific papers published between 2004 and mid-February 2007. He found "a tripling of the mean annual publication rate for papers using the search term 'global climate change', and, at the same time, a significant movement of scientific opinion away from the apparently unanimous consensus which Oreskes had found in the learned journals from 1993 to 2003. Remarkably, the proportion of papers explicitly or implicitly rejecting the consensus has risen from zero in the period 1993–2003 to almost 6% since 2004. Six papers reject the consensus outright." Using Oreskes' methodology and if her findings for the period 1993 to 2003 were accurate, then Schulte found that scientific publications in the more recent period of 2004–2007 showed a strong tendency away from the consensus that Oreskes claimed to have found.

The German scientists Dennis Bray and Hans von Storch conducted several surveys in 1996, 2003, 2008, and 2010. The results attracted relatively little attention, though they are listed in Table 1 of Cook et al. (2016). The reasons for this are twofold: (a) their reports were presented in publications that are not commonly read by many people and (b) instead of confining themselves to yes/no type questions they included a popular opinion survey technique of asking respondents to express on a scale (in their case from 1 to 7) their level of agreement with a statement (in this case about the adequacy of data availability for climate analysis). With this survey technique, it is more difficult to obtain a definite result than using single yes/no questions. Verheggen *et al.* (2014) and Strengers et al. (2015) reported the results of a survey they conducted in 2012 in which they asked specifically about agreement or disagreement with IPCC's claim in its *Fifth Assessment Report* (AR5) that it is "virtually certain" or "extremely likely" that net anthropogenic activities are responsible for more than half of the observed increase in global average temperatures in the past 50 years. A total of 7,555 authors were contacted, including contributors to IPCC Reports, and 1,868 questionnaires were returned. The survey found that fewer than half of the respondents agreed with the IPCC's then most recent claims. Idso et al. (2015a) comment "This survey shows IPCC's position on global warming is the minority perspective in this part of the science community. Since the sample was heavily biased toward contributors to IPCC reports and academics most likely to publish, one can assume a survey of a larger universe of scientists would reveal even less support for IPCC's position."

There have also been a number of surveys of meteorologists and environmental professionals and each of these surveys found a majority opposed to the alleged consensus for the IPCC's view; for details, see Idso et al. (2015b).

Finally, there is the Global Warming Petition Project which is a statement about the causes and consequences of climate change and which was signed by 31,478 American scientists, including 9,021 with PhDs. The full statement reads:

"We urge the United States government to reject the global warming agreement that was written in Kyoto, Japan in December, 1997, and any other similar proposals. The proposed limits on greenhouse gases would harm the environment, hinder the advance of science and technology, and damage the health and welfare of mankind.There is no convincing scientific evidence that human release of carbon dioxide, methane, or other greenhouse gases is causing or will, in the foreseeable future, cause catastrophic heating of the Earth's atmosphere and disruption of the Earth's climate. Moreover, there is substantial scientific evidence that increases in atmospheric carbon dioxide produce many beneficial effects upon the natural plant and animal environments of the Earth."

We repeat our view that surveys of public opinion play no legitimate role in establishing scientific truth and, even if they did, the overall result of these surveys would be inconclusive. However, we quote yet again from Idso et al. (2015b).

"This is a remarkably strong statement of dissent from the perspective advanced by IPCC. The fact that more than ten times as many scientists have signed it as are alleged to have "participated" in some way or another in the research, writing, and review of IPCC's *Fourth Assessment Report* is very significant. These scientists actually endorse the statement that appears above. By contrast, fewer than 100 of the scientists (and nonscientists) who are listed in the appendices to IPCC reports actually participated in the writing of the all-important *Summary for Policymakers* or the editing of the final report to comply with the summary and therefore could be said to endorse the main findings of that report.

The Global Warming Petition Project has been criticised for including names of suspected non-scientists, including names submitted by environmental activists for the purpose of discrediting the petition. But the organisers of the project painstakingly reconfirmed the authenticity of the names in 2007, and a complete directory of those names appeared as an appendix to *Climate Change Reconsidered: Report of the Nongovernmental International Panel on Climate Change (NIPCC)*, published in 2009."

Critics may claim that we have devoted too much space to the question of consensus or lack of consensus. In our view, this is justified given the space that we have devoted to the history of the IPCC and its various Assessments Reports. We have quoted extensively from chapter 1 of Idso et al. (2015a) but there are other serious writings from which we could also quote. According to Prof. Mike Hulme of the Universities of East Anglia and of Cambridge, and a contributor to IPCC reports, "What is causing climate change? By how much is warming likely to accelerate? What level of warming is dangerous? – represent just three of a number of contested or uncertain areas of knowledge about climate change" (Hulme, 2009, p. 75). He continues "Uncertainty pervades scientific predictions about the future performance of global and regional climates. And uncertainties multiply when considering all the consequences that might follow from such changes in climate" (p. 83). As to the IPCC, he admits it is "governed by a Bureau consisting of selected governmental representatives, thus ensuring that the Panel's work was clearly seen to be serving the needs of government and policy. The Panel was not to be a self-governing body of independent scientists" (p. 95). We have already noted a similar comment by Rupert Darwall (2017). A consensus cannot be said to exist – and does not exist – until it has been established. Temporarily we give the last word to Idso et al. (2015b, page uvw) "There is no scientific consensus on global warming."

Recently, Idso et al., (2015a) reconsidered climate science issues and have reached the conclusions presented below.

7.12.3.1   Controversies

- Reconstructions of average global surface temperature differ depending on the methodology used. The warming of the twentieth and early twenty-first centuries has not been shown to be beyond the bounds of natural variability.
- General circulation models (GCMs) are unable to accurately depict complex climate processes. They do not accurately hindcast or forecast the climate effects of anthropogenic greenhouse gas emissions.
- Estimates of equilibrium climate sensitivity (the amount of warming that would occur following a doubling of atmospheric CO_2 levels) range widely. The IPCC's estimate is higher than many recent estimates.
- Solar irradiance, magnetic fields, UV fluxes, cosmic rays, and other solar activity may have a greater influence on climate than climate models and the IPCC currently assume.

7.12.3.2   Climate impacts

- There is little evidence that the warming of the twentieth and early twenty-first centuries has caused a general increase in severe weather events. Meteorological science suggests a warmer world would see milder weather patterns.
- The link between warming and drought is weak, and by some measures, drought decreased over the 20th century. Changes in the hydrosphere of this type are regionally highly variable and show a closer correlation with multidecadal climate rhythmicity than they do with global temperature.
- The Antarctic ice sheet is likely to be unchanged or is gaining ice mass. Antarctic sea ice is gaining in extent, not retreating. Recent trends in the Greenland ice sheet mass and Arctic sea ice are not outside natural variability.
- Long-running coastal tide gauges show the rate of sea-level rise is not accelerating. Local and regional sea levels exhibit typical natural variability.
- The effects of elevated CO_2 on plant characteristics are net positive, including increasing rates of photosynthesis and biomass production.

7.12.3.3 Why scientists disagree

- Fundamental uncertainties and disagreements prevent science from determining whether human greenhouse gas emissions are having effects on Earth's atmosphere that could endanger life on the planet.
- Climate is an interdisciplinary subject requiring insights from many fields of study. Very few scholars have mastery of more than one or two of these disciplines.
- Many scientists trust the Intergovernmental Panel on Climate Change (IPCC) to objectively report the latest scientific findings on climate change, but it has failed to produce balanced reports and has allowed its findings to be misrepresented to the public.
- Climate scientists, like all humans, can have tunnel vision. Bias, even or especially if subconscious, can be especially pernicious when data are equivocal and allow multiple interpretations, as in climatology.

7.12.3.4 Appeals to consensus

- Surveys and abstract-counting exercises that are said to show a "scientific consensus" on the causes and consequences of climate change invariably ask the wrong questions or the wrong people. No survey data exist that support claims of consensus on important scientific questions.
- Some survey data, petitions, and peer-reviewed research show deep disagreement among scientists on issues that must be resolved before the man-made global warming hypothesis can be accepted.
- Some 31,000 scientists have signed a petition saying "there is no convincing scientific evidence that human release of carbon dioxide, methane, or other greenhouse gases is causing or will, in the foreseeable future, cause catastrophic heating of the Earth's atmosphere and disruption of the Earth's climate."
- Because scientists disagree, policymakers must exercise special care in choosing where they turn for advice.

7.13 CLIMATE MODELS, C.P., SCC, AND IAMS

7.13.1 C.P. OR CETERIS PARIBUS

Economists have used models for many years and a feature of their use of models is that they will often qualify their conclusions by adding "c.p." or "ceteris paribus" (other things being equal). The purpose is to make it clear that it has been assumed that certain factors remain unchanged and acknowledges that possible changes in those factors have not been taken into account. Consequently, a reader is expected to exercise caution when considering the output of an economic model. It is abundantly clear that individual professional climate modellers are very conscious of the limitations of their models and share the same cautionary attitude to the outputs of their models; this is evidenced by the report of the recent Fifth Workshop on Systematic Errors in Weather and Climate Models (WSE) held in Montreal 19-23 June 2017 (Zadra et al. 2018).

It is worthwhile to quote from the conclusions of this workshop:

All model evaluation efforts reveal differences when compared to observations. These differences may reflect observational uncertainty, internal variability, or errors/biases in the representation of physical processes. The following list represents errors that were noted specifically during the meeting:

- convective precipitation – diurnal cycle (timing and intensity); the organisation of convective systems; precipitation intensity and distribution; and the relationship with column-integrated water vapor, SST, and vertical velocity;
- cloud microphysics – errors linked to mixed-phase, supercooled liquid cloud, and warm rain;

- precipitation over orography – spatial distribution and intensity errors;
- MJO [Madden-Julian Oscillations] modelling – propagation, response to mean errors, and teleconnections;
- subtropical boundary layer clouds – still underrepresented and tending to be too bright in models; their variation with large-scale parameters remains uncertain; and their representation may have a coupled component/feedback;
- double intertropical convergence zone/biased ENSO (El Niño Southern Oscillation) – a complex combination of westward ENSO overextension, cloud–ocean interaction, and representation of tropical instability waves (TIW);
- tropical cyclones – high-resolution forecasts tend to produce cyclones that are too intense, although moderate improvements are seen from ocean coupling; wind–pressure relationship errors are systematic;
- surface drag – biases, variability, and predictability of large-scale dynamics are shown to be sensitive to surface drag; CMIP5 mean circulation errors are consistent with insufficient drag in models;
- systematic errors in the representation of heterogeneity of soil;
- stochastic physics – current schemes, while beneficial, do not necessarily/sufficiently capture all aspects of model uncertainty;
- outstanding errors in the modelling of surface fluxes; errors in the representation of the diurnal cycle of surface temperature;
- errors in variability and trends in historical external forcings;
- challenges in the prediction of midlatitude synoptic regimes and blocking;
- model errors in the representation of teleconnections through inadequate stratosphere– troposphere coupling; and
- model biases in mean state, diabatic heating, SST; errors in meridional wind response and tropospheric jet stream impact simulations of teleconnections.

The workshop also addressed the question of ways forward, but that is a matter of technical details that do not need to concern us here. The point is that the above account of the workshop makes it abundantly clear that professional climate modelers are very serious and honest about the limitations of their models. However, this cautionary approach is completely lost by the time the results have been through the "factory" of the IPCC process with its predefined agenda – in spite of the IPCC's protestations to the contrary.

The general public does not have the time or expertise to study the many climate models now in use and their constantly updated versions, but in practice has to receive the results after they have been digested and put into executive summaries by the IPCC, which we have already argued has a predetermined agenda to provide evidence to support the hypothesis of human-induced global warming and consequently the need to restrict anthropogenic CO_2 emissions. It would be a healthy discipline if the outputs of climate models were to be qualified by "c.p." or, in full "ceteris paribus." We referred earlier to the fact that when Kirill Kondratyev pointed out the fact that the models used in the early days of the IPCC did not take various factors into account he was quickly side-lined from the corridors of power in the IPCC. Since then, various other scientists who have made this point have received a similar treatment.

Prior to 2015, under the Kyoto Protocol, the international climate policy under the UNFCCC focused on the goal of keeping the global-mean temperature increase below 2 °C relative to pre-industrial levels. The Paris Agreement reset this long-term goal to holding the increase well below 2 °C and pursuing efforts to limit it to 1.5 °C. Rogelj et al. (2018) consider a number of scenarios of suggested reductions in GHG emissions for input to model calculations intended to lead to model output predictions that would achieve this limitation of global warming. We regard this as part of the well-financed climate modelling industry which lacks the humility to qualify its outputs by "ceteris paribus." Another fairly recent publication which sounds promising is an IPCC publication by Flato et al. (2013) which runs to over 80 pages. The 2 1/3 page executive summary starts off "Climate models have continued to be developed since the AR4. This too we regard as part of the well-financed climate modelling industry which lacks the humility to qualify its outputs by "ceteris paribus." There is no doubt that climate modelling is alive and well and a very active and well-resourced research activity. Our questions would be

(1) to what extent does it contribute to our understanding of the climate?

and

(2) to what extent does it contribute to the health, wealth, and happiness of mankind?

7.13.2 SCC, THE SOCIAL COST OF CARBON, CARBON TAX

One of the weapons available to governments in the war against global warming is to introduce a carbon tax, i.e. a tax to be paid for every ton of carbon emitted as CO_2. This is widely regarded by economists as being an effective way to contribute to the reduction of emissions of CO_2 and thereby reducing human-induced global warming. The question arises as to the level at which to set the tax. Values of $11 per ton (Nordhaus 2011) and $200 per ton (Stern 2007) and various other values have been proposed. This leaves aside the question of whether the money that is collected is used to abate the adverse effects of global warming or just as a general source of revenue for the governments concerned. One of the consequences of the Kyoto Protocol and the Paris Agreement is the idea that individual countries include a carbon tax among the tools to be utilised to achieve their targets for the reduction of CO_2 emissions. This has led governments, particularly that of the US, to use Integrated Assessment Models (IAMs) to calculate the social cost of carbon (SCC) as a basis on which to fix a carbon tax. We need to discuss IAMs which have been around for about 20 years and to distinguish them from what might be described as pure climate models which have been around for more like 50 years. We have already seen that climate models are based on attempting to predict future climate based on the equations governing the behaviour of the atmosphere and factors influencing the atmosphere.

A useful source on IAMs is the article by Pindyck (2017) on the "Use and Misuse of Models for Climate Policy." Pindyck should not simply be written off as a sceptic. We shall quote extensively from this and related articles, particularly so as not to lose the style of the message. This article starts off from the economists' point of view of 20 years or so ago that there was a need to integrate climate science with the economic effects of GHG emissions and so Integrated Assessment Models (IAMs) were born. These early modelling efforts of Nordhaus (1991) and others more than two decades ago helped economists to understand how GHG emissions accumulate in the atmosphere, how that accumulation can affect global mean temperatures, and how higher temperatures might affect GDP and consumption. By including a social welfare function that values the flow of consumption over time, these models can also be used to illustrate the possible welfare effects of different GHG abatement policies and how those welfare effects depend on various parameters. In effect, these early IAMs can be viewed as pedagogical devices. In his book, "The Climate Casino" Nordhaus (2013) uses his Dynamic Integrated Climate and Economy (DICE) model to help explain – at a textbook level – how unrestricted GHG emissions can cause climate change and lead to serious problems in the future. He also utilises the model to illustrate some of the uncertainties we face when thinking about the climate system and when trying to predict the changes to expect under different policies. The problem arises when we take these models so seriously that we use them to try to evaluate alternative policies, come up with an "optimal" (i.e. welfare-maximising) policy, or estimate the SCC. Economists often build and use models to help elucidate the interconnections among variables, but usually they understand and are clear about the limits of those models. They know that a model can help to tell a story in a logically coherent way, but the model might not be able to provide the numerical details of the story. In other words, the model might not be suitable for forecasting or quantitative policy analysis. This is the case for the various versions of the Nordhaus DICE model, as well as the plethora of IAMs (most of which are much more complex than DICE) that have been developed over the past couple of decades. As explained in Pindyck (2013a), many of the key relationships and parameter values in these models have no empirical (or even theoretical) grounding and thus the models cannot be used to provide any kind of reliable quantitative policy guidance.

Pindyck (2013b) argued that Integrated Assessment Models (IAMs) "have crucial flaws that make them close to useless as tools for policy analysis (page 860)" and continued (Pindyck 2017) "that the problem goes beyond their "crucial flaws": IAM-based analyses of climate policy create a perception of knowledge and prevision that is illusory and can fool policymakers into thinking that the forecasts the models generate have some kind of scientific legitimacy. Despite the fact that IAMs can be misleading as guides for policy, they have been used by the US government to estimate the social cost of carbon (SCC) and evaluate tax and abatement policies". The crucial flaws to which Pindyck (2017) refers include:

(1) Certain inputs – functional forms and parameter values – are arbitrary, but they can have huge effects on the results the models produce. One example is the discount rate.
(2) We know very little about *climate sensitivity,* i.e. the temperature increase that would eventually result from a doubling of the atmospheric CO_2 concentration, but this is a key input to any IAM.
(3) One of the most important parts of an IAM is the *damage function,* i.e. the relationship between an increase in temperature and gross domestic product (GDP; or the growth rate of GDP). When assessing climate sensitivity, we can at least draw on the underlying physical science and argue coherently about the relevant probability distributions. But when it comes to the damage function, we know virtually nothing – there is no theory and are no data that we can draw from. As a result, developers of IAMs have little choice but to specify what are essentially arbitrary functional forms and corresponding parameter values.
(4) IAMs can tell us nothing about "tail risk," i.e. the likelihood or possible impact of a catastrophic climate outcome, such as a temperature increase above 5 °C, that has a very large impact on GDP. And yet it is the possibility of a climate catastrophe that is (or should be) the main driving force behind a stringent abatement policy.

Pindyck (2017) raises the issues of scientific honesty and the veneer of scientific legitimacy. Scientific honesty refers to the argument which is sometimes made that we have no choice – that without a model we will end up relying on biased opinions, guesswork, or even worse. Thus, we must develop the best models possible and then use them to evaluate alternative policies. In other words, the argument is that working with even a highly imperfect model is better than having no model at all. This might be a valid argument if we were honest and up-front about the limitations of the model. But often we are not. By the veneer of scientific legitimacy is meant the fact that models sometimes convey the impression that we know much more than we really do. They create a veneer of scientific legitimacy that can be used to bolster the argument for a particular policy. This is particularly true for IAMs, which tend to be large and complicated and are not always well documented. IAMs are typically made up of many equations; these equations are hard to evaluate individually (especially given that they are often ad hoc and without any clear theoretical or empirical foundation) and even harder to understand in terms of their interactions in a complex system. In effect, the model is just a black box into which we put in some assumptions about GHG emissions, climate sensitivity, discount rates, etc., and we get out some results about temperature change, reductions in GDP, etc. And although it is not clear exactly what is going on, since the black box is "scientific," we are supposed to take those results seriously and use them for policy analysis. Pindyck considers a couple of examples might help to clarify this point but does acknowledge that:

Developers of IAMs generally do try to base their models' equations as much as possible on climate science and economic principles, and the models I am aware of have much more content than the "Limits to Growth" models. The problem is that climate science and economic principles are limited in what they can tell us about how to specify and parameterise an IAM's equations, which is why the models cannot tell us much about the design of climate policy. Unfortunately, IAM developers and users have sometimes failed to be clear about the models' inadequacies, thereby overselling the validity of the models. The result is that policy makers who rely on the projections of IAMs, and have little or no understanding of how the models are built and how they work, can be misled. ... I doubt that the developers of IAMs have any intention of using them in a misleading way. Nevertheless, overselling their validity and claiming that IAMs can be used to evaluate policies and determine the SCC can end up misleading researchers, policymakers, and the public, even if it is unintentional. If economics is indeed a science, scientific honesty is paramount. ... We must make it clear that IAMs are not climate models and that we should not attempt to discredit climate models by association with IAMs. The level of uncertainty associated with IAMs is much greater than that associated with climate models. Our point is that (1) there is, nevertheless, a level of uncertainty associated with climate models and (2) there is a widespread feeling that the IPCC is less than honest in its discussion of the uncertainties associated with climate models because of its (concealed) remit to establish that human-induced global warming exists.

To illustrate the difficulties associated with trying to use an IAM to establish the level of tax, we consider the work of the U.S. Interagency Working Group (2010, 2013) which used three IAMs to attempt to estimate the SCC. With a judicious choice of parameter values (varying the discount rate is probably sufficient), these models will yield an SCC estimate as low as a few dollars per ton, as high as several hundred dollars per ton, or anything in between. Thus, a modeler whose prior beliefs are that a stringent abatement policy is (or is not) needed can choose a low (or high) discount rate or choose other inputs that will yield the desired results. If there were a clear consensus on the correct values of key parameters, this would not be much of a problem. But "putting it mildly there is no such consensus". The Interagency Working Group did not try to determine the "correct" values for the discount rate. Instead, it used middle-of-the-road assumptions about the discount rate (setting it at 3 per cent) as well as other parameters and arrived at an estimate of around $33 per ton for the SCC (recently updated to $39 per ton). But, as we have already noted, Nordhaus (2011) obtained an estimate for the SCC of $11 per ton, while on the other hand, Stern (2007) considered an extremely stringent abatement to be optimal, a result that is consistent with an SCC of more than $200 per ton.

Pindyck (2013b) argued that, while other parameters have some effects, it is the value assumed for the discount rate which has the greatest effect on the calculated SCC and, moreover, that the actual choice of IAM is largely irrelevant. The discount rate is usually defined (and calculated) as the present value of future reductions in GDP resulting from one additional ton of CO_2 emissions today. The problem here is that there is no consensus regarding the "correct" discount rate. (The Interagency Working Group simply chose a midrange number – 3 per cent – that the members of the group could all live with; the group's reports never claimed that this number was in any sense "correct.") Because reasonable arguments can be made for a low discount rate or a high rate, the modeler simply has too much flexibility in the choice of discount rate. If the modeler is at all biased towards a more or less stringent abatement policy, he/she can choose a discount rate accordingly. If one believes that we should use market-based discount rates (i.e. the rates we actually observe in financial markets) and a relatively low value for climate sensitivity, then $11 might be roughly the right number for the SCC. But if instead one believes (perhaps based on some kind of ethical argument regarding intergenerational welfare comparisons) that we should use a very low discount rate, then $200 or so might be the right number, especially if one also uses a larger value for climate sensitivity. The point here is that there is hardly any need for a model; decide on the discount rate and climate sensitivity and you pretty much have an estimate of the SCC. The model itself is almost a distraction. Economists are sharply divided on the discount rate that should be used for the analysis of climate change policy and long-lived public projects such as the construction of dams and bridges. There is a large and growing literature on the discount rates (plural, because some argue that the rate should decline over time) that should be used for very long-time horizons. For an overview, see Gollier (2013).

Pindyck (2013a,b) goes on to discuss catastrophic outcomes and to discuss that what really matters for the SCC is the likelihood and possible impact of a catastrophic climate outcome: a much larger than expected temperature increase and/or a much larger than expected reduction in GDP caused by even a moderate temperature increase. IAMs, however, simply cannot account for catastrophic outcomes. They share this feature with climate models. Developers of IAMs sometimes claim that their models do account for catastrophic outcomes because they include "tipping points" in the damage functions, such that the loss of GDP increases very sharply when temperature reaches some threshold. But as with the rest of the damage function, the specification of the threshold and the extent to which GDP decreases when the threshold is crossed are arbitrary and not based on any theory or empirics, and thus they cannot tell us much about would happen if the temperature increase turns out to be very large. The damage function, with or without "tipping points," can do little more than reflect the beliefs of the modeler. Stern (2013) provides a detailed discussion of how IAMs grossly underestimate (or ignore) possible catastrophic outcomes. Kriegler et al. (2009) used expert elicitation to estimate tipping points in the climate system and thus potential catastrophic outcomes, see Pindyck (2011, 2012). See Weitzman (2011, 2013) for discussions of how "tail risk" affects the SCC, and also the discount rate that should be used to calculate the SCC.

We summarise Pindyck's conclusions. Economists need to be honest and forthcoming about what they do and do not know about climate change and its impact. Just as financial economists should not try to sell "technical analysis" to investors, environmental economists should not claim that IAMs can forecast climate change and its impact or that IAMs can tell us the magnitude of the social cost of carbon.

"Atmospheric scientists have made great progress in understanding how weather patterns develop and change, but they don't claim to be able to forecast next month's weather or when the next hurricane will arrive. There has also been great progress in our understanding of the drivers of climate, how GHG emissions can affect climate, and (to a much lesser extent) how changes in climate can affect GDP and other economic variables. But that progress still does not enable us to build and use IAMs as tools for forecasting and policy analysis, and we would be deluding ourselves if we thought otherwise.

This does not mean that IAMs are of no use. … IAMs can be valuable as analytical and pedagogical devices to help us better understand climate dynamics and climate – economy interactions, as well as some of the uncertainties involved. But it is crucial that we are clear and up-front about the limitations of these models so that they are not misused or oversold to policymakers. Likewise, the limitations of IAMs do not imply that we have to throw up our hands and give up entirely on estimating the SCC and analyzing climate change policy more generally."

After discussing the peculiarities of electricity as an economic good, we address a closely related question: what is, and what is not, special about variable renewable electricity sources, such as wind and solar power? Previous literature suggests that VRE have specific properties that lead to "integration costs" (e.g. Sims et al. 2011, Holttinen et al. 2011, Milligan et al. 2011, NEA 2012, Ueckerdt et al. 2013a, Hirth et al. 2015). This study relates integration costs to heterogeneity and offers a new definition of integration cost that has a welfare-theoretical interpretation. We argue that VRE are not fundamentally different from dispatchable power plants: all technologies are subject to integration costs. However, it turns out that what is special about wind and solar power is not the existence but the size of integration costs.

8 Climate change
Energy resources–nuclear accidents

8.1 BACKGROUND

Substantial changes in the environment caused by human industrial activities, and their possible climatic impact, have become matters of very great concern, because variations in climate can seriously affect agricultural productivity and many other aspects of human activity. Serious attention has recently been given, for example, to the possible effects of the increase of CO_2 and other greenhouse gases in the atmosphere.

The influence of nuclear tests on weather and climate has also been widely discussed (Crutzen and Birks 1982, Turco *et al.* 1983, Harwell 1984, London and White 1984, US National Research Council 1985, Bach 1986, 1987, Berger and Labeyrie 1987, Ginzburg 1988, Kondratyev 1988, Pittock *et al.* 1989) in KC98.

In addition, there have been many studies of the climate-forming processes in the Arctic, whose principal feature is determined by the isolation of the ocean from the atmosphere by the ice cover (American Meteorological Society 1995, Nagurny 1986, US Arctic Research Commission 1989, Kondratyev *et al.* 1996 in KC98). In this connection, of special interest is the programme of studies in the Arctic prepared by the US Interagency Committee on the Arctic Study Policy. This programme is based on the Act on the Arctic Study Policy approved by the US Congress in 1984. According to this Act: (1) the Arctic is of critical importance for national defense, (2) the Arctic is a natural laboratory to study the effect of extreme climatic conditions on man, which may play an important role in the solution of defense problems, (3) special atmospheric conditions in the Arctic determine its unique importance for testing telecommunication systems in high latitudes, which is extremely important from the viewpoint of defense problems, and (4) developments of the arctic marine technologies are very interesting for economical extraction and transportation of energy resources, as well as for national defense. According to the Act, the Arctic is defined as the territories north of the Polar circle which belong to the USA and other countries, as well as all the US territories located north and west of the boundary formed by the rivers Porcupine, Yukon, and Cuzcoquim, and by the adjacent seas, including the Arctic Ocean, the Bering, Beaufort, and Chukcha Seas, as well as the Aleutian Chain.

8.2 NUCLEAR WAR AND CLIMATE

A few decades ago, at the time of extensive above-ground nuclear testing programmes, we were concerned about the effect of those nuclear explosions on the weather and, possibly, the climate but there are several other factors which should be considered and which, in the long term, may be very important. Tropical deforestation, in a number of developing countries, is often referred to in this context. However, human beings and domestic animals have been affecting the surface of the Earth for a long time.

In recent years, assessments of the possible impact of a nuclear war on the atmosphere and climate have become of great concern (Chylek *et al.* 1983, 1987, Berger 1984, Chylek and

Ramaswamy 1984, Robock 1984, Sagan 1984, Cess 1985, Crutzen and Hahn 1986, Malone *et al.* 1986, Penner 1986, Thompson and Schneider 1986, Brühl and Crutzen 1987, Velikhov 1987, Kondratyev 1988, Singer 1988, Vupputuri 1988, Pittock *et al.* 1989 in KC98). The complexity of estimating the possible climatic impacts of post-nuclear strong disturbances of the gaseous and aerosol composition of the atmosphere necessitates a thorough analysis of the contributions from numerous interacting factors and the role of numerous feedbacks, using not only simulation techniques but also possible natural analogues and data of observations after the nuclear tests of the late 1950s and early 1960s. In this connection, results should also be analysed from studies on possible climatic effects connected with increasing concentrations of CO_2 and other greenhouse gases as well as volcanic eruptions. Despite their extremely limited similarity, natural analogues contain signatures that help reduce the uncertainties of numerical modelling caused by simplifying assumptions.

The results from numerical modelling of the climatic impact of a possible nuclear war, which substantiate a well-known concept of nuclear winter, have lately been criticised in connection with the simplifications and assumptions which can seriously affect the final results. Basic assumptions involve: (i) the neglect of the atmospheric greenhouse effect and the initial patchy distribution of smoke aerosol, (ii) a very schematic simulation of the effects of coagulation and removal of smoke aerosol, (iii) the neglect of various feedback mechanisms between the effect of smoke aerosol on the wind field and the effect of atmospheric motions on the distribution of smoke, and (iv) interaction between the atmosphere and the ocean.

A group of United States workers developed a programme of studies on possible climatic impacts of a nuclear war, aimed at the reduction of uncertainties in respective estimates, based on the nuclear winter concept (National Climate Program Office of Science 1985 in KC98). Bearing in mind the insufficient reliability of both climate models and input data, they concluded that at present this concept of nuclear winter can be neither rejected nor accepted.

Without sharing the excessive caution of the US workers, we believe that the nuclear winter concept is realistic but one-sided. This conclusion is based on the following (Kondratyev 1988 in KC98): (i) results from an analysis of the effects on the atmosphere and climate of the nuclear tests in the late 1950s and early 1960s and (ii) theoretical estimates of the climatic effect of the post-nuclear atmospheric composition. Nuclear testing involved the formation of enormous amounts of post-nuclear NO_2, which led to an increase of solar radiation absorption by the atmosphere and a global climate cooling that could reach 10 °C in the case of a nuclear war. Theoretical estimates of the transformation of the greenhouse effect in the zones of explosions show the possibility of warming and subsequent changes of the sign of climatic change.

Thus, it seems certain that a nuclear war would trigger a global ecological catastrophe which would manifest itself through a strong instability as well as spatial and temporal variabilities of climate. The catastrophic consequences would affect the ozone layer. All this would annihilate civilization. Further studies are needed for a more reliable and detailed substantiation of possible impacts of a nuclear war on the atmosphere and climate.

The effect of the dynamics of permafrost on the hydrological regime is very important. Estimates show, for example, that with permafrost destroyed in the regions of the Lena and Enissey rivers the run-off for these rivers would be halved, which will affect the salinity regime of the arctic seas and the annual change of the extent of the arctic ice cover (the process of ice formation would slow down). Changes in the albedo of snow-ice cover caused by contamination can play a substantial role. This effect must be particularly considered in the analysis of the effect of a possible nuclear war on climate. More accurate consideration of the processes in the polar regions is a fundamental aspect of further improvements on climate models.

The processing of the round-the-year data from acoustic sub-ice sounding from 13 nuclear submarines, covering practically the whole of the Arctic Ocean, has made it possible to draw maps of the ice cover thickness distribution in the Arctic and to characterise the fields of sea ice drift (Wadhams 1994 in KC98). As McLaren et al. (1991 in KC98) have shown, these data can also be used also to retrieve the ice cover roughness.

In this connection, of great concern is the problem of a possible climatic impact of soot particles ejected to the troposphere and the stratosphere by fires caused by nuclear explosions in the atmosphere. During recent years, the assessments of the climatic implications of fires on the oil fields of Kuwait have been of serious concern (Parungo et al. 1992 in KC98).

The surface air temperature increase over the ocean in both hemispheres is intensified from the tropics to mid-latitudes. This trend is similar to a latitudinal variation in excess concentrations of the ^{14}C isotope with respect to its quantity before nuclear tests, observed near the ocean surface. Over the last 16 years, the oceanic warming has penetrated much deeper in the sub-tropics and mid-latitudes and in the equatorial zone, which is also similar to the variability of the observed depth of excess ^{14}C penetration. The warming is reduced with depth in both hemispheres in the 60°-0° latitudinal belt and propagates downward, from a maximum in the near-surface sub-tropical waters towards high latitudes. The maxima observed near 65° N and 60° S at depths of 250–750 m is a very interesting feature of temperature variations with depth. The second of these maxima is not far from the region of shelf ice of the Ross Sea.

Since the 16 yr period of integration is not long enough to reach an equilibrium state, the results obtained were compared with data for the interactive energy balance climate model/box model of the ocean. It was found that the sensitivity of the OSU model to external factors constitutes 0.72 K $W^{-1}m^2$, the global mean effective coefficient of the ocean-atmosphere heat exchange is 80 $Wm^{-2}K^{-1}$, and the effective coefficient of heat diffusion in the ocean $K = 3.2, 3.8$ and 1.5 cm^2 s^{-1} (at depths of 50, 250, and 750 m, respectively). The K value averaged by mass (2.25 cm^2 s^{-1}) is consistent with the estimates from data on the penetration into the ocean of tritium and ^{14}C produced by nuclear tests. A climate model taking account of a box diffusion at $K = 2.25$–2.50cm^2 s^{-1} simulates well the evolution of respective differences between surface air temperature and the temperature of the oceanic surface layer, obtained in the OSU model. This suggests that the heat transport from the surface deep into the ocean, according to this model, occurs at a rate consistent with the observed intensity of diffusion of tritium and ^{14}C.

An analysis of the annual mean, zonal mean profile of warming due to CO_2 concentration doubling has demonstrated that the heating of the surface layer intensifies from the tropics to mid-latitudes in both hemispheres and gradually penetrates deep into the ocean, reaching deeper waters in the sub-tropics and mid-latitudes compared with the equatorial belt. The meridional profile of warming is very similar to the observed latitudinal distribution of excess ^{14}C concentration (compared with its values prior to nuclear tests in the atmosphere). The calculations of zonal mean profiles of variations in the ocean heat balance components showed that the CO_2 concentration-doubling-induced heating of the ocean takes place largely due to the transport of heat to deep waters, whereas the meridional heat transport is of low importance.

8.3 NUCLEAR ENERGY, NUCLEAR WINTER

8.3.1 NUCLEAR ENERGY

After World War 2 scientists were found to be at all points in the spectrum of opinions regarding the acceptability or otherwise of nuclear power. As two extreme examples we take James Lovelock and Martin Ryle. Martin Ryle has already been discussed in the previous section as being passionately opposed to nuclear power and to nuclear weapons. At the other end of the spectrum, James Lovelock supports belief in favour of nuclear power to provide our needs for energy as fossil fuels run out or lead to damage to the climate and the environment.

8.3.2 Nuclear winter

There is a great deal reported in the literature on the effect of volcanic eruptions on the climate. The material thrown up in the eruptions forms clouds that shield the land from the incident sunlight and so the temperature drops. There was also some work done by Kirill Kondratyev on the cooling effect of dust storms in the Karakum Desert showing that the presence of atmospheric aerosols leads to a cooling effect.

In 1969, Paul Crutzen discovered that oxides of nitrogen, NOx could be an efficient catalyst for the destruction of stratospheric ozone, i.e. of the ozone layer. Following studies of the potential effects of NO generated by the engines on supersonic aeroplanes flying in the stratosphere in the 1970s, it was suggested by John Hampson (1974) in Nature that due to the creation of atmospheric NO by nuclear fireballs, a full-scale nuclear war could result in the depletion of the ozone layer, possibly subjecting the Earth to solar ultraviolet radiation for a year or more. In 1975, Hampson's hypothesis "led directly" to the US National research council (NRC) reporting on the models of ozone depletion following nuclear war in the book Long-Term Worldwide effects of Multiple Nuclear-Weapons Detonations. The threat to the ozone layer from oxides of nitrogen turned out to be much less serious than the threats from CFCs (chlorofluorocarbons).

At one extreme James Lovelock (his books) and at the other end Martin Ryle Physics World article. So, Russia does appear to interfere in western politics. The FBI has charged 13 Russians with trying to influence the last American presidential election, including the whimsical detail that one of them was to build a cage to hold an actor in prison clothes pretending to be Hillary Clinton.

Meanwhile, it emerges that the Czech secret service, under KGB direction, near the end of the Cold War had a codename ("COB") for a Labour MP they had met and hoped to influence – presumably under the bizarre delusion that he might one day be in reach of power.

There is no evidence that Jeremy Corbyn was a spy, or of collusion by Trump campaign operatives with the Russians who are charged. Yet the alleged Russian operation in America was anti-Clinton and pro-Trump. It was also pro-Bernie Sanders, and pro-Jill Stein, the Green candidate – who shares with Vladimir Putin a strong dislike of fracking. The Keystone Cops aspects of these stories should not reassure.

The interference by Russian agents in western politics during the Cold War was real and dangerous. A startling example from the history of science has recently been discussed in an important book about the origins of the environmental movement.

In June 1982, the same month as demonstrations against the NATO build-up of cruise and Pershing missiles reached fever pitch in the West, a paper appeared in *AMBIO*, a journal of the Royal Swedish Academy of Sciences, authored by the Dutchman Paul Crutzen and the American John Birks. Crutzen would later share a Nobel Prize for work on the ozone layer. The 1982 paper, entitled *The Atmosphere after a Nuclear War, Twilight at Noon*, argued that, should there be an exchange of nuclear weapons between NATO and the Soviet Union, forests and oil fields would ignite and the smoke of vast fires would cause bitter cold and mass famine: "The screening of sunlight by the fire-produced aerosol over extended periods during the growing season would eliminate much of the food production in the Northern Hemisphere."

Carl Sagan, astronomer turned television star, then convened a conference on the "nuclear winter" hypothesis in October 1983, supported by leading environmental and anti-war pressure groups from Friends of the Earth to the Audubon Society, Planned Parenthood to the Union of Concerned Scientists. Curiously, three Soviet officials joined the conference's board and a satellite link from the Kremlin was provided.

In December1983, two papers appeared in the prestigious journal *Science*, one on the physics that became known as TTAPS after the surnames of its authors, S being for Sagan; the other on the biology, whose authors included the famous biologists Paul Ehrlich and Stephen Jay Gould as well as Sagan. The conclusion of the second paper was extreme: "Global environmental changes sufficient to cause the extinction of a major fraction of the plant and animal species on Earth are likely. In that event, the possibility of the extinction of *Homo sapiens* cannot be excluded."

Who started the scare and why? One possibility is that it was fake news from the beginning. When the high-ranking Russian spy Sergei Tretyakov defected in 2000, he said that the KGB was especially proud of the fact "it created the myth of nuclear winter." He based this on what colleagues told him and on research he did at the Red Banner Institute, the Russian spy school.

The Kremlin was certainly spooked by NATO's threat to deploy medium-range nuclear missiles in Europe if the Warsaw Pact refused to limit its deployment of such missiles. In Darwall's version, based on Tretyakov, Yuri Andropov, head of the KGB, "ordered the Soviet Academy of Sciences to produce a doomsday report to incite more demonstrations in West Germany." They applied some older work by a scientist named Kirill Kondratyev on the cooling effect of dust storms in the Karakurn Desert to the impact of a nuclear exchange in Germany.

Tretyakov said: "I was told the Soviet scientists knew this theory was completely ridiculous. There were no legitimate facts to support it. But it was exactly what Andropov needed to cause terror in the West." Andropov then supposedly ordered it to be fed to contacts in the western peace and green movement.

It certainly helped Soviet Propaganda. From the Pope to the Campaign for Nuclear Disarmament to the non-aligned nations, calls for NATO's nuclear strategy to be rethought came thick and fast. A Russian newspaper used the nuclear winter to inveigh against "inhuman aspirations of the US imperialists, who are pushing the world towards nuclear catastrophe." The award of the Nobel peace prize in 1985 to the prominent Russian doctor Evgeny Chazov specifically mentioned his support for the nuclear winter theory.

"Propagators of the nuclear winter thus acted as dupes in a disinformation exercise scripted by the KGB," concludes Darwall. We can never be entirely certain of this because Tretyakov's KGB colleagues may have been exaggerating their role and he is now dead. But that the KGB did its best to fan the flames is not in doubt.

It soon became apparent that the nuclear winter hypothesis was plain wrong. As the geophysicist Russell Seitz pointed out, "soot in the TTAPS simulation is not up there as an observed consequence of nuclear explosions but because the authors told a programmer to put it there." He added "The model dealt with such complications as geography, winds, sunrise, sunset and patchy clouds in a stunningly elegant manner – they were ignored!' The physicist Steven Schneider, concluded that "the global apocalyptic conclusions of the initial nuclear winter hypothesis can now be relegated to a vanishingly low level of probability."

The physicists Freeman Dyson and Fred Singer, who would end up on the opposite side of the global-warming debate from Schneider and Seitz, calculated that any effects would be patchy and short-lived, and that while dry soot could generate cooling, any kind of dampness risked turning a nuclear smog into, a warming factor and a short-lived one at that.

By 1986, the theory was effectively dead, and so it has remained. A nuclear war would have devastating consequences, but the impact on the climate would be the least of our worries.

The stakes were higher in the Cold War than today. The Soviet peace offensive secured the support of many western intellectuals and much of the media, and very nearly prevailed.

8.4   THE BIG MISTAKE SURROUNDING THE IPCC AND THE UNFCCC

The League of Nations was founded after the Great War of 1914-8 but was ineffective in preventing the Second World War. The United Nations has over the years passed many resolutions which have never been implemented. It is tempting to cite the numerous resolutions relating to the State of Israel and its treatment of the Palestinian people in the occupied territories but there are many others, for example referring to the recent occupation of the Crimea by Russia, where the resolutions are ignored by powerful states if the resolutions conflict with their own special interests. It is perhaps instructive to consider why it is that the Montreal Protocol was more or less successful while the Kyoto Protocol has been, if we are honest, a failure.

8.4.1 THE MONTREAL PROTOCOL AND THE KYOTO PROTOCOL

January 2019 was the 21st anniversary of the coming into force of the Montreal Protocol, which is concerned with limiting the production and use of ozone-destroying chemicals. It was discovered in the mid-1970s that some human-produced chemicals such as chlorofluorocarbons (CFCs) and halons could destroy ozone and deplete the ozone layer, which resides naturally in the stratosphere (see, for example, Kondratyev and Varotsos 2000). These chemicals are produced from industrial and agricultural activities and lead to increases in the concentrations of chlorine, bromine, nitrogen, and hydrogen radicals, which are involved in the destruction of ozone in the stratosphere. The resulting increase in ultraviolet (UV) radiation at the Earth's surface may increase the incidences of skin cancer and eye cataracts (Varotsos and Cracknell 2004, Varotsos 2005). Remote sensing has played a key role in identifying the cause of ozone depletion and especially the Antarctic ozone hole, which was first detected experimentally (Farman *et al.* 1985).

In response to the prospect of increasing ozone depletion, the governments of the world crafted the 1987 United Nations Montreal Protocol, which came into force in 1989, as an international approach to address this global issue. The Protocol, now ratified by over 190 nations, establishes legally binding controls on the national production and consumption of ozone-depleting gases. Production and consumption of all the principal halogen-containing gases by developed and developing nations will be significantly phased out before the middle of the 21st century.

As a result of the broad compliance with the Protocol and its Amendments and Adjustments and, of great significance, industry's development of "ozone-friendly" substitutes for the now-controlled chemicals, the total global accumulation of ozone-depleting gases has slowed down. This has reduced the risk of further ozone depletion. Now, with continued compliance, the recovery of the ozone layer is expected by the late 21st century (Varotsos *et al.* 2000). The results are shown in Figure 8.1 for various scenarios. In each case, production of a gas is assumed to result in its eventual emission to the atmosphere.

It is projected that without the Montreal Protocol and with continued production and use of CFCs and other ozone-depleting gases, effective stratospheric chlorine would have increased tenfold by the mid-2050s compared with the 1980 value. It is likely that such high values would have increased global ozone depletion far beyond that currently observed. As a result, harmful UV-B radiation would have increased substantially at the Earth's surface, causing a rise in skin cancer cases (Kondratyev *et al.* 1995 in AC98, Varotsos *et al.* 1995 in AC98).

Nowadays, there is a consensus in the scientific community that anthropogenic activities have resulted in the increase of tropospheric ozone and the decrease of stratospheric ozone (Varotsos *et al.* 1994). In particular, the stratospheric ozone depletion is considered as one of the strongest anthropogenic signals in the Earth's system.

Nearly two decades after the implementation of the Montreal Protocol, stratospheric ozone loss is as severe as ever over the Arctic and the timing of ozone recovery is uncertain (Reid *et al.* 1994, Varotsos and Kirk-Davidoff 2006). In this context, the "Ozone layer and UV radiation in a changing climate" (ORACLE-O3) project running during the International Polar Year (IPY) 2007–2008 is concerned with quantification of polar ozone losses in both hemispheres by using ground-based and satellite observations.

After a long and highly successful career in the use of remote sensing in atmospheric physics, Prof. Kondratyev turned his attention to ecological and sociological problems. In the editorial preface to a special issue edited by Cracknell and Varotsos (2008), it was noted that in those later years Prof. Kondratyev became extremely critical of the Intergovernmental Panel on Climate Change (IPCC) and its work and the reasons for this were identified. Following the publication of that special issue, the following question has been put: "Given Prof. Kondratyev's criticisms of the IPCC

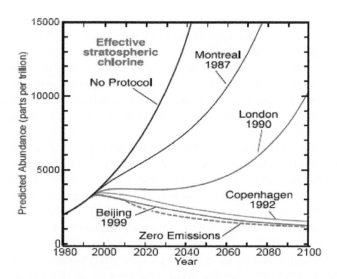

FIGURE 8.1 Long-term assessment of the abundance of the effective stratospheric chlorine for the following cases: (*a*) no Protocol and continued production increases of 3% per year (business-as-usual scenario); (*b*) continued production and consumption as allowed by the Protocol's original provisions agreed upon in Montreal in 1987; (*c*) restricted production and consumption as outlined in the subsequent Amendments and Adjustments as decided in London in 1990, Copenhagen in 1992, and Beijing in 1999; and (*d*) zero emissions of ozone-depleting gases starting in 2007 (WMO 2007).

and the Kyoto Protocol, was he also opposed to the Montreal Protocol?" The response of Cracknell and Varotsos (2008) was verbatim: "It is our view that the simple answer to this question is 'No'. But, of course, such a simple answer needs to be explained or justified."

Examination of the extensive writings of Prof. Kondratyev over the last 10–15 years of his life, that is up to 2006, reveals several examples of instances where he criticises the climate models used by the IPCC (e.g. Kondratyev 1997, 1998, 2004, Kondratyev and Cracknell 1998a) and where he was trying to influence the decisions of policy makers in Russia in the year or so before the Russian ratification of the Kyoto Protocol. Prof. Kondratyev also wrote extensively on atmospheric ozone and he co-authored an extensive monograph devoted entirely to atmospheric ozone (Kondratyev and Varotsos 2000). He discussed the Montreal Protocol in several publications (Kondratyev 1998, Kondratyev and Varotsos 2000). In his writings he recounted the science of anthropogenic causes of the depletion of stratospheric ozone. He was not involved in the politics surrounding the Montreal Protocol and we have found nothing in his writings that would support any suggestion that he opposed the Montreal Protocol.

This leads to another question, namely "Was the science of anthropogenic ozone depletion leading to the Montreal Protocol any more certain than the science of global warming leading to the Kyoto Protocol?" Views may differ on the answer to this question. But one important difference is that, in the case of ozone depletion, the issue in relation to the manufacture of a small group of chemical compounds that play a somewhat minor role in our lives, non-ozone-destroying substitutes, were known to be available and assistance was provided to developing countries to comply with the Montreal Protocol. A useful summary is given in box 21.2 of the Stern (2007) review. However, in the case of the Kyoto Protocol, the issues were more complex, the materials (fuels) involved play a much more important role in our lives, no simple substitute for the emission of greenhouse gases is easily available on the scale required to prevent global warming, and potential solutions were seen as severely damaging to national economies and their development (Kondratyev and Varotsos 1995).

The scientific investigations of total ozone changes (especially in high latitudes), in which remote sensing played a major role, put the total ozone variability (depletion) problem forward as a very important aim in studying global-scale environmental dynamics of great significance for humans and ecosystems. The recognition of such an environmental danger led to the signing of the Montreal Protocol and further amendments to the Protocol to avoid dangerous environmental consequences of ozone depletion, even though there were, and still are, unsolved scientific problems remaining. There are many lessons of the Montreal Protocol that can be applied to solving other global environmental issues.

The first lesson is the application of the "precautionary principle," that is taking necessary action in time to prevent damage rather than waiting until the damage has been proved, by which time the damage could have been extensive and irreversible. Another important lesson of the Montreal Protocol is on how to act on an issue when there is no scientific certainty (in 1987, when the Protocol was originally signed, many questions remained unanswered). The solution has been to undertake successive steps to phase out ozone-depleting substances and involve the scientific community in advising governments periodically on the further steps needed to protect the ozone layer and recommend alternate technologies. During the past 15 years, the governments changed the Protocol four times in accordance with relevant scientific advice. One more lesson of the Protocol is in promoting universal participation, including developing countries, in the Protocol by recognising "common and differential responsibility." Such a feature resulted in almost all countries committing themselves to protection of the ozone layer. Another important lesson is the integration of science, economics, and technology with politics, both in developing the control measures and in implementing them.

Nowadays, an urgent unsolved problem in ozone science is the poorly studied heterogeneous chemical processes taking place in aqueous aerosols or on solid aerosols. In particular, special attention must be paid to the fact that the chemical activity of solid particles at various altitudes (various pressures and temperatures) depends strongly on their physical properties and/or their composition (Varotsos et al. 2000; Varotsos 2005).

It should be stressed that "warming" and "ozone depletion" events are linked between them. A very recent investigation revealed that the same gases that caused holes in Earth's ozone layer in the past century are responsible for the rapid warming of the Arctic as well (Polvani et al., 2020). Looking at the effect of these gases in climate simulations between 1955 and 2005, Polvani et al., (2020) found that the gases accounted for up to half of the warming and sea-ice loss of the Arctic during that period. These so-called ozone-depleting substances (ODSs) are considered potent greenhouse gases and include organic chlorine and bromine compounds such as chlorofluorocarbons. Polvani et al., (2020) also concluded that the warming was caused directly by the gases and not because of their interactions with the ozone layer. ODSs in the atmosphere are declining since they were banned in the 1980s, following concerns over the ozone layer hole over Antarctica.

8.4.2 The United Nations and the Montreal: Kyoto Protocols

It is instructive to compare the Montreal Protocol and the Kyoto Protocol and to do so we need to understand one or two principles of international law. Basically the United Nations, in the form of UNFCCC proposed a treaty, the Montreal Protocol or the Kyoto Protocol, States are invited to ratify the treaty and when enough States have ratified it the treaty then passes into international law. However, the weakness is that States which did not ratify the treaty may choose to ignore it and moreover there is no international police force to enforce the treaty even on States that have ratified the treaty. Both the Montreal Protocol and the Kyoto Protocol have been ratified by the required number of States and have passed into international law. Remotely sensed data are used to monitor the extent to which countries are abiding by the Montreal Protocol. Remotely sensed data is also contributing input data to the computer models which are being used extensively to predict climate change/global warming (by the IPCC) for recommendations to the UNFCCC (United Nations Framework Convention on Climate Change). The UNFCCC led to the Kyoto Protocol which is

TABLE 8.1

The relative success of the Montreal Protocol and the Kyoto Protocol summarised

	Montreal Protocol	Kyoto Protocol
Scientific consensus	Yes	No
Scientific advice followed	Yes	IPCC Yes Sceptics No
Recommendations	Adequate, fit for purpose	Not adequate, i.e. Not fit for purpose
Government actions	Good	Limited *
Developing countries	Participation	Limited participation
Stimulate new technologies	Yes	Limited **

* Limited by vested interests and economic fears.

** e.g. carbon capture has not really taken off.

concerned with emissions of carbon dioxide (CO_2) from the burning of fossil fuels and all that. The Kyoto Protocol, which also is enshrined in international law, attempts to restrict the emissions of CO_2 by those countries that have signed up to it.

It is not unfair to say that the Montreal Protocol is probably being fairly successful in terms of stopping ozone depletion from getting much worse. On the other hand, it is also fair to say that the Kyoto Protocol has achieved almost nothing in terms of slowing down human-induced global warming. Why is this? Comparisons are sometimes made between the Kyoto Protocol and the Montreal Protocol, see Table 8.1.

The Montreal Protocol came about as a result of the scientific evidence for human-induced depletion of the ozone layer, and especially the famous "ozone hole" which appears in the Antarctic each spring. This was rapidly accepted to be a result of the escape of CFCs (chlorofluorocarbons) into the atmosphere. The world's leaders came together and, in the Montreal Protocol, agreed to phase out the production of CFCs and to replace them by other "ozone-friendly" (or perhaps less unfriendly) substances. The reasons for the relative success of the Montreal Protocol are neatly summarised in box 21.2 of the Stern (2007) Review, see Figure 7.stern. Twenty-four countries signed the original Protocol in 1987, and by October 2006, 74 countries had ratified the Protocol and this included the major developing countries. Emissions of CFCs have largely been brought under control, but of course, the ozone layer will not recover immediately; it is expected to take up to 100 years to do so. Comparison with the Kyoto Protocol shows that the Kyoto Protocol which attempts to halt, and hopefully even reverse, human-induced global warming is far less successful than the Montreal Protocol.

The philosophy behind the approach of the IPCC and the UNFCCC leading up to the Kyoto Protocol and all the subsequent meetings of the parties is based on the success of the Montreal Protocol and the idea that once an environmental problem has been identified then it can be solved by international legislation. But it has turned out that there are two fallacies in that. First international legislation can only be achieved by agreement by the various states which all have their own interests and agendas. Secondly, there is no international police force to enforce whatever international legislation is adopted. Recently there has been a turning away from the idea that climate change due to human-induced global warming can be reversed by international legislation in the way that ozone depletion was relatively successively dealt with.

8.5 WHAT CAN I DO?

We deliberately say "What can I do?" rather than "What can we do?" and there is a reason. If my wife, spouse, or partner says to me "I think we ought to xxx" she almost certainly means that she thinks that I should do xxx" where "xxx" is wash the car, call the plumber, take the dog for a walk, or one of another hundred domestic chores. So we use the singular.

The answer to the question "What can I do?" depends on who I am and, in theory, I could be any one of the approx. 7.3 billion people, rising rapidly to 10 billion people, who inhabit the Earth. We don't attempt to be any more accurate in giving the population of the Earth because by the time you are reading this it will have increased. Of this 7.3-10 billion people a very large number will not even ask the question because their first task is to stay alive, or to put a roof over their head, or to feed and clothe their family, or to defend their pathetically small patch from predatory neighbours, or to survive in a war zone. So how many people are not struggling with those problems and even have the time or energy to ask the question at all. Most of us will be from the countries described as Annex I countries to the Kyoto Protocol, the population of which we calculated from population figures of 2019 to be about 967 million. This is not to say that everyone in an Annex 1 country is sufficiently far from poverty to be able to contemplate the question "What can I do about climate change?" nor to say that there are not some people in another country who enjoy a sufficient degree of leisure to contemplate this question. So regarding 967 million as only an upper limit and considering that world population will soon be passing through 9 billion we can take it as true to say that no more than 1 in 10 of the world's population even has the leisure to contemplate this question. It is only those of us who are not struggling to stay alive and not struggling to put a roof over our heads or to feed and clothe our families who can indulge ourselves in the luxury of being able to ask the question at all. And it is only within the last few years that people have started to ask this question. Let us suppose that we are some of the few, the 10% or so of the world's population who are likely even in a position to be able to think about this question and we should acknowledge that only a small fraction of this 10% are actually likely to think seriously about the question.

So, what can I do? One answer is to be found in the title of Greta Thunberg's first book "No one is too small to make a difference" (Thunberg 2019).

David Mackay (2009) on page 3 of his book on "Sustainable Energy – without the hot air' comments on the use of numbers.

> Numbers are chosen to impress, to score points in arguments, rather than to inform. 'Los Angeles residents drive 142 million miles – the distance from Earth to Mars –every single day'

and several more. He continues:

> If all the ineffective ideas for solving the energy crisis were laid end to end, they would reach to the moon and back…

The result of this lack of meaningful numbers and facts? We are inundated with a flood of crazy innumerate codswallop. The BBC doles out advice on how we can do our bit to save the planet – for example "switch off your mobile phone charger when it's not in use"; if anyone objects that mobile phone chargers are not actually our number one form of energy consumption, the mantra "every little helps" is wheeled out. Every little helps? A more realistic mantra is:

> If everyone does a little, we'll achieve only a little.

The completely opposite point of view is expressed by Wagner and Weitzman (2015) "Climate Shock: the economic consequences of a hotter planet" in discussing the civic duty of voting for the right candidate. They quote the probability of the effect of one U.S. citizen making a difference to the outcome of a US presidential election as being 1 in 60 million and conclude "Your vote doesn't count". They then go on to say "Don't just vote for the sake of voting. Vote as an informed citizen. Vote well." Wagner and Weitzman have a whole chapter (Chapter 7) devoted to the question "What can you do?"

Mike Berners-Lee (2019), like David Mackay, concentrates on energy as the driver of our whole system. In many ways, these two books have no magic wand to wave but are constantly making points about numerous little ways I which we can be more energy efficient and more eco-friendly. When he comes to the "What can I do? Summary" near the end of his book "There is no planet B" he begins by saying "When the challenges are so global, and each one of us is so small, it can be tempting, but wrong, to think that there is nothing an individual can do to help humans to get a grip. To do so is a cop out. It is one form of human denial of the Anthropocene challenge. It is difficult to summarise the Berners-Lee's answer to the question "what can I do"; in a sense his answers permeate his whole book. Like Mackay his focus is very much on the efficient production and use of energy. Mackay's book is dedicated "to those who will not have the benefit of two billion years' accumulated energy reserves."

There is a whole chapter of (Wagner and Weitzman (2015) devoted to the question "What can you do?" Greta Thunberg's first little book of speeches claims "No one is too small to make a difference" and this leads us to a clue. The question is also addressed by Mike Berners-Lee (2019) in his "There Is No Planet B."

Extinction Rebellion is an international movement that uses non-violent civil disobedience in an attempt to halt mass extinction and minimise the risk of social collapse. The other book by Mike Berners-Lee that is worth mentioning is his "How Bad are Bananas?" (2010) which has the subtitle "The carbon footprint of everything." This book has a wealth of data to enable someone who wants to "do the right thing" in terms of minimising one's carbon footprint to choose between various actions in their lifestyle.

8.5.1 THE ANTHROPOCENE

This brings us (back) to the question of the relative importance of natural events and human activities in affecting the climate which we considered in section 1. It is perhaps an appropriate point at which to introduce the idea of the Anthropocene. Geologists divide the time from the creation of the Earth until the present into epochs. The term Anthropocene has been proposed as a geological epoch dating from the time when human beings begin to have a significant impact on the Earth's geology and ecosystems, in addition to the natural effects. The name Anthropocene is a combination of anthropo- from anthropos (Ancient Greek: ἄνθρωπος) meaning "human" and -cene from kainos (Ancient Greek: καινός) meaning "new" or "recent." It is claimed that as early as 1873, the Italian geologist Antonio Stoppani acknowledged the increasing power and effect of humanity on the Earth's systems and referred to an "anthropozoic era."

Although this is an interesting idea, the problem is to determine the beginning of this proposed epoch. Various start dates for the Anthropocene have been proposed, ranging from the beginning of the Agricultural Revolution 12,000–15,000 years ago, to as recent as the 1960s. An early concept for the Anthropocene was the Noosphere by Vladimir Vernadsky, who in 1938 wrote of "scientific thought as a geological force." Scientists in the Soviet Union appear to have used the term "anthropocene" as early as the 1960s to refer to the Quaternary, the most recent geological period. The biologist Eugene F. Stoermer is often credited with coining the term "anthropocene," it was in informal use in the mid-1970s and the term was widely popularised in 2000 by atmospheric chemist Paul J. Crutzen.

There is a whole chapter in the book "The Human Planet – How We created the Anthropocene (Lewis and Maslin 2018) devoted to trying to identify the date that should be used to define the beginning of the Anthropocene. Generally, the beginning and end of each geological era is not defined by a precise number of years before the present. Maybe seeking to define the precise start of the Anthropocene is misguided and it did not start in a precisely defined year anyway. Another view point is that in terms of geological time the period from the beginning of the Anthropocene (whenever that is taken to be) until its end, presumably when *Homo sapiens* becomes extinct is so short, in terms of geological time as not to justify being defined as a separate period in geological time, by which stage it will not matter because we shall be extinct and the geological timescale – which is a human invention – will be extinct too.

Rider and Harrison (2019) in their splendid book "Hutton's Arse" make some very perceptive comments about global warming. This book is fascinating. A review of the first edition, printed on the back cover of the book says "This book should be read by all: from the thoroughgoing special-ists to those who have been touched by the magnificence and awe-inspiring nature of Scotland's scenery and geology, and anyone who appreciates well-written work on an inspiring subject that is interspersed with wit, wisdom and emotion."

Some comments of particularly penetrating wisdom will be found in chapter 5 of Rider and Harrison (2019), entitled "The Coming Ice Age" They write that plagiarism is common, that every-thing has been written before and this title the 'The Coming Ice Age' has been taken from a book written in 1896 by an American sailor and scientist, called CAM Taber. Taber was writing 100 years or so ago when the next ice age was imminent. Rider and Harrison go on to comment that with global warming, today it is ice melt that is imminent. They make the point that science has fashions but that just because an idea is fashionable does not make it right – perhaps even the opposite. They go on to say that fashionable ideas are likely to be wrong and that science does not advance by dem-ocratic vote or public opinion poll, a point which we have made elsewhere in this book. They write that global warming has an even greater handicap than being fashionable as it now has a huge vested interest from environmentalists, scientists, politicians, software writers, technical and engineering companies, insurance brokers, and even bankers.

Riders and Harrison acknowledge that the Highlands of Scotland have their part to play in this debate. Windmills and small-scale hydro schemes have been established to try to defer the carbon release of fossil fuels, and the peat bogs hold on to a huge carbon reserve. But they query whether any of this helps and go on to comment that the number of books on the subject of climate change is huge but tend to be all the same: global warming the facts; global warming the human reasons; global warming the necessary remedies; and especially, global warming if we don't do the right thing.

They make the following point: "But are we giving ourselves a planetary importance that we do not have, first of all to be able to affect global climate for the 'worse' and then, when we have so decided, to affect it for the 'better'? (Note that better and worse are not scientific terms; they reflect what is supposedly better or worse for people). From a geological perspective, this is, of course, laughable. From a human perspective this is replacing the previous God of Noah's great flood with a new god, *Homo sapiens*, the species: both equally silly prospects. King Canute will always get his feet wet."

They comment on the limiting effect of dividing the world into specialisms and note that as science progresses, more and more specialists and specialist subjects are invented. They point out that Leonardo da Vinci was not only a supreme artist and sculptor, but also architect, engineer, and inventor of war machines, something which would simply not be possible today when art and sci-ence have become very far apart.

"Specialisation in geology, as in other sciences, has had the effect of creating sub-cultures all over the subject. Like street gangs, each specialist group has its own language which has to be understood to become a member. The Geological Society of London has 22 such groups, and in every university this is repeated: the palaeontology gang at this end of the corridor, the geomorphol-ogy gang at the other and the geophysicists in the middle. This is serious. Just as street gangs hinder proper social functions, so the specialists hinder the proper functions of science. And this is even more obvious when it comes to major subjects: cross the divide between geology and archaeology or astronomy or climatology, and you will be treated as a pitiful amateur. You don't have the right terms or the same language and you don't know the right people."

When considering ice ages and global warming, it can be seen that the study of ice ages cov-ers subjects such as geology, archaeology, history, climatology, even astronomy, as well as human anthropology. Humans have lived and developed through the ice ages and have left records in the form of artefacts, then written accounts, and now instrumental records. Riders and Harrison consider that humans were affected by the ice age climate and now the climate is being affected by humans, that geology, archaeology, history, and the future are all part of the same continuum although with the involvement of a huge number of different specialists. Their narrative is around the ice-world of the Highlands, looking at the past, demonstrating the present, and predicting the future. The rocks are the past, humans the present, and instruments and software the future. They claim that there will be no regard for subject limits, big business, environmentalists, politicians, or any other self-interest group.

Rider and Harrison go on to talk about the last 250,000 years and what has been discovered from ice cores and from ocean sediments. Fascinating stuff which does help to put our efforts in perspective in relation to nature. We must, however, leave our readers to find their own copies and read the book.

We cannot, however, resist the temptation to make one last quote from the same book:

Past and future climates in the Highlands

Winter and summer, spring and autumn come so regularly that we overlook their fragility, the delicate balance that climate is, being quite unaware of what even the slightest changes will do. If the average summer temperature in the Highlands were to drop by 4.0C, the difference between a tee-shirt and a sweater, all the Highland Munros would be permanently snow-covered, glaciers flowing from their flanks. Put another way, the height at which permanent ice would form in the Highlands today is only c.250-300m above present highest peaks. Even now, in some years small amounts of snow last well into summer in gullies on Ben Nevis. The balance is very delicate.

In the past, of course, such calculations were unnecessary, and during the depths of the last ice age 25,000 years ago, a huge ice-sheet covered all of Scotland and even northern England down as far as York. This idea is a familiar one today, though only 160 years ago such a proposition was unknown. The first person to propose an ice age, even for Scotland, was Louis Agassiz (1807–1873), a Swiss natural scientist who became convinced in the 1830s, while working on glaciers in his home country, that ice had once covered large parts of Northern Europe. He visited the Highlands in the autumn of 1840 with William Buckland (1784–1856), then President of the London Geological Society. The *Scotsman* newspaper published a letter from Agassia in October of that year, written from Fort Augustus:

"... at the foot of Ben Nevis, and in the principal valleys, I discovered the most distinct moraines and polished rocky surfaces, just as in the valleys of the Swiss Alps in the region of existing glaciers, so that the existence of glaciers in Scotland at early periods can no longer be doubted."

As Rider and Harrison comment, this is a rare example of a scientific discovery being communicated to the world in the letters column of a national newspaper.

8.5.2 Greta Thunberg; extinction rebellion

It would be negligent of us is we did not mention the Greta Thunberg movement and Extinction Rebellion. These are two different things, although they do have some features in common, namely public demonstrations and marches. Greta Thunberg is particularly concerned that targets for the reduction of CO_2 emission are not adequate to prevent human-induced global warming and even these targets are not being met. Extinction Rebellion is particularly concerned with the extinction of species as a result of human activities.

Greta Tintin Eleonora Ernman Thunberg was born on 3 January 2003 in Stockhom, Sweden, and sprang to fame in 2018 when she began her school strike and sat down outside the Swedish parliament with the placard shown in Fig. 8.11 "School strike for climate."

On 20 August 2018, Thunberg, who had just started the ninth grade in high school and who has Asperger's syndrome, decided not to attend school until the 2018 Swedish general election on 9 September and she protested by sitting outside the Riksdag, the Swedish Parliament, every day for three weeks during school hours with the sign Skolstreijk för klimatet (school strike for climate), see Figure 8.2. After the general elections, Thunberg continued to strike only on Fridays. She inspired

FIGURE 8.2   Greta Thunberg in August 2018 ("School strike for climate"). In late 2018, Thunberg began the school climate strikes and public speeches for which she has become an internationally recognised climate activist.

school students across the globe to take part in student strikes. After October 2018, Thunberg's activism evolved from solitary protesting to taking part in demonstrations throughout Europe; making several high-profile public speeches, and mobilising her growing number of followers on social media platforms. By March 2019, she was still staging her regular protests outside the Swedish parliament every Friday, where other students occasionally joined her. According to her father, her activism has not interfered with her schoolwork, but she has had less spare time.

As of December 2018, more than 20,000 students had held strikes in at least 270 cities. The school strikes for climate on 20 and 27 September 2019 were attended by over four million people, according to one of the co-organisers.

One feature of her campaign is her claim to be independent of any organisation or commercial supporter and another is that she eschews air travel on account of the carbon emissions. So in 2019, she sailed across the Atlantic Ocean from Plymouth, England, to New York, USA, in the 60-foot (18 m) racing yacht *Malizia II*, equipped with solar panels and underwater turbines. The trip was announced as a carbon-neutral transatlantic crossing serving as a demonstration of Thunberg's declared beliefs of the importance of reducing emissions. While not doubting her sincerity, all claims of carbon neutrality deserve to be the subject of close scrutiny.

Greta Thunberg attended and spoke at various high-profile meetings in Canada and the USA and had intended to remain in the Americas to travel overland to attend the United Nations Climate Change Conference (COP25) in Santiago, Chile, in December. However, it was announced at short notice that COP25 was to be moved to Madrid, Spain, because of serious public unrest in Chile. Since she refuses to fly, she hitched a lift from the USA to Lisbon, Portugal, arriving there on 3 December and then travelled on to Madrid to speak at COP25 and to participate with the local

Fridays for Future climate strikers. This was followed by other performances and an appearance at the January 2020 World Economic Forum held in Davos, Switzerland.

There appear to be four interwoven themes to Thunberg's message.

1. Humanity is facing an existential crisis because of climate change.
2. The current generation of adults is responsible for climate change.
3. Climate change will have a disproportionate effect on young people, and that too little is being done about the situation.
4. Politicians and decision-makers need to listen to the scientists.

With regard to (1), we and many others have argued that climate change is only one of several threats to our way of life. With regard to (2), this is only partly true. It goes back much earlier than that and, moreover, Greta Thunberg's generation in the rich west are enjoying the benefits of industrialisation which has been made possible by the consumption of fossil fuels and the consequent emissions of CO_2. Point (3) is true insofar as the life expectancy of young people is greater than that of their elders while the claim that too little is being done assumes that we know what needs to be done and we shall turn to this shortly. Point (4) overlooks the fact that "the scientists" don't all agree with the IPCC, in which Thunberg seems to have implicit faith.

Thunberg has given many speeches and she has published a collection of her climate action speeches, *No One Is Too Small to Make a Difference*, in May 2019 by Penguin Books, with the earnings being donated to charity, and subsequently an expanded volume with the same title, published by Allen Lane, a division of Penguin Books) later in the same year. We quote from just two of her speeches.

The first was at the 2019 UN Climate Action Summit on 23 September 2019, and subsequently an expanded volume with the same title Thunberg said to world leaders: "This is all wrong. I shouldn't be up here. I should be back in school on the other side of the ocean. Yet you all come to us young people for hope? *How dare you!* You have stolen my dreams and my childhood with your empty words. And yet I'm one of the lucky ones. People are suffering. People are dying. Entire ecosystems are collapsing. We are in the beginning of a mass extinction. And all you can talk about is money and fairy tales of eternal economic growth. How dare you!" "You are failing us." Thunberg said towards the end of her speech. "But the young people are starting to understand your betrayal. The eyes of all future generations are upon you. And if you choose to fail us, I say: We will never forgive you." OK, but does that solve the problem?

The second speech was to the European Economic and Social Committee 'Civil Society for rEUnaissance', Brussels, Belgium, 21 February 2019, illustrated by the then President of the EU, J-C Junker, giving her a condescending handshake. She said "We know that most politicians don't want to talk to us. Good, we don't want to talk to them either. We want them to talk to the scientists instead. Listen to them….." This overlooks the point that there is no consensus. One of the reasons for the relative failure of the Kyoto Protocol compared to the comparative success of the Montreal Protocol is that the scientists are not unanimous about human-induced global warming, whereas there was no such division about ozone depletion.

So, what did Greta Thonberg achieve? She showed that the IPCC, UNFCC, and the Kyoto Protocol had failed to achieve the reduction in CO_2 emissions that it was hoped to achieve when the IPCC was set up. For two reasons, first the reduction emission targets for greenhouse gases were not ambitious enough and secondly even these inadequate targets were not being met. And she struck a chord of sympathy with many young people and some not so young people who felt that they had been betrayed by the politicians and world leaders of the last 40 years. So far so good, but what solution has she offered? We have touched on the problem earlier, namely that the method adopted in the case of ozone depletion, i.e. that of the Montreal Protocol, is apparently not working for human-induced global warming because of the vested interests of individual nation states

FIGURE 8.3   Extinction Rebellion placard, with the "extinction symbol."

and the absence of any international policeman to enforce whatever measures have been agreed in international law. Greta Thonberg imagines that by delivering speeches to politicians, world leaders, etc., on behalf of young people whose future she sees to be at stake she will achieve something significant.

We turn now to Extinction Rebellion (XR) which burst on the scene at about the same time as Greta Thonberg. This is a global environmental movement with the stated aim of using nonviolent civil disobedience to compel government action to avoid tipping points in the climate system, biodiversity loss, and the risk of social and ecological collapse. The movement uses a circled hourglass, known as the extinction symbol, to serve as a warning that time is rapidly running out for many species, see Figure 8.3.

Taken from its website (https://rebelliion.earth accessed 28 January 2020), we note the aims and principles of Extinction Rebellion:

8.5.3 AIMS

- Government must tell the truth by declaring a climate and ecological emergency, working with other institutions to communicate the urgency for change.
- Government must act now to halt biodiversity loss and reduce greenhouse gas emissions to net-zero by 2025.
- Government must create, and be led by the decisions of, a citizens' assembly on climate and ecological justice.

When the movement expanded to the United States, a further demand was added demanding justice for underprivileged people.

8.5.4 Principles

- We have a shared vision of change – creating a world that is fit for generations to come.
- We set our mission on what is necessary – mobilising 3.5% of the population to achieve system change by using ideas such as "momentum-driven organising" to achieve this.
- We need a regenerative culture – creating a culture that is healthy, resilient, and adaptable.
- We openly challenge ourselves and this toxic system, leaving our comfort zones to take action for change.
- We value reflecting and learning, following a cycle of action, reflection, learning, and planning for more action (learning from other movements and contexts as well as our own experiences).
- We welcome everyone and every part of everyone – working actively to create safer and more accessible spaces.
- We actively mitigate for power – breaking down hierarchies of power for more equitable participation.
- We avoid blaming and shaming – we live in a toxic system, but no one individual is to blame.
- We are a non-violent network using non-violent strategy and tactics as the most effective way to bring about change.
- We are based on autonomy and decentralisation – we collectively create the structures we need to challenge power. Anyone who follows these core principles and values can take action in the name of Extinction Rebellion.

Extinction Rebellion has a decentralised structure. Providing that they respect the "principles and values," every local group can organise events and actions independently. Extinction Rebellion organises public demonstrations involving non-violent civil disobedience, leading to mass arrests in order to publicise its activities. Starting on Monday 15 April, Extinction Rebellion organised demonstrations in London, focusing on Oxford Circus, Marble Arch, Waterloo Bridge, and the area around Parliament Square and lasting several days. By 19 April, the police said 682 people had thus far been arrested in London. Activists gathered at Hyde Park at the end of the 11-day demonstrations in London, during which 1,130 people had been arrested.

As part of a two-week series of XR actions which was called "International Rebellion," to take place in more than 60 cities worldwide, events were planned around London from 7 to 19 October 2019 to demand the UK government take urgent action to tackle the climate crisis. Over 1000 arrests had been made by 11 October. Other cities involved included New York, Munich, Melbourne, Adelaide Sydney Brisbane, Berlin, Amsterdam Paris, Vienna, and Washington D.C.

Extinction Rebellion uses mass arrest as a tactic to try to achieve its goals. Extinction Rebellion's founders researched the histories of "the suffragettes, the Indian salt marchers, the civil rights movement and the Polish and East German democracy movements," who all used the tactic, and are applying their lessons to the climate crisis. It has been claimed that one of the founders of Extinction Rebellion claimed that letters, emailing, and marches don't work. You need about 400 people to go to prison and about two to three thousand people to be arrested. In London's April 2019 protests, 1130 arrests were made and during the two-week October 2019 actions in London as part of International Rebellion, 1832 arrests were made.

We need to ask how effective the impassioned speeches of Greta Thunberg and the massive marches and non-violent demonstration and disruption and mass arrests of Extinction Rebellion are likely to be. We consider two examples from recent history in democratic countries. Mass protests in other countries can end in tragedy, as for example in the June fourth Incident in Tiananmen Square in 1989.

Our first example is that of the Campaign for Nuclear Disarmament (CND) in the UK, see Fig. 8.13.

The single event that most put CND on the public map was the Aldermaston March of April 1958. The Easter march was to the Atomic Weapons Establishment at Aldermaston, Berkshire, the main location for the research, development, and production of Britain's nuclear warheads. CND members participated extensively in the event, and it was immediately, inextricably linked with the new-born CND in the public mind. It went on to become an annual event for many years

FIGURE 8.4   The 1958 CND march to Aldermaston.

and achieving considerable publicity. However, the U.K., over 60 years later, still has its nuclear weapons (Figure 8.4).

Our second example concerns the anti-Vietnam war protests in the USA in 1967, 1969, and 1970. It is estimated that over 15 million people took part in anti-Vietnam war marches in the USA. It is a matter of opinion as to how significant were these marches in contributing to the ending of the Vietnam war. One obviously important fact is that by the time President Richard Nixon took office in January 1969 about 34,000 Americans had been killed fighting in Vietnam and in 1969 approximately another 10,000 were killed. Or it could be argued that America simply lost the war militarily. Nevertheless, the three big marches of 1967, 1969, and 1970, along with a march in Melbourne in 1970 achieved some success.

Thus, while not discounting the efforts of Greta Thunberg and Extinction Rebellion, it seems that the message is clear that the IPCC/UNFCCC/Kyoto Protocol approach is not being spectacularly successful in terms of reducing human-induced global warming. It seems that the options are:

1. Acknowledging that the approach that has been tried is not working to the desired (by whom?) extent and that we should plan for increased temperatures.
2. Attempt some kind of geo-engineering solution – highly dangerous.
3. Acknowledge that individuals need to do their own little bit, but how to achieve this when it will certainly lead to a reduction in life style? This leads to the question "What can I do?" to which we shall turn in the next section.

Nathaniel Rich in his book "Losing Earth" (Rich 2019) began rather perceptively at the start of his chapter 1.

8.6 SUSTAINABILITY

Climate change is widely presented as being the only, or at least the most serious, threat to our way of life, or even to our survival as a species. We have already tried in section 1 to stress that it is only one of many threats to the survival of our way of life. In this book, we have acknowledged that human-induced global warming as a result of the emission of greenhouse gases is one threat to our way of life but we have tried to stress that it is only one of many threats. Climate change activists thrive on stirring up fear among the public who are urged to "do something" and thereby create the illusion that they (the activists or Jo Public) are "doing something" to solve the problem. What can I do? We introduced the idea of sustainability in section 1 at the beginning of this book and we now revisit the idea in the closing chapter. In 1988 or thereabouts, Kondratyev criticised the IPCC for relying too much on the results of computer modelling. Even more radical is the claim by Nathaniel Rich in his book "Losing Earth: the decade we could have stopped climate change" (2019). We summarise the introduction to that book.

In his book, Nathaniel Rich makes the point that almost everything about global warming was known and understood in 1979. He considers that it was almost better understood than today and notes that currently almost nine out of ten Americans are not aware that nearly all scientists agree that human beings have altered the global climate through the thoughtless burning of fossil fuels.

By 1979, the issue had been generally accepted, the "greenhouse effect" was a phrase in common usage, and attention had turned to predicting the consequences of climate change. Rich comments that the basic science was easily understood; the more carbon dioxide being put into the atmosphere, the warmer the planet would become. He notes that the world has warmed more than 1 degree Celsius since the Industrial Revolution and that the Paris climate agreement signed in 2016 aimed to restrict warming to 2 degrees Celsius. However, he also notes that this agreement was nonbinding, unenforceable, and unheeded and that a recent study had put the chances of this being achieved at less than one in twenty.

Rich comments that even if this was achieved, we would still have to negotiate the extinction of the world's tropical reefs, a sea level rise of several metres and the abandonment of the Persian Gulf. He quotes the climate scientist James Hansen who has called a two-degree warming "a prescription for long term disaster".

According to Rich, it would appear that long-term disaster is now the best-cast scenario. However, he warns that warming of three degrees would constitute short-term disaster with forests sprouting in the Arctic, the loss of most coastal cities and mass starvation. He notes that Robert Watson, a former chairman of the United Nations Intergovernmental Panel on Climate Change, has argued that a realistic minimum would be a three-degree warming, and that a four-degree warming would result in permanent drought in Europe with vast areas of China, India, and Bangladesh reduced to desert. Rich claims that it is this prospect of a five-degree warming that has prompted some of the world's pre-eminent climate scientists to warn of the collapse of human civilisation. This collapse, he suggests, will not be as a direct result of warming, but of the secondary effects. He quotes Red Cross estimates that already more refugees flee environmental crises than violent conflict and predicts that "starvation, drought, the inundation of the coasts, and the smothering expansion of deserts will force hundreds of millions of people to run for their lives." He goes on to suggest that "mass migrations will stagger delicate regional truces, hastening battles over natural resources, acts of terrorism, and declarations of war. Beyond a certain point, the two great existential threats to our civilization, global warming and nuclear weapons, will loose their chains and join to rebel against their creators."

If an eventual five- or six-degree warming scenario seems outlandish, it is only because we assume that we'll respond in time. We'll have decades to eliminate carbon emissions, after all, before we are locked into six degrees. But we've already had decades – decades increasingly punctuated by climate-related disaster – and we've done nearly everything possible to make the problem worse. It no longer seems rational to assume that humanity, encountering an existential threat, will behave rationally.

There can be no understanding of our current and future predicament without an understanding of why we failed to solve this problem when we had the chance. For in the decade that ran between 1979 and 1989, we had an excellent chance. The world's major powers came within several signatures of endorsing a binding framework to reduce carbon emissions – far closer than we've come since. During that decade, the obstacles we blame for our current inaction had yet to emerge. The conditions for success were so favourable that they have the quality of a fable, especially at a time when so many of the veteran members of the climate class – the scientists, policy negotiators, and activists who for decades have been fighting ignorance, apathy, and corporate bribery – openly despair about the possibility of achieving even mitigatory success. As Ken Caldeira, a leading climate scientist at the Carnegie Institution for Science in Stanford, California, recently put it, "We're increasingly shifting from a mode of predicting what's going to happen to a mode of trying to explain what happened."

So what happened? The common explanation today concerns the depredations of the fossil fuel industry, which in recent decades has committed to playing the role of villain with comic-book bravado. Between 2000 and 2016, the industry spent more than $2 billion, or ten times as much as was spent by environmental groups, to defeat climate change legislation. A robust subfield of climate literature has chronicled the machinations of industry lobbyists, the corruption of pliant scientists, and the influence campaigns that even now continue to debase the political debate, long after the largest oil and gas companies have abandoned the dumb show of denialism. But the industry's assault did not begin in force until the end of the eighties. During the preceding decade, some of the largest oil and gas companies, including Exxon and Shell, made serious efforts to understand the scope of the crisis and grapple with possible solutions.

We despair today at the politicisation of the climate issue, which is a polite way of describing the Republican Party's stubborn commitment to denialism. In 2018, only 42% of registered Republicans knew that "most scientists believe global warming is occurring," and that percentage has fallen. Scepticism about the scientific consensus on global warming – and with it, scepticism about the integrity of the experimental method and the pursuit of objective truth – has become a fundamental party creed. But during the 1980s, many prominent Republican members of Congress, cabinet officials, and strategists shared with Democrats the conviction that the climate problem was the rare political winner: nonpartisan and of the highest possible stakes. Among those who called for urgent, immediate, and far-reaching climate policy: Senators John Chafee, Robert Stafford, and David Durenberger; Environmental Protection Agency administrator William K. Reilly; and, during his campaign for president, George H. W. Bush. As Malcolm Forbes Baldwin, the acting chairman of Ronald Reagan's Council for Environmental Quality, told industry executives in 1981, "There can be no more important or conservative concern than the protection of the globe itself." The issue was unimpeachable, like support for the military and freedom of speech. Except the atmosphere had an even broader constituency, composed of every human being on Earth.

It was widely accepted that action would have to come immediately. At the beginning of the 1980s, scientists within the federal government predicted that conclusive evidence of warming would appear on the global temperature record by the end of the decade, at which point it would be too late to avoid disaster. The United States was, at the time, the world's dominant producer of greenhouse gases; more than 30% of the human population lacked access to electricity altogether. Billions of people would not need to attain the "American way of life" in order to increase global carbon emissions catastrophically; a light bulb in every other village would do it. A 1980 report prepared at the request of the White House by the National Academy of Sciences proposed that "the carbon dioxide issue should appear on the international agenda in a context that will maximize cooperation and consensus-building and minimize political manipulation, controversy and division." If the United States had endorsed the proposal broadly supported at the end of the eighties – a freezing of carbon emissions, with a reduction of 20% by 2005 – warming could have been held to less than 1.5 degrees.

A broad international consensus had agreed on a mechanism to achieve this: a binding global treaty. The idea began to coalesce as early as February 1979, at the first World Climate Conference in Geneva, when scientists from 50 nations agreed unanimously that it was "urgently necessary" to act. Four months later, at the Group of Seven meeting in Tokyo, the leaders of the world's wealthiest nations signed a statement resolving to reduce carbon emissions. A decade later, the first major diplomatic meeting to approve a framework for a treaty was called in the Netherlands. Delegates from more than 60 nations attended. Among scientists and world leaders, the sentiment was unanimous: action had to be taken, and the United States would need to lead. It didn't.

The inaugural chapter of the climate change saga is over. In that chapter – call it Apprehension – we identified the threat and its consequences. We debated the measures required to keep the planet within the realm of human habitability: a transition from fossil fuel combustion to renewable and nuclear energy, wiser agricultural practices, reforestation, and carbon taxes. We spoke, with increasing urgency and self-delusion, of the prospect of triumphing against long odds.

We did not, however, seriously consider the prospect of failure. We understood what failure would mean for coastlines, agricultural yield, mean temperatures, immigration patterns, and the world economy. But we did not allow ourselves to comprehend what failure might mean for us. How will it change the way we see ourselves, how we remember the past, how we imagine the future? How have our failures to this point changed us already? Why did we do this to ourselves? These questions will be the subject of climate change's second chapter. Call it the Reckoning.

That we came so close, as a civilisation, to breaking our suicide pact with fossil fuels can be credited to the efforts of a handful of people – scientists from more than a dozen disciplines, political appointees, members of Congress, economists, philosophers, and anonymous bureaucrats. They were led by a hyperkinetic lobbyist and a guileless atmospheric physicist who, at severe personal cost, tried to warn humanity of what was coming. They risked their careers in a painful, escalating campaign to solve the problem, first in scientific reports, later through conventional avenues of political persuasion, and finally with a strategy of public shaming. Their efforts were shrewd, passionate, and robust. And they failed. What follows is their story, and ours.

It is flattering to assume that, given the opportunity to begin again, we would act differently – or act at all. You would think that reasonable minds negotiating in good faith, after a thorough consideration of the science, and a candid appraisal of the social, economic, ecological, and moral ramifications of planetary asphyxiation, might agree on a course of action. You would think, in other words, that if we had a blank slate – if we could magically subtract the political toxicity and corporate agitprop – you'd think we'd be able to solve this.

Yet we did have something close to a blank slate in the spring of 1979. President Jimmy Carter, who had installed solar panels on the roof of the White House and had an approval rating of 46%, hosted the signing of the Israel-Egypt peace treaty. "We have won, at last, the first step of peace," he said. "A first step on a long and difficult road." The number one film in America was The China Syndrome; the number one song was the Bee Gees' "Tragedy." Barbara Tuchman's A Distant Mirror, a history of the calamities that befell medieval Europe after a major climatic change, had been near the top of the bestseller list all year. An oil well off Mexico's Gulf Coast exploded and would gush for nine months, staining beaches as far away as Galveston, Texas. In Londonderry Township, Pennsylvania, at the Three Mile Island nuclear plant, a water filter was beginning to fail. And in the Washington, D.C., headquarters of Friends of the Earth, a 30-year-old activist, a self-styled "lobbyist for the environment," was struggling through a dense government report, when his life changed.

If emissions are catastrophic, a light bulb in every other village would do it. A 1980 report prepared at the request of the White House by the National Academy of Sciences proposed that "the carbon dioxide issue should appear on the international agenda in a context that will maximize cooperation and consensus-building and minimize political manipulation, controversy and division." If the United States had endorsed the proposal broadly supported at the end of the eighties – a freezing of carbon emissions, with a reduction of 20% by 2005 – warming could have been held to less than 1.5 degrees.

8.6.1   THE UNITED NATIONS SUSTAINABLE DEVELOPMENT GOALS (SDGs)

The UN's Sustainable Development Goals (SDGs) have their origins in the eight Millennium Development Goals (MDGs) which were agreed on by world leaders at a UN summit in 2000 and set targets:

- to eradicate extreme poverty and hunger;
- to achieve universal primary education;
- to promote gender equality and empower women;
- to reduce child mortality;
- to improve maternal health;
- to combat HIV/AIDS, malaria, and other diseases;
- to ensure environmental sustainability; and
- to develop a global partnership for development.

From the Millennium Development Goals, progress has been made across the board, from combating poverty, to improving education and health, and reducing hunger, but there was still a long way to go. Fifteen years after the MDGs were created, they reached their expiry date and the UN created the Sustainable Development Goals (SDGs) to focus their attention on over the next 15 years.

The Sustainable Development Goals are a call for action by all countries – poor, rich, and middle-income – to promote prosperity while protecting the planet. They recognise that ending poverty must go hand-in-hand with strategies that build economic growth and address a range of social needs including education, health, social protection, and job opportunities, while tackling climate change and environmental protection. The SDGs, set in 2015 by the United Nations General Assembly are intended to be achieved by the year 2030.

The SDGs have 17 calls for action:

1. No Poverty
2. Zero Hunger
3. Good Health and Well-being
4. Quality Education
5. Gender Equality
6. Clean Water and Sanitation
7. Affordable and Clean Energy
8. Decent Work and Economic Growth
9. Industry, Innovation, and Infrastructure
10. Reducing Inequality
11. Sustainable Cities and Communities
12. Responsible Consumption and Production
13. Climate Action
14. Life Below Water
15. Life On Land
16. Peace, Justice, and Strong Institutions
17. Partnerships for the Goals.

We recall that in 1992, the first United Nations Conference on Environment and Development (UNCED) or Earth Summit was held in Rio de Janeiro, where the first agenda for Environment and Development, also known as Agenda 21, was developed and adopted. Twenty years later, in 2012, the United Nations Conference on Sustainable Development (UNCSD), also known as Rio+20, was held as a 20-year follow-up to UNCED. In the run-up to Rio+20, there was much discussion about the idea of the Sustainable Development Goals (SDGs). At the Rio+20 Conference, a resolution known as "The Future We Want" was reached by member states. The Rio+20 outcome document mentioned that "at the outset, the OWG [Open Working Group] would decide on its methods of work." On 19 July 2014, the OWG forwarded a proposal for the SDGs to the Assembly. After 13 sessions, the OWG submitted their proposal to the 68th session of the General Assembly in September 2014. On 5 December 2014, the UN General Assembly accepted the Secretary General's Synthesis Report, which stated that the agenda for the post-2015 SDG process would be based on the OWG proposals.

8.7   THE WORLD'S MOST DANGEROUS ANIMAL

As is well known, various food chains involve creatures higher up in the chain acting as predators on creatures lower down in the chain. But the weapons used are usually just their teeth and claws, although some have developed other "external" weapons, such as the spider's web or the traps of the pitcher plants, etc. *Homo sapiens* is different. We have developed all sorts of weapons to prey on other species, to extract minerals from the ground to generate energy that we use for all sorts of purposes to drive our whole way of life. This point has been made dramatically by Calgary Zoo, where in 1986 there was a cage with a label saying "World's most dangerous animal." Unlike other cages, it was open and visitors could go inside it and the photograph shown in Fig. 8.14 was taken. Though obviously this was a photo-opportunity for visitors, the management of the zoo probably had a more serious intention as well. They may have intended visitors to think about the relentless extinction of botanical and zoological species which human beings have been causing for hundreds of years. They may have been thinking about the many violent deaths of human beings caused by murders, by wars, and by traffic and industrial accidents. But apart from these direct causes of death, there is yet another way in which human beings are dangerous to one another and that is by the damage that we cause to the general environment in which we live. Why is our environment important? The simplest explanation is that, as humans, the environment-the Earth-is our homeland. If we damage the environment, we threaten harm to ourselves (Figure 8.5).

Today industrialisation and urbanisation have led to increasing affluence and a growing population. This places enormous pressure on the environment, and we need to transform and improve ourselves to tackle these challenges. We do not live in isolation; our entire life support system is dependent on the well-being of many other species living on Earth. The term biosphere was created to describe the totality of living things on Earth by Vladimir Vernadsky, a Russian scientist, in the 1920s. The biosphere refers to the one global ecological system in which all living things are interdependent to a greater or lesser extent. Within the overall biosphere, or ecosystem, there are smaller ecosystems like the rainforests, marine ecosystems, the desert, and the tundra. When any of these systems is threatened, the entire planet may be affected. The various anthropogenic activities that take place may have widespread and unpredicted consequences for the health of our planet.

There is a saying that we must think globally and act locally when we are handling issues related to the environment and natural resources.

8.8   ECONOMIC GROWTH IS NOT OUR SALVATION

The murmuring mantra of global markets – which prevailed between the end of the Cold War and the onset of the Great Recession, promising something like their own eternal reign – is that economic growth will save us from anything and everything.

But in the aftermath of the 2008 crash, a number of historians and iconoclastic economists studying what they call "fossil capitalism" have started to suggest that the entire history of swift economic growth, which began somewhat suddenly in the 18th century, is not the result of innovation or the dynamics of free trade, but simply our discovery of fossil fuels and all their raw power – a onetime injection of that new "value" into a system that had previously been characterised by unending subsistence living. This is a minority view, among economists, and yet the précis version of the

FIGURE 8.5   The world's most dangerous animal. An example of *Homo sapiens*, Arthur P. Cracknell in 1986 in Calgary Zoo.

perspective is quite powerful. Before fossil fuels, nobody lived better than their parents or grandparents or ancestors from 500 years before, except in the immediate aftermath of a great plague like the Black Death, which allowed the lucky survivors to gobble up the resources liberated by mass graves.

In the West especially, we tend to believe we've invented our way out of that endless zero-sum, scratch-and-claw resource scramble – both with particular innovations, like the steam engine and computer, and with the development of a dynamic capitalistic system to reward them. But scholars like Andreas Malm have a different perspective: we have been extracted from that muck only by a singular innovation, one engineered not by entrepreneurial human hands but in fact millions of years before the first ones ever dug at the earth – engineered by time and geologic weight, which many millennia ago pressed the fossils of Earth's earlier carbon-based life forms (plants, small animals) into petroleum, like lemon under a press. Oil is the patrimony of the planet's prehuman past: what stored energy the earth can produce when undisturbed for millennia. As soon as humans discovered that storehouse, they set about plundering it – so fast that, at various points over the last half century, oil forecasters have panicked about running out. In 1968, the labour historian Eric Hobsbawm wrote, "Whoever says Industrial Revolution, says cotton." Today, he would probably substitute "fossil fuel."

The timeline of growth is just about perfectly consistent with the burning of those fuels, though doctrinaire economists would argue there is much more to the equation of growth. Generations being as long as they are and historical memory as short, the West's several centuries of relatively reliable and expanding prosperity have endowed economic growth with the reassuring aura of permanence: we expect it, on some continents, at least, and rage against our leaders and elites when it does not come. But planetary history is very long, and human history, though a briefer interval, is long, too. And while the pace of technological change we call progress is today dizzying and may yet invent new ways of buffering us from the blows of climate change, it is also not hard to imagine those flush centuries, enjoyed by nations who colonised the rest of the planet to produce them, as an aberration. Earlier empires had boom years, too.

You do not have to believe that the economic growth is a mirage produced by fossil fuels to worry that climate change is a threat to it – in fact, this proposition forms the cornerstone around which an entire edifice of academic literature has been built over the last decade. The most exciting research on the economics of warming has come from Solomon Hsiang and Marshall Burke and Edward Miguel, who are not historians of fossil capitalism but who offer some very bleak analysis of their own in a country that's already relatively warm, every degree Celsius of warming reduces growth, on average, by about one percentage point (an enormous number, considering we count growth in the low single digits as "strong"). This is the sterling work in the field. Compared to the trajectory of economic growth with no climate change, their average projection is for a 23% loss in per capita earning globally by the end of this century.

Tracing the shape of the probability curve is even scarier. There is a 51% chance, this research suggests that climate change will reduce global output by more than 20% by 2100, compared with a world without warming, and a 12% chance that it lowers per capita GDP by 50% or more by then, unless emissions decline. By comparison, the Great Depression dropped global GDP by about 15%, it is estimated – the numbers weren't so good back then. The more recent Great Recession lowered it by about 2%, in a onetime shock; Hsiang and his colleagues estimate a one-in-eight chance of an ongoing and irreversible effect by 2100 that is 25 times worse. In 2018, a team led by Thomas Stoerk suggested that these estimates could be dramatic underestimates.

The scale of that economic devastation is hard to comprehend. Even within the post-industrial nations of the wealthy West, where economic indicators such as the unemployment rate and GDP growth circulate as though they contain the whole meaning of life in them, figures like these are a little bit hard to fathom; we've become so used to economic stability and reliable growth that the entire spectrum of conceivability stretches from contractions of about 15%, effects we study still in histories of the Depression, to growth about half as fast – about 7%, which the world as a whole last achieved during the global boom of the early 1960s. These are exceptional onetime peaks and troughs, extending for no more than a few years, and most of the time we measure economic fluctuations in ticks of decimal points – 2.9 this year, 2.7 that. What climate change proposes is an economic setback of an entirely different category.

The breakdown by country is perhaps even more alarming. There are places that benefit, in the north, where warmer temperatures can improve agriculture and economic productivity: Canada, Russia, Scandinavia, Greenland. But in the mid-latitudes, the countries that produce the bulk of the world's economic activity – the United States, China – lose nearly half of their potential output. The warming near the equator is worse, with losses throughout Africa, from Mexico to Brazil, and in India and Southeast Asia approaching 100%. India alone, one study proposed, would shoulder nearly a quarter of the economic suffering inflicted on the entire world by climate change. In 2018, the World Bank estimated that the current path of carbon emissions would sharply diminish the living conditions of 800 million living throughout South Asia. One hundred million, they say, will be dragged into extreme poverty by climate change just over the next decade. Perhaps "back into" is more appropriate: many of the most vulnerable are those populations that have just extracted themselves from deprivation and subsistence living, through developing-world growth powered by industrialisation and fossil fuel.

And to help buffer or offset the impacts, we have no New Deal revival waiting around the corner, no Marshall Plan ready. The global halving of economic resources would be permanent, and, because permanent, we would soon not even know it as deprivation, only as a brutally cruel normal against which we might measure tiny burps of decimal-point growth as the breath of a new prosperity. We have gotten used to setbacks on our erratic march along the arc of economic history, but we know them as setbacks and expect elastic recoveries. What climate change has in store is not that kind of thing – not a Great Recession or a Great Depression but, in economic terms, a Great Dying.

How could that come to be? The answer is partly in the preceding chapters – natural disaster, flooding, public health crises. All of these are not just tragedies but expensive ones, and beginning already to accumulate at an unprecedented rate. There is the cost to agriculture: more than three million Americans work on more than two million farms; if yields decline by 40%, margins will decline, too, in many cases disappearing entirely, the small farms and cooperatives and even empires of agribusinesses slipping underwater (to use the oddly apposite accountant's metaphor) and drowning under debt all those who own and work those arid fields, many of them old enough to remember the same plains' age of plenty. And then there is the real flooding: 2.4 million American homes and businesses, representing more than $1 trillion in present-day value, will suffer chronic flooding by 2100, according to a 2018 study by the Union of Concerned Scientists. Fourteen per cent of the real estate in Miami Beach could be flooded by just 2045. This is just within America, though it isn't only South Florida; in fact, over the next few decades, the real-estate impact will be almost $30 billion in New Jersey alone.

There is a direct heat cost to growth, as there is to health. Some of these effects we can see already – for instance, the warping of train tracks or the grounding of flights due to temperatures so high that they abolish the aerodynamics that allow planes to take off, which is now commonplace at heat-stricken airports Like the one in Phoenix. (Every round-trip plane ticket from New York to London, keep in mind, costs the Arctic three more square meters of ice.) From Switzerland to Finland, heat waves have necessitated the closure of power plants when cooling liquids have become too hot to do their job. And in India, in 2012, 670 million lost power when the country's grid was overwhelmed by farmers irrigating their fields without the help of the monsoon season, which never arrived. In all but the shiniest projects in all but the wealthiest parts of the world, the planet's infrastructure was simply not built for climate change, which means the vulnerabilities are everywhere you look.

Other, less obvious effects are also visible – for instance, productivity. For the past few decades, economists have wondered why the computer revolution and the Internet have not brought meaningful productivity gains to the industrialised world. Spreadsheets, database management software, email – these innovations alone would seem to promise huge gains in efficiency for any business or economy adopting them. But those gains simply haven't materialised; in fact, the economic period in which those innovations were introduced, along with literally thousands of similar computer-driven efficiencies, has been characterised, especially in the developed West, by wage and productivity stagnation and dampened economic growth. One speculative possibility: computers have made us more efficient and productive, but at the same time, climate change has had the opposite effect, diminishing or wiping out entirely the impact of technology. How could this be? One theory is the negative cognitive effects of direct heat and air pollution, both of which are accumulating more research support by the day. And whether or not that theory explains the great stagnation of the last several decades, we do know that, globally, warmer temperatures do dampen worker productivity.

The claim seems both far-fetched and intuitive, since, on the one hand, you don't imagine a few ticks of temperature would turn entire economies into zombie markets, and since, on the other, you yourself have surely laboured at work on a hot day with the air-conditioning out and understand how hard that can be. The bigger-picture perspective is harder to swallow, at least at first. It may sound like geographic determinism, but Hsiang, Burke, and Miguel have identified an optimal annual average temperature for economic productivity: 13 degrees Celsius, which just so happens to be the historical median for the United States and several other of the world's biggest economies. Today, the U.S. climate hovers around 13.4 degrees, which translates into less than 1% of GDP loss – though, like compound interest, the effects grow over time. Of course, as the country has warmed over the last decades, particular regions have seen their temperatures rise, some of them from suboptimal levels to something closer to an ideal setting, climate-wise. The greater San Francisco Bay Area, for instance, is sitting pretty right now, at exactly 13 degrees.

This is what it means to suggest that climate change is an enveloping crisis, one that touches every aspect of the way we live on the planet today. But the world's suffering will be distributed as unequally as its profits, with great divergences both between countries and within them. Already-hot countries like India and Pakistan will be hurt the most; within the United States, the costs will be shouldered largely in the South and Midwest, where some regions could lose up to 20% of county income.

Overall, though it will be hit hard by climate impacts, the United States is among the most well-positioned to endure – its wealth and geography are reasons that America has only begun to register effects of climate change that already plague warmer and poorer parts of the world. But in part because it has so much to lose, and in part because it so aggressively developed its very long coast-lines, the US is more vulnerable to climate impacts than any country in the world but India, and its economic illness won't be quarantined at the border. In a globalised world, there is what Zhengtao Zhang and others call an "economic ripple effect." They've also quantified it, and found that the impact grows along with warming. At degree Celsius, with a decline in American GDP of 0.88%, global GDP would fall by 0.12%, the American losses cascading through the world system. At two degrees, the economic ripple effect triples, though here, too, the effects play out differently in different parts of the world; compared to the impact of American losses at one degree, at two degrees the economic ripple effect in China would be 4.5 times larger. The radiating shock waves issuing out from other countries are smaller because their economies are smaller, but the waves will be coming from nearly every country in the world, like radio signals beamed out from a whole global forest of towers, each transmitting economic suffering.

For better or for worse, in the countries of the wealthy West, we have settled on economic growth as the single best metric, however imperfect, of the health of our societies. Of course, using that metric, climate change registers – with its wildfires and droughts and famines, it registers seismically. The costs are astronomical already, with single hurricanes now delivering damage in the hundreds of billions of dollars. Should the planet warm 3.7 degrees, one assessment suggests, climate change damages could total $551 trillion – nearly twice as much wealth as exists in the world today. We are on track for more warming still.

Over the last several decades, policy consensus has cautioned that the world would only tolerate responses to climate change if they were free – or, even better, if they could be presented as avenues of economic opportunity. That market logic was probably always short-sighted, but over the last several years, as the cost of adaptation in the form of green energy has fallen so dramatically, the equation has entirely flipped: we now know that it will be much, much more expensive to not act on climate than to take even the most aggressive action today. If you don't think of the price of a stock or government bond as an insurmountable barrier to the returns you'll receive, you probably shouldn't think of climate adaptation as expensive, either. In 2018, one paper calculated the global cost of a rapid energy transition, by 2030, to be negative $26 trillion – in other words, rebuilding the energy infrastructure of the world would make us all that much money, compared to a static system, in only a dozen years.

Every day we do not act, those costs accumulate, and the numbers quickly compound. Hsiang, Burke, and Miguel draw their 50% figure from the very high end of what's possible – truly a worst-case scenario for economic growth under the sign of climate change. But in 2018, Burke and several other colleagues published a major paper exploring the growth consequences of some scenarios closer to our present predicament. In it, they considered one plausible but still quite optimistic scenario, in which the world meets its Paris Agreement commitments, limiting warming to between 2.5 and 3 degrees. This is probably about the best-case warming scenario we might reasonably expect; globally, relative to a world with no additional warming, it would cut per-capita economic output by the end of the century, Burke and his colleagues estimate, by between 15 and 25%. Hitting four degrees of warming, which lies on the low end of the range of warming implied by our current emissions trajectory, would cut into it by 30% or more. This is a trough twice as deep as the deprivations that scarred our grandparents in the 1930s, and which helped produce a wave of fascism, authoritarianism, and genocide. But you can only really call it a trough when you climb out of it and look back from a new peak, relieved. There may not be any such relief or reprieve from climate deprivation, and though, as in any collapse, there will be those few who find ways to benefit, the experience of most may be more like that of miners buried permanently at the bottom of a shaft.

8.9   RECENT ADDITIONS TO THE CLIMATE CHANGE LITERATURE

There is a large body of climate change literature and we have cited quite a lot of it in this book. The purpose of this final section is just to mention a few new publications which have appeared recently and which we have not had time to digest properly to summarise their contributions to the literature. However, the literature is growing all the time and the danger in writing a book is that one never quite catches up with the latest papers and books to be published. We were anxious, towards the end of our period of writing, not to fall into the trap of what we might call the Mr Casaubon syndrome. Mr Casaubon was a pompous and ineffectual middle-aged fictional scholar in George Eliot's novel "Middlemarch" who marries the heroine, Dorothea Brooke, because he needs an assistant for his work. His "masterwork, "Key to All Mythologies" inevitably was never finished and so remained unpublished at his death.

We have collected half a dozen new books which contain interesting material which we have not had time to digest and so we present them to our readers with recommendations to read for themselves on the grounds that they have interesting and useful things to say. They are presented in no particular order.

1. John Browne "Make, Think, Imagine" 2019 London, Bloomsbury. We mention this book because it is by a representative of the Energy (oil) industry and the energy industry is often vilified by the "green" agenda as the people who are destroying the climate by extracting coal, oil, and gas from the bowels of the Earth. But what people forget is that this energy is extracted by the industry for the benefit of all of us who travel in cars, aeroplanes, and trains, who consume food and manufactured goods and who are the clients of these coal, oil, and gas companies.

2. Pope Francis "Laudato Si' – On care for our common home". 2015, London, Catholic Truth Society.

Though these two authors might seem unlikely bedfellows, we take these two books together. John Browne describes (pages 177–9) a meeting he attended in June 2018 at a unique summit hosted by Pope Francis at the Vatican. The meeting was attended by the leaders of many of the world's biggest oil and gas companies, investors who oversee trillions of dollars' worth of capital and many of the energy sector's leading thinkers and policymakers, along with various priests and bishops of the Catholic Church. The discussions focused on climate change and how we can tackle it without curtailing the world's economic growth. Browne comments that "The serene sixteenth-century chapels and palaces of the Vatican lend the appropriate gravitas to proceedings. The tone is one of humility, and the discussion is marked by a pragmatic optimism." He reflects that now more than two decades after he first urged the hydrocarbon sector to take heed of the grave reality of climate change, the terms of the debate have shifted – what he heard at the Vatican reaffirmed his conviction that we now have the pieces we need to solve this most pressing and complex of global puzzles.

We now have at our disposal the means to provide more energy to more people, while simultaneously extracting it at a lower economic, humanitarian, and environmental cost." Browne continues, "However, putting the right technology in place quickly will be expensive and will require that we have the courage to confront vested interests and design effective incentives, in order to create a market pull in the right direction. History shows us that energy transitions, from the use of only our muscles to wood to fossil fuels and beyond, are generally slow and incremental. The current transition is unlikely to be an exception, but this time we may not have the luxury of waiting to see what unfolds. We need to keep inventing better and more cost-effective energy sources, more efficient machines, and more intelligent ways to link our infrastructure together. "As Pope Francis summarised at the Vatican summit, 'Civilisation requires energy, but energy use must not destroy civilisation.'

As the world grows, we will need increasing amounts of energy. Even if the US and Europe become much more energy efficient, the rest of the world will consume more, as they strive to match our standard of living. This will only be possible with the use of fossil fuels, with efficient ways to capture and store or use carbon dioxide, in combination with the greater application of zero carbon sources such as renewables, as well as improved versions of nuclear fission. As things stand, nuclear fusion remains a distant prospect and is not the magical solution to all our energy needs. There are many forecasts regarding the future mix of energy sources, the cost of implementing the mix and the concentration of carbon dioxide in the atmosphere that will result. There remains a great deal of uncertainty, but almost all these forecasts indicate that the world must immediately increase its investment in energy by at least 60 per cent above the level seen in 2015. The amount of hydrocarbons in the mix will depend on effective incentives to reduce the use of carbon, and continued engineering and commercial breakthroughs in the production of all other forms of energy. How we set in motion these actions to provide all of humankind with the energy we need to progress, while also sustaining the biosphere's capability to nurture human life, is among the most profoundly consequential decisions that we will have to make."

In Laudato Si' Pope Francis said "I will begin by briefly reviewing several aspects of the present ecological crisis, with the aim of drawing on the results of the best scientific research available today, letting them touch us deeply and provide a concrete foundation for the ethical and spiritual itinerary that follows. I will then consider some principles drawn from the Judaeo-Christian tradition which can render our commitment to the environment more coherent. I will then attempt to get to the roots of the present situation, so as to consider not only its symptoms but also its deepest causes. This will help to provide an approach to ecology which respects our unique place as human beings in this world and our relationship to our surroundings. In light of this reflection, I will advance some broader proposals for dialogue and action which would involve each of us as individuals, and also affect international policy. Finally, convinced as I am that change is impossible without motivation and a process of education, I will offer some inspired guidelines for human development to be found in the treasure of Christian spiritual experience."

3. Bill McKibben "Falter – Has the Human Game Begun to Play Itself Out?" 2019. Wildfire, London. As the author notes in the introduction it is 30 years since a previous book of his "The End of Nature" which was an introduction to the human-induced greenhouse effect. As the title indicates, "The End of Nature" was not a cheerful book, and sadly its gloom has been vindicated. My basic point was that humans had so altered the planet that not an inch was beyond our reach, an idea that scientists underlined a decade later when they began referring to our era as the Anthropocene.

This volume "Falter" is bleak as well – in some ways bleaker, because more time has passed and we are deeper in the hole. It offers an account of how the climate crisis has progressed and of the new technological developments in fields such as artificial intelligence that also seem to me to threaten a human future. Put simply, between ecological destruction and technological hubris, the human experiment is now in question. The stakes feel very high, and the odds very long and the trends very ominous." McKibben opens "Falter" on a Note on Hope: Thirty years ago, in 1989, I wrote the first book for a wide audience on climate change – or, as we called it then – the greenhouse effect.

I know, too, that this bleakness cuts against the current literary grain. Recent years have seen the publication of a dozen high-profile books and a hundred TED talks devoted to the idea that everything in the world is steadily improving. They share not only a format (endless series of graphs showing centuries of decreasing infant mortality or rising income) but also a tone of perplexed exasperation that any thinking person could perceive the present moment as dark. As Steven Pinker, the author of the sanguine Enlightenment Now, explained, "None of us are as happy as we ought to be, given how amazing our world has become." People, he added, just "seem to bitch, moan, whine, carp and kvetch."

"I'm grateful for those books because, among other things, they remind us precisely how much we have to lose if our civilizations do indeed falter. But the fact that living conditions have improved in our world over the last few hundred years offers no proof that we face a benign future. That's because threats of a new order can arise – indeed, have now arisen. Just as a man or woman can grow in strength and size and wealth and intelligence for many years and then be struck down by some larger force (cancer, a bus), so, too, with civilizations. And – to kvetch and whine a little further – because of the way power and wealth are currently distributed on our planet, I think we're uniquely ill-prepared to cope with the emerging challenges. So far, we're not coping with them.

Still, there is one sense in which I am less grim than in my younger days. This book ends with the conviction that resistance to these dangers is at least possible. Some of that conviction stems from human ingenuity – watching the rapid spread of a technology as world changing as the solar panel cheers me daily. And much of that conviction rests on events in my own life over the past few decades. I've immersed myself in movements working for change, and I helped found a group, 350.org, that grew into the first planet-wide climate campaign. Though we haven't beaten the fossil fuel industry, we've organized demonstrations in every country on the globe save North Korea, and with our many colleagues around the world, we've won some battles. At the moment, we're helping as friends and colleagues push hard for a Green New Deal in the United States and similar steps around the world. (This book is dedicated to one of my dearest colleagues in that fight, Koreti Tiumalu, who died much too early, in 2017.) I've been to several jails, and to a thousand rallies, and along the way I've come to believe that we have the tools to stand up to entrenched power.

Whether that entrenched power can actually be beaten in time I do not know. A writer doesn't owe a reader hope – the only obligation is honesty – but I want those who pick up this volume to know that its author lives in a state of engagement, not despair. If I didn't, I wouldn't have bothered writing what follows."

4. Nathaniel Rich "Losing Earth – the decade we could have stopped climate change": 2019, Picador, London. We have already quoted most of the introduction to this book in Section 8.6. We can just reduce that section to a single paragraph:

"Nearly everything we understand about global warming was understood in 1979. It was, if anything, better understood. Today, almost nine out of ten Americans do not know that scientists agree, well beyond the threshold of consensus, that human beings have altered the global climate through the indiscriminate burning of fossil fuels. But by 1979 the main points were already settled beyond debate, and attention turned from basic principles to a refinement of the predicted consequences. Unlike string theory and genetic engineering, the "greenhouse effect"--a metaphor dating to the early twentieth century – was ancient history, described in any intro-to-biology textbook. The basic science was not especially complicated. It could be reduced to a simple axiom: the more carbon dioxide in the atmosphere, the warmer the planet. And every year, by burning coal, oil, and gas, human beings belched increasingly obscene quantities of carbon dioxide into the atmosphere."

5. Christiana Figueres and Tom Rivett-Carnac "The Future we Choose – Surviving the Climate Crisis". 2020, Manilla Press, London.

Figueres and Rivett-Carnac were involved in COP 21 in Paris in 2015 which led to the Paris Agreement which defined CO_2 emissions targets and their story, very briefly, of that Conference is in their Authors' Note:

Christiana Figueres and Tom Rivett-Carnac (The Future we Choose – Surviving the Climate Crisis) are from very different backgrounds, she being born in 1956 at what they describe as the end of the Holocene epoch of stable climate conditions and he in 1977 at the start of the Anthropocene epoch which is characterised by the breakdown of the conditions necessary for human survival. Christiana Figueres is from Costa Rica, a developing country with a long tradition of sound economic growth in keeping with the natural environment. Tom Rivett-Carnac is from the UK, the country which started the Industrial Revolution, with its inevitable reliance on coal.

They had very little in common, Christiana's father was three times president of Costa Rica and is considered "the father of modern Costa Rica." He was responsible for some of the most widespread environmental policies in the world and is the only head of state ever to have abolished a national army. Tom is directly descended from founders of the East India Company, an institution which operated a private army. However, they shared a deep conviction around the need to forge a better world for all children and decided to work together to this end.

From 2010 to 2016, Christina was Executive Secretary of the United Nations Framework Convention on Climate Change (UNFCCC), an organisation tasked with guiding the response of all governments to climate change. Assuming the highest responsibility for negotiations right after the dramatic debacle of the 2009 Copenhagen climate change conference, Christiana refused to accept that a global agreement was impossible.

In 2013 in New York, Christiana met Tom, who was then president and CEO of the Carbon Disclosure Project USA, and they discussed the possibility of his joining the UN. Although Christiana considered that Tom did not have the experience necessary for the job, she felt that he had "the humility to foster collective wisdom, and the courage to work within a complexity that is beyond any mapping."

Tom was invited to join the UN effort to advance the negotiations for the Paris Agreement as Christiana's Chief Political Strategist. In this role he designed and led the Groundswell Initiative, which mobilised support for an agreement from a wide range of stakeholders outside of national governments. A few years later the most far-reaching international agreement on climate change was finally achieved.

On 12 December 2015, at 7:25 P.M. the Paris Agreement was adopted to the delight of 5,000 delegates who had been eagerly awaiting this historical breakthrough. One hundred and ninety-five nations had unanimously adopted an agreement that would guide their economies for the next four decades.

A new global pathway had been established but pathways are only of value if they are used and in their book "The Future we Choose – Surviving the Climate Crisis" Christiana and Tom write that humanity has procrastinated for far too long on climate change and that now it is necessary to walk the path, or rather to run it. This book maps that route and the reader is encouraged to join them at www.GlobalOptimism.com

The final one of these six books is:

6. Anatol Lieven "Climate Change and the Nation State – the Realist Case". The dust jacket says "This is a book which offers realism in place of idealism, and national policy instead of global protest. It will provoke innumerable discussions." It is not immediately apparent what the Realist case is, but it includes the idea that we mentioned in Section 8.5.2 that school strikes,

big marches, and the like are not going to achieve very much. We believe that it is worth reading this book to try to find out what the author means by the Realist Case. According to the dust jacket:

> This is one of those rare books that have something really important to say. Anatol Lieven, one of the most original and independent-minded foreign policy thinkers, is telling his fellow realists that at this moment the world's great powers are far more threatened by climate change than they are by each other.

8.10 A NEW TOOL FOR ENVIRONMENTAL RISK ASSESSMENT: THE NATURAL TIME ANALYSIS

Many natural phenomena governed by physical laws allow their changes to be predicted. However, others, such as climate change, seem impossible. For centuries, these complex systems were considered random because the necessary mathematics were not available to understand their patterns. In the 1960s, after nearly a century, Edward Lorenz, while studying climate, came to the same conclusion as Henri Poincaré in 1887 about the unpredictability of dynamic systems. In the 1970s, Mitchell Feigenbaum, showed how the transition from order to chaos occurs through the existence of the constant 4.6692. In the mid-1980s, chaos was a thriving topic. Scientists, based on the fact that even complex systems contain an underlying order, began to look beyond the apparent random disorder of nature, finding links in the behaviour of meteorological phenomena, the evolution of ecosystems and climate, etc. (Lovejoy and Varotsos, 2016). In general, it is impossible to predict the long-term behaviour of these systems since in them, small differences in the initial conditions produce very divergent results (Varotsos 2020).

The results of the new analysis termed as natural time analysis (NTA) showed that novel dynamical features hidden behind time series in complex systems can emerge upon analyzing them in the new time domain of natural time, which conforms to the desire to reduce uncertainty and extract signal information as much as possible. The analysis in natural time enables the study of the dynamical evolution of a complex system and identifies when the system approaches a critical point. Hence, natural time plays a key role in predicting impending catastrophic events in general (Varotsos *et al.* 2011 and references therein). Data analysis in natural time have appeared to date in diverse fields, including Biology, Earth Sciences (Geophysics, Seismology), Environmental Sciences, Physics and Cardiology.

As a first example, impressive results emerged from the study of the depletion of the atmospheric ozone layer by applying NTA to ozone observations since 1979 (Varotsos and Tzanis 2012). More precisely, the NTA has been shown to capture key features of the dynamics of the ozone hole complex system, especially before the unprecedented event of a major, sudden stratospheric warming and the subsequent break-up of the Antarctic ozone hole into two holes in September 2002 (Varotsos 2002). In particular, precursor changes in entropy in natural time under time reversal, have been identified from 2000 to 2001. There was also a demonstration of NTA potential by studying the unusual dynamic conditions in the Arctic during 2019–2020 that led to the lowest ozone level in much of central Arctic, with an area about three times the size of Greenland (Varotsos *et al.* 2020a). Although this first ozone hole over the Arctic during the winter-spring 2019–2020 was attributed to the colder stratospheric temperatures, the winter-spring temperature in 2019–2020 was very close to that of 2010–2011, when no such ozone hole occurred. The latter demonstrated via the NTA that the ozone hole phenomenon is complex and does not depend solely on the temperature regime of the polar stratosphere (von der Gathen *et al.* 1995). In addition, the advanced nowcasting model that is mainly based on NTA showed much earlier that Arctic ozone depletion in 2019–2020 would reach the limits of the ozone hole and that it would disappear in the last week of April 2020, as did happen. A striking agreement between real and nowcasted stratospheric ozone and temperature extreme values was found in the NTA domain. As it turns out, the problem of ozone depletion remains and sometimes grows unexpectedly even though a lot of excellent work has been done to address it, which won the Nobel Prize in Chemistry in 1995. It was awarded to three Professors Mario J. Molina of the MIT University, Sherwood Rowland of the University of California at Irvine and Paul Crutzen of the Max Planck Institute for Chemistry in Mainz, Germany, who proposed the thinning of the ozone layer as a consequence of chlorofluorocarbons and other chemicals used in a range of products.

A second example of using NTA, is the study of the El-Niño/La Niña phenomenon that is closely related to extreme weather events worldwide (e.g. heat waves, tornadoes, floods, and droughts), incidence of epidemic diseases (e.g. malaria), severe coral bleaching, etc. The El Niño/La Niña Southern Oscillation (ENSO) phenomenon is a composite oceanic–atmospheric phenomenon which is quasi-periodic (with a period of 3–7 years). It results from the interaction of the ocean and the atmosphere with many effects on climatic and weather conditions not only in the tropical Pacific but also in many regions of the world. The investigation of the temporal evolution of ENSO since January 1876 by means of NTA reveals that the major ENSO events display precursors that are maximized over a period of almost two years (Varotsos *et al.* 2016). This finding improves the accuracy of the short-term prediction models of the ENSO extreme events, thus preventing disastrous effects in advance. Among the devastating effects of the El Niño events are: increased and intense forests fires (emitting greenhouse gases—CO_2 and CH_4—that trap heat in the atmosphere and contribute to global warming), reduction in phytoplankton, severe droughts in Australia, Indonesia, India and southern Africa and heavy rains in California, Ecuador, and the Gulf of Mexico. The negative impacts of La Niña are abnormally heavy monsoons in India and Southeast Asia, cool and wet winter weather in southeastern Africa, wet weather in eastern Australia, cold winter in western Canada and the northwestern United States and winter drought in the southern United States.

A third example of using NTA, is the study of extreme weather events. There is growing evidence that extreme weather events, such as frequent and intense cold spells and heat waves, are causing unprecedented deaths and diseases in both developed and developing countries. Thus, they require extensive and urgent research to reduce the risks. Average temperatures in Europe in June–July 2019 were the hottest ever measured and attributed to climate change. The problem, however, of a thorough study of natural climate change is the lack of experimental data from the long past, where anthropogenic activity was then very limited. Today, this problem can be successfully resolved by using, inter alia, biological indicators that have provided reliable environmental information for thousands of years in the past. An interesting problem of the scientific community was to determine if a slight temperature change in the past could trigger current or future sharp temperature change or changes. When NTA was used for this purpose, it showed that temperature fluctuations were persistent. That is, they exhibit long memory with scaling behavior, which means that an increase (decrease) in temperature in the past is always followed by another increase (decrease) in the future with multiple amplitudes. Therefore, the increase in the frequency, intensity, and duration of extreme temperature events due to climate change will be more pronounced than expected. This will affect human well-being and mortality more than is estimated in today's modeling scenarios. The scaling property detected can be used to more accurately forecast monthly to decadal extreme temperature events. Thus, it is possible to develop improved early warning systems that will reduce public health risk locally, nationally and internationally (Varotsos, and Mazei 2019).

A fourth example of applying NTA refers to the current debate on the role of carbon dioxide in global warming. The NTA tool enables the examination of the intrinsic dynamics of the temporal evolution of the carbon dioxide amount and temperature by analyzing instrumental measurements data and paleo-reconstructed data to see if this dynamics remains stable in terms of its internal properties. A recent study (Varotsos *et al.* 2020b) used the paleoecological data that were based on over 20,000 sea surface temperature point reconstructions obtained from ocean sediment cores using alkenone unsaturation indices, applying ratios of Mg/Ca in planktonic foraminifera. The NTA results concerning the instrumental data fluctuations in land-ocean temperature and carbon dioxide, in the period March 1958–April 2017, showed that the long-term dynamics of the fluctuations of these two quantities exhibit persistent power-law type behavior. This type of behavior also emerges from the same analysis of the reconstructed global mean sea temperature and carbon dioxide amount fluctuations over the past 805 thousand years. In addition, the time series of both parameters are characterised by multifractality for time scales of less than 4 years. Consequently, the intrinsic properties of carbon dioxide and global temperature have not changed over the past nearly one million years [16]. This result is very important for the theory of global warming which is of crucial importance for humanity. It is worth mentioning that the UN's Intergovernmental Panel on Climate Change (IPCC) and former US Vice President Al Gore were awarded the Nobel Peace Prize in 2007 for their efforts to build up and disseminate greater knowledge about man-made climate change, and to lay the foundations for the measures that are needed to counteract such change.

A fifth example of using NTA, is the reconstruction of historic climate parameters. In this regard, atmospheric pollutants and environmental indicators are often used to reconstruct historic atmospheric pollution from peat, as it is accumulated over time by decomposing plant material, thus leading to a history of air pollution. Very recently, three key parameters related to the peat bogs' surface wetness dynamics in European Russia during the Holocene were investigated using modern statistical analysis. These parameters are: (i) the water table depth (WTD) in relation to the surface, which is reconstructed based on the community structure of the subfossil testate amoeba assemblages; (ii) the peat humification estimated as absorption of alkaline extract that directly reflects moisture at the time the peat was formed; (iii) the Climate Moisture Index (CMI) and the Aridity Index derived from pollen-based reconstructions of the mean annual temperature and precipitation and classifying moisture conditions as the ratio between available annual precipitation and potential land surface evapotranspiration. All these parameters provide useful information about the paleoclimate (atmospheric moisture component) dynamics. According to a recent investigation [17] high values of WTD and peat humification appear to comply with the Gutenberg-Richter law for earthquakes, which interestingly results from NTA just by considering the maximum entropy principle [9]. It is noteworthy that this law also seems to reproduce the high values of the modeled climate moisture and aridity indices. The validity of this new result is checked by applying NTA. On this basis, a new nowcasting tool was developed to more accurately estimate the average waiting time for the extreme values of these climate parameters. This has helped to understand better climate variability to address emerging development needs and priorities by implementing empirical studies of the interactions between climatic effects, mitigation, adaptation, and sustainable growth (Varotsos et al. 2020c).

The above-mentioned examples of the application of the breakthrough NTA research achievements contribute substantially to the human well-being for the following reasons:

(1) The climate is a unique complex system and from an historical viewpoint, man is an element of it. However, at present, the problem of co-evolution of human society and nature has arisen. The impact of human activity on natural systems has reached a global scale, and it is important to try to make a conditional division between man-made and natural processes. A typical description of this division can be obtained using the tool of nature-society system analysis. Usually, there are two interacting systems: human society with technologies, sciences, economics, sociology, agriculture, industry, etc., and nature with climatic, ecological, biogeochemical, hydrological, geophysical and other natural processes. Parameterisation and investigation of the interaction of these two systems is the main objective of current investigations.

   A system, like the climate system, is defined by its behaviour and structure. The behaviour of such a system is intended to provide uninterrupted functioning by means of a correspondingly organised structure and behaviour. This characteristic of the complex system to actively interact with an external medium is referred to as survivability. The key questions to be answered within the numerous investigations of global ecodynamics are: (i) What are the levels, interactions and significance of the 'human dimension' in the development of society and its role in global environmental change? (ii) What are the current and potential future impacts of global environmental variability on economic development, what factors determine the capability of society to respond to changing events, what are the opportunities to provide sustainable development and reduce man's sensitivity to forcings? (iii) What are the possible methods of decision-making in sustainable development in view of the Nature-Society system complexity and high-level uncertainties regarding the global environmental variability?

(2) Climate Summits like Paris – 2015 (COP 21), Marrakech – 2016 (COP 22), Bonn – 2017 (COP 23), Katowice – 2018 [AC1] (COP 24), Madrid UN – 2019 (COP 25), or the International Arctic Forum 2019 in St. Petersburg did not fulfil the overall understanding of the global environmental crisis and its future consequences. However, they have determined the need to substantiate the priorities related to the interaction of society and nature, creating a set of Sustainable Development Goals (SDGs). This has substantially contributed to the transition from the Millennium Development Goals in the period 2000–2015 to the United Nations 2030 Agenda for Sustainable Development. This agenda through its SDGs addresses key global challenges including climate change and environmental degradation, with a plan for achieving all these goals by 2030, among which are: (i) Take urgent action to combat climate change and its impacts and (ii) Conserve and sustainably use the oceans, seas and marine resources for sustainable development (Varotsos and Cracknell 2020). However, further discussion on the priorities of global change problems is needed, but it must be done taking into account the fact that the Earth System has been operating in different quasi-stable states for the last half million years and that global change cannot be understood from the standpoint of a simple cause-effect paradigm. Do human activities actually have the ability to change the Earth System in ways that can turn out to be irreversible? Has the Earth System moved far beyond the range of natural variability? In the case of changing climatic issues, are recent estimates relevant to the facts? Such general and extremely important questions need to be addressed urgently using the Earth observations, whose recent research advances in both methodological development and technological solutions can make a significant contribution.

References

Abbott, B. W., Bishop, K., Zarnetske, J. P., Minaudo, C., Chapin, F. S., Krause, S., ... and Plont, S. (2019). Human domination of the global water cycle absent from depictions and perceptions. *Nature Geoscience*, 12, 533–540.

Adamenko V. N., Ya Kondratyev, K., (1990) Global climate changes and their empirical diagnostics. Anthropogenic impact on the nature of the North and Its ecological implications. Izrael Y. A., G. V. Kalabin and V. V. Nikonov. *Apatity: Kola Scientific Centre, Russian Academy of Sciences.* p. 17–34 (in Russian).

Akhtar, M. K., Simonovic, S. P., Wibe, J., & MacGee, J. (2019). Future realities of climate change impacts: an integrated assessment study of Canada. *International Journal of Global Warming*, 17(1), 59–88.

Albino, V., Ardito, L., Dangelico, R. M., and Petruzzelli, A. M. (2014). Understanding the development trends of low-carbon energy technologies: A patent analysis. *Applied Energy*, 135, 836–854.

Al-Fattah, S. M. (2020). Non-OPEC conventional oil: Production decline, supply outlook and key implications. *Journal of Petroleum Science and Engineering*, 107049, 1–16.

Allen, G. H. and Pavelsky, T. M. (2018a). Global extent of rivers and streams. *Science*, 361(6402), 585–588.

Allen, G. H., & Pavelsky, T. M. (2018b). Global extent of rivers and streams. *Science*, 361(6402), 585–588.

Alley, R. B., Brook, E. J., & Anandakrishnan, S. (2002). A northern lead in the orbital band: North–south phasing of Ice-Age events. *Quaternary Science Reviews*, 21(1-3), 431–441.

American Meteorological Society, 1995, *Fourth Conference on Polar Meteorology and Oceanography, 15–20 January 1995*, American Meteorological Society: Dallas, Texas, J1–J12.

Anderegg, W.R.L., Prall, J.W., Harold, J., and Schneider, S.H. (2010). Expert credibility in climate change. *Proceedings of the National Academy of Sciences* 107: 27. 12107–12109.

Angell, J. Y. (1986). Annual and seasonal global temperature changes in the troposphere and low stratosphere, 1960–85. *Monthly Weather Review*, 114(10), 1922–1930.

Anyamba, A. and Tucker C.J. (2005). Analysis of Sahelian vegetation dynamics using NOAA-AVHRR NDVI data from 1981–2003. *J Arid Environ* 63:596–614.

Arrhenius, S. (1896). "On the influence of carbonic acid in the air upon the temperature of the ground" *The London, Edinburgh, and Dublin Philosophical Magazine and Journal of Science*. 41 251, 237–276.

Bach, W., 1986, Nuclear war: the effects of smoke and dust on weather and climate. *Progress in Physical Geography*, 10, 315–363.

Bach, W., 1987, *Kann die Menschheit einen Atomkrieg überlegen? In Verantwortung für den Frieden Naturwissenschaftler – Initiative.* Köln.

Bagla, P. (2009). No sign yet of Himalayan meltdown, Indian report finds. *Science*. 326, p. 326.

Ball, P. (2004). *The Elements: A Very Short Introduction*. OUP: Oxford. p. 33.

Balsamo, G., Agustí-Panareda, A., Albergel, C., Arduini, G., Beljaars, A., Bidlot, J., ... and Buizza, R. (2018). Satellite and in situ observations for advancing global Earth surface modelling: A review. *Remote Sensing*, 10(12), 2038.

Barnes, J. (1987). *Early Greek Philosophy*. London: Penguin.

Barraclough, G. (1989). *The Times Atlas of World History*. London: Times Books.

Barrage, L. (2019). The Nobel Memorial Prize for William D. Nordhaus. *The Scandinavian Journal of Economics*, 121(3), 884–924.

Bate, L. F. (1998). *El proceso de investigación en arqueología*. Barcelona: Crítica.

Bate, R. and Morris, J. (1994). *Global Warming: Apocalypse or Hot Air?*, London: IEA Environment Unit.

Bell, R. E., A.F. Banwell, L.D. Trusel, and J. Kingslake. (2018). Antarctic surface hydrology and impacts on ice-sheet mass balance. *Nature Climate Change*, 8, p. 1044–1052.

Benedick, R.E. (1991). *Ozone Diplomacy: New Directions in Safeguarding the Planet*. Harvard University Press: Cambridge, MA p. 300.

Benedick, R. E. (2007). Avoiding gridlock on climate change. *National Academy of Sciences: Issues in Science and Technology*, Winter, 37–40.

Benedick, R. E. (2009). Science inspiring diplomacy: the improbable Montreal Protocol. *Proceedings of the symposium for the 20th anniversary of the Montreal Protocol*, eds. C. Zerefos, G. Contopoulos, and G. Skalkeas, Springer Berlin, 13–19.

Bentham, J., G.M. Singh, G. Danaei, R. Green, J.K. Lin, G.A. Stevens, … and M. Ezzati. (2020). Multidimensional characterization of global food supply from 1961 to 2013. *Nature Food*, 1(1), 70–75.

Bereiter, B., S. Eggleston, J. Schmitt, C. Nehrbass-Ahles, T.F. Stocker, H. Fischer, S. Kipfstuhl and J. Chappellaz. (2015). Revision of the EPICA Dome C CO_2 record from 800 to 600-kyr before present. *Geophysical Research Letters*, 42(2), 542–549, doi:10.1002/2014gl061957.

Berger, A., 1984, Hiver nucléaire ou les conséquences climatiques d'un conflit nucléaire généralisé. *Revue des Questions Scientifiques*, 155, 461–493.

Berger, W. H. and Labeyrie, L. D. (eds.), 1987, *Abrupt Climatic Change: Evidence and Implications*. (Dordrecht : Reidel).

Berger, A., Gallee, H., Fichefet, T., Marsiat, I., & Tricot, C. (1990). Testing the astronomical theory with a coupled climate—ice-sheet model. *Palaeogeography, Palaeoclimatology, Palaeoecology*, 89(1-2), 125–141.

Berger, A., T. Fichefet, H. Gallee, C. Tricot, and J.P. Van Ypersele. (1992). Entering the glaciation with a 2-D coupled climate model. *Quaternary Science Reviews*, 11(4), 481–493.

Bergquist, A. K., and K. Söderholm. (2017). Business and Green Knowledge Production in Sweden 1960s-1980s. *Harvard Business School Research Paper Series*. 18-050.

Bernauer, T., and T. Böhmelt. (2020). International conflict and cooperation over freshwater resources. *Nature Sustainability*, 3(5), 350–356.

Berners-Lee, M. (2010). *How Bad are Bananas? The Carbon Footprint of Everything*. London: Profile Books.

Berners-Lee, M. (2019). *There is no Planet B: A Handbook for the Make or Break Years* Cambridge: Cambridge University Press.

Best, K. (2019). A Data-Driven Analysis of Environmental Migration in Coastal Bangladesh (Doctoral dissertation, Vanderbilt University).

Beven K. (2002) Towards a coherent philosophy for modeling the environment *Proc. Roy. Soc. London* vol. 458, N2026, p. 2465–2484.

Biehl, J. and P. Staudenmaier. (1995). "*Ecofacism: Lessons from the German Experience*" Edinburgh: AK Press.

Björkman, M. and S. Widmalm. (2010). Selling Eugenics: The Case of Sweden, Notes and Records of the Royal Society, 18 August 2010.

Boden, T. A., Andres, R. J., & Marland, G. (2017). Global, regional, and national fossil-fuel co2 emissions (1751-2014)(v. 2017). Environmental System Science Data Infrastructure for a Virtual Ecosystem; Carbon Dioxide Information Analysis Center (CDIAC), Oak Ridge National Laboratory (ORNL), Oak Ridge, TN (United States).

Boehmer-Christensen, S. (1994a). Global climate protection policy: The limits of scientific advice. Part 1. *Global Environmental Change*. 4, 140–159.

Boehmer-Christensen, S. (1994b). Global climate protection policy: The limits of scientific advice. Part 2. *Global Environmental Change*. 4, 189–200.

Boehmer-Christensen, S.A. and A.P. Cracknell. (2009). Kirill Kondratyev and the IPCC: His opposition to the Kyoto Protocol. In A.P. Cracknell, V.F. Krapivin and C.A. Varotsos *Global Climatology and Ecodynamics: Anthropogenic Changes to Planet Earth*. Berlin, Chichester, Springer, Praxis.

Boehmer-Christiansen, S. (1996). 10 The international research enterprise and global environmental change. The environment and international relations, 171. *Global Environmental Change Series* Edited by Michael Redclift, Martin Parry, Timothy O'Riordan, Robin Grove-White, Routledge.

Bolch, T., J.M. Shea, S. Liu, F.M. Azam, Y. Gao, S. Gruber, W.W. Immerzeel, A. Kulkarni, H. Li, A.A. Tahir, and G. Zhang. (2019). Status and Change of the Cryosphere in the Extended Hindu Kush Himalaya Region. In *The Hindu Kush Himalaya Assessment* (pp. 209–255). Springer: Cham.

Bolin, B. (2007). "*A History of the Science and Politics of Climate Change: The Role of the Intergovernmental Panel on climate Change*." Cambridge University Press: Cambridge.

Bond, W.J. and G.F. Midgley, 2012. Carbon dioxide and the uneasy interactions of trees and savannah grasses. *Philosophical Transactions Royal Society London, B, Biological Sciences*, 367, (1588) 601–612.

Bony, S., B. Stevens, D.M.W. Frierson, C. Jakob, M. Kageyama, R. Pincus, T. G. Shepherd et al. "Clouds, circulation and climate sensitivity." *Nature Geoscience* 8, no. 4 (2015): 261.

Boucher, O., Randall, D., Artaxo, P., Bretherton, C., Feingold, G., Forster, P., … and Rasch, P. (2013). Clouds and aerosols. In *Climate Change 2013: The Physical Science Basis. Contribution of Working Group I to the Fifth Assessment Report of the Intergovernmental Panel on Climate Change* (pp. 571–657). Cambridge University Press, Vienna, Austria.

Broecker, W.S. and G.H. Denton. (1990). What drives glacial cycles? *Scientific American*. 262, 43–50.

Brough, W. and M. Kimenyi. (2004). "Desertification of the Sahel". *PERC –Reports*, 22, (2), 2004.

Browne, J. (2019). *Make, Think, Imagine*. London, Bloomsbury.

Bruckmann, G. (1980). *Input-Output Approaches in Global Modeling*; *Proceedings of the Fifth IIASA Symposium on Global Modeling*, September 26–29, 1977 (Vol. 9). Pergamon Press.

Brühl, C. and Crutzen, P., 1987, Scenarios of possible changes in atmospheric temperatures and ozone concentrations due to man's activities, estimated with a 1-D coupled photochemical climate model. *Climate Dynamics*, 2, 173–203.

Brun, E., D. Six, G. Picard, V. Vionnet, L. Arnaud, E. Bazile, A. Boone, A. Bouchard, C. Genthon, V. Guidard, et al. (2011). Snow/atmosphere coupled simulation at Dome C, Antarctica. *Journal of Glaciology*, 57, 721–736.

Brundtland Report. (1987). see World Commission on Environment and Development, 1987.

Budyko, M.I. (1948). *"Isparenie v estestvennykh usloviyakh (Evaporation Under Natural Conditions)"* Leningrad: Gidrometeorologicheskoe izdatel'stvo.

Budyko, M. I. (1969). The effect of solar radiation variations on the climate of the Earth. *Tellus*, 21(5), 611–619.

Budyko, M. I. (1973). *Climate and Life*. (trans. Z. Uchijima), Tokyo University Press, Tokyo.

Budyko, M.I. (1974). *Climate and life*. edited by David Miller. New York: Academic Press.

Budyko, M.I. (1977). *Global'naya ekologiya (Global Ecology)*, Moscow: Mysl.

Budyko, M. I., and Izrael, Y. A. (1987). *Anthropogenic Change of Climate*. Leningrad, Russia, Gidrometeoizdat.

Burns, P.C., R.C. Ewing and A. Navrotsky. (2012). Nuclear fuel in a reactor accident. *Science*, 335 (6073) 1184–1188.DOI: 10.1126/science.1211285

Butos, W.N. and T.J. McQuade. (2015). Causes and consequences of the climate science boom. *The Independent Review* 20: 2 (Fall), 165–196.

Buznikov, A.A. (2008). "Creative career of Kirill Kondratyev." In *"Academician Kirill Ya Kondratyev. A Person from the Generation of Victors."* Russian Academy of Sciences: Levsha St Petersburg, p. 7–20.

Callendar, G.S. (1938). The artificial production of carbon dioxide and its influence on temperature. *Quarterly Journal of the Royal Meteorological Society*, 64, 223–240.

Canby, T.Y. (1984). El Niño's ill wind. *National Geographical Magazine*, 165, 144–183.

Carson, R. (1962). *"Silent Spring"*. Boston, MA: Houghton Mifflin.

Cess, R. D., 1985, Nuclear war: illustrative effects of atmospheric smoke and dust upon solar radiation. *Climatic Change*, 7, 237–252.

Chaisson, E. J. (2001). *COSMIC EVOLUTION: The Rise of Complexity in Nature*. Harvard Univ. Press. 274 pp.

Chaisson, E.J. (2008). Long-term global heating from energy usage, *Eos*, 89, 253.

Chaisson, E. J. (2011). Energy rate density as a complexity metric and evolutionary driver. *Complexity*, 16(3), 27–40.

Charlson, R.J., J. Lovelock, M. Andreae, and S. Warren. (1987). Oceanic phytoplankton, atmospheric sulfur, cloud albedo, and climate. *Nature*, 326, 655–661.

Chattopadhyay, G., P. Chakraborthy, and S. Chattopadhyay. (2012). Mann-Kendall trend analysis of tropospheric ozone and its modeling using ARIMA, *Theor. Appl. Clim.*, 110, 321–328, doi:10.1007/s00704-012-0617-y.

Chu, S., and A. Majumdar. (2012). Opportunities and challenges for a sustainable energy future. *Nature*, 488(7411), 294–303.

Chubachi, S. (1984). Preliminary result of ozone observations at Syowa Station from February, 1982 to January, 1983. *Memoirs of National Institute of Polar Research Japan*, special Issue, 34, 13–20.

Chubachi, S. (1985). A special ozone observation at Syowa Station, Antarctica, from February 1982 to January 1983, in C.S. Zerefos and A. Ghazi (eds.) *Proceedings of the Quadrennial Ozone Symposium held in Halkidiki, Greece, September 3-7, 1984*. D. Reidel: Dordrecht, The Netherlands, pp. 285–289.

Chubachi, S. (1993). Relationship between total ozone amounts and stratospheric temperature at Syowa, Antarctica. *Journal of Geophysical Research: Atmospheres*, 98(D2), 3005–3010.

Chylek, P. and Ramaswamy, V., 1984, Effect of graphitic carbon on the albedo of clouds. *Journal of Atmospheric Science*, 41, 3076–3094

Chylek, P., Ramaswamy, V. and Srivastava, V., 1983, Albedo of soot-contaminated snow. *Journal of Geophysical Research*, 88, 10837–10843.

Chylek, P., Srivastava, V., Cahenzli, L., Pinnick, R. G., Dod, R. L., Novakov, T., Cook, T. L. and Hinds, B. D., 1987, Aerosol and graphitic carbon content of snow. *Journal of Geophysical Research*, 92, 9801–9810.

Climate Change Science Program. (2003). Vision for the Program and Highlights of the Science Strategic Plan. A Report by the Climate Change Science Program and the Subcommittee on Global Change Research. Washington, D.C., July 2003, 34

Coal. (n.d.) On the economics of electricity, *The Energy Journal* (3) 1–27.

Cogley, J.G., J.S. Kargel, G. Kaser, and N.C. Van de Veen. (2010). Tracking the source of glacier misinformation. Science, published E-Letter responses (20 January 2010).

Collins, T. (2018). *"Can the Sahel recapture its lost glory?"* New African, August/September

Cook, J. D. Nuccitelli, S.A. Green, M. Richardson, B. Winkler, R. Painting, R. Way, P. Jacobs, and A. Skuce. (2013). Quantifying the consensus on anthropogenic global warming in the scientific literature. *Environmental Research Letters* 8: 2.

Cook, J., Oreskes, N., Doran, P. T., Anderegg, W. R., Verheggen, B., Maibach, E. W., ... and Nuccitelli, D. (2016). Consensus on consensus: A synthesis of consensus estimates on human-caused global warming. *Environmental Research Letters*, 11(4), 048002.

COSPAR (1978) Proceedings of the Open Meeting of the Working Group on Space Biology of the Twenty-First Plenary Meeting of COSPAR, Innsbruck, Austria, 29 May-10 June 1978.

Cottey, A. (2018). "Martin Ryle: An energy visionary." *Physics World*, 31: (9), 36–40.

Cox, J.D. (2005) *Climate Crash – Abrupt Climate Change and What it Means for Our Future*. Joseph Henry Press, Washington D.C. p. 215.

Cracknell, A. P. (1997). *Advanced very high resolution radiometer AVHRR*. CRC Press.

Cracknell, A.P. (2009)."Sustainability – no hope!" or "Sustainability – no hope?" in "Global Climatology and Ecodynamics – Anthropogenic Changes to Planet Earth".

Cracknell, A.P., and C.A. Varotsos. (2012). *Remote Sensing and Atmospheric Ozone: Human activities versus natural variability*. Praxis, Chichester, Springer, Berlin.

Cracknell, A.P. and C.A. Varotsos, (2007a) The IPCC fourth assessment report and the fiftieth anniversary of Sputnik. *Environmental Science and Pollution Research*, 14(6): 384–387.

Cracknell, A.P., and C.A. Varotsos, (2007b). Editorial and cover: Fifty years after the first artificial satellite: From Sputnik 1 to Envisat. *International Journal of Remote Sensing*, 28(10), 2071–2072.

Cracknell, A. P., and C. A. Varotsos, (2008). Editorial comment-the Montreal protocol. *International Journal of Remote Sensing*, 29(19), 5455–5459.

Cracknell, A. P., and C.A. Varotsos. (2011). New aspects of global climate-dynamics research and remote sensing. *International Journal of Remote Sensing*, 32(3), 579–600.

Cracknell A.P., V.F. Krapivin and C.A. Varotsos (eds.) (2009). *Global Climatology and Ecodynamics: Anthropogenic Changes to Planet Earth*. Springer/Praxis: Chichester, U.K. p. 518.

Croll, J. (1867). On the change in the obliquity of the ecliptic, its influence on the climate of the polar regions and on the level of the sea. *Philosophical Magazine*, 33, 426–445.

Crutzen, P., & Birks, J. (1982). Twilight at noon: The atmosphere after a nuclear war. *Ambio*, 11(2–3), 114–125.

Crutzen, P. and Hahn, J., 1986, *Schwarzer Himmel, Auswirkungen einer Atomkrieges auf Klima und Globale Umwelt* (Frankfurt: Fischer Verlag).

D'Odorico, P., L. Rosa, A. Bhattachan, and G.S. Okin. (2019). Desertification and Land Degradation. In *Dryland Ecohydrology* (pp. 573–602). Springer: Cham.

Daly, H. (1990). Towards some operational principles of sustainable development. *Ecological Economics*, **2** (1), 1–6.

Darwall, R. (2013). *"The Age of Global Warming – A History"* Quartet Books, London.

Darwall, R. (2017). *False claims on low-carbon energy are damaging UK*, http://www.telegraph.co.uk/business/2017/03/23/false-claims-low-carbon-energydamaging-uk/, accessed on 06.07.2017.

Darwall, R. (2019). *"Green Tyranny – Exposing the totalitarian roots of the climate industrial complex."* Encounter, New York, London.

Davis, K. F., A. Bhattachan, P. D'Odorico, and S. Suweis. (2018). A universal model for predicting human migration under climate change: Examining future sea level rise in Bangladesh. *Environmental Research Letters*, 13(6), 064030. doi:10.1088/1748-9326/aac4d4

de Leon Barido, D. P., N. Avila, and D.M. Kammen. (2020). Exploring the enabling environments, inherent characteristics and intrinsic motivations fostering global electricity decarbonization. *Energy Research & Social Science*, 61, 101343.

Dedić, N., C. Stanier. (2017). *Towards Differentiating Business Intelligence, Big Data, Data Analytics and Knowledge Discovery*. 285. Berlin; Heidelberg: Springer International Publishing.

DeFoor, T. E., Robinson, E., and Ryan, S. (1992). Early lidar observations of the June 1991 Pinatubo eruption plume at Mauna Loa Observatory, Hawaii. *Geophysical Research Letters*, 19(2), 187–190.

Deshler, T., Hofmann, D. J., Johnson, B. J., & Rozier, W. R. (1992). Balloonborne measurements of the Pinatubo aerosol size distribution and volatility at Laramie, Wyoming during the summer of 1991. *Geophysical Research Letters*, 19(2), 199–202.

Deshler, T., Johnson, B. J., and Rozier, W. R. (1993). Balloonborne measurements of Pinatubo aerosol during 1991 and 1992 at 41 N: Vertical profiles, size distribution, and volatility. *Geophysical Research Letters*, 20(14), 1435–1438.

Dial, R.J., G.Q. Ganey, and S.M. Skiles. (2018). What color should glacier algae be? An ecological role for red carbon in the cryosphere. *FEMS Microbiology Ecology*, 94(3), p. fiy007.

Diamond, J. (2005). *Collapse: How Societies Choose to Fail or Survive*. Allen Lane, London, p. 575.

Donchenko, V.K. (2008). In *"Academician K.Ya. Kondratyev"* Russian Academy of Sciences, Levsha, St Petersburg, pp. 21–31.

Doran, P.T. and Zimmerman, M.K. (2009). Examining the scientific consensus on climate change. *EOS* 90: 3, 22–23. DOI:10.1029/2009EO030002.

Duan, L., L. Cao, and K. Caldeira. (2018). Estimating contributions of sea ice and land snow to climate feedback. *Journal of Geophysical Research: Atmospheres*, 124(1), 199–208.

Duchesne, L., Houle, D., Ouimet, R., Caldwell, L., Gloor, M., and Brienen, R. (2019). Large apparent growth increases in boreal forests inferred from tree-rings are an artefact of sampling biases. *Scientific Reports*, 9(1), 1–9.

Einstein, A. (1996). *Quoted in A. Calaprice, The Quotable Einstein*. Princeton, MA: Princeton University Press, p. 224.

Eklundh, L. and Olsson, L. (2003). "Vegetation index trends for the African Sahel 1982-1999" *Geophysics Research Letters*, 30(8):13.1–13.4.

Ellingson, R.G., R.D. Cess, and G.L. Potter. (2016). The atmospheric radiation measurement program: Prelude. *Meteorological Monographs*, 57, 1–1.

Emsley, J. (1996). *The Global Warming Debate: The Report of the European Science and Environment Forum. European Science and Environment Forum*. Bournemouth: Bourne Press Limited

Farman, J. C., B.G. Gardiner and J.D. Shanklin. (1985) Large losses of total ozone in Antarctica reveal seasonal ClO_x/NO_x interaction, *Nature*, 315, 207–210.

Federal Ministry of Economics and Technology. (2010). *"Energy Concept for an Environmentally Sound, Reliable and Affordable Energy Supply"*. Federal Ministry of Economics and Technology (BMWi): Berlin.

Fensholt, R. and K. Rasmussen. (2011). "Analysis of trends in the Sahelian 'rain-use efficiency' using GIMMS NDVI, RFE and GPCP rainfall data." *Remote Sensing of Environment*, 115 (2):438–451.

Fensholt, R., T. Langanke, K. Rasmussen, A. Reenberg, S.D. Prince, C. Tucker, R.J. Scholes, Q. Le Bao, A. Bondeau, R. Eastman, H. Epstein, A.E. Gaughan,U. Hellden, C. Mbow, L. Olsson, J. Paruelo, C. Schweitzer, J. Seaquist, and K. Wessels. (2012). "Greenness in semi-arid areas across the globe 1981-2007----an earth observing satellite based analysis of trends and drivers." *Remote Sensing of Environment*, 121:144–158.

Fensholt, R., K. Rasmussen, P. Kaspersen, S. Huber, S. Horion, and E. Swinnen. (2013). "Assessing land degradation/recovery in the African Sahel from long-term earth observation based primary productivity and precipitation relationships." *Remote Sensing*, 5 (2):664–686.

Ferguson, C.W. and D.A. Graybill. (1983). Dendrochronology of bristlecone pine: A progress report. *Radiocarbon*, 25 (2) 287–288.

Ferraro, R and T. Smith. (2015). Global precipitation monitoring. *Satellite-Based Applications on Climate Change*. 81–93. doi: 10.1007/978-94-007-5872-8_6.

Ferrey, S. (2019). Against the wind-sustainability, migration, presidential discretion. *Columbia Journal of Environmental Law*, 44, 341.

Figueres C. and T. Rivett-Carnac. (2020). *The Future we Choose: Surviving the Climate Crisis*. (London: Manila Press.

Flato G, Marotzke J, Abiodun B, Braconnot P, Chou SC, Collins W, Cox P, Driouech F, Emori S, Eyring V, Forest C, Gleckler P, Guilyardi E, Jakob C, Kattsov V, Reason C, Rummukainen M. (2013). Evaluation of climate models. In *Climate Change 2013: The Physical Science Basis. Contribution of Working Group I to the Fifth Assessment Report of the Intergovernmental Panel on Climate Change*, in TF Stocker, D Qin, GK Plattner, M Tignor, SK Allen, J Boschung, A Nauels, Y Xia, V Bex, PM Midgley (eds.). Cambridge University Press: Cambridge, UK and New York, NY, 741– 866, doi: 10.1017/CBO9781107415324.020.

Forrester, J.W. (1973). *World Dynamics*. Cambridge, MA: Wright-Allen Press.

Foster, G. L., D.L. Royer, and D.J. Lunt. (2017). Future climate forcing potentially without precedent in the last 420 million years. *Nature Communications*, 8, 14845.

Foukal. P., and J. Lean. (1990), An empirical model of total solar irradiance variation between 1874 and 1988. *Science*, 247, .556–558.

Fowler, C., Emery, W. J., & Maslanik, J. (2004). Satellite-derived evolution of Arctic sea ice age: October 1978 to March 2003. *IEEE Geoscience and Remote Sensing Letters*, 1(2), 71–74.

Francis, Pope. (2015). *Laudato Si' On Care for our common Home*. London: Catholic Truth Society.

Friedrich M, S. Remmele, B. Kromer, J. Hofmann, M. Spurk, K.F. Kaiser, C. Orcel, and M. Küppers. (2004). "The 12,460-year Hohenheim oak and pine tree-ring chronology from central Europe — A unique annual record for radiocarbon calibration and paleoenvironment reconstructions". *Radiocarbon*. 46 (3): 1111–1122.

Galindo, I. G. (1965). Turbidometric estimations in Mexico city using the Volz sun photometer. *Pure and Applied Geophysics*, 60(1), 189–196.

Geist, H. (2005). *"The Causes and Progression of Desertification."* Ashgate, Aldershot

Geist, H. and E.F. Lambin. (2004). "Dynamic causal patterns of desertification." *Bioscience*, 54 (9):817–829.

Giannetti, F., Reggiannini, R., Moretti, M., Adirosi, E., Baldini, L., Facheris, L., ... & Vaccaro, A. (2017). Real-time rain rate evaluation via satellite downlink signal attenuation measurement. *Sensors*, 17(8), 1864.

Gibney, E. (2015). Why Finland now leads the world in nuclear waste storage. Nature News. Springer Nature, Date: Dec 2, 2015

Gillis, J. (2014). Panel's warning on climate risk: Worst is yet to come. *New York Times* (March 31).

Ginzburg, A. S., 1988, *The Planet Earth in 'Afternuclear' Epoch* (Moscow: Nauka).

Glazner, A.F., C.R. Manley, J.S. Marron, and S. Rojstaczer. (1999). Fire or ice: Anticorrelation of volcanism and glaciation in California over the past 800,000 years. *Geophysical Research Letters*, 26(12), pp. 1759–1762.

Goldman, A., Murcray, F. J., Rinsland, C. P., Blatherwick, R. D., David, S. J., Murcray, F. H., & Murcray, D. G. (1992). Mt. Pinatubo SO2 column measurements from Mauna Loa. *Geophysical Research Letters*, 19(2), 183–186.

Goldsmith, E., R. Allen, M. Allaby, J. Davoll and S. Lawrence. (1972). *"A Blueplrint for Survival"* London: Penguin.

Gollier, C. (2013). *Pricing the Planet's Future: The Economics of Discounting in an Uncertain World*. Princeton University Press, Princeton, NJ.

Gong, P., J. Wang , L. Yu , Y. Zhao , Y. Zhao , L. Liang , Z. Niu , X. Huang , H Fu , S. Liu , C. Li , X. Li , W. Fu, C. Liu , Y. Xu , X. Wang , Q. Cheng , L. Hu , W. Yao , H. Zhang , P. Zhu , Z. Zhao , H. Zhang , Y. Zheng, L Ji , Y. Zhang , H. Chen , A. Yan , J. Guo, L. Yu , L. Wang , X. Liu , T. Shi , M. Zhu , Y. Chen , G. Yang, P. Tang, B. Xu , C. Giri, N Clinton, Z Zhu, J Chen and J. Chen. (2013). Finer resolution observation and monitoring of global land cover: first mapping results with Landsat TM and ETM+ data, *International Journal of Remote Sensing*, 34:7, 2607–2654

Goody R. (2001) Climate benchmarks: Data to test climate models. *Studies of the Earth from Space*, N6, p. 87–93 (in Russian).

Goody R. (2002). Observing and thinking about the atmosphere. *Annu. Rev. Energy Environ.*, 27, p. 1–20.

Goody R., J. Anderson, and G. North. (1998). Testing climate models: An approach *Bulletin of the American Meteorological Society* 79, p. 2541–2549.

Goody R., J. Anderson, T. Karl, R. B. Miller, G. North J. Simpson, G. Stephens and W. Washington. (2002). Why monitor the climate? *Bulletin of the American Meteorological Society*, 83, p. 873–878.

Gore, A. (1992). *Earth in the Balance – Forging a New Common Purpose*. Earthscan, London, p. 407.

Gore, A. (2006). *An Inconvenient Truth*. Bloomsbury, London, Eamus, PA: Rodale Press, p. 325.

Gorshkov, V.G. (1990). *Energetics of the Biosphere and Environmental Stability*. ARISTI, Moscow, p. 237.

Gorshkov, V.G. (1995). *Physical and Biological Bases of Life Stability - Man, Biota, Environment*. Springer-Verlag, Berlin.

Gribbin, J. (1988). *"The Hole in the Sky - Man's Threat to the Ozone Layer"* London: Corgi.

Gribbin, J. (1990). Assault on the climate consensus[describes opposing views to greenhouse effect theory]. *New Scientist*, 128, 26–27.

Griggs D. J., and M. Noguer (2001) Climate Change 2001: The scientific basis. Contribution of working group I to the third assessment report of the intergovernmental panel on climate change. *Weather*, vol. 57, p. 267–269.

Grigoryev Al.A. and Kondratyev K.Ya. (2001a). *Ecological Disasters*. St. Petersburg Scientific Center of RAS. St. Petersburg, 206. (in Russian).

Grigoryev Al.A. and Kondratyev K.Ya. (2001b). *Ecological Catastrophes*. The St.-Petersburg Scientific Centre of RAS, St. Petersburg, p. 661 (in Russian).

Grigoryev Al.A. and Kondratyev K.Ya. (2001c). *Natural and Anthropogenic Ecological Disasters*. St. Petersburg, SPb Sci. Centre RAS, St. Petersburg. p. 688. (in Russian).

Grubb, M., Vrolijk, C. and Brack, D. (1999) *The Kyoto Protocol. A guide and Assessment*. Royal Institute of International Affairs. Earthscan, London.

Sun Guangqi. (1992). Zheng He's expeditions to the western ocean and his navigation technology. *Journal of Navigation*, 45, 329–343.

Hall. N. (ed.) (1991). *"The New Scientist, Guide to Chaos"* London: Penguin.

Hall, C. A. (2017). Methods and Critiques for EROI Applied to Modern Fuels. In *Energy Return on Investment* (pp. 119–143). Springer, Cham.

Hampson, J. (1974). Photochemical war on the atmosphere. *Nature*, 250(5463), 189–191.

Han, F., Westover, A. S., Yue, J., Fan, X., Wang, F., Chi, M., ... & Wang, C. (2019). High electronic conductivity as the origin of lithium dendrite formation within solid electrolytes. *Nature Energy*, 4(3), 187–196.

Haque, U., Da Silva, P. F., Devoli, G., Pilz, J., Zhao, B., Khaloua, A., ... and Yamamoto, T. (2019). The human cost of global warming: deadly landslides and their triggers (1995–2014). *Science of the Total Environment*, 682, 673–684.

Harrison, A. M. (2019). *Constitution and Specification of Portland Cement*. Edited by Peter Hewlett, Martin Liska Butterworth-Heinemann, Lea's Chemistry of Cement and Concrete, 87.

Harwell, M. A., 1984, *Nuclear Winter* (Berlin: Springer).

Hayashi, M., and L. Hughes. (2013). The policy responses to the Fukushima nuclear accident and their effect on Japanese energy security. *Energy Policy*, 59, 86–101.

Heath, M., F. Werner, F. Chai, B. Megrey, and P. Monfray. (2004). Challenges of modeling ocean basin ecosystems. *Science*, 304(5676), 1463–1466.

Hecht, A. D., and Döös, B. R. (1988). Climate change, economic growth and energy policy: A recommended US strategy for the coming decades an editorial. *Climatic Change*, 13(1), 1–3.

Heinberg, R. (2003). *The Party's Over – Oil, War and the Fate of Industrial Societies*. Clairview Books: Forest Row, East Sussex, p. 306.

Heinberg, R. (2005). *The party's over: oil, war and the fate of industrial societies*. New Society Publishers. p. 288, ISBN 9781550923346

Heinberg, R. (2006). *The Oil Depletion Protocol – A Plan to Avert Oil Wars, Terrorism and Economic Collapse*. Clairview Books: Forest Row, East Sussex. p. 194.

Heinberg, R. (2007). Out of time? The end of oil. *Public Policy Research*, 14(3), 197–203.

Henry, D.O., C.E. Cordova, M. Portillo, R- M. Albert, R. DeWitt, and A. Emery-Barbier. (2017). "Blame it on the goats? Desertification in the Near East during the Holocene." *The Holocene*, 27, 625–637.

Herman, A. (2016). Discrete-element bonded-particle sea ice model DESIgn, version 1.3. Model description and implementation. *Geoscientific Model Development* 9: 1219–1241

Herrmann, S.M. and G.G. Tappan. (2013). "Vegetation impoverishment despite greening: A case study from central Senegal." *Journal of Arid Environment*, 90:55–66.

Herrmann, S. M., Anyamba, A., & Tucker, C. J. (2005). Recent trends in vegetation dynamics in the African Sahel and their relationship to climate. *Global Environmental Change*, 15(4), 394–404.

Heumann, B.W., J.W. Seaquist, L. Eklundh, and P. Jonsson (2007). "AVHRR derived phenological change in the Sahel and Soudan, Africa, 1982–2005." *Remote Sensing of Environment*, 108:385–392.

Heyen, D.A. and F. Wolff. (2019). Drivers and barriers of sustainability transformations: A comparison of the "Energiewende" and the attempted transformation to organic agriculture in Germany. *GAIA – Ecological Perspectives on Science and Society*, 28 (S1), 226–232.

Higgins, S. and S. Scheiter. (2012). "Atmospheric CO_2 forces abrupt vegetation shifts locally, but not globally." *Nature*, 488 (7410): 209–212.

Hirth, L. (2013). The market value of variable renewables: The effect of solar wind power variability on their relative price. *Energy Economics*, 38, 218–236.

Hirth, L. (2014). "The Economics of Wind and solar Variabiity – How the Variability of Wind and Solar Power affects their Marginal Value, Optimal Deployment and Integration Costs". PhD thesis, Technischen Universität Berlin."

Hirth, L. (2015). "The Optimal Share of Variable Renewables", *The Energy Journal*, 36(1)

Hirth, L. and F. Ueckerdt. (2013). "Redistribution effects of energy and climate policy: The electricity market", *Energy Policy* , 62, 934–947.

Hirth, L. and I. Ziegenhagen. (2015). "Balancing power and variable renewables: Three links", *Renewable and Sustainable Energy Reviews*, 50, 1035–1051.

Hirth, L., F. Ueckerdt and O. Edenhofer. (2014). "Why Wind is not coal: on the economics of electricity. FEEM Working Paper 2014.039.

Hirth, L., F. Ueckerdt and O. Edenhofer. (2015). "Integration costs revisited – An economic frame- work of wind and solar variability", *Renewable Energy*, 74, 925–939.

Hoff, R. M. (1992). Differential SO2 column measurements of the Mt. Pinatubo volcanic plume. *Geophysical Research Letters*, 19(2), 175–178.

Hofmann. D., B.J. Johnson, and S.J. Oltmans. (2009). Twenty-two years of ozonesonde measurements at the South Pole, *International Journal of Remote Sensing*, 30, 3995–4008.

Högberg, L. (2013). Root causes and impacts of severe accidents at large nuclear power plants. *Ambio*, 42(3), 267–284.

Holbrook, N. J., H. A. Scannell, A.S. Gupta, J.A. Benthuysen, M. Feng, E.C. Oliver, ... P.J. Moore, (2019). A global assessment of marine heatwaves and their drivers. *Nature Communications*, 10(1), 2624.

Holttinen, H., P. Meibom, A. Orths, B. Lange, M. O'Malley, J.O. Tande, A. Estanqueiro, E. Gomez, L. Söder, G. Strbac, J.C. Smith and F. van Hulle. (2011). 'Impacts of large amounts of wind power on design and operation of power systems', *Wind Energy*, 14(2), 179–192

Hooper, N. J., and Sherman, J. W. (1986). *Temporal and spatial analyses of civil marine satellite requirements.* NOAA Technical Report NESDIS 16, Washington DC.

Houghton, John T., ed. (1985). The global climate. CUP Archive.

Houghton, J.T. (1991). The Bakerian Lecture 1991, the predictability of weather and climate. *Philosophical Transactions of the Royal Society of London*, A337, 521–572.

Houghton, J.T., Jenkins, G.J. and Ephraums, J.J. (1990). *Climate Change – the IPCC Scientific Assessment.* Cambridge University Press, Cambridge. p. 365.

Houghton, J. T., Meira Filho, L. G., Griggs, D. J., and Maskell, K. (eds.). (1997). An introduction to simple climate models used in the IPCC Second Assessment Report. WMO.

Hubbert, M.K. (1956). "Nuclear Energy and the Fossil Fuels" Presented before the Spring Meeting of the Southern District, American Petroleum Institute, Plaza Hotel, San Antonio, Texas, March 7–8–9, 1956 "Archived copy" (PDF). Archived from the original (PDF) on 2008-05-27. Retrieved 2014-11-10.

Hulme, M. (2009). *Why We Disagree About Climate Change: Understanding Controversy, Inaction and Opportunity.* New York, NY: Cambridge University Press.

Huntford, R. (1971). *"The New Totalitarians".* Allen Lane, Penguin, London.

Huss, M. and R. Hock. (2015). A new model for global glacier change and sea-level rise. *Frontiers in Earth Science*, 3, p. 54.

Huybers, P. and C. Langmuir. (2009). Feedback between deglaciation, volcanism, and atmospheric CO_2. *Earth and Planetary Science Letters*, 286(3-4), pp. 479–491.

IAEA. (2004). Annual Report for 2004, GC(49)/5, Vienna, Austria (https://www.iaea.org/sites/default/files/anrep2004_full.pdf)

IAEA. (2008). Advisory Material for the IAEA Regulations for the Safe Transport of Radioactive Material, No. TS-G-1.1 (Rev. 1), Vienna, Austria https://www-pub.iaea.org/MTCD/publications/PDF/Pub1325_web.pdf.

IAEA. (2009). Annual Report for 2009, GC(54)/4, Vienna, Austria.

IAEA. (2016a). *Nuclear Power and Sustainable Development*, International Atomic Energy Agency, Vienna, Austria.

IAEA. (2016b). *Climate Change and Nuclear Power*, International Atomic Energy Agency, Vienna, Austria.

IAEA (2019) 63rd IAEA General Conference 16 – 20 2019, Vienna International Centre, Vienna https://www.iaea.org/about/policy/gc/gc63/gc-at-glance and https://www.iaea.org/gc-archives/gc

Idso, C.D., R.M. Carter and S.F. Singer. (2015a). *Why Scientists Disagree about Global Warming.* Chicago: Heartland Institute.

Idso, C. D., Carter, R. M., & Singer, S. F. (2015b). Why scientists disagree about global warming. The Heartland Institute. Non Profit Research Organization. " NIPCC Report, 135 p. Arlington Heights: Heartland Institute.

Ilčev S.D. (2019) *Global Satellite Meteorological Observation (GSMO) Applications.* Springer, Cham. https://doi.org/10.1007/978-3-319-67047-8

Ilyinskaya, E., Mobbs, S., Burton, R., Burton, M., Pardini, F., Pfeffer, M. A., ... & Bergsson, B. (2018). Globally significant CO2 emissions from Katla, a subglacial volcano in Iceland. *Geophysical Research Letters*, 45(19), 10–332.

IPCC. (1990) *Climate Change, The IPCC Scientific Assessment.* Edited by J.T. Houghton, G.J. Jenkins and J.J. Ephraums. New York: Cambridge University Press.

IPCC. (1992). Climate Change 1992: The Supplementary Report to the IPCC Scientific Assessment by J.T. Houghton (eds.), et al. , Cambridge, UK: Cambridge University Press. p. 218.

IPCC. (2007a). *Climate Change 2007, the Fourth Assessment Report (AR4) of the United Nations Intergovernmental Panel on Climate Change (IPCC)* (Cambridge: Cambridge University Press).

IPCC. (2007b). *Climate Change 2007: Synthesis Report. Contribution of Working Groups I, II and III to the Fourth Assessment Report of the Intergovernmental Panel on Climate Change.* [Core Writing Team, Pachauri, R. K. and Reisinger, A. (eds.)]. IPCC, Geneva, Switzerland, p. 104.

IPCC. (2013). Climate Change 2013. *The Physical Science Basis. Working Group I Contribution to the Fifth Assessment Report of the Intergovernmental Panel on Climate Change*. Technical Report, WMO/UNEP. Cambridge University Press.

IPCC. (2014). *Climate Change 2014: Synthesis Report. Contribution of Working Groups I, II and III to the Fifth Assessment Report of the Intergovernmental Panel on Climate Change*. [Core Writing Team, R.K. Pachauri and L.A. Meyer (eds.)]. IPCC: Geneva, Switzerland, p. 151.

IPCC. (2018). Global Warming of 1.5° C: An IPCC Special Report on the Impacts of Global Warming of 1.5° C Above Pre-industrial Levels and Related Global Greenhouse Gas Emission Pathways, in the Context of Strengthening the Global Response to the Threat of Climate Change, Sustainable Development, and Efforts to Eradicate Poverty. Intergovernmental Panel on Climate Change.

IPCC. (2019). Climate Change. Land: An IPCC Special Report on Climate Change, Desertification, Land Degradation, Sustainable Land Management, Food Security, and Greenhouse Gas Fluxes in Terrestrial Ecosystems.

IPCC 1. (1990). Houghton, J. T. (ed.). Climate Change: The IPCC Scientific Assessment; Report Prepared for IPCC by Working Group 1.

IPCC 2. (1992). Houghton, J.T. et al. (eds.), *Climate Change 1992. The Supplementary Report to the IPCC Scientific Assessment*. Cambridge University Press, Cambridge.

IPCC CCS. 2005. Metz, B., O. Davidson, H. C. de Coninck, M. Loos, and L.A. Meyer (eds.) *IPCC special report on Carbon Dioxide Capture and Storage*. Cambridge University Press, Cambridge, United Kingdom and New York, NY, USA. p. 442. Available in full at www.ipcc.ch (PDF - 22.8MB).

IPCC Third Assessment Report (2001). Vol. 1. Climate Change 2001. The Scientific Basis Cambridge Univ. Press. p. 881.

IPCC X. (1996) J.T. Houghton, et al. (eds.), *Climate Change 1995. The Science of Climate Change*. Cambridge University Press: Cambridge.

Irwin, S., and Good, D. (2017). EPA interpretation of the "inadequate domestic supply" waiver for renewable fuels ruled invalid: Where to from here?. *Farmdoc daily*, 7. ISBN 1118897390, 9781118897393.

Izrael, Yu, and B. Kuvshinnikov. (1975). SSSR-SSHA: sotrudnichestvo v oblasti okhrany prirodnoi sredy, Mezhdunarodnaya Zhizn' 2 34.

Jacobs, D. (2012): The German – History, Targets, *Policies and Challenges. – Renewable Energy Law and Policy Review (RELP)*, 4, p. 223–233.

Jäger, H. (1992). The Pinatubo eruption cloud observed by lidar at Garmisch-Partenkirchen. *Geophysical Research Letters*, 19(2), 191–194.

Jankó, F., Á. Drüszler, B. Gálos, N. Móricz, J. Papp-Vancsó, I. Pieczka, … and O. Szabó. (2020). Recalculating climate change consensus: The question of position and rhetoric. *Journal of Cleaner Production*, 254, 120–127.

Jellinek, A.M., M. Manga, and M.O. Saar. (2004). Did melting glaciers cause volcanic eruptions in eastern California? *Probing the mechanics of dike formation: Journal of Geophysical Research*, v. 109, B09206, doi:10.1029/2004JB002978.

Jennings, H. (ed.), (1993). *Atmospheric Aerosols* (Tucson, AZ: Arizona Press).

Jouzel, J. (2013). A brief history of ice core science over the last 50 yr. *Climate Past*, 9, 2525–2547.

KC98. Kondratyev K. Ya. and A. P. Cracknell. (1998). *Observing Global Climate Change*. London: Taylor and Francis, p. 562.

Kellogg, W. W. (1992, May). *What the Greenhouse Skeptics Are Saying*. In *Climate Change and Energy Policy: Proceedings of the Conference October 21–24 1991*, Los Alamos, NM (p. 430). Springer Science & Business Media.

Kelly, P. M., and Sear, C. B. (1982). The formulation of Lamb's dust veil index. Atmospheric effects and potential climatic impact of the 1980 eruptions of mount St. *Helens*, 2240, 293–298.

Keong, C. Y. (2018). From Stockholm Declaration To Millennium Development Goals: The United Nation's Journey To Environmental Sustainability. Developmental State And Millennium Development Goals// Country Experiences, 209–256.

Keys, D. (1999). *Catastrophe – An Investigation into the origins of the Modern World*. London, Arrow Books

Khayal, O. M. E. S., and M.I. Osman, (2019). The technical and economic feasibility of establishing ethanol fuel plant in kenana sugar company. *GPH-International Journal of Applied Science*, 2(10), 01–36.

Khodri, M., T. Izumo, J. Vialard, S. Janicot, C. Cassou, M. Lengaigne, J. Mignot, G. Gastineau, E. Guilyardi, N. Lebas, A. Robock and M.J. McPhaden. (2017). Tropical explosive volcanic eruptions can trigger El Niño by cooling tropical Africa. *Nature Communications*, 8, 778, DOI: 10.1038/s41467-017-00755-6.

Kinne, S., Schulz, M., Textor, C., Guibert, S., Balkanski, Y., Bauer, S. E., ... & Tie, X. (2006). An AeroCom initial assessment–optical properties in aerosol component modules of global models. *Atmospheric Chemistry and Physics*, 6(7), 1815–1834.

Kiyosugi, K., C. Connor, R.S.J. Sparks, H.S. Crosweller, S.K. Brown, L. Siebert, T. Wang, and S. Takarada. (2015). How many explosive eruptions are missing from the geologic record? Analysis of the quaternary record of large magnitude explosive eruptions in Japan. *Journal of Applied Volcanology*, 4(1), p. 17.

Kleiner, K. (2008). Nuclear energy: Assessing the emissions. *Nature Reports Climate Change*, 130–131.

Kondratyev, K.Ya. (1956). *Radiant Sun Energy*. Hydrometeoizdat: Leningrad. p. 600. (in Russian).

Kondratyev, K. Y. (1961). Soviet Investigations in Actionmetry and Atmospheric Optics. *Weather*, 16(6), 180–186.

Kondratyev K.Ya. (1965). *Radiative Heat Exchange in the Atmosphere*. Pergamon Press: New York. p. 350.

Kondratyev K.Ya. (1969). *Radiation in the Atmosphere*. Academic Press, New York. p. 912.

Kondratyev K.Ya. (1972a). Radiation in the Atmosphere. WMO Monograph No. 309, Geneva, p. 214.

Kondratyev, K. Ya. (1972b). Complex Atmospheric Energetics Experiment. WMO Bulletin No. 12.

Kondratyev, K.Ya. (1976), Aerosol and climate. *Trudy GGO, Issue* 381, pp. 3–66.

Kondratyev K.Ya. (1982). *The World Climate Research Programme: The State and Perspectives, and the Role of Spaceborne Observational Means*. ARISTI: Moscow. p. 274. (in Russian).

Kondratyev, K. Ya., 1983, *Satellite Climatology* (Leningrad: Gidrometeoizdat).

Kondratyev, K. Y. (1985). Nuclear war, the atmosphere, and climate. *Optics News*, 11(11), 20–21.

Kondratyev, K. Ya., 1988, *Climate Shocks: Natural and Anthropogenic* (New York: Wiley).

Kondratyev K.Ya. (1989). *Global Ozone Dynamics*. ARISTI: Moscow. p. 212. (in Russian).

Kondratyev K.Ya. (1990a). *Key Problems of Global Ecology*. ARISTI: Moscow. p. 454. (in Russian).

Kondratyev K.Ya. (1990b). *Planet Mars*. Hydrometeoizdat: Leningrad. p. 368. (in Russian).

Kondratyev, K.Ya. (1991). *Atmospheric Aerosols* (Leningrad: Gidrometeoizdat).

Kondratyev K.Ya. (1997). Key issues in global change. *Energy and Environment*, 8, 5–9.

Kondratyev K.Ya. (1998). *Multidimensional Global Change*. Wiley/PRAXIS, Chichester, U.K. p. 761.

Kondratyev K.Ya. (2001). Biogenic aerosol in the atmosphere. *Atmospheric and Oceanic Optics C/C of Optika Atmosfery I Okeana*, 14(3), 153–160.

Kondratyev K.Ya. (2004). Key aspects of global climate change. *Energy and Environment*, 15, 469–503.

Kondratyev, K. Ya and Binenko, V.I. (eds.). (1981). *First Global GARPExperiment V.2. Polar aerosol heavy cloudiness and radiation*. Moscow: Leningrad Gidrometeoizdat Publishing. p. 152. (in Russian).

Kondratyev, K. Ya., and V. I. Binenko. (1984). Impact of Clouds on Radiation and Climate (in Russian). Gidrometeoizdat. p. 240.

Kondratyev, K.Ya. and A.P. Cracknell, (1998a). *Observing Global Climate Change*. Taylor & Francis, London, 562.

Kondratyev, K. Y., & Cracknell, A. P. (1998b). *Observing global climate change*. CRC Press.

Kondratyev, K.Ya. and A.P. Cracknell, (2001). "Global climate change. Socioeconomic aspects of the problem.". In *Remote Sensing and Climate Change. Role of Earth Observation*, Edited by: Cracknell, A. p. 37–79. Chichester: Springer-Praxis.

Kondratyev K.Ya., and I. Galindo (1997). *Volcanic Activity and Climate*. A. Deepak: Hampton, Virginia, 382 pp.

Kondratyev K. Ya., and Galindo I. (2001). *Global Change Situations: Today and Tomorrow*. Universidad de Colima, Colima, Mexico. pp. 164.

Kondratyev, K.Ya. and A.A. Grigoryev. (2002). *Environmental Disasters. Anthropogenic and Natural*. Springer, Praxis: Chichester, U.K. p. 484.

Kondratyev, K. Y., and Krapivin, V. F. (2004). Monitoring and prediction of natural disasters. *Nuovo Cimento C Geophysics Space Physics C*, 27, 657.

Kondratyev K.Ya. and Moskalenko N.I. (1984). *Greenhouse Effect of the Atmosphere and Climate*. ARISTI, Moscow, p. 262. (in Russian).

Kondratyev, K.Ya. and Yu. M. Timofeyev, (1970). *Thermal Sounding of the Earth from Space*. Hydrometeoizdat: Leningrad. p. 421. (in Russian).

Kondratyev, K.Ya. and C. Varotsos, (1995). Atmospheric greenhouse effect in the context of global climate change. *Nuovo Cimento della Societa Italiana di Fisica C-Geophysics and Space Physics*, 18: 123–151.

Kondratyev, K.Ya. and C. Varotsos, (2000). *Atmospheric Ozone Variability: Implications for Climate Change*. Human Health and Ecosystems, Chichester: Springer-Praxis.

Kondratyev, K. Ya., and Zhvalev, V. F. (1981). *The First GARP Global Experiment Aerosol and Climate*, Gidrometeoizdat, Leningrad.

Kondratyev, K.Ya., O.M. Pokrovsky, and C.A. Varotsos. (1995). Atmospheric ozone trends and other factors of surface ultraviolet radiation variability. *Environmental Conservation*, 22: 259–261.

Kondratyev K.Ya., Johannessen O.M., and Melentyev V.V. (1996). *High Latitude Climate and Remote Sensing.* Wiley/Praxis: Chichester, U.K. p. 200.

Kondratyev, K.Ya., Fernando Moreno Pena, and Galindo, I. (1997). *Sustainable Development and Population Dynamics.* Universidat de Colima: Mexico. p. 128.

Kondratyev K.Ya., Krapivin V.F., and Phillips G.W. (2002a). *Global Environmental Change: Modelling and Monitoring.* Springer: Berlin. p. 319.

Kondratyev K.Ya., V.F. Krapivin G.V. Phillips (2002b). *Problems of the High-latitude Environmental Pollution.* Saint-Petersburg State Univ. Publ: Saint Petersburg. p. 280. (in Russian).

Kondratyev K.Ya., V.F. Krapivin and V.P. Savinykh (2003a). *Prospects of Civilization Development. Multidimensional analysis.* Moscow: Logos Publ. p. 575. (in Russian)

Kondratyev K.Ya., V.F. Krapivin and C. A. Varotsos. (2003b). *Global Carbon Cycle and Climate Change.* Springer/PRAXIS: Chichester, U.K., p. 370.

Kondratyev K. Ya., K.S. Losev, M.D. Ananicheva and I.V. Chesnokova. (2003c) *Natural-Scientific Basis for Life Stability.* Moscow: VINITI. p. 240. (in Russian)

Kondratyev K.Ya., Krapivin V.F., Savinykh V.P., and Varotsos C.A. (2004a). *Global Ecodynamics: A Multidimensional Analysis.* Springer/Praxis: Chichester, U.K. p. 658.

Kondratyev K.Ya., K. S. Losev, M.D. Ananicheva and I.V. Chesnokova. (2004b). *Stability of Life on Earth Springer.* Praxis: Chichester, U.K., p. 165.

Kondratyev K.Ya., Krapivin V.F., Lakasa H., and Savinikh V.P. (2005). *Globalization and Sustainable Development: Ecological Aspects.* Science: St. Petersburg. p. 240. (in Russian)

Kondratyev, K.Ya., Krapivin, V.F., and Varotsos, C.A. (2006). *Natural Disasters as Components of Ecodynamics.* Springer/PRAXIS: Chichester, U.K. p. 625.

Kopp, G. (2016). Magnitudes and timescales of total solar irradiance variability. *Journal of Space Weather and Space Climate*, 6, A30.

Koromyslova, A. V., Baraboshkin, E. Y., & Martha, S. O. (2018). Late Campanian to late Maastrichtian bryozoans encrusting on belemnite rostra from the Aktolagay Plateau in western Kazakhstan. *Geobios*, 51(4), 307–333.

Kramer, H. J. (2002). *Observation of the Earth and its Environment: Survey of Missions and Sensors.* Springer Science & Business Media, Gilching, Germany

Krapivin, V.F. and Kondratyev K.Ya. (2002). *Global Changes of the Environment.* St. Petersburg Univ. Publ.: St. Petersburg. p. 724. (in Russian).

Krapivin, V. F., & Varotsos, C. A. (2019). *Geoecological information-modeling system and its implication for the biocomplexity and survivability assessment of the Okhotsk Sea ecosystem.*

Krapivin, V.F., C.A. Varotsos, and V. Yu. Soldatov (2015). *New Ecoinformatics Tools in Environmental Science: Applications and Decision-making.* Springer: London, U.K. p. 903.

Krause, Florentin; Bossel, Hartmut; Müller-Reißmann, Karl-Friedrich (1980). *Energie-Wende: Wachstum und Wohlstand ohne Erdöl und Uran [Energy transition: growth and prosperity without petroleum and uranium] (PDF) (in German).* Germany: S Fischer Verlag. ISBN 978-3-10-007705-9. Retrieved 2016-06-14.

Kreidenweis, S. M., Petters, M., and Lohmann, U. (2019). 100 years of progress in cloud physics, aerosols, and aerosol chemistry research. *Meteorological Monographs*, 59, 1114.

Kriegler, E., Hall, J. W., Held, H., Dawson, R., and Schellnhuber, H. J. (2009). Imprecise probability assessment of tipping points in the climate system. *Proceedings of the national Academy of Sciences*, 106(13), 5041–5046.

Kulkarni, A.V., B.P. Rathmore, S.K. Singh and I.M. Bahuguna. (2011). Understanding changes in the Himalayan cryosphere using remote sensing techniques. *International Journal of Remote Sensing*, 32, pp. 601–615.

Kushnir, Y. (2000). Solar Radiation and the Earth's Energy Balance. Published on The Climate System, complete online course material from the Department of Earth and Environmental Sciences at Columbia University. Accessed December 12, 2008

Kutterolf, S., Jegen, M., Mitrovica, J.X., Kwasnitschka, T., Freundt, A. and Huybers, P.J. (2013). A detection of Milankovitch frequencies in global volcanic activity. *Geology*, 41(2), pp. 227–230.

Kutterolf, S., Schindlbeck, J.C., Jegen, M., Freundt, A. and Straub, S.M. (2019). Milankovitch frequencies in tephra records at volcanic arcs: The relation of kyr-scale cyclic variations in volcanism to global climate changes. *Quaternary Science Reviews*, 204, pp. 1–16

Lamb, H.H. (1970). Volcanic dust in the atmosphere, with a chronology and assessment of its meteorological significance. *Philosophical Transactions Royal Society London Ser. A*, 266, 425–533.

Lamb, H.H. (1977). Supplementary volcanic dust veil index assessments. *Climate Monitor*, 6, 57–67.

Lamb. H.H. (1983a). Update of the chronology of assessments of the volcanic dust veil index Climate. *Monitor*, 12, 79–90.

Lamb, Peter J. Sub-saharan rainfall update for 1982; continued drought. *J. Climatol.*, 1983b, 3.4: 419–422.

Lamb, H.H. (1995). *Climate History and the Modern World*. 2nd Edition. London: Routledge.

Lamb, H. H. (2002). *Climate, history and the modern world*. Routledge, London, 410

Langway, C.C. (2008). The history of early polar ice cores. *Cold Regions Scence and Technology*, 52, 101–117.

Lee, D. (1977). *Plato*. Timaeus and Critias (translated), Revised edition. Penguin: London.

Lee, K.Y. (2000) *From Third World to First: The Singapore Story:1965–2000*. Times Media Pte Ltd.: Singapore. p. 778.

Leggett, J.K. (1990). *Global Warming. The Greenpeace Report*. Oxford: Oxford University Press.

Leggett, J. (2005). *Half Gone: Oil, Gas, Hot Air and the Global Energy Crisis*. (Portobello: London).

Lelieveld, J., D. Kunkel, and M. G. Lawrence. (2012). Global risk of radioactive fallout after major nuclear reactor accidents. *Atmospheric Chemistry and Physics*, 12(9), 4245.

Leroy, S.A.G., F. Marret, S. Giralt, and S.A. Bulatov. (2006). Natural and anthropogenic rapid changes in the Kara-Bogaz Gol over the last two centuries reconstructed from polynological analyses and a comparison to instrumental records. *Quaternary International* 150, 52–70.

Letcher, T. M. (2019). Why do we have global warming? In *Managing Global Warming* (pp. 3–15). Amsterdam, the Netherlands: Academic Press.

Levathes, Louise (1996). *When China Ruled the Seas: The Treasure Fleet of the Dragon Throne*, 1405–1433. Oxford University Press. ISBN 978-0-19-511207-8.

Lewis, S.L. and M.A. Maslin (2018). *The Human Planet: How We Created the Anthropocene*. Pelican: Penguin, London.

Li, R., and R. Jiang. (2019). Is carbon emission decline caused by economic decline? Empirical evidence from Russia. *Energy & Environment*, 30(4), 672–684.

Lieven, A. (2020). *Climate Change and the Nation State*. The Realist Case: London, Allen Lane.

Lin, M., and Huybers, P. (2019). If rain falls in India and no one reports it, are historical trends in monsoon extremes biased?. *Geophysical Research Letters*, 46(3), 1681–1689.

Lipp JS (2008) Intact membrane lipids as tracers for microbial life in the marine deep biosphere. *Dissertation zur Erlangung des Doktorgrades der Naturwissenschaften, Am Fachbereich Geowissenschaften der Universität Bremen* (https://d-nb.info/989330281/34).

Liu, Z., H. Hong, C. Wang, W. Han, K. Yin, K. Ji, Q. Fang, and T. Algeo. (2019). Oligocene-Miocene (28–13 Ma) climato-tectonic evolution of the northeastern Qinghai-Tibetan Plateau evidenced by mineralogical and geochemical records of the Xunhua Basin. *Palaeogeography, Palaeoclimatology, Palaeoecology*, 514, pp.98–108.

Logan, L.A. (1994). Trend in the vertical distribution of ozone: an analysis of ozonesonde data. *Journal of Geophysical Research*, 99. 25553–25585.

Lohmann, U. (2006). Aerosol effects on clouds and climate, *Space Science Review*, this volume. doi: 10.1007/s11214-006-9051-8.

Lohmann, U. and J. Feichter. (2005). Global indirect aerosol effects: A review, *Atmospheric Chemistry and Physics*, 5, 715–737.

Lomborg, B. (2010). *Cool It – The Sceptical Environmentalist's Guide to Global Warming*. London: Marshall Cavendish.

London, J. and White, G. F. (eds.), 1984, *The Environmental Effects of Nuclear War* (Boulder, CO : Westview Press).

Lorenz, E. (1963). Deterministic non-periodic flow. *Journal of the atmospheric sciences*, 20, 130–141.

Lovejoy, S. (2019). *Weather, Macroweather, and the Climate: Our Random Yet Predictable Atmosphere*. New York: Oxford University Press.

Lovejoy, S. and C. Varotsos. (2016). Scaling regimes and linear/nonlinear responses of last millennium climate to volcanic and solar forcings. *Earth System Dynamics*, 7(1), pp. 133–150.

Lovelock, J. (2006). *The Revenge of Gaia: Why the Earth is Fighting Back – and How We Can Still Save Humanity*. London: Penguin.

Lovelock, J. (2009). *The Vanishing Face of Gaia, a Final Warning*. London: Allen Lane, Penguin.

Lovering, J.R., A. Yip, and T. Nordhaus. (2016). Historical construction costs of global nuclear power reactors. *Energy Policy*, 91, 371–382.

Lüthi, D., M Le Floch, B. Bereiter, T. Blunier, J-M Barnola, U. Siegenthaler, D. Raynaud, J. Jouzel, H. Fischer, K. Kawamura and T.F. Stocker. (2008). High-resolution carbon dioxide concentration record 650, 000–800,000 years before present. *Nature* 453. p. 379–382.

Machin, A., & Ruser, A. (2019). What counts in the politics of climate change? Science, scepticism and emblematic numbers. In *Science, numbers and politics* (pp. 203–225). Palgrave Macmillan, Cham.

Mackay, D.J.C. (2009). "Sustainable energy – without the hot air" UIT, Cambridge Ltd. 368 page. Downloadable free from www.withouthotair.com.

Malone, R. C., Auer, L. H., Glatzmaier, G. A., Wood, M. C. and Toon, O. B., 1986, Nuclear winter: three-dimensional simulations including interactive transport, scavenging and solar heating of smoke. *Journal of Geophysical Research*, 91, 1039–1054.

Mandelbrot, B.B. (1982). *The Fractal Geometry of Nature*. Salt Lake City: W.H. Freeman.

Mankin, W. G., Coffey, M. T., and Goldman, A. (1992). Airborne observations of SO2, HCl, and O3 in the stratospheric plume of the Pinatubo volcano in July 1991. *Geophysical research letters*, 19(2), 179–182.

Marchuk G.I. and Kondratyev K.Ya. (1992). *Priorities in Global Ecology*. Science, Moscow, p. 264. (in Russian).

Marcott, S.A., J.D. Shakun, P.U. Clark, and A.C. Mix. (2013). A reconstruction of regional and global temperature for the past 11,300 years. *Science*, 339(6124), 1198–1201, doi:10.1126/science.1228026

Martinez, S. and D. Mollicone. (2012). "From land cover to land use: a methodology to assess land use from remote sensing data." *Remote Sensing*, 4(4):1024–1045.

Martinez, D. M., B. W. Ebenhack, and T. P. Wagner. (2019). *Energy Efficiency: Concepts and Calculations*. Amsterdam, the Netherlands: Elsevier.

Masson-Delmotte, V. et al. (2013). Information from Paleoclimate Archives. In: *Climate Change 2013: The Physical Science Basis. Contribution of Working Group I to the Fifth Assessment Report of the Intergovernmental Panel on Climate Change* [Stocker, T.F., D. Qin, G.-K. Plattner, M. Tignor, S.K. Allen, J. Boschung, A. Nauels, Y. Xia, V. Bex, and P.M. Midgley (eds.)]. Cambridge University Press: Cambridge, United Kingdom and New York, NY, USA, pp. 383–464.

Matoba, S., T. Shiraiwa, A. Tsushima, H. Sasaki, and Y.D. Muravyev. (2011). Records of sea-ice extent and air temperature at the Sea of Okhotsk from an ice core of Mount Ichinsky, Kamchatka. *Annals of Glaciology*, 52 (58): 44–50.

Mbow, C., R. Fensholt, K. Rasmussen, and D. Diop. (2013) Can vegetation productivity be derived from greenness in a semi-arid environment ? Evidence from ground-based measurements. *Journal of Arid Environment*, 97: 56–65.

McCormick, M. P. (1992). Initial assessment of the stratospheric and climatic impact of the 1991 Mount Pinatubo eruption: Prologue. *Geophysical Research Letters*, 19(2), 149–149.

McCormick, M., and Veiga, R. E. (1992). SAGE II measurements of early Pinatubo aerosols. *Geophysical Research Letters*, 19(2), 155–158.

McKibben, B. (2019). *Falter, Has the Human Game Begun to Play itself Out?* London: Headline.

Mclaren, A. S., Bourke, R. H. and Weaver, R. L. S., 1991, *Contour mapping of Arctic Basin ice roughness parameters. International Conference on the Role of the Polar Regions in Global Change. Proceedings of a Conference held June 11–15, 1990, at the University of Alaska Fairbanks*, edited by G. Weller, C. L. Wilson and B. A. B. Severin (Fairbanks: Geophysical Institute and Center for Global Change and Arctic Systems Research), p. 79.

Meadows, D.H., D.L. Meadows, J. Randers, and W.W. Behrens. (1972). *The Limits to Growth: A report or the Club of Rome's project on the predicament of mankind*. Universe Books: New York. p. 205.

Meadows, D.H., D.L. Meadows, J. Randers, and W.W. Behrens III. (1974). *The Limits to Growth: A Report for the Club of Rome's Project on the Predicament of Mankind*. New York: Universe Books.

Meadows, D.H., D.L. Meadows, and J. Randers. (1992). *Beyond the Limits*. Chelsea Green: Post Mill, VT. p. 300.

Meadows, D.H., J. Randers, and D.L. Meadows. (2005) *Limits to Growth: the 30-year Update*. Earthscan: London. p. 338.

Meleshko, V.P. and R.T. Wetherald. (1981). The effect of a geographical cloud distribution on climate: A numerical experiment with an atmospheric General Circulation Model. *Journal of Geophysical Research* 86:C12 11995–12014.

Menzies, G. (2004), *1421: The Year China Discovered America*, Harper Perennial, New York.

Menzies, G. (2008), *1434: The Year a Magnificent Chinese Fleet Sailed to Italy and Ignited the Renaissance*, William Morrow, New York.

Menzies, G., & Hudson, I. (2013). Who Discovered America?: The Untold History of the Peopling of the Americas. Harper Collins (9780062236777). Australia Books.

Michener, W.K., T.J. Baerwald, P. Firth, M.A. Palmer, J.L. Rosenberger, E.A. Sandlin and H. Zimmerman. (2001). Defining and unraveling biocomplexity. *BioScience*, 51 (12): 1018–1023.

Micklin, P.P. (1987). The fato 'Siberial': Soviet water politics in the Gorbachev era. *Journal Central Asian Survey*, 6(2), 67–88.

Micklin, P.P. (1988). Desiccation of the aral sea: A water management disaster in the soviet union. *Science* 241, 4870, 1170–1176.

Micklin, P.P. (2014). The Siberian water transfer scheme. In: P.P. Micklin et al. (eds.) *The Aral Sea. Springer Earth System Sciences*. Springer-Verlag: Berlin Heidelberg. 2014, 381–404.

Micklin, P.P. (2016). The future Aral Sea: Hope and despair. *Environmental Earth Sciences*, 75 (9), 1–15.

Milankovitch, M.M. (1920). *Theorie Mathématique des phenomènes thermiques produits par la radiation solaire. Academie Yougoslave des Sciences et des Arts de Zagreb*. Paris: Gauthier-Villars.

Miller, D.H. (1965). The heat and water budget of the Earth's surface, *Advances in Geophysics*, 11: 176–302.

Miller, D. H. (1978). *The factor of scale: ecosystem, landscape mosaic, and region. Sourcebook on the Environment: A Guide to the Literature*. University of Chicago Press, Chicago, 63–88.

Milligan, M., E. Ela, B.-M. Hodge, B. Kirby, D. Lew, C. Clark, J. DeCesaro and K. Lynn. (2011). 'Integration of variable generation, cost-causation, and integration costs', *Electricity Journal*, 24(9), 51–63.

Mills, M.P. (2019). *"The 'New Energy Economy' An exercise in Magical Thinking"*. Manhattan Institute: Manhattan.

Ming, Y. and I.M. Held. (2018). Modeling water vapor and clouds as passive tracers in an idealized GCM. *Journal of Climate*, 31(2), pp. 775–786.

Mobiot, G. (2006). *Heat: How to stop the Planet Burning*. Allen Lane: London. p. 304.

Molina, M., D. Zaelkeb, K.M. Sarma, S.O. Andersen, V. Ramanathan, and D. Kaniaru. (2009). Reducing abrupt climate change risk using the Montreal Protocol and other regulatory actions to complement cuts in CO_2 emissions. Proceedings of the National Academy of Scicnces of the U.S.A. 106, 20616–20621.

Naafs, B.D.A., M. Rohrssen, G.N. Inglis, O. Lähteenoja, S.J. Feakins, M.E. Collinson, E.M. Kennedy, P.K. Singh, M.P. Singh, D.J. Lunt, and R.D. Pancost. (2018). High temperatures in the terrestrial mid-latitudes during the early Palaeogene. *Nature Geoscience*, 11(10), p. 766.

Nagurny, A. P., 1986, *Modelling the continental and marine ice in climate models. Progress in Science and Technology. Meteorology and Climatology, Volume 13 Itogi Nauki i Tekhniki (Meteorologia i Gidrologia)* (Moscow: VINITI).

Nairobi, Kenya http://www.unep.org/ozone or http://www.unep.ch/ozone.

NASA (2015) *Global Climate Change*, Scientific Consensus: Earth's Climate Is Warming, https://climate.nasa.gov/scientific-consensus/

NEA. (2012). *Nuclear Energy and Renewables - System Effects in Low-carbon Electricity Systems*. Nuclear Energy Agency: Paris.

NEA. (2015a). *Nuclear New Build: Insights into Financing and Project Management*, OECD: Paris.

NEA. (2015b). *Climate Change: An Assessment of the Vulnerability of Nuclear Power Plants and the Cost of Adaptation*. OECD: Paris.

Neumann, J. E., Willwerth, J., Martinich, J., McFarland, J., Sarofim, M. C., and Yohe, G. (2020). Climate damage functions for estimating the economic impacts of climate change in the United States. *Review of Environmental Economics and Policy*, 141: 25–43.

Newfield, T.P. (2018). *"The Climate Downturn of 536-50"* In *"The Palgrave Handbook of Climate History"* Ed. S. White, C. Pfister and F. Mauleshagen. Palgrave MacMillan: London. pp. 447–493.

Newhall, C.G. and S. Self. (1982). The volcanic explosivity index (VEI): An estimate of explosive magnitude for historical volcanism. *Journal of Geophysical Research*, 87, 1231–1238.

Nielsen JO, S.A.L. D'haen, and A. Reenberg. (2012) Adaptation to climate change as a development project: A case study from Northern Burkina Faso. *Climate Development* 4(1):16–25.

Nikkei. (3 August, 2018). (https://asia.nikkei.com/Opinion/SoutheastAsian-fund-can-complement-Chinese-investment-by-boosting-self-reliance).

Noilhan J., Planton S. (1989). A simple parameterization of land surface processes for meteorological models. *Mon. Weather Rev.* 117: 536–549.

Nordhaus, W.D. (1973. World dynamics: Measurement without data. *Economic Journal* 83: 1156–1183.

Nordhaus, W. D. (1991). The cost of slowing climate change: A survey. *The Energy Journal*, 12(1).

Nordhaus, W.D. (1992). Lethal model 2: The limits to growth revisited. *Brookings Papers on Economic Activity* 2: 1–59.

Nordhaus, W. D. (2011). Estimates of the social cost of carbon: background and results from the RICE-2011 model (No. w17540). *National Bureau of Economic Research*, 5(2), 240–257.

Nordhaus, W. D. (2013). *The Climate Casino: Risk, Uncertainty, and Economics for a Warming World*. NEw USA: Yale University Press.

Ohshima, K.-I. and S. Martin. (2004). Introduction to special section: Oceanography of the okhotsk sea. *J Geophys. Res.* 109 (C09S01): 1–3.

Ohshima, K.-I., S. Nihashi, E. Hashiya, and T. Watanabe. (2006). Interannual variability of sea ice area in the Sea of Okhotsk: Importance of surface heat flux in fall. *Journal of the Meteorological Society of Japan*, 84 (5): 907–919.

Oldfield, J.D. (2016). "Mikhail Budykos (1920–2001) contributions to Global Climate Science: from heat balances to climate change and global ecology." *WIREs Climate Change*, 7, 682–692.

Oldfield, J. D. (2018). Imagining climates past, present and future: Soviet contributions to the science of anthropogenic climate change, 1953–1991. *Journal of Historical Geography*, 60, 41–51.

Olsson, L., L. Eklundh, and J. Ardo. (2005). "A recent greening of the Sahel—trends, patterns and potential causes." *Journal of Arid Environment*, 63:556–566.

Olwig, M.F. (2013) Beyond translation: reconceptualizing the role of local practitioners and the development 'interface'. *European Journal of Development Research* 25(3):428–444. doi:10.1057/ejdr.2013.9

Oreskes, N. (2004). Beyond the ivory tower: the scientific consensus on climate change. *Science* 306: 5702 (December) 1686. doi: 10.1126/science.1103618.

Page, J. (2010a). Sorry, says UN, our Nobel prize-winning climate change report got key fact wrong. *The Times (London)*, Thursday 21 January 2010, p. 37.

Page, J. (2010b). Climate change chief admits report may have more errors. *The Times (London)*, Saturday 23 January 2010, p. 54.

Pahl, G. (2005). *Biodiesel. Growing a New Energy Economy*. Chelsea Green: White River Junction, VT. p. 281.

Parsons, D.;, M. Beland, D. Burridge, P. Bougeault, G. Brunet, J. Caughey, S. Cavallo, M. Charron, H. Davies, A.D. Niang, et al. (2017). THORPEX Research and the Science of Prediction. *Bulletin American Meteorological Society*. 98, 807–830.

Parungo, F., Kopcewicz, B. and Nagamoto, C., 1992, Aerosol particles in the Kuwait oil fire plumes : Their morphology, size distribution, chemical composition, transport, and potential effect on climate. *Journal of Geophysical Research*, 97, 15867–15882.

Paterne, M., J. Labeyrie, F. Guichard, A. Mazaud, and F. Maitre. (1990). Fluctuations of the Campanian explosive volcanic activity (South Italy) during the past 190,000 years, as determined by marine tephrochronology: *Earth and Planetary Science Letters*, 98, p. 166–174, doi:10.1016/0012-821X(90)90057-5.

Pearce, F. (2006) *The Last Generation: How Nature Will Take Her Revenge for Climate Change. Eden Project Book*. London: Transworld Publishers. p. 324.

Pearce, F. (2018) When the Rivers Run Dry. The global water crisis and how to solve it. Granta, London. 312 pp.

Pearce, J. (2019). Teaching science by encouraging innovation in appropriate technologies for sustainable development. HAL Id: hal-02120521, https://hal.archives-ouvertes.fr/hal-02120521

Penner, J. E., 1986, Uncertainties in the smoke source term for 'nuclear winter' studies. *Nature*, 324, 922–926.

Petit, J.R., J. Jouzel, D. Raynaud, N. I. Barkov, J.-M. Barnola, I. Basile, M. Bender, J. Chappellaz, M. Davis, G. Delaygue, M. Delmotte, V. M. Kotlyakov, M. Legrand, V. Y. Lipenkov, C. Lorius, L. PÉpin, C. Ritz, E. Saltzman and M. Stievenard. (1999). Climate and astmospheric history of the past 420,000 years from the Vostok ice core in Antarctica. *Nature*, 399, 429–436.

Pindyck, R. S. (2011). Fat tails, thin tails, and climate change policy. *Review of Environmental Economics and Policy*, 5(2), 258–274.

Pindyck, R. S. (2012). Uncertain outcomes and climate change policy. *Journal of Environmental Economics and management*, 63(3), 289–303.

Pindyck, R. S. (2013a). Climate change policy: What do the models tell us? *Journal of Economic Literature*, 51(3), 860–872.

Pindyck, R. S. (2013b). The climate policy dilemma. *Review of Environmental Economics and Policy*, 7(2), 219–237.

Pindyck, R.S. (2017). The Use and Misuse of Models for Climate Policy *Review of Environmental Economics and Policy*, 11(1), p. 100–114.

Pittock, A. B., Walsh, K. and Frederiksen, J. S., 1989, General circulation model simulation of mild nuclear winter effects. *Climatic Dynamics*, 3, 191–206.

Polvani, L. M., Previdi, M., England, M. R., Chiodo, G., and Smith, K. L. (2020). Substantial twentieth-century Arctic warming caused by ozone-depleting substances. *Nature Climate Change*, 10(2), 130–133.

Popper, K. (1959). "The Logic of Scientific Discovery", reprinted 2002. Routledge Abingdon. English edition based on the original German "Logik der Forschung. Zur Erkenntnistheorie der modernen Naturwissenschaft" of 1934.

Popper, Karl. *The logic of scientific discovery*. Routledge, 2005.

Porter, J.R., A.J. Challinor, C.B. Henriksen, S.M. Howden, P. Martre, and P. Smith. (2019). Invited review: Intergovernmental panel on climate change, agriculture, and food—A case of shifting cultivation and history. *Global Change biology*, 25(8), 2518–2529.

Pöschl, U. and M. Shiraiwa. (2015). Multiphase chemistry at the atmosphere–biosphere interface influencing climate and public health in the anthropocene. *Chemical Reviews*, 115(10), pp. 4440–4475.

Post, M. J., Grund, C. J., Langford, A. O., and Proffitt, M. H. (1992). Observations of pinatubo ejecta over Boulder, Colorado by lidars of three different wavelengths. *Geophysical Research Letters*, 19(2), 195–198.

Prata, A.J. (1990) Satellite-derived evaporation from Lake Eyre, *South Australia, International Journal of Remote Sensing*, 11, 2051–2068.

Prăvălie, R., & Bandoc, G. (2018). Nuclear energy: between global electricity demand, worldwide decarbonisation imperativeness, and planetary environmental implications. *Journal of Environmental Management*, 209, 81–92.

Prince, S. D., Wessels, K. J., Tucker, C. J., & Nicholson, S. E. (2007). Desertification in the Sahel: a reinterpretation of a reinterpretation. *Global Change Biology*, 13(7), 1308–1313.

Prueher, L.M., and D.K. Rea. (2001). Tephrochronology of the Kamchatka–Kurile and Aleutian arcs: evidence for volcanic episodicity. *Journal of Volcanology and Geothermal Research*, 106(1–2), 67–84.

Rabier, F., Bouchard, A., Brun, E., Doerenbecher, A., Guedj, S., Guidard, V., … and Genthon, C. (2010). The CONCORDIASI project in Antarctica. *Bulletin of the American Meteorological Society*, 91(1), 69–86.

Randall, D.A., R.D. Cess, J.P. Blanchet, G.J. Boer, D.A. Dazlich, A.D. Delganio,. M. Deque, V. Dymnikov, V. Galin, S.J. Ghan, A.A. Lacis, H. LeTreut, Z.D. Li, X.Z. Liang, B.J. McAvaney, V.P. Meleshko, J.F.B. Mitchell, J.J. Morcrette, G.L. Potter, L. Rikus, Roeckner, E., Royer, J.F., Schlese, U., Sheinin, D.A., Slingo, J., Sokolov, A.P., K.E. Taylor, W.M. Washington, R.T. Wetherald, I. Yagali, and M.H. Zhang. (1992). Intercomparison and interpretation of surface energy fluxes in atmospheric general circulation models. *Journal of Geophysical Research*, 97, 3711–3724.

Rao, K.R. (2019). Wind Energy and Power Generation Options. Socioeconomic Factors. In *Wind Energy for Power Generations* (pp 703–828. Berlin: Springer.

Rasmussen, K. (1999). Land degradation in the Sahel-Sudan. *The conceptual basis Danish Journal of Geography*, Special Issue, 2:151–159.

Rasmussen, K., S. D'haen, R. Fensholt, B. Fog, S. Horion, J.O. Nielsen, L.V. Rasmussen, and A. Reenberg. (2015). "Environmental change in the Sahel: Reconciling contrasting evidence and interpretations." *Regional Environmental Change*. DOI 10.1007/s10113-015-0778-1.

Raval, A., & Ramanathan, V. (1989). Observational determination of the greenhouse effect. *Nature*, 342(6251), 758–761.

Ravishankara, A.R., J.S. Daniel, and R.W. Portmann. (2009). Nitrous oxide (N_2O): The dominant ozone-depleting substance emitted in the 21st century. *Science*, 326 (5949), 123–125. (Oct 2)

Ravishankara, A. R., Y. Rudich, and J.A. Pyle. (2015). "Role of chemistry in Earth's climate." *Chemical Reiews*.115, 10, 3679–3681.

Rees M. (2004). *Our Final Century: Will the Human Race Survive the Twenty-First Century?* Heinemann: Oxford. p. 266.

Reid, S. J., G. Vaughan, N. J. Mitchell, I. T. Prichard, H. J. Smit, T. S. Jorgensen, C. Varotsos, and H. De Backer. (1994). Distribution of the ozone laminae during EASOE and the possible influence of inertia-gravity waves. *Geophysical Research Letters*, 21: 1479–1482.

Ribrant, J., and L. Bertling. (2007, June). Survey of failures in wind power systems with focus on Swedish wind power plants during 1997–2005. IEEE Power Engineering.

Rich, N. (2019). *"Losing Earth: The Decade We Could Have Stopped Climate Change"*. London: Picador.

Richardson, L.F. (1922). *Weather Prediction by Numerical Process*. Cambridge University Press: Cambridge.

Rider, M. (2019). *Hutton's Arse: 3 Billion Years of Extraordinary Geology in Scotland*. Edinburg, UK: Dunedin Academic Press Ltd.

Rider, M. and Harrison P., (2019). *Hutton's Arse: 3 billion years of extraordinary geology in Scotland*. Dunedin Academic Press Ltd.

Roberts P. (2004): *The End of Oil: The Decline of the Petroleum Economy and the Rise of a New Energy Order*. London: Bloomsbury. p. 399.

Robinson, N.A. and G.R. Waxmonsky. (1988). "The US-USSR agreement to protect the environment: 5 years of cooperation." *Environmental Law*, 18, 436–440.

Robock, A., 1984, Snow and ice feedbacks prolong effects of nuclear winter. *Nature*, 310, 667–670.

Robock, A. (2000). Volcanic eruptions and climate. *Reviews of Geophysics*, 38, 191–219.

Robock, A., and M.P. Free. (1995). Ice cores as an index of global volcanism from 1850 to the present. *Journal of Geophysical Research*, 100, 11, 549–567.

Robock, A., and M.P. Free. (1996). The volcanic record in ice cores for the past 2000 years. In *Climatic Variations and Forcing Mechanisms of the last 2000 Years*, edited by P.D. Jones, R.S. Bradley and J. Jouzel. (pp 533:546). Springer-Verlag: New York.

Rogelj, J.,, A. Popp, K.V. Calvin, G. Luderer, J. Emmerling, D. Gernaat, S. Fujimori, J. Strefler, T. Hasegawa, G. Marangoni, V. Krey, E. Kriegler, K. Riahi, D. P. van Vuuren, J. Doelman, L. Drouet, J. Edmonds, O. Fricko, M. Harmsen, P. Havlík, F. Humpenöder, E. Stehfest and M. Tavoni. (2018). Scenarios towards limiting global mean temperature increase below 1.5 °C. *Nature Climate Change*, 8, 325–332.

Rosa, E.A., S.P. Tuler, B. Fischhoff, T. Webler, S.M. Friedman, R.E. Sclove, K. Shrader-Frechette, M.R. English,, R.E. Kasperson,, R.L. Goble, T.M. Leschine,, W. Freudenburg,, C. Chess, C. Perrow, K. Erikson, J.F. Short and T.M. Leschine. (2010). Nuclear waste: Knowledge waste?. *Science*, 329(5993), 762–763.

Rosso, O.A., H.A. Larrondo, M.T. Martin, A. Plastino and M.A. Fuentes. (2007). Distinguishing noise from chaos. *Physical Review Letters*, 99, pp. 154102-1–154102-4.

Sagan, C., 1984, Nuclear war and climatic change – guest editorial. *Climatic Change*, 6, 1–4.

Sarma, M.K., and K.N. Taddonio. (2009). The Role of Financial Assistance by the Multilateral Fund in Technology Change to Protect the Ozone Layer. In Twenty years of ozone decline, *Proceedings of the Symposium for the 20th Anniversary of the Montreal Protocol*, eds. C. Zerefos, G. Contopoulos and G. Skalkeas, 441–458.

Scheiter, S. and , S. Higgins. (2009). "Impacts of climate change on the vegetation of Africa: An adaptive dynamic vegetation modelling approach." *Global Change Biology*, 15(9):2224–2246.

Schindlbeck, J.C., S. Kutterolf, S.M. Straub, G.D. Andrews, K.L. Wang, and M.J. Mleneck-Vautravers. (2018). One Million Years tephra record at IODP Sites U 1436 and U 1437: Insights into explosive volcanism from the Japan and I zu arcs. *Island Arc*, 27(3), e12244.

Schlesinger, W. H.; Bernhardt, E. S. (2013). *Biogeochemistry: An Analysis of Global Change*. Academic Press: New York.

Schneider, S.H. and J. Lane. (2005). *The Patient From Hell: How I Worked with My Doctors to Get the Best of Modern Medicine and How You Can Too*. Cambridge, MA: Perseus.

Schneider, M., A. Froggatt, S. Thomas, J. Hazemann, and L. Mastny. (2011). The World Nuclear Industry Status Report 2010-2011. Nuclear Power in a Post-Fukushima World. 25 years after the Chernobyl accident.

Scholz, Y. (2019). Cooperative Renewable Energy Expansion in Europe: Cost Savings and Trade Dependencies. In *The European Dimension of Germany's Energy Transition* (pp. 353–361). Springer: Cham.

Schulte, K. M. (2008). Scientific consensus on climate change?. *Energy & Environment*, 19(2), 281–286.

Schumann, G.J.P. and A. Domeneghetti, (2016). Exploiting the proliferation of current and future satellite observations of rivers. *Hydrological Processes*, 30 (16), 2891–2896. doi:10.1002/hyp.10825

Scott, M., and Lindsey, R. (2016). NOAA: "Do Volcanoes Emit More Carbon Dioxide Than Humans?" https://snowbrains.com/noaa-do-volcanoes-emit-more-carbon-dioxide-than-humans/.

Scovronick, N., Budolfson, M., Dennig, F., Errickson, F., Fleurbaey, M., Peng, W., ... and Wagner, F. (2019). The impact of human health co-benefits on evaluations of global climate policy. *Nature communications*, 10(1), 1–12.

Sharma, P. R. (2019). Bayesian Analysis of Banking Policy and Regulation (Doctoral dissertation, University of California, Irvine).

Sigl, M., M. Winstrup, J.R. McConnell, K.C. Welten, G. Plunkett, F. Ludlow, U. Büntgen, M. Caffee, N. Chellman, D. Dahl-Jensen, H. Fischer, S. Kipfstuhl, C. Kostick, O.J. Maselli, F. Mekhaldi, R. Mulvaney, R. Muscheler, D.R. Pasteris, J.R. Pilcher, M. Salzer, S. Schüpbach, J.P. Steffensen, B.M. Vintner, and T.E. Woodruff. (2015). Timing and climate forcing of volcanic eruptions for the past 2,500 years. *Nature*, 523, 543–549.

Simkin, T and L. Siebert. (1994). *Volcanoes of the World*, 2nd ed. p. 349. Geoscience Press: Tucson, Ariz.

Sims, R., P. Mercado, W. Krewitt, G. Bhuyan, D. Flynn, H. Holttinen, G. Jannuzzi, S. Khennas, Y. Liu, M. O'Malley, L. J. Nilsson, J. Ogden, K. Ogimoto, H. Outhred, Ø. Ulleberg and F. V. Hulk. (2011). 'Integration of Renewable Energy into Present and Future Energy Systems'. In: *IPCC Special Report on Renewable Energy Sources and Climate*. O. Edenhofer, R. Pichs-Madruga, Y. Sokona, K. Seyboth, P. Matschoss, S. Kadner, T. Zwickel, P. Eickemeier, G. Hansen, S. Schlömer and C. V. Stechow, Eds. Cambridge University Press: Cambridge, United Kingdom and New York, NY, USA.

Singer, S. F., 1988, Re-analysis of the nuclear winter phenomenon. *Meterological and Atmospheric Physics*, 38, 228–239.

Singer, S.F. (1989). (ed.) *Global Climate Change*. New York, NY: Paragon House.

Singer, S. F. (1996). Climate change and consensus. *Science-AAAS-Weekly Paper Edition*, 271(5249), 579.

Singer, S.F., (ed.) (2008). *Nature, Not Human Activity, Rules the Climate: A Critique of the UN-IPCC Report of May 2007*. The Heartland Institute: Chicago.

Sklyarov, E. V., Simonov, V. A., Buslov, M. M., Coleman, R. G., & Juvigné, E. H. (1994). *Ophiolites of the southern Siberia and Northern Mongolia*. Reconstruction of the Palaeo-Asian Ocean. VSP, Utrecht, 85–98.

Slingo, A. (1988). Can plankton control climate? *Nature*, 336, 421.

Smith, D. M., Dunstone, N. J., Scaife, A. A., Fiedler, E. K., Copsey, D., and Hardiman, S. C. (2017). Atmospheric response to Arctic and Antarctic sea ice: The importance of ocean–atmosphere coupling and the background state. *Journal of Climate*, 30(12), 4547–4565.

Soon, W., Baliunas, S. L., Robinson, A. B., & Robinson, Z. W. (1999). Environmental effects of increased atmospheric carbon dioxide. *Climate Research*, 13(2), 149–164.

Sorge, E. (1935). Glaziologische Untersuchungen in Eismitte (Glacio- logical research at Eismitte), in: *Wissenschaftliche Ergebnisse der Deutschen Groenland Expedition Alfred Wegener 1929 und 1930–31*. F. A. Brokaus: Leipzig, Germany. p. 270.

Sovacool, B. K. (2008). Valuing the greenhouse gas emissions from nuclear power: A critical survey. *Energy Policy*, 36(8), 2950–2963.

Sparrow, C. (1982). *The Lorenz Equations: Bifurcations Chaos and Strange Attractions* New York: Springer.

Steffen, W. and P. Canadell. (2005). "Carbon Dioxide Fertilisation and Climate Change Policy." Australian Greenhouse Office, Department of the Environment and Heritage, Commonwealth of Australia.

Stephens, G.L. and T. L'Ecuyer. (2015). The Earth's energy balance. *Atmospheric Research*, 166, pp. 195–203.

Sterling, C. (1974). The Making of the Sub-Saharan Wasteland. *Atlantic Monthly*, May, 98–105.

Stern, N. (2007). *The Economics of Climate Change: the Stern Review*. Cambridge: Cambridge University Press. p. 692.

Stern, N. (2013). The structure of economic modeling of the potential impacts of climate change: grafting gross underestimation of risk onto already narrow science models. *Journal of Economic Literature*, 51(3), 838–859.

Stockholm (1972),The United Nations Conference on the Human Environment (UNCHE), held in Stockholm, Sweden, in 1972. In: Chatterjee D.K. (eds.) *Encyclopedia of Global Justice*. Springer, Dordrecht. https://doi.org/10.1007/978-1-4020-9160-5_655.

Stowe, L. L., Carey, R. M., and Pellegrino, P. P. (1992). Monitoring the Mt. Pinatubo aerosol layer with NOAA/11 AVHRR data. *Geophysical Research Letters*, 19(2), 159–162.

Strahan, D. (2011). *The last oil shock: A survival guide to the imminent extinction of petroleum man*. Hachette UK, London, UK.

Strengers, B., B. Verheggen, and K. Vringer (2015). Climate science survey questions and responses (April 10). PBL Netherlands Environmental Assessment Agency. http://www.pbl.nl/sites/default/files/cms/publica-ties/pbl-2015-climate-science-su rvey-questions-and-responses_01731.pdf.

Stuiver M., B. Kromer, B. Becker and C.W. Ferguson. (1986). Radiocarbon age calibration back to 13,300 Years BP and the ^{14}C age matching of the German Oak and US Bristlecone pine chronologies. *Radiocarbon*. 28 (2B): 969–979.

Sugiyama, M., I. Sakata, H. Shiroyama, H. Yoshikawa, and T. Taniguchi. (2016). Research management: Five years on from Fukushima. *Nature*, 531(7592), 29–30.

Summerhayes, C. P. (2015). *Earth's Climate Evolution*. Chichester, UK: John Wiley & Sons. p 410.

Sun, Guangqi "Zheng He's expeditions to the western ocean and his navigation technology." *The Journal of Navigation* 45.3 (1992): 329–343.

Suni, T., A. Guenther, H.C. Hansson, M. Kulmala, M.O. Andreae, A. Arneth, P. Artaxo, E. Blyth, M. Brus, L. Ganzeveld, and P. Kabat. (2015). The significance of land-atmosphere interactions in the Earth system—iLEAPS achievements and perspectives. *Anthropocene*, 12, pp. 69–84.

Swift, J. (1977). Sahelian Pastoralists: Underdevelopment, desertification, and famine. *Annual Review of Anthropology*, 6: 457–478.

Tanzer M. (1980). *The Race for Resources: Continuing Struggles Over Minerals and Fuels*. London: Heinemann. p. 285.

Thompson, S. L. and Schneider, S. H., 1986, Nuclear winter reappraised. *Foreign Affairs*, Summer, 981–1005.

Thunberg, G. (2019). *No One Is Too Small to Make a Difference: Illustrated Edition*. Penguin UK.

Tian, J., Ma, X., Zhou, J., Jiang, X., Lyle, M., Shackford, J., & Wilkens, R. (2018). Paleoceanography of the east equatorial Pacific over the past 16 Myr and Pacific–Atlantic comparison: High resolution benthic foraminiferal δ18O and δ13C records at IODP Site U1337. *Earth and Planetary Science Letters*, 499, 185–196.

Tickell, C. (1977). *Climatic Change and World Affairs*. Harvard: Center for International Affairs.

Tickell, C. (1986). Climatic change and world affairs. *Eos, Transactions American Geophysical Union*, 67(17), 425–425.

Toman, M.A., Chakravorty, U., and Gupta, S. (2003). *India and Global Climate Change. Perspectives on Economics and Policy from a Developing Country*. Resources for the Future: Washington DC. p. 366.

Trenberth, K.E. (ed). (1992). *Climate System Modeling*. Cambridge University Press: Cambridge.

Trenberth, K., J. Fasullo, and J. Kiehl. (2009). Earth's global energy budget. *Bulletin of the American Meteorological Society.*

Tsareva, O. O., Zelenyi, L. M., Malova, H. V., Podzolko, M. V., Popova, E. P., and Popov, V. Y. (2018). What humankind can expect with an inversion of Earth's magnetic field: threats real and imagined. *Physics-Uspekhi,* 61(2), 191.

Turco, R. P., Toon, O. B., Ackerman, T., Pollack, J. B. and Sagan, C., 1983, Nuclear winter: global consequences of multiple nuclear explosions. *Science,* 222, 1283–1292.

Turner, B.L. II and W.B. Meyer. (1994). "Global land-use and land-cover change: an overview." In: Meyer, W.B. and B.L. Turner, II (eds.) *Change in Land Use and Land Cover: A Global Perspective.* Cambridge University Press, Cambridge, pp 3–10

Ueckerdt, F., Hirth, L., Luderer, G., and Edenhofer, O. (2013a). System LCOE: What are the costs of variable renewables?. *Energy,* 63, 61–75.

Ueckerdt, F., Hirth, L., Müller, S., and Nicolosi, M. (2013b). *Integration costs and Marginal value. Connecting two perspectives on evaluating variable renewables.* In *Proceedings of the 12th Wind Integration Workshop.*

UNEP (2005) Production and Consumption of Ozone Depleting Substances under the Montreal Protocol 1986 – 2004. *Published by Secretariat for The Vienna Convention for the Protection of the Ozone Layer & The Montreal Protocol on Substances that Deplete the Ozone Layer.*

UNEP Production and Consumption of Ozone Depleting Substances under the Montreal Protocol. (1986–2004). Ozone Secretariat, UNEP, 2005, Nairobi, p. 79.

UNESCO. (2019). World Water Assessment Programme The United Nations World Water Development Report 2019: Leaving No One Behind (UNESCO).

United Nations Scientific Committee on the Effects of Atomic Radiation (UNSCEAR). (2008). Sources, Effects and Risks of Ionizing Radiation. Report to the General Assembly with Scientific Annexes VOLUME II Scientific Annexes C, D and E. UNITED NATIONS, New York, 2011.

United Nations Scientific Committee on the Effects of Atomic Radiation (UNSCEAR). (2013). Sources, Effects and Risks of Ionizing Radiation. Report to the General Assembly with Scientific Annexes VOLUME II Scientific Annex B, New York, 2013.

Valero, F. P., & Pilewskie, P. (1992). Latitudinal survey of spectral optical depths of the Pinatubo volcanic cloud-derived particle sizes, columnar mass loadings, and effects on planetary albedo. *Geophysical Research Letters,* 19(2), 163–166.

Varentsov, M. I., Grishchenko, M. Y., & Wouters, H. (2019). Simultaneous assessment of the summer urban heat island in Moscow megacity based on in situ observations, thermal satellite images and mesoscale modeling. *Geography, Environment, Sustainability,* 12(4), 74–95.

Varotsos, C. (2002). The southern hemisphere ozone hole split in 2002. *Environmental Science and Pollution Research,* 9(6), 375–376.

Varotsos, C. (2005). Airborne measurements of aerosol, ozone, and solar ultraviolet irradiance in the troposphere. *Journal of Geophysical Research,* 110: D09202.1–D09202.10.

Varotsos, C. (2020). Weather, macroweather, and the climate our random yet predictable atmosphere, *Physics Today,* 73(1), 54. doi:1063/PT.3.4392

Varotsos, C. and A. P. Cracknell, (2004). New features observed in the 11-year solar cycle. *International Journal of Remote Sensing,* 25: 2141–2157.

Varotsos, C. and D. Kirk-Davidoff, (2006). Long-memory processes in ozone and temperature variations at the region 60°S-60°N. *Atmospheric Chemistry and Physics,* 6: 4093–4100.

Varotsos, C.A., and A.P. Cracknell, (1993). Ozone depletion over Greece as deduced from Nimbus-7 TOMS measurements. *Remote Sensing,* 14, 2053–2059.

Varotsos, C. A. & A. P. Cracknell. (2020). Remote Sensing Letters. Contribution to the success of the Sustainable Development Goals-UN 2030 agenda, *Remote Sensing Letters,* 11(8): 715–719.

Varotsos, C. A., M. N. Efstathiou, & J. Christodoulakis. (2020a). The lesson learned from the unprecedented ozone hole in the Arctic in 2020; A novel nowcasting tool for such extreme events. *Journal of Atmospheric and Solar-Terrestrial Physics,* 207, 105330.

Varotsos, C.A., and V.F. Krapivin, (2019). Modeling the state of marine ecosystems: A case study of the Okhotsk Sea. *Journal of Marine Systems,* 194, 1–10.

Varotsos, C.A., and V.F. Krapivin, (2020). Microwave Remote Sensing Tools in Environmental Science. Springer International Publishing. ISBN: 978-3-030-45767-9

Varotsos, C. A. & Y. A. Mazei. (2019). Future temperature extremes will be more harmful: A new critical factor for improved forecasts. *International Journal of Environmental Research and Public Health,* 16(20), 4015.

Varotsos, C., Kalabokas, P., & Chronopoulos, G. (1994). Association of the laminated vertical ozone structure with the lower-stratospheric circulation. *Journal of Applied Meteorology,* 33(4), 473–476.

Varotsos, C.A., K. Ya Kondratyev, and S. Katsikis. (1995). On the relationship between total ozone and solar ultraviolet radiation at St Petersburg, Russia. *Geophysical Research Letters*, 22, 3481–3484.

Varotsos, C.A., K. Ya. Kondratyev, and A.P. Cracknell. (2000). New evidence for ozone depletion over Athens, Greece. *International Journal of Remote Sensing*, 21(15), 2951–2955.

Varotsos, C.A., M.N. Efstathiou, and J. Christodoulakis. (2019a). Abrupt changes in global tropospheric temperature. *Atmospheric Research*, 217, pp. 114–119.

Varotsos, C.A., V.F. Krapivin, and A.A. Chukhlantsev, (2019b). Microwave polarization characteristics of snow at 6.9 and 18.7 GHz: Estimating the water content of the snow layers. *Journal of Quantitative Spectroscopy and Radiative Transfer*, 225, 219–226.

Varotsos, C.A., V.F. Krapivin, and F.A. Mkrtchyan. (2019c). On the recovery of the water balance. *Water, Air, & Soil Pollution* (in press).

Varotsos, C., Y. Mazei, & M. Efstathiou. (2020b). Paleoecological and recent data show a steady temporal evolution of carbon dioxide and temperature. *Atmospheric Pollution Research*, 11(4), 714–722.

Varotsos, C., Y. Mazei, E. Novenko, A. N. Tsyganov, A. Olchev, T. Pampura, … M. Efstathiou. (2020c). A new climate nowcasting tool based on paleoclimatic data. *Sustainability*, 12(14), 5546.

Varotsos, C. A. & C. Tzanis. (2012). A new tool for the study of the ozone hole dynamics over Antarctica. *Atmospheric Environment*, 47, 428–434.

Varotsos, C. A., C. Tzanis, & A. P. Cracknell. (2016). Precursory signals of the major El Niño Southern Oscillation events. *Theoretical and Applied Climatology,* 124(3-4), 903–912.

Varotsos, P., N. V. Sarlis, & E. S. Skordas. (2011). *Natural time analysis: The new view of time: precursory seismic electric signals, earthquakes and other complex time series*. Springer Science & Business Media.

Varoufakis, Y. (2017) "Talking to My Daughter about the Economy: A Brief History of Capitalism". *The Bodley Head London.*

Vaughan, R.A. (2001). "Cooling off on global warming – the continuing debate" In *"Remote Sensing and Climate Change"* ed. A.P. Cracknell: Springer, Berlin, Praxis, Chichester, pp 205–215.

Velders, G. J. M., S.O. Andersen, J. S. Daniel, D.W. Fahey, and M. McFarland. (2007). The importance of the Montreal Protocol in protecting climate. *Proceedings of the National Academy of Sciences of the U.S.A., 2007, 104*, 4814–4819.

Velikhov, E. E. (ed.), 1987, *Climate and Biological Consequences of a Nuclear War* (Moscow: Nauka).

Verdes, P.F. (2007). Global warming is driven by anthropogenic emissions: A time series analysis approach. *Physical Review Letters*, 99, 048501-1-048501-4.

Verheggen, B., Strengers, B., Cook, J. van Dorland, R., Vringer, K., Peters, J., Visser, H., and Meyer, L. (2014). Scientists' views about attribution of global warming. *Environmental Science & Technology* 48: 16. 8963–8971, DOI: 10.1021/es501998e

Viana, M., Hammingh, P., Colette, A., Querol, X., Degraeuwe, B., de Vlieger, I., & van Aardenne, J. (2014). Impact of maritime transport emissions on coastal air quality in Europe. *Atmospheric Environment*, 90, 96–105.

Victor B. G., K. Raustiala and E.B. Skolnikoff (eds.) (1998) *The Implementation and Effectiveness of International Environmental Commitments: Theory and Practice*. MIT Press: Cambridge, MA, and London. p. 737.

Villach. (1986). "Report of the International Conference on the Assessment of the Role of Carbon Dioxide and of Other Greenhouse Gases in Climate Variations and Associated Impacts" World Climate Programme, WMP - No 661.

Vivoda, V. (2016). *Energy security in Japan: challenges after Fukushima*. New York: Routledge.

von der Gathen, P., M. Rex, N. R. Harris, D. Lucic, B. M. Knudsen, G. O. Braathen, … C. Varotsos. (1995). Observational evidence for chemical ozone depletion over the Arctic in winter 1991–92. *Nature*, 375(6527), 131–134.

Vupputuri, R. K. R., 1988, The interactive effects of large injections of smoke, dust and NOx on atmospheric temperature and ozone structure and surface climate. In *Natural and Man–Made Hazards*, edited by M. I. El-Sabh and T. S. Murthy (Dordrecht: Reidel), pp. 643–668.

Wackernagel, M., N.B. Schulz, D. Deumling, A. Callejas Linares, M. Jenkins, V. Kapos, C. Monfreda, J. Loh, N. Myers, R. Norgaard, and J. Randers. (2002). Tracking the ecological overshoot of the human economy. *Proceedings of the Academy of Science*, 99 (14), 9266–9271.

Wadhams, P., 1994, Remote sensing of snow and ice and its relevance to climate change processes. In *Remote Sensing and Global Climate Change*, edited by R. A. Vaughan and A. P. Cracknell (Berlin: Springer), pp. 303–339.

Wadhams, P. (2017). *A Farewell to Ice*. London: Penguin.

Wagner, G., & Weitzman, M. L. (2015). *Climate shock: the economic consequences of a hotter planet*. Princeton University Press, 264 p., ISBN 0691159475, 9780691159478

Walker, M. (2013). "*6.2.3 Dendrochronological Series. Quaternary Dating Methods*. New York, John Wiley and Sons.

Walker, G. and D. King. (2008). *The Hot Topic – How to Tackle Global Warming and Still Keep the Lights On*. Bloomsbury: London.

Wallace-Wells, D. (2019). *The Uninhabitable Earth*. London: Penguin. p. 320.

Wang, H., Shi, G. Y., Zhang, X. Y., Gong, S. L., Tan, S. C., Chen, B., ... & Li, T. (2015). Mesoscale modelling study of the interactions between aerosols and PBL meteorology during a haze episode in China Jing–Jin–Ji and its near surrounding region–Part 2: Aerosols' radiative feedback effects. *Atmospheric Chemistry and Physics*, 15(6), 3277–3287.

Ward, B. and R. Dubos. (1972). "*Only One Earth – The Care and Maintenance of a Small Planet*" London: Penguin.

Washington, W. M., & Parkinson, C. (2005). *Introduction to three-dimensional climate modeling*. University Science Books. 2nd Edition, 354 pages,

WCRP. (2009). "WCRP Implementation Plan 2010-2015". WCRP, WMO TD No 1503.

Weart. S.R. (2008). *The Discovery of Global Warming*. Harvard University Press: Cambridge, MA.

Weitzman, M. L. (2011). Additive damages, fat-tailed climate dynamics, and uncertain discounting. In *The Economics of Climate Change: Adaptations Past and Present* (pp. 23–46). Cambridge, MA: University of Chicago Press.

Weitzman, M. L. (2013). Tail-hedge discounting and the social cost of carbon. *Journal of Economic Literature*, 51(3), 873–882.

White, S., C. Pfister and F. Mauelshagen. (2018). *The Palgrave Handbook of Climate History*. London: Palgrave Macmillan.

Wild, M., D. Folini, C. Schär, N. Loeb, E.G. Dutton, and G. König-Langlo. (2013). The global energy balance from a surface perspective. *Climate Dynamics*, 40(11–12), pp. 3107–3134.

Williams, S. E., A.J. Hobday, L. Falconi, J. M. Hero. N. J. Holbrook, Capon, S ... and L. Hughes. (2020). Research priorities for natural ecosystems in a changing global climate. *Global Change Biology*, 26 (2), 410–416.

Wilson, R. C. and H.S. Hudson. (1988). Solar luminosity variations in solar cycle 21. *Nature*, 332. 810–812.

Winchester, S. (2004). *Krakatoa: The day the world exploded*. Penguin UK, 448p

Winchester, S. (2008). *The Day the World Exploded: The Earthshaking Catastrophe at Krakatoa*. New York: Harper Collins.

Winker, D. M., and Osborn, M. T. (1992a). Preliminary analysis of observations of the Pinatubo volcanic plume with a polarization-sensitive lidar. *Geophysical Research Letters*, 19(2), 171–174.

Winker, D. M., and Osborn, M. T. (1992b). Airborne lidar observations of the Pinatubo volcanic plume. *Geophysical Research Letters*, 19(2), 167–170.

WMO (1987) *The Influence of Climate Change and Climatic Variability on the Hydrologie. Regime and Water Resources (Proceedings of the Vancouver Symposium, August 1987)*. IAHSPubl. no. 168, 1987. http://hydrologie.org/redbooks/a168/iahs_168_0421.pdf.

WMO (1988) *UNEP: 1988, Climate Change: The IPCC Scientific Assessment*, Houghton, J., Jenkins, G. J., and Ephraums, J. J. Cambridge University Press, London, 365 pp.

WMO (1998). World Meteorological Organisation, Geneva, Switzerland, 324.

WMO. 2007. The Role of Climatological Normals in a Changing Climate. WCDMP-No. 61, WMO-TD/No. 1377, World Meteorological Organization.

WMO (2018) *WMO statement on the state of the global climate in 2017*. World Meteorological Organisation, Geneva

WMO, UNEP. (2018). National Oceanic and Atmospheric Administration. National Aeronautics and Space Administration, European Commission.

WMO-UNEP. (1989). Proceedings of the World Conference on the Changing Atmosphere: Implications for Global Security. Toronto, June 27–30, 1988, WMO DOC 710 91989).

Wold, H. (1938). *A Study in the Analysis of Stationary Time Series*. Uppsala Sweden: Almqvist

World Commission on Environment and Development. (1987). Our Common future. Oxford University Press, Oxford, i.e. the Brundtland Report.

World Meteorological Organization (WMO). (1986). Report of the International Conference on the Assessment of the Role of Carbon Dioxide and of Other Greenhouse Gases in Climate Variations and Associated Impacts, Villach, Austria, 9-15 October 1985. WMO Technical Note, No. 661.

World Meteorological Organization (WMO), 1988, Water resources and climatic change; sensitivity of water-resource systems to climate change and variability. WCAP-4 (WMO/TD-N247), Geneva

Worldwatch. (2007). *Biofuels for Transport: Global Potential and Implications for Sustainable Energy and Agriculture*. Earthscan: London.

Xu, R., H. Hu, F. Tian, C. Li, and M.Y.A. Khan. (2019) Projected climate change impacts on future streamflow of the Yarlung Tsangpo-Brahmaputra River. *Global and Planetary Change*, 175, 144–159.

Yan, Y., M. L. Bender, E.J. Brook,. H.M. Clifford, P.C. Kemeny, Kurbatov A.V.,... and J.A. Higgins. (2019). Two-million-year-old snapshots of atmospheric gases from Antarctic ice. *Nature*, 574 (7780), 663–666.

Yin, Z., Zhu, L., Li, S., Hu, T., Chu, R., Mo, F., ... and Li, B. (2020). A comprehensive review on cultivation and harvesting of microalgae for biodiesel production: environmental pollution control and future directions. *Bioresource Technology*, 122804, 1–19.

Zadra, A., K. Williams, A. Frassoni, M. Rixen, Á.F. Adames, J. Berner, F. Bouyssel, B. Casati, H. Christensen, M.B. Ek, G. Flato, Y. Huang, F. Judt, H. Lin, E. Maloney, W. Merryfield, A. van Niekirk, T. Rackow, K. Saito, N Wedi and P. Yadav. (2018). Systematic errors in weather and climate models. Nature, Origins and Ways Forward, Bulletin of the American Meteorologial Society, ES67–ES70.

Zavialov, P. (2005) *Physical Oceanography of the Dying Aral Sea*. Praxis/Springer: Chichester/Berlin. p. 140.

Zillmann, J.W. (2007). Some observations of the IPCC assessment process 1988–2007. *Energy and Environment*, 18, 869–891.

Index

Page numbers in *Italics* refer to figures; **bold** refer to tables

A

absorbed solar radiation, 15, 26, 112, 274, 312
acid rain, 31, 35–38, 42–43, 173, 175, 217, 304, 308,
    342–343, 345, 347
Advanced Earth Observation Satellite (ADEOS), **219**
Advanced Microwave Sounder (AMSU-A), 214
Advanced Radiometer to Measure Thermal Emission/
    Reflection (ASTER), 214
aerosol; *see also* condensation nuclei; gas-to-particle
    reactions
    anthropogenic, 78–80, 93–96, 217, 234–238, 286,
        296, 299
    atmospheric, 14, 42, 78–80, 85, 93–96, 98–99, 102,
        209, 217, **219**, 221, 223, 228, 234–236,
        240–241, 264, 277–278, 297, 299, 313–314,
        323, 362, 364
    optical thickness (AOT), 221, 237, 239, **240**,
        248–250, 256
    pollution, 82, 217, 286–287
    radiation experiments, 234
    radio-sonde, 253
    stratospheric, 238–242, **240**
    sulphate, 237–238
    tropospheric, 237
    volcanic, **162**, 228, 238, 245, 247, 251, 253, 255–256
agriculture, 23, 26, 31, 35, 50, 64, 72, 112, 137, 150, 152,
    169, 171, 226, 259, 284, 290, 298, 304,
    306–308, 332, 345, 385–386
Agung eruption, 239, 244, 246
air temperature, 89, 99, 113, 119–120, 124, 126, 165, 221,
    223, 232, 242, 244–246, 249, 256–257, 299,
    327–329, 363
albedo, 121, 157, 168–169, 286, 301, 322, 362; *see also*
    surface albedo
    effect (clouds), 12, 81, 94, 99
    feedback, 92, 159, 166, 322
    feedback (ice), 81, 159
American Meteorological Society, 288, 361
Antarctic
    circumpolar current
    climate-forming processes, 91
anthropogenic
    aerosol, *see* aerosol
    effects, 29, 48, 100, 234
    emissions 31, 90, 93–94, **95**, 150, 234, 238, 296,
        331–332, 342, 345, 349, 355
    WCRP, 166, 311
anticyclones, 249
Applied Climatology Programme
Aral Sea, 124–125, 128–139, *134*, **136**, *138*, **139**
Arctic
    climate-forming processes, 361
    haze, 234

atmosphere

atmosphere
    aerosol, 93–96, 250
    boundary layer, 124, **240**
    cryosphere system, 160, 278
    ocean interaction, 15, 113–114, 121, 166
    pollution, *see* pollution
    top of (TOA), 94
atmospheric chemistry, 79, 151, 345
atmospheric general circulation, 114, 123–124, 160, 216,
    224, 276
Atmospheric Infra-red Sounder, 209, 211, 214
Atmospheric Radiation Measurement, 241
atmospheric temperature, 49, 88–89, 96, 223, 299
atmospheric transparency, 317
attenuation, 9, 116–117, 239–240
aurora/auroral zone
available potential energy (APE), 183
AVHRR, 116, 205
    data, 209, 221, 247–249, 271

B

backscattering, 240, 250–251, 254–255
balloon observations, 211, 216, 252–253
baroclinic layer
barotropic waves
biogenic emissions, 93, 95, 217, 235
biogeochemical cycles, 91–92, 125, 154, **219**
biomass, 16, **30–32**, **86–87**, 90, **92**, **95**, 143, **149**, 172, 174,
    235, 321, 353
biome model, 143, 146
bioproductivity, 17
biosphere, 9, 16–18, 42–43, 50, 52, 83, 85, 110, 112, 121,
    141–170, 212, *215*, 221, 269, 271, 284,
    286–287, 297, 313, 315–316, 318–320,
    322–323, 346, 383, 389
blackbody radiation
blocking, 54, 93, **162**, 279, 355
boundary layer, 15, 100, 113–114, 124, 225, **240**, 272, 279,
    355
box model, 144, 363
brightness, 82, 115, 170, **219**, 226, 255, 271
    temperature, 82, 170, **219**, 255
bromine, 366, 368
bromoform

C

CAENEX programme, 234
calibration, 104, 221, 225, 232–233, 276, 311
carbon, 16, *17*, *19*, 23–24, 35, 42, 63, 66, 75–80, 83, 93–94,
    114, 121, 142, **143**, 144–149, 152, 154,
    168–169, 175, 178–179, 187–188, 192, *193*,

194–195, 214, 228, *231*, 232, 238, 243, 277,
290, 293, 297, 301, 306, *309*, 315, 317–319,
321–323, 325–328, 330–331, 334, 336,
339–341, 352, 354, 356–357, 359, **369**,
371–372, 374, 377, 379–381, 384–385, 389–391
carbon cycle, 16–17, 114, 142–146, 148, 152, 154, 169,
297, 330
  global-scale climate changes, 41, 162, 189, 195,
  217–218, 236, 239, 368
carbon dioxide
  aerosols/climate, 234–242
  cloud-radiation interaction, 168, 225
  CO₂ cycle, 18, 49, 65, 164, 182, 277
  concentration, 4, 17–19, 26, 29–31, 33, 42, 47, 49, 73,
  75, *76*, 77–78, 82–83, 85, 114, 124, **143**,
  144–147, *147*, 148, 150, 152, 154, 157,
  160–161, 192, 230, 244, 273, 275–276, 278,
  282, 286, 299, 306, 315, 323–324, 331, 336,
  339–340, 345, 349, 357, 362–363, 366
  detecting greenhouse signal, 121, 145, 150, 161,
  278, 299
  impact on Arctic/Antarctic, 160–165, 168, 224, 336
  impact on ocean, 18, 112–117, 121
  internal variability, 118, 142, 269, 279, 299, 354
  numerical modelling, 98–99, 119–120, 124, 141, 143,
  145, 148, 216, 223, 225, 238, 241, 256–257,
  266, 268–269, 313, 328–329, 362
  planetary comparison, 104, 208, 313, 372, 381, 384
  response experiments, 142, 269
  simple models, 7, 89
  solar-terrestrial interrelations, 14, 145, 148–149, 156,
  211, 275, 308, 314, 336
  volcanic eruptions, 14, 26, 33, 95, 156, 224, 226, 228,
  234–235, 238, 241–246, 248, 250, 253, 255,
  257, 317, 362, 364
carbon monoxide, 214
carbonyl sulphide
chlorine, 23, 210, 252, 366, *367*, 368
chlorocarbon compounds
chloroffuoromethanes
chlorofluorocarbons (CFCs)
  greenhouse effect, 18, 362, 379, 389
  impact, 42, 329, 365
chlorophyll, **213**
cirrus clouds, 98, *99*, 216
clear-sky conditions, 88–89, 99, 249
climate carbon dioxide
  diagnostics, 12, 98, 157, 225, 279
  forming factors, 85, 150, 211, 217, 316, 323
  observed regularities, 124, 146, 254, 316
  predictability, 7, 42, 141, 183–184, 276–277, 279, 355
  studies (future research), 187–188
  system parameters, 7, **163–164**
  world ocean and, 42, 138, 290
climate change background
  factors involved, 70, 260
  future studies, 31
  global-scale, 41, 162, 189, 195, 217–218, 236, 239, 368
  natural factors affecting, 165, 217, 298, 317
  nuclear war (impact), 165, 217, 298, 317
  research directions, 188
  World Climate Programme, 41

climate models development
  nested, 119, 124, 143
  simple (further studies), 121–122, 246, 257, 362
Climate of the Past Programme, 41, 162, 281, 287
Climate System Modelling Programme
climatically important processes aerosols and climate
  Arctic and Antarctic, 160–165, 168, 224, 336
  climate models, 329
  cloudiness and radiation, 98, **163**, 255
  land surface processes, 121, 213, 297
  minor gaseous components, 83–93, 154, 217, 237, 248
  processes in the ocean, 96, 121, 142, 152, 237, 277,
  361–362
  solar-terrestrial interrelations, 62–63, 145, 336
climatic system external impacts, 141, 154, 237, 269
  internal variability, 118, 142, 269, 279, 299, 354
climatology of planets, 48, 83, 354, 372
cloud
  albedo effect, 99, 240
  condensation nuclei, 96–97, 234, 236–237, 253, 352
  cover, 98–99, 117, 204, 206, 212–213, 221, 224–225,
  247, 302
  feedback, 238, 273–274
  formation, 29, 94, 97, 234, 267
  liquid water content, 221
  optical feedback, 275
  optical thickness, 221, 237, 239, 248, 250, 256
  radiation forcing, 224
  radiation interaction, 168, 225, 238, 329
Clouds and the Earth's Radiant Energy System (CERES),
  100, 102, 212
Coastal Zone Color Scanner
collision-induced heating, 363
combustion process, 16–17, 30–31, 75, 93, 174
condensation nuclei (CN), 96–97, 234, 236–237, 252–253
convection, 7, 115, 124, 267
convective clouds, 115–116, 212
corpuscular radiation, 227
correlation radiometer, **219**
correlation spectrometer, 247, 250–252
coupled general circulation model, 100, 119–120, 122, 141,
  146, 155, 160, 274, 276, 283, 291, 301, 353
cryosphere; *see also* ice; permafrost; snow
cumulus clouds, 75, 98
cyclogenesis, 121
cyclones, 162, 242, 244, 279, 355

D

Darwin (Australia)
Defense Meteorological Satellite Program (DMSP), 204, **205**
Deforestation, *see* forests
delta-Eddington approximation, 345
dendroclimatic data, 271, 287, 311
depolarization ratio
desert, 6, 14, 26–27, 72, 128–131, 146, 149, **150**, 156, 207,
  211, **240**, 249, 257–260, 277, 306, 364–365,
  379, 383
desertification, 128, 137, 257–261, 263–264, 306–307, 310
diabatic heating, 279, 355
dimethyl sulphide, 31, 94, 234, 327
dipole heating/structure

diurnal cycle, 272, 275, 279, 354–355
Dobson stations data, 20–21, 90
Doppler lidar, **219**
Doppler radars, **219**
droughts, *16*, 41, 79, 112, 226–228, 233, 259–260,
264–265, 301, 306, 309, 341, 353, 379, 387
dust clouds, 9, 14, 156, 246
dust veil index (DVI), 242–245

E

Earth, 1, 4–6, 9–12, 14–15, 18–19, 24, 26, 28–29, 31, 35,
37, 39, 41, 43, 47–50, 52, 57–59
comparative climatology, 10, 48, 83, 85, 100, 201, 212,
218, 233, 257, 276, 282, 288, 292–293, 354, 372
Earth Observing System, 102, 211–212, *215*
Earth Radiation Budget
components, 99, 221, 313
Experiment (ERBE), 82, 89, 100–104, 208
interannual variability, 101, 121, 144, 146, 159, 166,
218, 242, 244, 256
satellite, 82, 89, 99–102, 208–210, 221, 313
solar constant, 10, 81, 103–104, **163**, **219**, **222**
teleconnections, **162**, 279, 355
water vapour, 100, **210**
ecosystems, 44, 111–112, 114, 125, 127, **143**, 145, 148,
157, 196, 290, 304, 308, 321, 330, 333, 342,
349, 368, 371, 375, 383
eddy heat, **92**
El Chichón eruption, 239, 246–247, 249, 251–252,
254–256
Electronic Scanning Microwave Radiometer (ESMR)
El Niño, 15–16, 101, 121, 169, 218, 241, 256, 306, 355,
393
Southern Oscillation (ENSO), 121, 146, 169, 241, 355
emissions
anthropogenic biogenic, 93, 95, 217, 235
natural, 93–94, 238
pre-industrial, 30, 90, **143**, 148, 169, 171, 238, 294,
324, 341
thermal, 18, 88, 122, 214, 239, 278
energetically active zones of the ocean (EAZO), 114
energy
available potential, 174, 183, 185
cycles, 221
fluxes, 12, 144, 151, 158, 160, 221
kinetic, 47, 62, 174, 185
solar, 9, 62, 81, 100, 172, 175, 178, 180, 182–184,
187–188
transport, 153
energy balance
land surface, 14, 154
models, 14, 154
energy cycles, 221
GEWEX, 42, 150, 221
equilibrium response experiments, 142, 269
ERS-1 118, 159, 212, 218, 220
ERS-2 159, 212, 218, 220
euphotic zone
Eurasia, 161
European Centre for Medium-Range Weather Forecasts
(ECMWF), 146, 225, 268

European Space Agency (ESA), 159, 207, 216, 222
evaporation, 12, 15, 28, 61, 82, 95–96, 110–112, 117–118,
120–122, 124, 133–134, **136**, 137, 139, 153,
174, 235–236, 238, 267, 272–273, 276, 305, 316
evapotranspiration
carbon dioxide, 122, 153, 272–273, 301
volcanoes, 156

F

Fabry-Perot interferometer, **219**
feedback
albedo, 81, 159, 166
cloud, 238, 273–274
strength, first-order, 238, 299
fertilizers/fertilization, 291
floods, *16*, 41, 112, 124, 166, 169, 227, 265, 303, 309
force-restore method, 273
forests
deforestation, 64, 70, 82, 90, 112, 144–145, 150, 259,
305, 324, 361
fossil fuels, 16–18, 29–30, 37, 40, 47, 50, 54, 56, 59, 65–66,
68, 73, 78–79, 93, 124, 131, 140, 145, 172, 175,
178, 182–184, 195, 268, 278, 282, 320, 322,
331, 334, 337, 340, 342, 345, 347, 350, 363,
369, 372, 375, 379, 381, 383–385, 388–390
Fourier-analysis, **219**, 247, 250–251, 253
Fourier spectrometer, **219**, 247, 250–251, 253
frontogenesis

G

GARP Atlantic Tropical Experiment (GATE), 115
gas-to-particle reactions, 95–96, 234–236, 246, 250, 252, 257
Gaussian filter, 116, 145, 246
general circulation models (GCMs); *see also* atmospheric
general circulation
oceanic (OGCMs), 123–124, 146, 155, 160, 270,
274, 276
geochemical cycles, 144, **219**
geodynamic laser range-sounder
geophysical fields, 126, 154, 169, 214, **220**, 224, 283, 346
geopotential flux, 16, 81, 144, 151, 160, 211, 221, 223, 239,
242, 244, 250, 270, 353
GEOS-3 radiometer
GEOSAT satellite, 118
geosphere (lithosphere), 41, 43, 150–151, 221, 269, 271,
297, 315
geostationary satellites, 115, 117, 204, 206–207, 267
geostrophic wind
GEWEX programme, 42, 150, 221
glaciers, 2, 6, 12, 107–111, 132, **136**, 157, **158**, 159,
164–166, 213, 220, 231, 301, 303, 321, 373
Global Energy and Water Cycle Experiment (GEWEX), 42,
150, 221
Global Ocean Observing System (GOOS), 329
Global Precipitation Climatology Project (GPCP), 218, 311
global system of observations, 203, 211, 216, 283
Global Terrestrial Observing System, 329
global vegetation index (GVI), 16–17, 28, 52, **86–87**, 93,
121–122, 142–146, 148–149, 153–154, 217,
219, 222, 232, 260, 262–263, 270–273, 306

global warming, 1–73, 75, 91, 94, 112, 124, 172–173, 175,
    178, 243, 256, 264–265, 273–275, 277–278, 281,
    289, 291, 293–295, 297, 301, 304, 309–311, 314,
    320–323, 325, 327–328, 330–337, 340, 342,
    344–346, 348–356, 358, 365, 367–369, 371–373,
    375, 378–380, 390, 393–394
    CO2 signal, 121, 145, 150, 161, 269, 278, 299–300,
        366, 387, 392
global water and energy cycles, 221
    GEWEX, 42, 150, 221
Global Weather Experiment, 12
GLRS-A/GLRS-R
GOS instrument
gradient ratio, 113, 120, 123, 186, 269
grassland, **150**, 262, 306
Great Lakes, 107, 120, *138*
Great Plains (USA), 365, 386
greenhouse effect
    atmospheric, 18, **30**, 42, 87, 104, 123, 152, 213, 223,
        314, 328, 336, 362
    climatically-important processes, 195, 241
    climatic impacts (external), 152, 236, 245–246,
        361–363
    CO_2 cycle, 16, *17*, 114, 142–143, **143**, 144–146, 148,
        152, 154, 330, 334
    depletion of ozone layer, *see* ozone
    detecting CO_2 signal, 150
    nuclear testing/war, 26, 361–362
    World Climate Programme, 41
greenhouse gases; *see also* carbon dioxide; nitrogen oxides;
    sulphur dioxide
Greenland, 12, 69, 110–111, 121, 157, 159, **162**, 169,
    228–230, 245, 265, 321, 353, 385, 392
Greenland Sea, 12, 157, 159, 169, 228–229, 265, 321,
    353, 392
Gulf Stream, 121

H

Hadley circulation cells, 117
halocarbons, 78, 85–86
halogenized hydrocarbons, 78, 85–86
halogens, 23, **85–86**, 217, 366
haze (Arctic), 234
heat; *see also* latent heat; sensible heat
    balance, 118, 157, 160, 237, 278, 284, 286, 312,
        316–317, 363
    content, upper-ocean, **163**, 327, 339
    fluxes, *115*, 242, 244
    transport, 42, 106, 112–113, 117, 121, 153, 276, 363
Henry's law constant (for DMS), 257–258, *258*
high latitudes, 89, 100, 114, 121, 123–124, 154, 159,
    166, 228, 234, 243–244, 256, 272, 317, 328,
    361, 363, 368
HIRDLS, 215
HIRIS, **219**
HIRS, 209, 223
Holocene warming, 391, 394
homeostasis
human activities, 6, 9–10, 18, 26, 28–29, 31–33, 41–42, 48,
    75, 77–78, 82, 93–94, 112, 118, 124–125, 128,
    131, 139–140, 144–146, 154–155, 201,

    233–234, 261, 269, 274, 277, 286, 295, 299,
    311–312, 315–316, 320–321, 326, 330,
    332–334, 337, 343, 346, 348–349, 351, 361,
    371, 373, 394–395
humidity, 1–2, 88, 119–120, 144, **163**, 202–203, 206, 209,
    214, 224–225, 237, 275
humus, **143**
hydrocarbons, 78–79, 389
hydrodynamic models, 113
hydrogen sulphide, 23, 35, 65, 183, 185, 336, 366
Hydrological Atmospheric Pilot Experiment (HAPEX)
hydrological cycle, 12, 14, 16, 107, 109–112, 124, 134,
    151, 154, 157, 174, 239, 276
hydrology, 28, 72, 109–111, 115, 118, 122, 133, 155, **163**,
    166, 290, 317
hydrosol
hydrosphere, *see* oceans
hydrothermodynamics, 267
hydroxyl OH, 218
hygroscopic nuclei, 31

I

ice
    albedo feedback, 81, 159, 166
    glaciers, 108–110, 157, 159
    icebergs, 14, 108, 158, 272
    on lakes, 109–110, 161
    sea ice, 12, 113, 115, 123, 127, 156–161, *161*, **164**, 166,
        168, 211, 224–225, 269, 272, 276, 321, 341,
        353, 363, 368
    sheets, 12, 107, 110, 153, 157, **158**, 159, 165, 169, 232,
        269, 296–297, 321
ice cover
    characteristics, 168
    CO_2 signal, 82, 121, 124, 160, 161, 269, 327
industrial pollution
infra-red radiation, 276
Infra-red Scanning Radiometer (IRSR)
interannual variability of ERB, 121, 144, 146, 159, 166,
    242, 244, 256
interferometric spectrometer
interglaciation period, 10, 157, 229, 336
Intergovernmental Panel on Climate Change (IPCC), 41,
    43–44, 193, 275, 281–282, 315, 318–323, 333,
    338–340, 342–343, 346–347, 349–350, 354,
    366, 379
internal variability of climatic system climate diagnostics
    nested climate models, 119, 124, 143
    Sections programme, 114
    short-term changes (theory), 10, 114, 118, 142, 155
    world ocean and climate, 138, 290
International Association of Meteorology and Atmospheric
    Physics (IAMAP), 314
International Council of Scientific Unions (ICSU), 42–43,
    290, 298
International Date Line 68, 370–371
International Geosphere-Biosphere Programme (IGBP), 41,
    150–151, 221, 271, 315, 347
Inter-Tropical Convergence Zone (ITCZ), 111, 115
ionization
ionosphere, 80

IR radiometer, **219**
irrigation 28, 61, 118, 128, 130–133, **136**, 137
IR satellite radiometry
ISLSCP Retrospective Analysis Program (IRAP)
isostatic factors
isotopic composition, 144

J

JERS-1, 218, 220
jet streams, **162**, 242, 244, 249
Joint Global Ocean Flux Study

K

Kalahari Desert, 146
katabatic winds, 161
Kelvin waves
kinetic energy, 47, 62, 174, 185
Krakatoa eruption, 226, 228, **242**, 244–246

L

lake ice, 157, 161
lakes, 28, 96, 108–111, 118–121, 124–125, 128, **136**, 138, **160**, 161
land areas, development, 26–29
Landsat satellite, 212, 216, 271, 277
Landsat Thematic Mapper (TM), 271
land surface
    characteristics, **212**
    energy balance, 14, 65, 104, *105*, 106, 154, 225, 243, 290, 363
    processes, 297
    temperature, 75, 120, 212, 222, 224
Langley method, 250
La Niña, 101, 218
latent heat
    climatically-important processes
    climatic system (external impacts)
    climatic system (internal variability)
lidars/lidar sounding, 254–255
lightening limb-sounding, 215–216, **219**
liquid water content
    of clouds, 221–223
liquid water path, 96, 108, 116, 168, 221
lithosphere (geosphere), 16, *17*, 83, 112, 141–170, 269, 300
Little Ice Age, 9, 150, 230
long-range forecasts, 15, 113
long-wave radiation, 275
    outgoing, 88, 93, 100, 104, 221, **222**, 224, 250, 328
low-orbit satellites, **116**
Lyman-alpha line

M

magnesium
marine bioproductivity, 114, 146
Mars, 155, 313, 370
mathematical climate models, 265
Mauna Loa observatory, 146, 250, 253

Maunder Minimum
maximum entropy technique
mediaeval warming, 230
Mediterranean Sea, 27, *156*
mesoscale energy, 225, 317
mesoscale model, 119–120, 124
mesoscale structure, 124
mesoscale vortices, 212
meteor showers, 14, 156
Meteor/TOMS instrumentation, 223
Meteosat, 206, *206*, 207, **207**, 223
methane, 23, 31, **32**, 42, 57, 61, 83, 87, 90, **92**, 228, 341, 352, 354
methyl chloroform
Michelson interferometer
micro-meters, 9, 14
microphysical processes (clouds), 96, 236
microwave data, 116, 159
Microwave Humidity Sounder (MHS), **205**
microwave limb sounder (MLS), 216, **219**
microwave passive sounding, **116**, 159, 170, 209, 212, 214
microwave radiometer, 159, 212, **219**, 220, **220**
microwave remote sensing, 9, 122, 135, 159, 166, 169, 201, 210, **212**, 216, 218, 241–244, 248, 250, 253, 262, 272, 277, 283–287, 300, 311–314, 319, 366, 368
Microwave Sounding Unit (MSU), 209, 214, 221, 223
mid-resolution video-spectrometers
Mie formulae
Milankovitch theory, 10
minor gaseous components (MGCs), 83–92
    optically-active, 83–92
mixing (in clouds), 96, 126
mixing ratios, *77*, **152**, 252
modelling aerosol properties, 235–236
MODIS-N, **219**
MODIS-T, **219**
MOGUNTIA model
moist-adiabatic adjustment
moist convective adjustment, 124
moisture cycle
monsoons, 155, 335, 386, 393
MOPITT, 214
Multi-angle Spectroradiometer (MISR), 214
multi-channel flux radiometer (MFR), 247, 250
multispectral satellite based radiometry
    HIRS data, 209, 223
    NOAA-4, **205**, **207**
    NOAA-5, **205**, **207**, 224
    NOAA-6, **207**
    NOAA-7 satellite, **207**
    NOAA-9 satellite, 101, 103, **207**, 223
    NOAA-10 satellite, 101, 103, **207**
    NOAA-11 satellite, **207**, 223
    TOVS, 209, 221, 223–225

N

natural emissions, 93–94, 238
natural events on/within Earth, 9–18
natural factors affecting climate change, 165, 217, 298, 317
natural gas, **58**, 95, 171–172, 174, 176, **179**, 190, 193, 235
near surface air temperature, 165

Neptune
net primary productivity (NPP), 17, 144
Nimbus-3, 224
Nimbus-4
Nimbus-5
Nimbus-6
Nimbus-7, 20–21, 91, 103, 210, 217, 223, 248
Nimbus-TOMS data, 21, 248
nitrogen, 19, 30, **30**, 31, 76, 83, 142, 152, 232, 257, 366
nitrogen cycle, 142, 152
nitrogen oxides, 29–30, **30**, 31, 78, 96, 236, 327, 364
noise experiment, 62, 118, 141, 145, 150, 299–300
non-convective clouds, 115–116, 212
non-linearity, 100, 329
non-marine sulphate aerosol, 95, 237–238, 246, 251–253,
    299, 330
normalized backscattering (NRBS), 154
Normalized Difference Vegetation Index (NDVI), 262
North Atlantic Oscillation, 114–115, 121
Northern Oscillation
North Pacific Oscillation, 15, 113
nuclear energy, 56, 59, 85, 173, 175, 189–191, *191*,
    193–195, 197, 199, 363, 381
nuclear testing, 26, 361–362
nuclear war, 33, 121, 283, 287–288, 343, 361–362,
    364–365
nucleation model, 253

O

observational system (GEWEX programme), 218
observed regularities of climate, 241, 244, 328, 333, 341
occultation measurements (SAGE), 248
oceanic general circulation models (OGCMs)
oceanographic information, 114, 213, **213**
oceans
    atmosphere-cryosphere system, 160, 278
    atmosphere interaction, 120, 213
    boundary conditions, 120, 124, 207, 209, 211, 270,
        274–275, 277–278
    boundary layer, 15, 100, 113–114, 124, 225, **240**, 272
    climate, 117
    climatically-important processes
    detecting CO_2 signal, 150
    energetically active zones (EAZO), 114
    hydrosphere, 83, 107–141, 155, 213, 341, 353
    mixed layer models, 276
    pollution, 332, 345
OH hydroxyl, 218
optically-active minor gaseous components, 83–92, 217
optical thickness of aerosols, 221, 237, 239, **240**,
    248–250, 256
    of clouds, 221–224
organic sulphur, **85**
orography, 279, 355
outgoing longwave radiation (OLR), 88, 104, 221, **222**,
    224, 250, 328
outgoing shortwave radiation (OSR), **222**, 249–250
ozone
    content, total (TOC), **164**, 223–224, 247, 251
    layer (depletion), 79, 217
    sondes, 202

    stratospheric, 14, 26, 30, 46, 90, 209–210, **210**, 223,
        322, 364, 366–367, 392
    TOMS, 20–21, 223–224, 247–248, **249**, 251, 311
    tropospheric, 77, 79, 366

P

Pacific Ocean; *see also* El Niño; Southern Oscillation
    ITCZ, 111, 115
    North Pacific, 15, 113, 196
palaecyclones
palaeoclimatic changes, 114, 257, 264–265, 287, 291–292
'pancake ice,'
parametrizations
    climate-forming (key region), 85, 99, 150, 156, 162,
        211, 217, 323, 328–329, 361
    climate models, 141
    climatically-important processes, 141
permafrost, 12, 90, 112, 121, 157–158, **158**, *158*, 160, 166,
    169, 321, 341, 362
peroxyacetylnitrate (PAN), 96, 236
perspective concept (climate change studies), 44, 46
perturbations theory, 113, 144, 199, 234, 289
photodissociation
photometric sun-spot index, 9, 138, 212, 271, 285, 310
photo-oxidation, 78–79, 250
photosphere emissions
photosynthesis, 16–18, 29, **72**, 144–145, 148, 150, 172,
    174, 232, 306, 353
phytomass, 144–145
phytoplankton, 393
Pinatubo eruption, 247–248, 250–252, 255–256
planetary albedo, 81, 208
planetary boundary layer, 15, 100, 113–114, 124, 225, **240**,
    272, 279, 355
planetary waves, 65
planets, 7, 9, 141, 287, 345
plankton, 31, 114, 123, 153, 231, 269
plate tectonic movements, 10–12
Pleistocene period, 4, 155
polar ice sheets, 6, 121, 124, 166, 296–297, 317
polarization ratio, 170, **219**, 251
polar-orbiting satellites, 7, 103, 203–204, **205**, 207–209,
    211–212, 222–223, 313
polar platforms, 104, 374
polar stratosphere, 392
pollution aerosol
    atmospheric, *93*, 234–235, 237, 240–241, 264, 323, 364
    oceanic, 287, 347, 394
    thermal, 313
    tropospheric, 96, **164**, 236–238, **240**
precipitation; *see also* rainfall; rain rate
    acidification, 39, 72, **85**, **87**
    dynamics, 121
    GEWEX, 42, 150, 221
    internal variability, 118, 142, 269, 279, 299, 354
pyrheliometric observations, 104

Q

quasi-biennial oscillation (QBO), 239, 246, 256
quasi-periodic oscillation, 393

R

radar altimeters, 218, **219–220**, 223
radar echo, 116
radiation
    cloud (forcing)
    cloud interaction, 94, 238
    solar , *see* solar radiation
radiative-convective models, 239
radiative effects, 100, 209, 238, 275
radiative equilibrium, 154
radiative forcing, 78–79, 88, 90, **90**, *94*, 240, 273, 276,
        296–297, 299
radiative heat flux divergence, 89, **89**, 99, 221
radiative heating, 221
radiative processes (clouds), 87
radiocarbon data, 202, 227, 232–233
radiometry, 169, 232
radio-sonde, 160, 252–253
rainfall, 1–2, 16, 27, 82, 102, 117, 122, 131, 137, 140, 142,
        146, 244, 258–260, 262, 268, 287, 292,
        305–306, 310–311, 335
rain rate, 115–116, **116**, 117, 224
Raman lidar, **162**, **219**, 223–224, 247, 250–256
Rayleigh scattering, 250–251
relative humidity (RH), **163**, 202, 225, 275
remote sensing, 9, 122, 135, 159, 166, 169, 201, 210–211,
        212, 216–218, 241–244, 248, 250, 253, 262,
        272, 277, 283–287, 300, 311–314, 319, 366, 368
remote sounding, 159, 223–224, 245, 247, 250
research future directions
    future studies
response experiments, 142, 269
river run-off, 115, 121–124, **163**, 272

S

SAFIRE (spectrometer)
SAGE, 208–210, 214, 217, 245, 247–248
Sahara, 128, 143, 249, 257, 259, 308–309
satellite (global system), 117
Satellite Infra-red Spectrometer, 223
satellite lidar/lidar sounding, **162**, **219**, 224, 250–254, 256
satellite microwave data, 159
satellite multispectral radiometry, 209, 214
satellite observations (of volcanic eruptions), 88, 114, 203,
        208, 213, 217, 221, 224, 226, 241, 247, 267,
        311, 313, 315, 328, 366
satellite radars, 116, 220
satellite radiation budget data, 80–83, 89, 99–102, 113–115,
        160, **163**, 208–210, **210**, 248
Saturn, 313
scanning spectropolarimeter
scattering ratio, 250–251, 254
sea ice, 12, 19, 112–113, 115, 123, 127, *128*, 152, 156–161,
        164, 166, 168, 211, 224, 269, 272, 276, 321,
        341, 353, 363, 368
Seasat satellite, 212
seasonal boundary layer (of oceans), 15, 113–114
sea state spectrum, 206, 236, 300, 346, 349, 363, 385
sea surface temperature
    climatically-important processes, 195, 241

climatic system (external impacts), 141, 154, 237, 269
climatic system (internal variability), 141, 154,
    237, 269
    detecting CO_2 signal, 150
    global system of observation, 203
    variations, 123
sensible heat
    climatic system (internal variability), 141, 154,
        237, 269
    external impacts on climatic system, 141, 154, 237, 269
    outgoing (OSR), **222**
Siberia, 107, 146, **162**, 228, 304
side-looking radars (SLR)
signal/noise ratio, 150
silicate aerosol, 246, **247**
simple climate models, 7, 29, 87, 89, 99–100, 104, 112,
        121, 139, 141, 148, 150, 152, 156, 159
single-scattering albedo, 239–240
smoke aerosol, 252, 362
snow
    climatically-important processes, 195, 241
    climatic system (external impacts), 141, 154, 237, 269
    climatic system (internal variability), 141, 154,
        237, 269
    observed regularities of climate, 124, 146, 254
soil, 26, **30**, 38, 69–70, **95**, 108–110, 112, 122
    moisture, 14, 110–111, 122, 124, 142, 153–154,
        162–163, 212, 264, 272–273, 292, 296, 304
solar activity (impacts), 242, 244, 255, 338, 353
solar backscattered ultraviolet radiation (SBUV), 223
solar constant
    climatically-important processes, 195, 241
    variability, 83, 166, 223
solar energy, 9, 62, 81, 100, 172, 175, 178, 180, 182–184,
        187–188
solar protons
solar radiation
    absorption (ASR), 239, 256, 362
    limb sounding, 215–216, **219**
    measurement of intensity, *see* solar constant
    UV, **163**, 217, 223
solar-terrestrial interrelations, 316–317
solar wind, 184–188, 359, 365
SOLSTICE instrument
Somali current, **162**
soot particles, 236, 363
Soufrière eruption, 346
Southern Oscillation, 121, 146, 169, 241, 355, 393
spectral radiometry, 169, 209, 214, 248, 250
spectrometry spectroscopy
SPOT, 212, 271, 285, 310
Stefan-Boltzmann constant, 81, 88
storms, 26, 98, 123, 131, 227, 234, 306, 364–365
stratiform clouds, 234, 254–255
stratocumulus clouds, 98, *99*
stratosphere, 9, 14, 18–19, 21, 26, 29–31, 42, 89–91, 96,
        156, 209–210, 215, 223–224, 236, 238–240,
        243, 246–248, 250–252, 254, 256, 279, 355,
        363–364, 366, 392, 397
Stratospheric Aerosol and Gas Experiment, 102, 208–209
stratospheric aerosols, *see* aerosols SAGE
stratospheric chemistry, 216

Stratospheric Measurement Mission (SMM), 103
stratospheric ozone, *see* ozone
stratospheric temperature, 223–224, 239, **247**, 248,
            255–256, 392
stratus clouds, 98, 117, 121, 238
sulphate, 31, 93–95, 228, 235, 237–238, 246, 251–253,
            299, 330
      non-marine, 231, 308, 321
sulphur cycle, 29, 31, 38, 94–95, 190, 235, 253, 257
sulphur dioxide, 38, 96, 234, 236, 243
sulphuric acid, 31, 77, 96, 236, 243
Sun
      atmosphere correlation, **219**
      spots, 9, 285
      synchronous satellites, 117, 247
surface air temperature
      climatically-important processes, 195, 241
      climatic system (external impacts), 141, 154, 237, 269
      climatic system (internal variability), 141, 154,
            237, 269
      detecting CO_2 signal, 150
surface albedo, 82, **160**, **163**, 166, **212**, 237, 250, 272, 302
surface-atmosphere system, 99, 160, 234, 250
surface energy budgets, 276
surface pressure (geostrophic wind), *8*, 120, **163**, 268
surface radiation budget (SRB), 115, 221–222
surface temperature, 16, 75, 81, 88–89, 94, 114, 120,
            122–124, 126–127, 142, 150, 153, 161, 206,
            212, 214, 218, 220–222, 224–225, 231, 238,
            243, 246, 249, 265, 274, 276, 279, 299, 302,
            327, 333, 335, 349, 353, 355, 393
surface-troposphere system, 239
swamp ocean models, 112
SWIRLS (infra-red limb wind sounder)
synergism, 85
synthetic-aperture radar (SAR), 159, **219–220**

T

taiga 20, 146
Tambora eruption, 239, **242**
tectonic plates, 10–12, 118, 155
teleconnections, **162**, 279, 355
temperature lapse rate (TLR), 239
temporal climatic variability, 158, 362
terrestrial planets
terrestrial-solar interrelations, 275
TES (emission spectrometer), 216
thematic mapper (TM), 271
thermal emissions, 18, 88, 122, 239, 278
thermocline, 15, 113, 123, 159
thermodynamic models
      clouds
      polar regions
      short-term climate changes
thermohaline circulation, 12, 112–114, *113*, 157, 169
thermohydrodynamics, 113
3-D climate models, 156, 323
thunderstorms, 98
TIROS-1, 7, 204, 244
TIROS-N, **205**, 209, 223–224
TOPEX/POSEIDON, 118, 218, 220
total ozone content, *see* ozone

Total Ozone Monitoring Spectrometer, 20–21, 223–224,
            247–248, **249**, 251, 314
TOVS, 209, 215, 221, 223–225
trace gases, 77, 81, 169, 203, 210, 217–221, 223, 277–278
trade winds
transient response experiments, 142, 269
tropical forests, 68, 70, 153
tropical ocean and global atmosphere (TOGA), 42, 221
Tropical Rain Measuring Mission (TRMM), 102, **116**
tropopause, 239, 251–252, 254
troposphere, 78, **89**, 96, 100, 152, **152**, 214–218, **219**,
            224–226, 236, 239–240, 243, 245–246, 250,
            254, 335, 363
      emission spectrometer, 216
tropospheric aerosols, *see* aerosols
tropospheric chemistry
tropospheric moisture
tropospheric ozone, *see* ozone
tropospheric pollution, 214
tropospheric temperature, 246, **247**
tundra, **32**, 146, 149, **150**, 383
2-D models, 156

U

ultraviolet solar radiation, 18, 90, **163**, 216–217, 223, 364
Umkehr technique
United Nations Environment Programme (UNEP), 22–23,
            24, 26, *27–28*, 42–43, 290, 298, 301, 343,
            346–347
University Corporation on Atmospheric Research
            (UCAR), 217
Upper Atmosphere Research Satellite (UARS), 209–210, 223
Uranus
US Arctic Research Commission, 159, 361
US Center for Climate Analysis
US Department of Energy, 278
US Environmental Protection Agency, 289, 319, 331
US Geological Service, **52**
US National Academy of Sciences, 278
UV solar radiation, 18, 90, **163**, 217, 223

V

vanadium
vegetation
      biome model, 143, 146
      cover, 120, 122, 144, 149, 153, 217, **219**, 222, 306
      index, 262
Venus, 313
Visible Infra-red Scanning Radiometer (VISR)
volcanoes
      aerosols, 93, 228, 234, 238, 243, 245, 247, 250–251,
            253, 255–256
      climate eruptions, 226, 238–239, 241–243, 255, 257,
            317, 362, 364

W

Walker circulation cells
Water; *see also* liquid water content
      balances, 26, 114–117, 133–137, 169, 282, 290, 316

content, 117, 225, 274–275
  phase transformation, 234
water cycle, 101, 109, *110*, 111–112, 122, 133, 218, 221
  GEWEX, 42, 150, 221
water vapour, 18, 29, *92*, 96, *97*, 98, 100, 108–109, 111,
    134, 141, 151, **210**, **219–220**, 221, 224–225,
    236, 238, 241, 243, 252–253, 267, 273,
    275–279, 295, 297, 327, 340
wave energy
weather forecast models
  climate models, 7, 29, 89, 209, 211, 216, 266, 268, 274,
    290, 302
  future studies, 266, 268
  long-range, 15, 113–114, 290
weather-forming processes, 85, 162, 217, 328–329, 361
wind
  direction, 134–135, *137*, 202, **213**, **220**, 252, 266
  katabatic winds, 161
  shear, **219**, 221
  speed, 34, 114–115, **163**, 202, **213**, **219–220**, 221,
    266–267
  stress, **163**, **212**
  trade winds

World Climate Application Programme (WCAP), 417
World Climate Research Programme (WCRP), 41–42, 101,
    121, 150, 162, 166, 211, **212**, 217–218, 221,
    297, 311, 314–315
  TOGA, 42
World Meteorological Organization (WMO), 41–43,
    151, **164**, 221, 283, 287–292, 298, 301,
    306, 312, 314–315, 318, 320, 323, 343,
    347, *367*
World Ocean Circulation Experiment (WOCE), 42
World Weather Watch (WWW), 283

X

X-ray emission

Z

Zaporozhye experiment
zenith-looking lidar, 247, 250
zonal available potential energy
zonal flow

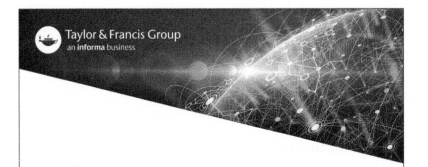

Printed and bound by CPI Group (UK) Ltd, Croydon, CR0 4YY

17/10/2024

01775698-0009